MW01480330

BIT–STRING PHYSICS

A Finite and Discrete Approach
to Natural Philosophy

KE Series on Knots and Everything — Vol. 27

BIT–STRING PHYSICS

A Finite and Discrete Approach to Natural Philosophy

H. Pierre Noyes

Stanford University, USA

edited by

J. C. van den Berg

Wageningen University
The Netherlands

World Scientific

Singapore • New Jersey • London • Hong Kong

Published by

World Scientific Publishing Co. Pte. Ltd.

P O Box 128, Farrer Road, Singapore 912805

USA office: Suite 1B, 1060 Main Street, River Edge, NJ 07661

UK office: 57 Shelton Street, Covent Garden, London WC2H 9HE

Library of Congress Cataloging-in-Publication Data
Noyes, H. Pierre.
   Bit-string physics : a finite and discrete approach to natural philosophy / H. Pierre
Noyes ; edited by J.C. van den Berg.
   p. cm. -- (K & E series on knots and everything ; vol. 27)
   Includes bibliographical references.
   ISBN 9810246110
   1. Physics--Philosophy. I. Berg, J. C. van den, 1944– II. Series.

QC5.58 .N69 2001
630'.01--dc21   2001026980

British Library Cataloguing-in-Publication Data
A catalogue record for this book is available from the British Library.

The editors and publisher would like to thank the following organisations and publishers of the various
journal for their permission to reproduce the selected reprints found in this volume:

Alternative Natural Philosophy Association	Plenum Publishing Corporation
Elsevier Science	Physics Essays Publication
Indian Statistical Institute	The Royal Society

Printed in Singapore by Mainland Press

To Mary

'Fair, kind and true' have often liv'd alone
Which three till now never kept seat in one
Sonnet CV — Shakespeare

CONTENTS

Introduction

Major scientific revolutions are rarely, if ever, started deliberately. They can be "in the air" for a long time before the first recognizable paradigm appears. I have believed for some time that this is the kind of intellectual environment in which we now find ourselves. For a while I even thought that some of the work collected here might, in fact, provide a few of the first steps toward a new natural philosophy alternative to the prevailing world view.) Since this has not come to pass, I offer these papers to the community of scientists in the hope that some younger physicist may find clues in them that help him or her to succeed where I have proved wanting.

Two of the fields which attracted me while I was in high school were history and political science. World War II intervened and diverted me into physics, a change in direction which I have never regretted. But this early interest, which had been reinforced while I was a freshman at Harvard by my good fortune in having Carl Schorske as my section man in medieval and modern history, was revived when my classmate Tom Kuhn asked me to review an early draft of *The Structure of Scientific Revolutions*[4]. His analysis convinced me that an historical approach to the problems of science could greatly aid in understanding those problems in a deeper way, and lead one into philosophical issues of great importance. I have continued to read and think about the origins of modern science, and in particular about the work of Galileo, ever since.

This interest of mine in scientific revolutions remained casual, until I heard Ted Bastin talk about the *combinatorial hierarchy* in 1973. This remarkable construction, "discovered" by Fredrick Parker-Rhodes[12] in 1961, yields algorithmically the sequence 3, 10, 137($\approx \hbar c/e^2$), $2^{127} + 136 (\approx 1.7 \times 10^{38} \approx \hbar c/Gm_p^2)$ and cannot be extended past the fourth term for reasons intrinsic to the construction. Why a simple mathematical algorithm should have anything to do with two of the fundamental dimensionless constants of modern physics remained unexplained, and so far as I am concerned remains unexplained to this day. It could — as the prevailing paradigms in theoretical physics seem to require — just be a coincidence, like the "prediction" by Swift that mars has two satellites. To make it plausible that, although still mysterious, the fact that the number of entities calculated for the third and fourth levels of the combinatorial hierarchy correspond closely to the two dimensionless numbers which characterize the two long range, macroscopic forces observed in nature (electromagnetism and gravitation) is probably something *more* than a coincidence is a main objective of this book.

When I had first met Ted the year before and heard casual mention of the combinatorial hierarchy, my immediate reaction was that it must be dangerous, mystical nonsense. When a year later (in 1973) I went to a seminar by Ted organized by Pat Suppes (then head of the Philosophy Department at Stanford) my intent was to start the question period with "How can a mathematical argument arrive at an integer approximation to dimensionless combinations of numbers which, individually, have to be determined empirically?" But part way through the seminar I was struck by the thought that my question might have a partial answer if the dimensionless *integers* themselves had a direct physical interpretation. In this connection I realized that Dyson's contention[2] that the renormalized perturbation series of QED (quantum electrodynamics) is not uniformly convergent and must diverge beyond 137 terms is relevant. As I discussed in more detail in my first paper about the combinatorial hierarchy[7], in a theory in which like charges attract rather than repel, Dyson's argument shows that 137 is the maximum number of charged particle pairs which can be quasi-localized within the compton wavelength of a mass corresponding to that number of pairs. Looked at in this way, since in quantum gravity like "charges" *do* attract, the same argument fixes the maximum number of particles of protonic mass one can discuss within a volume whose linear dimension is the proton Compton wavelength. These form an ephemeral black hole! This realization was enough for me to suspend judgment, for a while, on the significance of the combinatorial hierarchy. I have tried several times over the years to get this idea into the mainstream literature. The latest unsuccessful attempt[8] was a comment on Zurek and Thorne's work on rotating, charged black holes.

If we could actually "calculate" the low energy fine structure constant, and (given the velocity of light, Plank's constant, and the proton mass) the low energy value of Newton's gravitational constant in a way that could be made convincing to physicists following the currently acceptable paradigms, I think some of them would agree that this fact might become the starting point for a "scientific revolution". The proviso that we are "given" c, $\hbar$ and m_p is already met by the conventional physics paradigm in the following way. The laboratory standards of mass, length and time are, to begin with, arbitrary. But they can be calibrated against theory by identifying any three independent dimensional combinations of specified theoretical constants which are accepted in the dominant theoretical physics formalism. Maxwell theory, and its relativistic successors, require a fixed dimensional constant that can be identified with "c" and measured directly or in combination with other elementary constants in a number of different ways. Quantum theory fixes $\hbar$, again in many ways. Relativistic quantum field theory relates the mass of any particle and its antiparticle to the energetic threshold for pair creation. Consequently any *stable* elementary particle mass can be chosen as a standard. Since the proton is known to be stable for times a factor of at least 10^{25} longer than the current age of the universe, m_p is also "given" in the required sense, completing the argument.

This implies that the theory in question also has a way of identifying masses and of calculating dimensionless mass ratios. We have already given one instance in that our interpretation of the Dyson argument gives a first approximation to the electron-pion mass ratio. We will see later that a second piece of work by Parker-Rhodes provides a calculation of the electron-proton mass ratio to eight significant figures. Clearly these two mass ratios can be the starting point for meeting this mass ratio requirement in bit-string physics. Since there is currently no empirical evidence for proton decay, and current physics does not even suggest that the electron is unstable, we are free to pick the proton as the mass which is singled out as our standard. Finally the combinatorial hierarchy relations then allows us to calculate a low energy approximation to the charge on the electron and Newton's gravitational constant from

$$e^2 \approx \hbar c/137; \quad G_N \approx \hbar c \times 10^{-38}/1.7m_p^2 \tag{1}$$

This much is straightforward, conventional dimensional analysis in MLT (mass-length-time) physics. But it still leaves untouched the mystery of *why* the Parker-Rhodes algorithm can — let alone *should* — lead to a replacement for the conventional understanding of the foundations of physics. I emphasize the fact that I am still frustrated by this mystery, and find the published works of Amson, Bastin, Kilmister and Parker-Rhodes unhelpful on this vital point. I mention this so that the reader will understand why I place little emphasis on those works and try to follow a different route.

What I have ended up doing is to try to understand in my own terms why a finite and discrete approach to the foundations of physics might make sense. One aspect of the program bothered me from the start. According to Kuhn a significant paradigm shift in a mature science like physics usually arises from some recognized anomaly in the explanations provided by "normal science". It is true that the very existence of significant dimensionless numbers like $e^2/\hbar c$ and $G_N m_p^2/\hbar c$ which (so far as we know) can only be uncovered by experiment has proved deeply disturbing to some physicists. Among them are some of the greatest, such as Einstein and Pauli. Yet I have never heard of — let alone heard — any physics department colloquium devoted to this topic. This seems to rule out identifying such dimensionless numbers as a Kuhnian anomaly likely to lead to a scientific revolution.

My worry as to whether I might be wasting my time in pursuing the understanding of the combinatorial hierarchy was compounded when I had occasion, some years ago, to discuss Kuhn's model for scientific revolutions with Freeman Dyson. Recall that Dyson was one of the four creators of renormalized QED (quantum electrodynamics), the paradigm shift which was the starting point for most of the work in theoretical elementary particle physics for the last half century. Also Schweber's history of QED[13] informs us that Freeman's paper on why the renormalized perturbation theory does not converge[2] convinced Dyson that QED could *not* be the starting point for a new fundamental theory. I have yet to persuade Freeman that the use I have made of his paper, discussed in various places in this volume, *could* be a clue pointing toward a theoretical breakthrough. Therefore I had to take particularly seriously his contention that scientific revolutions start from technological advances rather than from intellectual puzzles. My rather cursory reading in the history of science made this a very plausible alternative to the Kuhnian model. In a recent discussion Dyson [3] does allow some scope for Kuhnian paradigm shifts. However, most of his examples reinforce his contention that tools rather than ideas are usually the driving force behind scientific revolutions.

At an earlier stage I had already found comfort rather than discouragement in Dyson's view. This was after I had started working with two far seeing computer scientists: Mike Manthey and David McGoveran. In very different ways they taught me that digital computers actually provide an intellectual background for thinking about the finite and discrete events which lie at the core of quantum mechanics, and which differs profoundly from the immersion in continuum mathematics that most older theoretical physicists like myself took for granted. This gives Dyson's point of view an unexpected twist: the revolutionary technological shift which *might* form the starting background out of which a contemporary scientific revolution could arise may be the computer revolution. This is what Mike called "the computer metaphor" a couple of decades ago. David obviously understood long ago that the paradigm shift for which we yearn is already implicit in starting from finite and discrete events[5]. Once I had thoroughly digested this idea of finite and discrete events modeled by bit-string operations as lying at the core of my research program, I began to see that this could tie in with Bridgman's operational approach. I had been attracted to Bridgman's way of thinking as a graduate student, but it took me most of my professional life to begin to meld it into the finite and discrete approach. The launching of a new Indian journal gave me the opportunity to make a preliminary excursion into this realm[9].

Since I have already written a brief autobiographical memoir about the history of my involvement with the combinatorial hierarchy, and this is included in one of the selections in this volume[10], I will not duplicate that material in this introduction. However, it is important to stress that I was not convinced that my research program along these lines could lead anywhere until David McGoveran showed that it was possible to extend the coupling constant and mass ratio calculations into new areas and in so doing also improve the fit to experiment of *all* the predictions. The first of these calculations was his study of the fine structure spectrum of hydrogen from the finite and discrete physics point of view[6]. This gave both the usual Dirac formula and four more significant figures in the calculation of the fine structure con-

6

stant. Also the work with Mike Manthey which gave a simple way to generate the bit-string universe algorithmically has recently been vindicated by the latest cosmological observations[11] a decade after the predictions were made. I hope that all of this excitement will be made clearer and more compelling by reading the papers themselves.

References

[1] K.Bowden, ed. " The Emergence of Physical Structure from Information Theory", special issue of the *International Journal of General Systems* **27**, No.'s 1-3 (1998).

[2] F.J.Dyson, *Phys Rev*, **85**, 631 (1952).

[3] F.J.Dyson, *The Sun, the Genome and the Internet*, Oxford (1999), esp. pp 13-16.

[4] T.S.Kuhn, *The Structure of Scientific Revolutions*, University of Chicago Press, 1962.

[5] D.O.McGoveran and H.P.Noyes, "FOUNDATIONS OF A DISCRETE PHYSICS", SLAC-PUB-4526 rev, Jun 1989, 93pp. Published in *Proc. ANPA 9*, pp 37-104 (1987).[Ch.4; see also Comment by DMcG in Ch.6]

[6] D.O.McGoveran and H.P.Noyes, *Physics Essays*, **4**, 115-120 (1991).[Ch.6]

[7] H.P.Noyes, "Non-Locality in Particle Physics" prepared for an unpublished volume entitled *Revisionary Philosophy and Science*, R.Sheldrake and D.Emmet, eds; available as SLAC-PUB-1405 (Revised Nov. 1975).[Ch.1]

[8] H.P.Noyes, "Comment on 'Statistical Mechanical Origin of the Entropy of a Rotating, Charged Black Hole", SLAC-PUB-5693, November 1991 (unpublished).[Ch.7]

[9] H.P.Noyes, "Operationalism Revisited", *Science Philosophy Interface*, **1**, 54-79 (1996).[Ch.10]

[10] H.P.Noyes, "A Short Introduction to BIT-STRING PHYSICS", in *Mereologies: Proc. ANPA 18 (1996)*, T.L.Etter, ed., pp 41-61.[Ch.14]

[11] H.P.Noyes, "Program Universe and Recent Cosmological Results", in *Proc. ANPA 20.*[Ch.16]

[12] A.F.Parker-Rhodes, "Hierarchies of Descriptive Levels in Physical Theory", Cambridge Language Research Unit, internal document I.S.U.7, Paper I, 15 January 1962. Included in [1], pp 57-80.

[13] S.Schweber, *QED and the Men Who Made It*, Princeton, 1994, p.565.

NON-LOCALITY IN PARTICLE PHYSICS[*]

Pierre Noyes

Stanford Linear Accelerator Center
Stanford University, Stanford, California 94305

Submitted to the Cambridge University Press as a chapter for

Revisionary Philosophy and Science

Rupert Sheldrake and Dorothy Emmet, Editors

[*] Work supported in part by the United States Atomic Energy Commission.

I. INTRODUCTION

The quantum mechanical description of systems of electrons and nuclei –
atoms, crystals, metals, molecules, semiconductors, plasmas, ... - has
turned out to be enormously successful. Not only can the experimental properties of simple systems be calculated to high accuracy using only four universal
constants (c, $\hbar$, m_e, e^2) and the masses of the nuclei $M(A, Z)$, but the extension
of the calculations by means of a modest number of empirical constants referring
to specific systems allows the quantitative prediction of many of the properties
of quite complicated structures. Few physicists doubt that these empirical
constants could also be calculated from the basic set given above if a proposal
to do so generated sufficient enthusiasm and adequate financial support. These
quantum mechanical and empirical ingredients support a detailed physical
description of the DNA double helix – the organ of heredity and the instruction
tape for protein synthesis within living cells. Biologically important mutation
phenomena happen due to the quantum uncertainties in the positions of the
hydrogen atoms which zip the two strands of the helix together. Therefore, there
is a significant transition region between the quantal description of particle
phenomena and the "classical" physio-chemical descriptions of molecular
biology and cell metabolism; physicists usually believe that the two regions can
be joined without conceptual conflict.

Yet quantum mechanics was born during a period of raging scientific controversy and philosophical doubt, a time when many physicists were deliberately
seeking for acausal physical phenomena, in order to break the chains of
"classical" physical determinism.[1] Many philosophers still do not accept

quantum physics. Neither do some physicists who cling to Einstein's critique.[2]
Contemporary physicists who grew up using quantum mechanical paradigms
seldom question the validity of those models. As students they may well have
been uneasy about how and whether classical and quantum mechanical descriptions join. Yet once they succeed in solving some specific quantum mechanical
problem, familiarity starts to breed contempt. Subsequently they are usually
content to ignore the basic paradoxes and get on with what they consider to be
the main job.

High energy particle physicists are not so fortunate. They study the materialization and disappearance of "particles", usually as "counts" in detectors.
They also investigate the "virtual" effects of such particulate degrees of freedom
at energies which do not allow the additional particles to be materialized and
isolated from the initial system. That new particles could be created from
energy, and that they can have measureable effects below the energetic "threshold"
for their creation was demonstrated by Wick[3] in a very brief but profound analysis of Yukawa's[4] meson theory of nuclear forces. Both predictions are inescapable consequences of theories which include the Heisenberg uncertainty principle
and the mass-energy equivalence of Einstein's special theory of relativity. The
successful artificial production of Yukawa particles (pions) in 1948 was one of
the great triumphs of experimental and theoretical particle physics, and of
accelerator technology.

Theoretical physics has had precious few, if any, comparable triumphs
since. Experiments of increasing subtlety, precision, and cost have revealed
detailed and intricate systems of ephemeral particles with intriguing characteristics. Many beautiful regularities and partial symmetries have been developed
by theorists to describe these results; within broad areas these theoretical

structures have demonstrated great predictive power. Yet there is no unified theory exhibiting overall consistency, let alone reliable and quantitative predictive accuracy. The conjuring act by which success appears to be achieved by some practitioners of high energy physics does not lead to that body of "public knowledge" which many see to be the aim of science.[5] Indeed, the tale is told of one theorist (I fear his name is legion) who prepares a separate model to "predict" any conceivable outcome of proposed key experiments, and files the stack in a locked drawer. His task is easy thanks to the ambiguities in basic theories. The time for making the calculations is ample because particle experiments are often major engineering enterprises that can take several years to bring to fruition. When our theorist learns the preliminary results of some experiment from the grapevine, he hurries to his drawer, extracts the "correct" prediction, and mails it off for publication. With luck, his paper can be in print before the results of the experiment are common knowledge.

A scientific community which tolerates the type of behavior just described creates ephemeral theories. There are frantic rushes from one fashion to the next. This situation provides some experimental physicists grim satisfaction and even sadistic delight in shooting down the flimsy structures that pass for predictions. But all experimentalists share to some extent the frustrations of their theoretical brethren, particularly as natural selection weeds out theorists who have not learned the skill of concealing a face-saving ambiguity behind the facade of what appears to be a clear prediction.

In spite of this unhappy situation, there have been surprisingly few attempts to attack the fundamental ambiguities. Epistemological tools developed during the somewhat similar periods when relativity and quantum mechanics were gestating have yet to be effectively employed. The reason is not far to seek. Although particle physics lacks a rigorous paradigm, it has what passes for one in the "local quantum field theory", which with skill can be steered around various ambiguities to arrive at genuinely useful predictive results. Few would challenge the assertion that the elementary version of this theory is quite ambiguous. The basic reason for the ambiguity is quite simple. Local quantum field theory takes over, unexamined, the continuously infinite four-dimensional Minkowski space-time of point events, and uses this framework for the definition of dynamical "field amplitudes" at each point. It is only in this sense that the field theory is "local"; the predictions derived from it definitely are not. Since this is a quantum theory, each field amplitude is subject to the uncertainty principle ($\delta E \delta t \geq \hbar$, $\delta p_i \delta x_i \geq \hbar$). But this means that whenever (as it always must) the theory requires a limit to be taken in which the volume $\delta x_1 \delta x_2 \delta x_3 \delta t$ surrounding a point shrinks to zero, the energy and the momentum carried by the field at that point must go to infinity.

Various clever ways have been found to avoid this apparent disaster. For the interaction of charged particles with the electromagnetic field (quantum electrodynamics or QED) Tomonoga, Schwinger, and Feynman showed that these infinities can be removed by a redefinition ("renormalization") of the charge and mass of the particles, provided the consequences of the theory are calculated to some finite order in a power series in the fine structure constant $e^2/\hbar c \approx 1/137$. For particles such as the electron and muon which exibit no "strong interactions", many of the properties that can be very precisely measured at low energy have been computed and confirmed by experiment[6] to the fantastic accuracy of one part in $(137)^3$. At very high energy still other properties can be computed and measured; conventionally interpreted, these results show that quantum electrodynamics is empirically "local" down to dimensions at least a factor of 10 shorter than the characteristic nuclear dimension of 1.4×10^{-13} cm.

The paradigm can also be used for systems of pions and nucleons and other "strongly interacting particles" called hadrons. This requires some care as the parameter analogous to the fine structure constant $G^2/\hbar c \approx 14$ does not allow a sensible power series expansion. The alternative of expanding in powers of $\hbar c/G^2$ is not available, because no one knows how to construct the "strong coupling limit" with which the series would have to start. So progress has been made by considering situations which introduce a second parameter that cuts down the effective interaction strength. For example, if two hadrons of mass M are so far apart that the uncertainty principle and the mass-energy relation only allow a single pion to be exchanged between them with any great likelihood, the characteristic parameter becomes $f^2/\hbar c = (m_\pi/2M)^2 G^2/\hbar c \approx 0.08$. Thus the leading term in the series has an a priori accuracy of about 10%. Since the next term can rarely be computed unambiguously, discrepancies of 10 to 30% between theory and experiment often count as a "validation" of the theory.

If this were all that was available to test theories of hadrons, one would expect considerably more pressure for fundamental revision than actually exists. But there are other trials that can be carried out to high accuracy which test general features a solution of the field equations must exhibit (if it exists) rather than predictions of numbers for specific dynamical situations. One test connects the probability amplitude for scattering in the forward direction with a specified integral over the probability that there will be scattering at all angles and all energies (total cross section). Such mathematical relationships are called "forward dispersion relations". It is often claimed that any local relativistic field theory in which effects cannot propagate faster than the speed of light (a property often called "causality" in this context) must predict amplitudes which satisfy the forward dispersion

relations. Since theorists are really willing to stick their necks out on this prediction, experimental tests have been many and have been pushed to the highest energies available in the accelerator laboratories. Both forward scattering amplitudes and total cross sections show rapid fluctuations in energy (resonances) over the energy region up to a couple of thousand million electron volts and then smooth out, making the tests very restrictive. No "counterinstance" to any forward dispersion relation has been uncovered for any system of particles so far tested. For many physicists this string of successes provides the strongest argument for "local field theory".

Other general properties of particle systems which were first predicted via local field theory and have been shown empirically to hold to very high accuracy provide much of the psychological underpinning for those who resist revisionary thinking in particle theory. Among these were the prediction of antiparticles (both the positron and the anti-proton, for example) electron-positron pair creation, neutrinos, Yukawa's prediction of the meson.... Actually all these predictions can now be viewed as necessary requirements of certain symmetry properties connected with relativistic transformations and the possibility of particle creation which comes from coupling relativity to quantum mechanics (the Wick-Yukawa mechanism already mentioned above, and to be discussed in greater detail below). Another great triumph in the early days of quantum field theory was the proof that particles with integral spin (bosons) must have symmetric wave functions (i.e. wave functions which do not change sign on the interchange of any pair of particle coordinates) and that particles with half-integral spin (fermions) must have wave functions which do change sign (are anti-symmetric) on such an interchange. For this proof Pauli received his Nobel prize. That integral spin particles must obey Bose-Einstein statistics (which is why they are called bosons), and half-integral spin particles must obey Fermi-Dirac statistics had been known (empirically) before, but Pauli provided a fundamental explanation. Further, his proof required that particles be treated as matter fields, not as the relativistic generalization of non-relativistic particle wave functions. Closely connected with this proof is the general proposition (CPT theorem) that the theory must be invariant if particles and antiparticles are interchange (C = charge conjugation), coordinates are mirrored (P = parity operation) and the direction of all motions are reversed (T = time reversal operation). But Stapp[7] has subsequently shown that once particles have separated from each other sufficiently so that their individual energies and momenta can be measured (an epistemological requirement in S-matrix theory), it is possible to describe both particles and antiparticles as having positive energies, thus avoiding the negative energy states which had forced elaborate constructions onto Dirac and Pauli. Further, since these amplitudes must lead to real probabilities lying between zero and one, he was able to show that very mild assumptions about the structure of these observable amplitudes suffice to establish both the connection between spin and statistics and the CPT theorem. Similarly, the successes of current algebra and the like, which are often cited as evidence for local field currents could probably be restated in terms of particle wave functions without invoking fields.

This summary of current ideas in particle theory could be extended to provide much evidence that, in spite of ambiguities at the fundamental level, and the somewhat questionable scientific ethics of some of its devotees, the field seems to be enjoying one of Kuhn's[8] periods of normal science rather than a crisis situation. There are no obvious "anomalies", let alone "counter-instances" which might immediately serve as a rallying point for a concerted attack. Thus, if we are to attempt revisionary activity, we know in advance that the going is going to be tough. We should learn from our comrades in Southeast Asia that we must "know our enemy" and attack where he is weak, not where he is strong. The strongest point in the defense of local field theory is obviously QED, so we should leave this to the last, and try to outflank it by finding weaker points. The analysis indicates, as should the title of this paper, that we believe one such weak point could lie in the reliance on locality in the formulation of the theory. S-matrix theorists[9] have already chosen this as a point of attack in their own revisionary efforts, and we should try to sieze any ground they have already gained in mounting our own attack. But we believe it possible to go deeper than they have done into the foundations of the theory without giving up (as they tend to) the objective of including both electromagnetism and gravitation within the quantum particulate description.

I was led to the idea of using non-locality as a starting point for theoretical reconstruction in particle physics, not by the analysis presented above, but because I was forced to recognize that the conventional theory already is inescapably non-local. This happened in terms of quite specific research problems, which I will review in the next three sections before turning to the main revisionary discussion. After noting a more conventional operational analysis presented elsewhere[10] we attempt to build up the known structure of particle physics using as the fundamental postulate the principle of rational discrimination. Both methodologies are based on a metaphysical presupposition that the quantum particles and their relationships must, in time, be capable of leading to the physicists who now are discussing the particles. A scientific retrodiction of our past in terms of the particles and their historical consequences which might make this proposition plausible, has been presented elsewhere.[11]

II. THE ETERNAL TRIANGLE EFFECT

Any quantum physicist knows that in order to solve a dynamical problem, he must, explicitly or implicitly, describe the system of interest throughout all of space-time. Any attempt to localize space-time regions within this infinite space-time volume can only be fuzzy because of the uncertainty principle, and this "fuzz" can extend to arbitrarily large distances. Yet this elementary fact about quantum mechanics is often ignored by working physicists. One reason is that for many (but not all) macroscopic (compared to atomic) situations the underlying atomic systems can be treated as extended Euclidean volumes (viz. the DNA double helix mentioned in the introduction). In such situations the probabilistic aspects of quantum mechanics can often be kept in the subliminal background of the calculations. A second reason for ignoring non-locality is that the theory makes use of the coordinates of <u>particles</u>, which are treated as space time <u>points</u>. Although the "actual" positions and velocities of the particles can only be computed in terms of probability amplitudes (wave functions), it is taken for granted that the points to which these distributions are referred have themselves a precise meaning. It can therefore be somewhat shocking when a specific problem forces a physicist to accept the <u>necessity</u> of extremely non-local effects within the framework of quantum mechanics; at least this was my own experience.

The specific problem which first forced me to take quantum mechanical non-locality seriously was the quantum mechanical problem of three particles. I had been led to study this problem because my own field of specialization - the "nuclear forces" between two nucleons (i.e. neutron + proton, neutron + neutron, or proton + proton) - had been pretty well worked out experimentally and pretty well correlated theoretically with current ideas in particle physics. It thus seemed that the time had come to try to calculate the properties of systems of three nucleons from first principles. I was encouraged in this task by the fact that Faddeev[12,13] had constructed a rigorous mathematical theory of the quantum mechanics of three non-relativistic particles interacting via "local potentials", i.e. potentials which depend only on the relative distance between two particles. I knew from the start that I would have to make some modification in the treatment, since we know on general grounds that the Wick-Yukawa mechanism generates a non-local interaction (see below), and I had demonstrated a specific non-local effect in the course of my own work.[14] Further, the details of the non-local interaction are particularly uncertain at short distances. I therefore wanted to recast the problem in such a way that it could be made as insensitive as possible to these sources of uncertainty. Because I understood these problems best in terms of the relative distance, r, between the nucleons, while Faddeev had presented the whole theory in a highly abstract form using a description in terms of their relative momenta, my first step was, naturally, to convert his description from momentum space to coordinate space.

The first step was easy. I showed that the dynamical driving terms in the Faddeev equations which come from a description of the nuclear force contain three pieces: (1) the two nucleon observable amplitudes ("phase shifts") directly available from experiment, (2) a factor multiplying these amplitudes which describes the probability distribution (wave function) of the nucleons at short distance (i. e. $\gamma \lesssim \hbar/m_\pi c$), and (3) a third function which can be constructed from the first two[15,16,17]. But when I tried to put this description into the Faddeev equations, I found that even if the nuclear "potential" is strictly zero for all separations r greater than some finite distance R (e. g. 10^{-13} cm), each pair of particles generates an effect which perturbs the motion of the third particle at arbitrarily large distances! Superficially this new interaction falls off only like 1/r, as is very easy to show.[17] At first I thought I had simply made a mathematical mistake in transcribing Faddeev's description into configuration space, and I puzzled over the result for two years before I could make significant progress. A 1/r "potential" has the same dependence on distance as the classical electric force between two charges (Coulomb potential), so would have observable macroscopic effects. Actually, if we use a little more care, it can be shown that the coefficient of the 1/r term is zero, and hence that the leading term in an expansion in powers of 1/r goes like $1/r^2$; this term has no need to vanish and in one case (the Efimov effect discussed below) has been rigorously proved to survive. Thus two structureless particles whose "interactions" classically described as a function of their separation vanish outside some finite radius R can affect the motion of a third particle (whose pairwise interactions are similarly bounded) at arbitrarily large distances from the pair! This rigorously proved result demonstrates the extreme non-locality predicted by ordinary non-relativistic quantum mechanics.

The physical origin of this effect is relatively easy to understand. When two quantum particles scatter, the emerging particles do not come out in precisely defined directions, as they would classically, but as probability waves. In the Faddeev equations, the interaction between each pair contains not only the "potential" which would exist if they were isolated (and then would be the whole story) but two additional terms representing the scattering of the other two possible pairs. So long as the pair in question are within the range of forces, their energy and momentum are not connected in the way they would be if free, and consequently they can pick up momentum from the outgoing waves in the other two channels, changing the effective interaction. Geometrically, the target presented by this region is proportional to R/y, where y is the distance to the third particle, thus explaining the long-range character of the effect.

Once I had understood this effect, I was struck by an analogy to a well-known phenomenon in behavioral science. If two people in a room with a door come to think that there is a third person outside that door who might enter (in the quantum mechanical analog this is the third particle, out of range but "virtually" present), their behavior changes in ways that could not be readily predicted from their previous communications in the room. This analogy looks superficial, but is in fact profound. If we follow the time-dependent development of a system of three particles (cf. the Figure, next page) which were initially isolated, the first interaction generates probability waves which distort the wave function in subsequent interactions and changes them from those which would occur in a system which contained only two particles. Thus to understand the present state of the system we must know not only the forces momentarily acting between the "isolated" pairs, but also the entire past history of all relevant particles in the system. Similarly, to understand why and how the two people change their behavior, we would need to know their individual histories before entering the room, the histories of their cultures, the evolutionary development of communication systems on this planet, ... For obvious reasons, I have dubbed this "the etermal triangle effect", and first so referred to it in an after-dinner speech.[18] I subsequently presented a paper on this subject at an American Physical Society meeting.[19]

As has already been mentioned, Efimov[20] discovered a specific example of this effect, also using a configuration space treatment, in a system of three identical spinless particles with no (relative or total) angular momentum. Actually, as we have seen, the existence of the effect does not require this high degree of symmetry, but assuming such symmetry simplified his original treatment, which he has subsequently generalized.[21] For two spinless particles with no relative angular momentum the cross section (i. e. the area of a beam of particles which is scattered out when the beam is incident on a target composed of the second particles) goes to a constant value, $4\pi a^2$, at low energy. The constant a, which is called the "scattering length" can be arbitrarily large compared to the range of forces R, and in the limit when there is a bound state of zero binding energy, a goes to infinity. Since this implies an infinite cross section, and the eternal triangle effect is due to the scattering of the third particle from this pair, it is not to surprising that in this limit the three particle system has an infinite number of bound states of arbitrarily large size (the number of such states approaches infinity like $\pi \ln(a/R)$ and corresponds in the limit to the bound state spectrum of a "potential" with radial dependence const. $/r^2$). The Efimov effect provides a rigorously demonstrated example of non-locality generated by finite (and arbitrarily short) range quantum mechanical forces between pairs.

Because of the way in which I arrived at the result, I initially believed that the effect depended on the wave functions of the pairs inside the range of forces, and tried for some time to manipulate the equations into a form that would allow these wave function effects to be observed in three particle systems. I should have been warned by the Efimov effect (which depends only on the measurable two-body scattering length, and not on the details of the force) that this effort might fail. In fact, as we will discuss below, it is impossible using particles which "interact" only at short range to "measure" the wave functions at short distances, no matter how many particles there are in the system. But I was led to that result, which has also been rigorously proved[22] in the case of three particles, by an entirely different line of thought, to which we now turn.

III. FIXED PAST AND UNCERTAIN FUTURE

As can be guessed from the Introduction, I have succeeded in retaining some of the fundamental doubts about quantum mechanics which plague most physics students when they first try to come to grips with the subject. Like many other theorists, my doubts were kept subliminal, or at best preconscious, by the successes of the renormalized perturbation theory of Quantum Electrodynamics. But one paper by Thomas Phipps[23], which eventually got published in an emasculated version by the Physical Review, kept me interested in more than the usual game of shooting down speculations in conflict with experience. He is much concerned with the fact that the conventional route used to pass from classical to quantum mechanics throws away the "constants of the motion" $\underline{X}_k$, $\underline{P}_k$ used to describe the initial state of the system and retains only the "dynamical variables" $\underline{x}_k$, $\underline{p}_k$. To quote a recent communication[24]

"There is nothing to be happy about in a theory that claims to embody a formal "Correspondence", yet absent-mindedly mislays half the classical canonical variables in the process, then covers its nakedness with a fog of blather about "mind", which could just as well be the "God" whose sensorium provided Newton with such convenient cover in circumstances of like embarrasment. I'm pretty absent-minded myself, but when it comes to counting parameters I'll take on any performing horse (or nonperforming physicist). " As he showed[25], the Hamilton-Jacobi equations of classical physics can be interpreted as operator equations acting on a state-vector ψ with classical physics in the limit $\psi \rightarrow$ const. and quantum physics in the limit in which the action $S \rightarrow \hbar/i =$ const. The quantum limit concerns us here, but Phipps showed that there are more general solutions ("Class III") in which neither ψ nor S are constant. Using the one-particle Dirac equation, he showed that these could be related to effects at nuclear dimensions, thanks to the (still unexplained) "coincidence" between half the classical electron "radius" $e^2/2m_e c^2 = 1.4 \times 10^{-13}$ cm and the pion "Compton wavelength" $\hbar/m_\pi c = 1.4 \times 10^{-13}$ cm. If unpublished results[26] which counterindicate Bell's inequality[27,28], and hence would allow hidden variables (in theoretical conflict with the Freedman-Clauser result[29]) are confirmed, this theoretical possibility should be explored with vigor. It would indeed be a start of unquestionably "revisionary" physics.

Lacking clear need to abandon the quantum limit, we use below only the fact that Phipps' prescription supplies the conventional Schroedinger wave function $\phi(\underline{x}_k, t)$ with a phase factor $\exp i \Sigma_k \underline{P}_k \cdot \underline{X}_k$, converting it into his ψ. Since physicists compute the probability of future events from $|\phi|^2$, which is identically equal to $|\psi|^2$ (in the quantum limit), the two theories are observationally indistinguishable in that limit. Nevertheless, Phipps' phase factor can still be interpreted as the description of a definite former condition which persists throughout all "virtual" (reversible, uncompleted) processes, yet changes irreversibly (and discontinuously in the quantum limit)when these processes are completed and join the fixed past. This way of looking at quantum mechanics has two conceptual advantages. It removes any subjective element ("the observer") from the physical theory and allows any describable process to take place whether it is "observed" or not. Thus no esoteric element such as the "collapse of the wave function" (other than the non-locality discussed in this article) need enter the theory.

The second conceptual advantage of Phipps' version of quantum mechanics[30,31] is that the irreversible changes in the "constants" of the phase factor define a unique sequence, irreversible as time progresses, independent of the observer. In contrast, both classical statistical mechanics and the conventional interpretation of quantum mechanics rest on laws which are time-reversible at the microscopic level. Classically, the "unidirectionality" of heat flow in time is a statistical prediction applying only to systems of large numbers of particles, while in quantum mechanics irreversibility is a direct consequence of the uncertainty principle. Retrodiction starting from classical microstates leads to hypothetical states that do not correspond to the actual preconditions of systems where the theory is used predictively, while starting from quantum mechanics retrodiction leads into an inceasingly chaotic past that looses contact with present experience. These paradoxes can be "avoided" by fiat, as recently suggested[32], by simply accepting that the time-irreversibility arises from the statement of the "boundary conditions" by the physicist who poses the problem. This makes the "observer" as much a part of classical physics as he is of conventional quantum physics. One way to avoid "humanizing" science in this apparently arbitrary way is to attribute irreversibility directly to some unidirectional (in time) cosmological model for the universe.[33] But even in the Einstein

models for expanding universes, we must still supply "boundary conditions"

either a priori or on the basis of current observations. I find Phipps' alter-

native of making time-irreversibility a microscopic property of the solution

to the equations of motion preferable to either bringing in the "observer" in

the disguised form of boundary conditions or of tying local problems of heat

flow to an overall cosmology.

In a deeper sense, I accept the necessity of admitting that the theoretical

physics we discuss is only possible among physicists at a certain cultural

level, which in turn presupposes a long period of both biological and cultural

evolution. The advantage of Phipps' approach is that we can now find it easy

to understand how the irreversibility of quantum processes and the consequent

increasing complexity of systems in time (already mentioned in connection with

the eternal triangle effect) could, given time, lead to just such an evolutionary

development. We are simply trying to state laws (which conceptually speaking

require beings vaguely describable as physicists to state those laws)

in such a way that those laws entail an evolutionary development of the physicists

who eventually state them.

Although I was familiar with Phipps' ideas for some time, and discussed

them casually with colleagues in several institutions, this thinking did not bear

fruit until I was questioned by Tom Phipps about the current status of the rela-

tivistic quantum mechanical two-body problem. (He had in mind trying to

generalize his 1960 covariant treatment of the Dirac equation in a Class III

theory[23]). The usually accepted mechanism for the "strong interactions" (the

generalized nuclear force problem) is the materialization of massive particles which transfer momentum from one system to another. Such particles ("mesons"), first predicted by Yukawa[4], are for most physicists the inescapable consequence of the coupling of quantum mechanics to special relativity, as was argued very simply by Wick.[3] We will describe this process in more detail in the next section, and use it as one of the basic postulates in approaching relativistic quantum mechanics. What Phipps' theory enabled me to do was to see that particle dynamics could be described using only the Wick-Yukawa process. Conceptually, the result is similar to S-matrix theory, and abandons ideas such as "potentials", "forces", "interactions" and "fields". Technically, the possibility of such a theory is obvious once any standard theory of scattering (such as that of Goldberger and Watson) is examined with an eye to grafting on the Phipps phase factor. The result is intuitively obvious[24], and has since been demonstrated in detail[10]; the phase factor represents particles which disappear from the initial state and (non-locally) appear in the final state. Thus the dynamics of calculating the transition matrix can be unambiguously separated from the description of the quantum scattering process.

IV. THE PRIMACY OF PARTICLE NUMBER

The atoms of Leucippus and Democritus had no "natural" or "original" motion; their random collisions were strikingly similar to the nineteenth century model for the kinetic theory of gases. Epicurus assumed that the atoms were falling in straight lines and that it was necessary to postulate that some of them "swerve" in order to initiate the processes which lead to the generation (and decay) of worlds. His random element in atomic theory has been criticized as foreign to the basically materialistic and deterministic focus of this natural philosophy. In recent years we have learned from the success of quantum mechanics that determinism does not inhere in the individual atomic events. The approximate validity of determinism stems from the flow of probability amplitudes from the past up to some event in which the massive particles at least potentially observable in that event manifest their individual particulate behavior. The random character of these individual events is an integral part of quantum mechanics, and supplies a possible answer to the puzzle raised by ancient atomism and its critics.

The conceptual revolution implied by this view is still in process. The existence of physicists (not just philosophers) concerned about the question of "hidden variables" at the experimental and not just the theoretical level[27,28,29,34] shows that the issue is by no means settled. But the concept of quantum fluctuations can be used to unify an enormous range of superficially disparate phenomena, and in particular to account for the existence of "forces" between structureless particles. This possibility is counter-intuitive for many people, but Wick[3] showed long ago that once one accepts both special relativity and quantum mechanics, the existence of "short-range forces" is inevitable. The argument goes as follows. Suppose there is a particle with mass m which is <u>finite,</u> and we bring together two other particles (whose masses may be as large as we wish, but also finite) close together. If the distance between them is r, and this close approach persists for a time interval δt, their energy must, because of the Heisenberg principle, be uncertain by an amount $\delta E \gtrsim \hbar/\delta t$. Thus for short enough times the uncertainty of the energy in the whole system must (because of the Einstein relation $\delta E \geq mc^2$ for a particle of mass m) exceed the rest energy mc^2 of the particle we postulated. This particle could, in principle, appear anywhere, but during the time δt it can only move a distance less than $c\delta t$, where c is the limiting velocity for particle motion. If we further assume that momentum (but not energy) is conserved (i.e. Newton's third law) when the "particle" appears and "disappears", this means that in instances

when it appears near one of the two systems, that system must change its

momentum; after the particle travels the distance r to the other system and

disappears, the second system must acquire the momentum lost by the first.

Thus momentum is transferred from one system to the other, or in the language

of Newton, there is a "force" between them. The distance over which this force

acts can now be calculated easily as $r \leq c\delta t$ (limiting velocity) $\lesssim c\hbar/\delta E$ (uncer-

tainty principle) $\lesssim c\hbar/mc^2 = \hbar/mc$ (mass-energy relation). We have proved

that provided only there is some particle of mass m <u>any</u> two systems which

can be brought close together, and which are "coupled" to this particle, will

experience a "force" of "range" $\hbar/mc$. Further, if the two particles initially

present are brought together sufficiently violently so that both can emerge from

the collision after loosing an amount of energy greater than mc^2 while still

conserving momentum, we expect in some instances to find a particle of mass

m emerging, along with the two particles in various ways, but <u>total</u> momentum

will be conserved between the initial and the final situations.

The actual history of the verification of this prediction was complicated by

the fact that the Yukawa particle inferred from the 1.4×10^{-13} cm range of

nuclear forces (which has a rest energy $m_\pi c^2$ of 140 Million electron volts) was

not the first particle discovered intermediate in energy between the electron

$(m_e c^2 = 0.51$ MeV) and the proton $(M_p c^2 = 938$ MeV). The first "mesotron" found

was the muon, a "heavy electron" now known to have a mass-energy $m_\mu c^2 = 106$ MeV.

Three Italian physicists hiding out from the Gestapo in the basement of the

University of Rome showed that this particle interacts with nuclei $\sim 10^{-13}$ times

more weakly than would be required if it were to serve as the particle predicted by Yukawa to generate nuclear forces.[35] But the Yukawa particle - now called the pion - was eventually discovered and shown to account quantitatively for the longest range part of nuclear forces. Unfortunately from the point of view of simplicity, the same Wick-Yukawa mechanism predicts that at distances of $\hbar/2m_\pi c$ we can expect some of the time to encounter two pions in a nuclear system, at distances less than $\hbar/3m_\pi c$ we can expect to encounter 3 pions, and so on up to indefinitely larger numbers of particles as we refine our spacial description. Experimentally, the uncertainty principle requires us to use particles of higher and higher energy as we try to "peel the nuclear onion" in to shorter and shorter distances, and indeed as we do so, we produce more and more pions. If we use indirect methods to refine our distance measurements, such as electromagnetic fields, we also find (as is required by consistency) phenomena which can be attributed to "virtual" pions, even though we do not use enough energy to produce them as free outgoing particles. Because of the identity of various types of particles, we cannot distinguish at short distance which particles were "initially present" and which were there because of quantum fluctuations. In other words, the Wick-Yukawa mechanism necessarily generates an extremely non-local description of particulate systems at short distance.

These general arguments took concrete form for me when I first heard in a seminar by Ted Bastin at Stanford that the sequence 3, 10, 137, $\sim 10^{38}$ results from a simple hierarchical construction starting from the basis 0,1. This sequence, for a physicist, is the (inverse) numerical sequence of the super-strong, strong, electromagnetic, and gravitational interaction strengths. Current physics leaves the ratios arbitrary, discoverable and corrigible by

means of experiment. That they are given by hierarchical construction using
more basic elements, represented by $3 = 2^3 - 1$, $10 - 3 = 7 (= 2^3 - 1)$,
$137 - 3 - 7 = 127 (= 2^7 - 1)$, $\sim 10^{38} - 3 - 7 - 127 = \sim 10^{38} (\sim = 2^{127} - 1)$ was
tanatlizizing but not illuminating. Interactions are dimensional concepts, tied
in conventional physics to the units of mass, length and time, and hence have
no logical connection to "pure numbers" defined by any mathematical sequence.
The clue, for me, came in a later statement by Bastin in the same seminar
that he viewed "quantization" as quantization of mass rather than of action
(quantization of "action" was the historical route to quantum mechanics). If this
\"pure number" sequence represented numbers of particles, its dimensionless
character could be established using non-dimensional concepts. This reminded
me of an old paper of Dyson's which showed that quantum electrodynamics
changes its character for systems with more than 137 particle-antiparticle
pairs. Once this non-dimensional way of describing where an interaction concept
fails is suggested, a uniform description of "interactions" might be given in
terms of where the number of particles describable by each concept "becomes
inoperative", to steal the immortal phrase of Ronald Zeigler.

Dyson's argument[36] was constructed to meet a different problem. He was
concerned with the problem of how many terms are meaningful in the "renormalized perturbation theory of quantum electrodynamics" (QED), - a series in
powers of the fine structure constant $e^2/\hbar c \approx 1/137$. He noted that if we replace
e^2 by $-e^2$ in this series the result should still be meaningful (converge) if the
series is absolutely convergent. Physically this amounts to replacing QED by
a theory in which like charges attract and unlike charges repel. The original
series corresponds, term by term, to including as many electron-positron

pairs as there are terms in the series; by the 137th term in e^2 there are 137

pairs. If all particles of the same charge happen to be within their own Compton

wavelength ($\hbar/m_e c$), they generate within that volume an electrostatic energy

of $\sim 137\ e^2/r = 137\ e^2/\hbar/m_e c = 137\ (e^2/\hbar c)m_e c^2 \approx m_e c^2$. This is highly improb-

able in the "real world", since all these like charges would repel each other,

and the system would rapidly disassemble. But in a theory with $e^2 \rightarrow -e^2$

(accomplished by the imaginary replacement $e \rightarrow ie$) the system would implode

rather than explode, with enough gained energy to keep sucking mass out of

the quantum fluctuations particle by particle until the whole theory collapsed.

Therefore the QED series becomes meaningless after 137 terms. This is a

specific application of the Wick-Yukawa mechanism to a problem about the

self-consistency of renormalized quantum electrodynamics.

But this same calculation supports a description using different language.

We can restate the result as saying that it is not possible to isolate more than

137 individual electrically charged particles (with the universal charge e) within

a region as small as their own Compton wavelength. Since the electron is the

least massive charged particle, this also says that we cannot meaningfully

define what we mean by space-time volumes in regions smaller than (1/137)

($\hbar/m_e c$) by means of electromagnetic measurements. Gravitational definition

is still conceptually (though not practically) meaningful down to much shorter

distance, but once we try to push it down to distances of the Schwartschild

radius, which we could do by trying to describe $\sim 10^{38}$ protons within their

own Compton wavelength, the mathematical singularity discussed by Dyson

becomes physical - the particles disappear down a black hole and loose their

particulate identity. Thus the pure number 137 is simply the maximum number of charged particles we can identify individually within their own Compton wavelength, while the pure number $\sim 10^{38}$ is the maximum number of gravitating protons we can identify individually within their own Compton wavelength.

There is a technical point about "black holes" which needs explanation here. Dyson points out that the density of "particles", even when there are 137 of them within their common quantum wavelength, is so low that, although the total assemblage has an electrostatic energy of mc^2, the calculation requires the use of the Coulomb potential e^2/r for the individual particles only in an energy region which is well known and does not involve general relativistic effects. This is also true for the corresponding gravitational case where the Coulomb potential is replaced by the Newtonian potential Gm^2/r. For the problem of interest here, the "Schwartschild", or "black hole", radius can be estimated using only Newtonian gravitational concepts and special relativity.[32] We do not have to invoke "curved space time" or any of the complicated technical apparatus (and postulates about continuous space-time coordinates) of the general theory of relativity in order to get, qualitatively, particulate quantum systems that have such intense gravitational fields that no particle or quantum can escape from them. Perhaps it would be better to call these particulate "black holes" generated by the Dyson mechanism by another name, but they have the most important property of a "black hole", and we think the term should be retained. It will be an interesting technical problem if these ideas ever generate a quantitative theory to see whether all the properties of the "black holes" predicted by the general theory can be reproduced, or whether there will be experimentally detectable quantitative differences.

With this clue, the pure numbers 3 and 10 are also interpretable. The electrostatic interaction we used above was e^2/r, while the corresponding

Yukawa interaction is $f^2 \exp(-m_\pi c/\hbar r)/r$. Since $f^2 = 0.08$, were it not for the exponential the same argument says that we cannot (e.g. by tracing back the trajectories of energing particles to their production volume) meaningfully describe the presence of more than ~ 12.5 pions within their own Compton wavelength; hopefully a more exact calculation including the exponential would bring this down to ~ 10. In any case we do not expect this number to be an exact integer (or $e^2/\hbar c$ to be <u>exactly</u> 1/137) because in some sense quark, pionic, electromagnetic, and gravitational effects all occur within any system.

To understand the number 3 as the maximum number of quarks we can meaningfully define requires a slightly different argument. In the quark model, massive bosons (pions, kaons, ...) are quark-antiquark pairs with quark number zero, while hadronic fermions (protons, neutrons, sigmas, lambdas, ...) are bound states of three quarks. No free quarks have ever been observed, and the quark quantum numbers can be assigned to the individual hadronic systems only in experiments where the two combinations mentioned above can be isolated. Thus the quark theory (if, as it often does, it excludes free quarks as a possibility), has precisely the required character of saying that the only quark numbers which make sense for isolatable individual particles are zero or three.

A number of problems remain before this dimensionless description can be meaningfully connected up to the Bastin approach. One of these is trivial, and was solved during a conference on the chapters in this volume. This is simply to reduce the relativistic quantum particle theory to dimensionless form. As any physicist knows, any dimensional result (i.e. any result that depends on the units in which we express our measurement) can be expressed in terms of three basic units for mass, length, and time. Thus all we need do is to show that the theory <u>requires</u> three fundamental units which can be defined independent of any dimensional considerations. It then is a matter of experimental convenience how we choose to assign numerical values to these units, and is not a matter of any fundamental significance. Our unit of mass is obvious, as the lightest massive particle is the electron (i. e. , empirically mass is quantized). All mass ratios with respect to the electron are pure numbers; whether these ratios are real or rational numbers is left open for the theory to decide. Such a question can never be settled by experiment, except for theories which require a specific rational number as an exact consequence (this is how Eddington's theory of the fine structure constant became experimentally counter-indicated). The second fundamental unit is the limiting signal velocity of special relativity. Whitehead[36] has argued that any theory which uses events in space-time volumes (not necessarily at points) as a basic conceptual tool should have such a limiting velocity. That this is also the ratio between electrostatic and electromagnetic units (and hence the "velocity of light") and the square root of the conversion factor between mass and energy is then a requirement of the theory; the theory is therefore frangible if these approximate equalities turn out not to be exactly true. The third fundamental constant comes from the wave-particle duality of quantum mechanics, and is the conversion factor between the (reduced) deBroglie wavelength of the particle (λ) and its momentum (p) namely $\hbar = p\lambda$; $\hbar$ is Planck's constant divided by 2π. In fact these last two "natural" units ($\hbar$ and c) are already customarily set equal to unit in high energy particle physics; only the masses of the particles (rather than ratios to the electron mass) are used dimensionally.

The much more difficult problem is to construct a theory in which the hierarchical steps of Bastin's construction lead successively from quarks to pions to electrons to black holes. Starting as I do from existing particle theory, there is no apparent reason why the 10 should not have turned out to be 5 or 15, or the 137 to be 100 or 150, for example. Thus all that can be presented here is a program for constructing, step by step, a theory which is conceptually compatible with Bastin's, but whose success or failure must lie in the uncertain future. This program will be presented in the next section.

V. ATOMS AND THE VOID SUFFICE

Although I believe that the program for creating a Democritean quantum mechanics presented in this section has points of similarity with the self-generating "computer program" type of approach advocated by Parker-Rhodes, Bastin, Amson, and Kilminster[39], the methodology used is different. To me it is obvious that discussions of physics (and metaphysics) such as this can only take place in a culture with a long history of linguistic communication. I therefore find it silly to ignore what is already current experiential "knowledge" in that community, and will make free use of it in what follows. But I would also insist that "circularity" in the argument can be avoided if the physical theory which we construct can be shown to be capable of producing, in time, the community in which this discussion is taking place. Of course to make that statement "scientifically plausible" would take many volumes of detailed argument covering various aspects of physical cosmology, evolutionary biology, the evolution of communication systems, and the class struggles that have led to the current world crisis. An outline of this evolutionary development has been sketched out, and will be published elsewhere.[11]

To set the philosophical tone of the discussion, I take exception at the outset with the famous mathematician who said that "God made the integers; all else is the work of man". The integers, like any other part of mathematics, are a human creation, and subject to limitations than come from our finite nature. This is made clear by Godel's Theorem[40], which shows that any postulate system capable of generating the integers necessarily creates an infinite class of (arithmetically true) propositions which cannot be proved to be true within the system. By suitable additional postulates, these undecidable propositions can be made provable, but that process necessarily creates a new infinite class of undecidable propositions. Thus, the indefinitely extendable characteristic of the integers extends to the propositions which they allow us to state, and never closes. This is quite compatible with the methodology I adopt. I recognize from the outset that any system of mathematical or scientific propositions can always be indefinitely extended, and that "closure" can never be more than approximate.

Although I do not aim at "closure" I still aim at the maximum generality I can achieve at any (necessarily finite) stage in the development, and try to find in any aspect of experience "counterinstances" - either factual or conceptual - to the construction which is being developed. That is, part of my methodology is to deliberately try to force contradictions, and to try to meet them. This is how I understand the dialectical process as described by Chairman Mao.[41] But to make the contradictions as clean as possible I also invoke the methodology of the spiritual godfather of the positivists - William of Occam. I try to use his razor to pare away the excressences which have grown up historically around the physical concepts invoked, and more specifically to use an "operational" analysis similar in spirit to that of Mach, Einstein, Heisenberg, and

Bridgman. This approach gets rather technical so only an outline is given here. Details will be presented elsewhere. [10]

I start with the (initially unexamined) concept of "particle detectors" such as are actually employed in high energy particle physics, and which can to some finite "accuracy" be used to assign macroscopic space and time coordinates (which concept also is not initially examined) and hence to define particle velocities. Postulating a limiting signal velocity, and adopting the Einstein convention for the simultaneity of distant events, then gives the proper Lorentz transformations. Postulating homogeneity and isotropy extends this to the Poincare transformations. To define the mass of particles I assume that there are devices which can change either the energy or the direction of the momentum of a particle; together these define a ratio of charge to mass and either, in conjunction with a velocity measurement, then defines an invariant mass. Adding the concept of a "grating" (or discussing finite aperture diffraction) then allows me to define the deBroglie wavelength, establish the wave-particle duality, and hence the uncertainty principle. This suffices to define covariant free-particle wave functions, and thanks to the Wick-Yukawa mechanism, a complete theory of particle scattering experiments. Since this aparatus entails the concept of (quantized) angular momentum, I can also introduce a dicotomic spin function. The simplest representations of a particle with this internal coordinate are the Dirac spinors, which provides an argument for the spacial coordinates being three-dimensional. It also provides us with the possibility of anti-particles.

Having defined particle wave functions, it is possible now to introduce external electromagnetic fields (defined in terms of macroscopic measurements

of charge and current) and hence to exibit explicitly the energy and momentum
changing devices invoked above to define mass. Further, this construction
yields the Dirac equation, and in the non-relativistic limit the Schroedinger
equation, thus recovering the conventional results of quantum mechanics to
lowest order in the external field. By assuming that density and velocity
distributions computed from the wave functions of <u>charged</u> particles are <u>also</u>
the sources of electromagnetic fields in Maxwell's equations, we can then bring
the external fields within the framework of description, and justify the use of
energy and momentum changes to measure mass. By using "quantum transitions"
generated by external field (specifically the photoelectric effect), we can then
also explain how a "particle detector" works, and thus close the logical circle
to an accuracy of $e^2/\hbar c$. The theory at this level will stand or fall, in the eyes
of most physicists, on whether the motion of <u>two</u> charged particles, each
acting as the source of the field for the other, can be computed to order
$(e^2/\hbar c)^3$ in agreement with experiment and the renormalized perturbation
series for QED. A second critical test will be whether the properties of pion
and nucleon systems can be computed, not from phenomenological T-matrices
but from T-matrices "bootstrapped" from the 3π and 4π systems and the Yukawa
coupling constant. Assuming such successes can be achieved (which will take
time), it will be interesting to see how far the theory can be pushed into
the less well understood regions of very short times and high energies where
quarks, neutrinos, and gravitational effects become dominant.

So much for "normal science" and operationalism. In the remainder of
this paper I will attempt instead to build up the same picture of particle
quantum mechanics starting from the minimum number of abstract postulates,

introducing additional ones only when they appear to be required by known experimental "facts". This methodology is not so different as it might seem at first glance - building up a minimal set of postulates in order to achieve agreement with an extant theory is just as much an application of "Occam's Razor" as the reverse procedure of paring them down to a minimal set. In both cases we are still guided by the current status of physics; the requirement that we ultimately be able to explain the development of human culture and science in terms of the theory remains. But the constructive approach is much closer to the methodology of Parker-Rhodes, et. al.[39] and was deliberately undertaken for the purposes of this volume in the hopes that it might aid their program. I was quite surprised, and a little frightened, to find out how far one can go toward recreating, at least qualitatively, most of the ideas currently being pursued at the frontiers of particle physics using so few assumptions.

The basic postulate adopted here is similar in spirit to that of Parker-Rhodes, Bastin, Amson, and Kilminster.[39] They start with the dicotomic pair (0, 1), but I would prefer to leave my starting point even vaguer and claim that a minimal requirement for rational thought be that one can distinguish something from nothing. I believe that this is the idea behind the basic materialistic postulate of Leucippus and Democritus that there are only atoms and the void; I therefore call it the "Democritean" postulate. An immediate corrolary for me is that any refinement of this idea must not allow the construction of an undifferentiated continuum. For me, the idea of the undifferentiated continuum, the common goal of many mystical traditions[42] also called Nirvana, is the antithesis of rational thought, --although I am quite prepared to admit that the attempt to attain that goal might be a rational activity. Of course we allow discrete approximations that can approach "points in the continuum" in the mathematical sense; but we avoid making use of the limit points themselves.

According to our basic postulate, there must be "particles" which we can distinguish from the void, and which are denumerable. The necessity for distinguishing them requires an additional concept, which I take to be relative motion. For the moment, we will assume that the particles can be distinguished from one another only by their "motions", and that the possibility of zero motion exists for some sets of the particles. But if these motions could be arbitrarily rapid, we could use them to define a continuum background space, a concept we have ruled out above. Therefore we must assume that there is an upper limit to relative motion. Once we accept this we can, roughly speaking assume that motions, relative to some set of particles that have zero motion, can be ordered between zero motion and the upper limit; for the moment we assume that this ordering does not change. But if the motions do not change, something must, or we would be back in a static Nirvana   with no means of distinguishing our particles. What changes when two particles are in constant relative motion with respect to some set of particles that have zero relative motion is the "distance" between the pair. We distinguish two cases: the distance first decreases to zero and then starts increasing, or it keeps on increasing; we save discussion of how long distances can keep on increasing till later. Unless we are willing to introduce "structure" into this description of particles some of which are moving and some of which are at rest, we now have all the ingredients for describing particles in uniform relative motion with respect to each other along a line; further if only relative motions are to be meaningful and there is a limiting velocity, the whole system can be made invariant with respect to which particles are assumed at rest and which in motion only by means of the Lorentz transformations.

At this point the thoughtful reader will note that, although we have introduced relative motion, we have nowhere introduced the concept of direction, so in conventional terms all we have are particles moving along a line. This produces a peculiar result. If we assign numerical values to some description, and then replace all velocities and distances by their negatives, we seem to have a description of the same situation but with different (negative for positive and positive for negative) numbers assigned to each particle. We could either assume that this artificial character of our numerical description has no significance (in which case we would have to describe only the symmetric systems in which this interchange produced no change) or add a new descriptive element. This can take the form of an "internal coordinate" for each particle that tells us whether, relative to some convention for the "direction" of the line, the particle is moving in the positive or negative direction. The simplest interpretation of this "spin coordinate" is that it represents a rotation about the line in a space with at least two directions[10] perpendicular to each other and to the line. If we confined ourselves to a plane we would do no better of than on a line ; two directions perpendicular to the line are needed to form non–superposable objects that distinguish "left" from "right". This possibility is a basic empirical requirement. This allows us to extend our concepts to a 3+1 dimensional Minkowski space - simply by requiring that this space be homogeneous and isotropic. Note that the space we invoke was constructed from finite particle motions assumed to be specified to some approximate numerical accuracy; we do not <u>have</u> to assume that this accuracy of measurement can be increased indefinitely (in fact this is ruled out by our anti-continuum corrolary to the Democritean postulate).

Given the particles in uniform motion, we note that once the description is established, all past and future motions are determined; our apparent motions are simply the description of a static four-dimensional world and we are back in Nirvana. So we assume that both directions and velocities can <u>change</u>. At this point, so far as I can see, we need a new concept that does not follow from the Democritean postulate in any obvious way. This is the basic mechanical postulate of Newton's third law, that action and reaction are equal and opposite, or that <u>momentum</u> is conserved. This allows us to introduce a Lorentz invariant descriptive of each particle by noting that the Newtonian concept applies only to the three spacial components p, and that for Lorentz invariance we must have a time-like component ϵ. The mass in then <u>defined</u> by the scalar invariant $m^2 = \epsilon^2 - p^2$, the same in all coordinate systems. All we need do now is postulate some law that specifies how the motions of the particles affect each other, conserving momentum and mass in a Lorentz-invariant way, and we have a full blown relativistic particle mechanics. But this is again a <u>deterministic</u> system, just as much a static four-dimensional world as the one ruled out above for "free particles", and again in conflict with our basic requirements.

The basic assumption that we make in order to avoid our essay into rational thought collapsing back into Nirvana should be clear from the last section: we assume that the <u>number</u> of particles can change if we try to count them in a small enough volume. In order to make this concept universal, we say that for any (finite) mass, this volume has a radius of $\hbar/mc$ where c is the limiting velocity already introduced and $\hbar$ is a universal constant whose numerical value depends on how we ultimately choose to relate this theory to our experiences. The reason we need a fundamental length at this point is also Democritean; if our particles could be arbitrarily small we would get back a continum at short distance just as surely as if their number could be arbitrarily large. <u>How many</u> particles we can still distinguish within this volume depends on the <u>means</u> we use to describe them - quark, pionic, electromagnetic or gravitational, and is dimensionless. Our remaining task is to try to derive, again wherever possible from our discriminatory postulate, the necessity for these different aspects of what are by now quantum particles satisfying a Lorentz-invariant wave-particle dual description, and whose number can change thanks to the Wick-Yukawa mechanism.

The fact that we have introduced the concept of changing particle number, and also rejected determinism on the grounds that it would lead back to a static four-dimensional universe, suggests that the means by which particle number changes is random. This basic method of avoiding Nirvana immediately provides a model for time in accord with experience - past events can be considered fixed, but due to the random character of particulate events, only the probabilities of future occurances can be predicted from this knowledge of the past. This means that our basic description of particles is statistical. Thanks to Max Born we know how to accomplish such a description: wave functions whose amplitudes satisfy a "causal" law are interpreted as predicting the probability of finding a (the) particle(s) represented by the wave function in one (or more) specifiable space-time volume(s). As has been discussed at length in other contexts in this paper, these wave functions need only represent the motion of "free particles". The phase velocity of these waves is ϵ/p and hence lies between the limiting velocity and infinity, while the velocity with which a group of waves known at some time in the past to have been (approximately) localized in some finite volume (the "group velocity") is $d\epsilon/dp = p/\epsilon$, and lies between zero and the limiting velocity. The necessity of constructing such groups of waves to describe localizable particulate events automatically introduces the "uncertainty principle" into our theory. Since we already have the mass-energy relation, this immediately allows us to identify the "Wick-Yukawa mechanism" as the means by which particle number changes, and hence allows us to have scattering, particle production, and particle annihilation (change) without invoking the concept of "interaction".[10] In the limit in which we can ignore effects due to the finite wave lengths of the particles, the wave theory can be reinterpreted as describing particles moving on trajectories perpendicular to the wave fronts, and we recover the relativistic mechanics of a system of particles whose velocities are the "group velocity" of the waves.

The expert will note that, since we have already argued for the necessity of "spin" in order to discriminate direction of motion, the construction sketched in the last paragraph allows us to write down conventional "Dirac spinor wave functions" for one or more particles. This raises an immediate problem. Although our "fixed past-uncertain future" point of view allows us to distinguish a unique meaning for the algebraic sign of the time parameter in our theory (conventionally, negative times are called past and positive times future), the corresponding uncertainty in the conjugate coordinate, the energy, is $\underline{not}$ resolved. Thus the theory seems to require negative as well as positive energies - which would have disasterous consequences. For instance of two particles of the same $(mass)^2$ value and equal but opposite momenta came together, they could annihilate each other leaving behind not just the "undifferentiated continuum" of Nirvana, but quite literally $\underline{nothing}$. We conclude that we $\underline{cannot}$ allow negative energies in the theory. Another reason we cannot allow negative masses is that they would lead to "anti-gravity", for which there is no empirical evidence. The simplest way to avoid them is to assume that (consistent with our Democritean framework), $\underline{mass}$ is an intrinsically positive concept; then, thanks to the mass-energy relation, only positive energies can occur. We still must require that whatever our lightest mass is, there is a $\underline{maximum}$ number of particles of that mass which can be meaningfully described within their own Compton wavelength. Since this ultimate limit refers $\underline{only}$ to mass, this is the basic $\underline{gravitational}$ concept in our particle theory. Empirically the lightest massive particle is the electron, and the maximum meaningful number of electrons within an electron Compton wavelength is $\sim 10^{44}$. Also empirically, the maximum number of baryons (protons, neutrons, sigmas,...) which can be described within a baryon Compton wavelength is $\sim 10^{38}$. The fact that there are two basic numbers here rather than one poses a still unsolved problem - where does the dimensionless mass ratio $M_p/m_e \sim 1837$ come from ? The fact that the sequence $3, 10, 137$ ends at 10^{38} rather than 10^{44} shows that it should be read up from quarks (which is the baryon sequence) and does not refer to electrons (leptons). We will see below that there are other reasons for believing that, at some very deep level, there must be two basic types of particle and not just one. The origin of this fact is left as a problem for future research.

Although we have succeeded in arguing that the only interpretation of Dirac spinors which we can allow is one corresponding to particles of positive mass and energy, this does not eliminate the "negative energy states" from the picture. Instead we are now required to assign a second dicotomic variable (in addition to spin). This new quantum number is conserved for systems of particles of one type, and the difference between the numbers of the two types is conserved in mixed systems. Since this is a distinguishing characteristic separate from mass or spin, there is some maximum number of particles which can still be discriminated by this characteristic when they are packed within their own Compton wavelength. If we assume that this number is (approximately) 137, we can identify this new quantum number with electric charge. Why the actual value should be 137 is left for future research; hopefully it can be "derived" along the lines of Parker-Rhodes, et.al.[39]. But whatever this number is, the conservation law already mentioned requires it to be the same

for <u>all</u> massive spin 1/2 particles, independent of their mass. From this point
of view, conservation of charge is therefore a reflection of a dicotomic property
of spinor wave functions. Since we also have positive mass values, and scattering
via the Wick-Yukawa mechanism, Stapp's derivation[7] shows that reasonable
assumptions about the conservation of probability suffice to establish "the
connection between spin and statistics" and the Pauli exclusion principle for
spinors (i.e. that no two half-integral spin particles can occupy the same state)
which is crucial for what follows.

Since it is only the <u>difference</u> between the charges which is conserved, our
theory allows for electrically neutral systems, and hence for transitions between
a state consisting of a positively and a negatively charged particle (e.g. an
electron and a positron or a proton and an anti-proton) to such states. One
possibility is that such states are massive, of which a specific empirical
example would be a neutron and an anti-neutron, but it is also conceivable that
the transition leads to two "particles" of negligible mass traveling at the
limiting velocity and with their spin aligned along or opposite to their direction
of motion. These are obviously <u>neutrinos</u>. We would like to have them available
for empirical reasons, but would like to get them out of the framework already
established in order to avoid having to apply Occam's sharp tool at a later stage.
One possibility is that they simply spinning neutral black holes. This would
explain why at high enough energy so that the <u>gravitational</u> mass of the relative
energy of motion of the electron and positron becomes comparable to the mass
energy, the cross section for producing neutrinos becomes enormous (in the
phenomenological Fermi theory of "weak interactions" it becomes so large
that the probability exceeds one - i.e. the theory has to break down - in this

limit), yet at low energy the effect is extremely weak. This would also explain why 10^{13} (the conventional measure for "weak interactions" in the $3, 10, 137, 10^{38}$ sequence) is not primary; it becomes a derived concept which must eventually be calculated from a more detailed articulation of electromagnetic and gravitational concepts. Another problem is that there are not one kind of lepton but two - muons and anti-muons as well as electrons and positrons - and each carries with it an associated type of neutrino. The ratio of 207 between the muon and electron mass is the only other distinguishing characteristic between the two types of leptons. Hopefully it is not an independent quantity, but can be linked up to the other large mass ratio of baryon to electron mass. Qualitatively, the m_μ/m_e ratio is the m_π/M_p ratio times the M_p/m_e ratio (to the extent that 273 and 207 are approximately the same), so one place to start looking is whether we can understand the m_π/m_e ratio. Before we do this, however, we must say a few words about electromagnetism.

Historically, the first "zero mass particles" used in quantum theory were light quanta (photons) not neutrinos. Like neutrinos, they have two spin states parallel or antiparallel to their momentum, but these states have spin 1 rather than 1/2. Therefore any number of them can be packed into a volume (at least until their energy begins to produce gravitational effects - Wheeler's geons), and as is characteristic of Bose statistics, the more there are, the more likely this is. Outside of their angular momentum and energy they have no other defining characteristics (unlike neutrinos which carry both lepton number and muon or electron number); indeed the type of "quantum" used to describe transitions of charged particle systems depends on mathematical convenience and is not uniquely dictated by the problem. From a Democritean point of view, I would therefore hesitate to call light quanta "particles", and I have a strong urge (already expressed above) to eliminate them from the description altogether. While the program undertaken here must, ultimately, be able to lead to an understanding of <u>classical</u> electromagnetic and gravitational fields, we have already seen that these are rather special limits in a particle theory. As far as "field quantization" goes, Bohr and Rosenfeld showed long ago[44] that to lowest order in $e^2/\hbar c$, any material system of sources and sinks of the "field" which satisfies the uncertainty principle leads to the <u>same</u> restrictions on the measurability of the fields as does "field quantization". This is a problem which must be tackled to higher order when discussing QED; as was already noted above, that problem is beyond the horizon of this paper.

One of the oldest ideas about the electron is that its mass may not be an intrinsic property, but simply a reflection of the energy due to its electric charge and mass-energy equivalence. Classically, if we assume that the charge is packed into a small enough radius to accomplish this, we find a "classical electron radius" of $e^2/m_e c^2 = 2.8 \times 10^{-13}$ cm, 137 times shorter than the Compton wave length of the electron $\hbar/m_e c = (\hbar c/e^2)(e^2/m_e c^2) = 137\,(e^2/m_e c^2)$. We have also seen that we can define what we mean by up to 137 charged particles of one sign within their own Compton wavelength. Such a system would be highly unstable, but if we combined it with a system of 137 particles of <u>opposite</u> charge the resulting neutral system although electrostatically unstable is not completely ephemeral. It would have a <u>finite</u> lifetime against decay into neutrinos or electromagnetic radiation, and is the most massive system of electrons and positrons we can define. If electrons and positrons have intrinsic mass, this system could not decay purely into radiation, but if they have only electrostatic energy, this system could go into 2 γ-rays, and would have a Compton wave length of

1.4×10^{-13} cm. Such a model <u>predicts</u> the mass of the observed neutral pion, and its decay ($\pi^0 \rightarrow 2\gamma$). Systems with smaller numbers of electrons and positrons are conceivable, but we gain electrostatic energy by adding an electron (or positron) and the appropriate neutrino to such a system; since they are stable against electromagnetic decay (because of charge conservation), they have a longer lifetime. Systems with two electrons or two positrons are presumably electrostatically unstable, thus explaining why the pion is a charge triplet $\pi^+ \pi^0 \pi^-$. If indeed the pion is composed of 137 $e^+ e^-$ pairs (with an electron-neutrino pair or positron-antineutrino pair added for the charged members), the pion would have the requisite odd spacial parity.

Whether or not this model for the pion works, once we have a charge-triplet, pseudoscalar entity of the right mass, a great deal of "strong interaction physics" can be "bootstrapped" out of this. All one has to do is to require the pion to be a bound state of three pions (a particular application of the Wick-Yukawa mechanism, where we need three rather than two because even and odd pion number systems do not freely transform into each other). One approximate way of doing this has recently been presented by Brayshaw[45] and can be interpreted as showing that the probability of the scattering of two pions at low energy is not a free parameter but is determined by the pion mass. Earlier work by Gore[46] and others showed that given the right low energy scattering, and the analytic structure required by the charge-triplet pseudoscalar structure of the pion, one of the "vector mesons" (resonances in the two and three pion systems) has to follow in terms of this single parameter. If, for each system, the single parameter is indeed determined by the pion mass itself, as Brayshaw's calculation suggests, there may well be a complete low energy dynamics of pionic systems with no explicit "empirical" input.

The construction of a parameter-free theory of "strong-interactions" need not stop with the pions. Provided only we can see why no more than about 10 pions can be meaningfully described within their own Compton wave length, we can "derive" what is conventionally called the (pseudovector) Yukawa coupling constant. But again, given the correct dynamics of two and three pion systems, it has already been shown that this one number plus the existence of spin 1/2 neutrons and protons can be used to couple the nucleonic dynamics to the known pionic dynamics and predict correctly the very complicated structure observed experimentally in the scattering of a pion by a nucleon (including the production of a second pion) and, given that, the scattering of two nucleons (including the production of a pion). Thus all of the "nuclear force" picture up to the point where "strange particles" enter can be brought, at least conceptually, within this basic schema.Bringing strange particles into the act in turn requires only going on up the structural hierarchy to the three quarks of which the neutron and proton, and the "strange" baryons are sometimes thought to be composed.

The entrance of the nucleons (or underlying quarks) at this point in the construction suggests a possible answer to the question of why there are, apparently, two basic masses for spin -1/2 particles - leptonic and baryonic - and not just one unit of mass. We have seen that it is at least conceivable that the "mass" of the electron and positron is simply a reflection of its electromagnetic properties, and not a "mechanical mass". Further, we have seen that there is a possibility of using electrons and positrons to construct the pions and other heavy bosons without introducing any new concepts or constants. The difficulty with this "universe" is that it is not stable, if it starts out electrically neutral. Eventually the electrons and positrons annihilate each other

leading to neutrinos (spinning black holes) and perhaps to ordinary black holes.
If the neutrinos are indeed spinning black holes, and there is no overall angular
momentum to the universe, they to will end up as spinless black holes. And a
universe consisting only of black holes with zero angular momentum strikes
me as having so few observable properties that it may well be yet another
version of Nirvana.   Another way of putting it is that so long as we have
spin -1/2 neutrinos we still have a lever for bootstrapping us back to the world
of quantized masses, but once these disappear we might have left only the black
holes of the general theory of relativity, which can have continuous mass values
and hence violate our basic Democritean postulate. We need a basic mass value
in the theory, which we can take to be the quark mass. Given that we can then
try to construct the sequence: 3-quarks, 10 quark-antiquark pairs (pions and
other bosons), 137 charged particles, 10^{38} hadrons, as already suggested.
Once we have electric charge, the possibility of systems with charge and only
electromagnetic mass occurs naturally and gives us electrons, positrons, and
their associated neutrinos. But since we now have a basic mass, the muon
might carry the remnant of this quark mass, explaining the large mass ratio
to the electron, and some trace of it might even persist in the muon neutrino
(provided it is small enough) without doing violence to any currently known
experimental facts.

One conceptual loose end remains, and at a level that could bring this
whole scheme down in ruins. We found the logical necessity for the existence
of electrical charge as a consequence of our basic postulates, once we had
succeeded in constructing positive energy spinors, but then went on to use
other electromagnetic properties of the conventional theory (basically Coulomb's

law) without constructing them.

From the point of view of "classical" physics, it is the long-range gravitational and electrostatic forces which knit the world together and provide the solid base on which all these speculations rest. Newton's laws were brought to the fore by gravitational problems and illustrated by him with examples drawn from the solar system. Electricity was investigated by Coulomb in analogy with Newton's gravitational theory, and both were mathematicized by Laplace in a single fertile paradigm. Rutherford provided the experimental facts for Bohr's "planetary" model of the atom, and Bohr quantized that model, using this mathematical paradigm as the known limit at large distances. Even Schroedinger's equation and Heisenberg's matrix mechanics used the inverse square law as the exemplar of the theory and the test case for comparison with experiment. From that point of view, quantum mechanics provides an explanation of the stability of atoms needed to complete the picture and (with the exclusion principle, electrons and nuclei) the details needed to understand chemistry, rods and clocks, evolution, and the phenomenal world. The ultimate justification of physics as a human creation that leads us back to these concepts by understandable historical and retrodictable stages relies on that consistency.[11] But the link connecting these long-range, classical phenomena with the Democritean picture discussed so far has not been provided.

The missing link in the argument may not be too hard to supply. As has already been noted, electromagnetic "quanta" (photons) differ from the other particles we have been discussing in that they have no quantum numbers intrinsic to their description. Loosely speaking they have "spin one", but the same mathematical property is expressed by the classical equations of Maxwell and Lorentz. We can bring them into the picture developed above by putting a neutrino and an anti-neutrino together. We cannot simply use two neutrinos or two anti-neutrinos, as then the system would have lepton (or muon) number ±2, while photons cannot be allowed particulate quantum numbers. If we fuse a neutrino and an anti-neutrino travelling in the same direction, their spin's and lepton (or muon) numbers cancel, so the "spin one" character of photons cannot be intrinsic if this model is to apply. If we give the neutrino and anti-neutrino one unit of relative angular momentum in space, we provide the missing quantum number, and the two states needed with respect to the direction of the photon. As already noted, this is all we need for describing other ways photons are used (i.e. we can use these circularly polarized states to construct linearly polarized states or visa versa).

There are a number of advantages in thinking about photons as zero mass states of neutrinos constructed in such a way as to eliminate any quantum numbers which come from their particulate substructure. To begin with this "explains" why classical physics could get started without using particles as a basic and unavoidable concept. The momentum and energy of such a system are not separate but equal (in a dimensional system with c=1), and continuously variable. The zero mass, given a limiting velocity gives the inverse square law for forces (Coulomb's law) in the static limit. But this continuity breaks down once particles with electromagnetic mass (electrons) or intrinsic mass (baryons) come up over the horizon. Whether this neutrino model of photons can meet the problem of introducing electromagnetism into a Democritean theory remains for the future to decide. In the meantime, we take comfort from the fact that more conventional theorists like Steve Weinberg are (for different reasons, but presumably making use of the same symmetries used

above) trying to identify weak and electromagnetic "interactions" as two aspects
of related phenomena.

Putting together a neutrino and an anti-neutrino to form a spin one photon
suggests that the other long-range classical "field" could be constructed in a
similar way. Putting them together with zero relative angular momentum would
yield a zero-spin zero-mass field congruent with Newton's gravitational theory.
So, at least in terms of quantum numbers, our particulate point of view can
recover two different long-range effects that mimic the needed aspects of the
Newtonian and the Maxwell fields. By adding a limiting velocity (special relativity) we get photons with all the symmetry properties needed, and (within a
factor of two) the properties of "black holes" needed at the level of precision
of the arguments presented in this paper. To construct the full tensor "field"
of Einstein's general theory from neutrinos would require at least two kinds,
and hence might provide an additional clue to the baryon-lepton and muon-
electron puzzles which are, from the point of view of this author, the least
understood problems confronting modern Democriteans.

Two basic ideas emerge from this discussion as extraordinarily fertile:
the atomic or particulate requirement that allows us to start a rational discussion and the random fluctuations of particle number which prevent us from
slipping back into a deterministic world; instead we find a fixed past from which
we can at present deduce only the probabilities of future events. These basic
concepts entail a limiting velocity, wave-particle duality and a unit of mass,
thus removing dimensional constants from the theory. Once we add the basic
mechanical postulate of the conservation of momentum, these ideas can be
articulated quantitatively, and by further application of the discriminatory

postulate require two types of limiting particle number which we can identify with the short-range breakdown of the space-time description of the classical electromagnetic and gravitational fields. By invoking technical details about Dirac spinors we arrive at neutrinos and (possibly) the massless quanta of the Maxwell and Newtonian fields. But this world is unstable, implying the necessity of an intrinsic unit of mass, which we identify with the baryons. By viewing bosons as assemblages of electrons and positrons we provide the link between the hadronic and the leptonic worlds. This whole construction would not be possible without the empirical knowledge of the classical fields and quantum numbers of the "elementary particles" which guided each step. But the classical and quantum pictures of earlier theories, when applied to known cosmological phenomena provide a possible route from the particles to the physicists who discuss them. [11] Thus the logical loop "closes" at the point when physicists can logically reconstruct the particles, but necessarily only as an approximation to the fixed past. The novelty which has already emerged by this route, together with the basic fact that at best we can only predict probabilities for future events, warns us to anticipate still more novelty in the future.

REFERENCES

1. Paul Forman, "Weimar Culture, Causality, and Quantum Theory, 1918-1927: Adaptation by German Physicists and Mathematicians to a Hostile Intellectual Environment", Historical Studies in the Physical Sciences, 3, 1 (1971).

2. P. Schlipp, ed. Albert Einstein, Philosopher Scientist, Library of Living Philosophers, Evanston, 1949.

3. G. C. Wick, Nature, 142, 993 (1938).

4. H. Yukawa, Proc. Phys. Math. Soc. Japan, 17, 48 (1935).

5. J. M. Ziman, Public Knowledge, Cambridge University Press, London, 1968.

6. S. J. Brodsky and S. D. Drell, Ann. Rev. Nucl. Sci., 20, 147 (1970).

7. H. P. Stapp, Phys. Rev., 125, 2139 (1962).

8. T. S. Kuhn, The Structure of Scientific Revolutions, University of Chicago Press, Chicago, 1962.

9. G. F. Chew, S-Matrix Theory of Strong Interactions, Benjamin, New York, 1961.

10. H. Pierre Noyes, "A Democritean Phenomenology for Quantum Scattering Theory", to be submitted to Foundations of Physics.

11. Pierre Noyes, "A Scientific Retrodiction of Our Past", submitted to Theoria to Theory.

12. L.D. Faddeev, Zh. Eksperim. i. Teor. Fiz., $\underline{39}$, 1459 (1960) (English transl. Soviet Physics - JETP, 12, 1014 (1961).)

13. L.D. Faddeev, Mathematical Aspects of the Three Body Problem in Quantum Scattering Theory, Davey, New York, 1965.

14. H.P. Noyes, Proc. 2nd Int. Symp. on Polarization Phenomena of Nucleons, P. Huber and H. Schopper, eds., Birkhauser Verlag, Basel und Stuttgart, 1966, p. 238.

15. H.P. Noyes, Phys. Rev. Lett., $\underline{15}$, 538 (1965).

16. K.L. Kowalski, Phys. Rev. Lett., $\underline{15}$, 798 (1965).

17. H.P. Noyes and H. Fiedeldey, Three Particle Scattering in Quantum Mechanics, J. Gillespie and J. Nuttall, eds., Benjamin, New York, 1968, p. 195.

18. H.P. Noyes, "Some Spontaneous Remarks", The Three Body Problem in Nuclear and Particle Physics, J.S.C. McKee and J.M. Rolph, eds., North Holland, Amsterdam, 1970, p. 434.

19. H.P. Noyes, Bull Am. Phys. Soc., $\underline{16}$, 20 (1971).

20. V. Efimov, Phys. Letters, $\underline{33B}$, 563 (1970); Yadern. Fiz., $\underline{12}$, 1080 (1970) (English transl. Societ J. Nucl. Phys., $\underline{12}$, 589 (1971).)

21. V. Efimov, Nuclear Physics, $\underline{A210}$, 157 (1973).

22. H.P. Noyes, "Solution of the On-Shell Faddeev Equations" (submitted to Phys. Rev. Letters).

23. T. E. Phipps, Jr., Phys. Rev., 118, 1653 (1960).

24. T. E. Phipps, private communication, quoted in SLAC-PUB-1351 "Fixed Past and Uncertain Future: an exchange of correspondence between Pierre Noyes, John Bell, and Thomas E. Phipps, Jr."

25. T. E. Phipps, Jr., Dialectica, 23, 189 (1969).

26. This is a rumor of an experiment by an unlucky graduate student whose results of several years standing lead to the opposite conclusion to that of Ref. 29; there is no obvious experimental contradiction between the two results, but the hesitation to publish is understandable, if regretable.

27. J. S. Bell, Physics, 1, 195 (1965).

28. E. P. Wigner, American Journal of Physics, 38, 1005 (1970).

29. S. J. Freedman and J. F. Clauser, Phys Rev. Letters, 28, 938 (1972).

30. T. E. Phipps, Jr., Foundations of Physics, 3, 435 (1973).

31. T. E. Phipps, Jr., Nature, 195, 1088 (1962).

32. B. Gal-Or, Science, 176, 11 (1972).

33. T. Gold and D. L. Schumacher, eds., The Nature of Time, Cornell Univ. Press, Ithaca, 1967.

34. J. M. Jauch, Are Quanta Real ?, Indiana Univ. Press, Bloomington, 1973.

35. M. Conversi, E. Pancini, and O. Piccioni, Phys. Rev. 71, 209L (1947).

36. F.J. Dyson, Phys. Rev., $\underline{85}$, 631 (1952).

37. P.S. Laplace, Exposition du Système du Monde, Paris, 1795 (Vol. II, p.305); English trans. W. Flint, London 1809.

    I am indebted to E. Guth for finding this reference for me. It is quoted in C.W. Meisner, K.S. Thorne, and J.A. Wheeler, Gravitation, Freedmon, San Francisco, 1970 as follows:

    "A luminous star, of the same density as the earth, and whose diameter should be 250 times larger than that of the Sun, would not, in consequence of its attraction, allow any of its rays to arrive at us; it is therefore possible that the largest luminous bodies in the universe may, through this cause, be invisible".

38. A.N. Whitehead, The Concept of Nature, Cambridge Univ. Press, Cambridge, 1920.

39. A.F. Parker-Rhodes, E.W. Bastin, J. Amson, and C.W. Kilminster, "On the Origin of the Scale Constants of Physics", E.W. Bastin, Studia. Phil. Gandensia, $\underline{4}$, 77 (1966).

40. E. Nagel and J.R. Newman, Godel's Proof, New York Univ. Press, New York, 1958.

41. Mao Tsetung, On Contradiction, (August 1937); Reprinted in Selections from the Writings of Mao Tsetung, Foreign Language Press, Peking, 1971.

42. F.S.C. Northrup, The Meeting of East and West, Mac Millan, New York, 1946.

43. H. P. Noyes, "On the Origin of Molecular Handedness in Living Systems", to be published in the proceedings of the conference on "The Generation and Amplification of Molecular Asymmetry, Jülich, Kernforschung Anlage, 1973.

44. N. Bohr and L. Rosenfeld, Det. Kgl, Dansk, Vid. Selskab, XII, 8 (1933).

45. D.D. Brayshaw, private communication.

46. B. F. Gore, Phys. Rev. D6, 666 (1972).

Comment on *"Non-Locality in Particle Physics"*
H. Pierre Noyes (2000)

The only serious error in this paper occurs on p. 21, where I wrote

$$r \leq c\delta t \ (\textit{limiting velocity}) \ \leq c\hbar/\delta E \ (\textit{uncertainty principle})$$

I should have written

$$r \leq c\delta t \sim c\hbar/\delta E$$

as Wick (*Nature* **142**, 993 (1938)) was careful to do. My mistake is that when the uncertainty principle in energy, $\delta E \delta t \geq \hbar$, is divided by δE to find δt, the result reads $c\delta t \geq c\hbar/\delta E$ instead of the way it is written in the first equation displayed above. This apparent triviality conceals a more serious conceptual error. If one could connect a radius in space to virtual particle energy in the way it is done in the first (incorrect) equation, and then compute in this way the number of virtual particles of finite mass one can have inside that radius, one could think of elementary particle structure as spacially ordered, like an onion, with one virtual particle of mass m_x inside a sphere of radius $r_1 = \hbar/m_x c$, two virtual particles inside $r_2 = \hbar/2m_x c$, ... n virtual particles inside a sphere of radius $r_n = \hbar/n \ m_x c$... This is the misleading visualization which may occur when a theorist says that, for example, "QED has been tested down to $10^{-15} cm$" or whatever. I am here confessing that I have fallen into this trap in many discussions of nuclear forces I have presented in the past. I hope that some of my colleagues in elementary particle physics will recognize that they sometimes make a similar conceptual error, and that it can seriously mislead not only the public but conscientious science journalists when they are trying to understand the connection between relativistic particle physics and the foundations of quantum mechanics.

Another point is that Wick introduced the limitation on the size of the region ($r \leq c\delta t$) so that the whole region in which the meson appears can act *coherently*. But the argument does not say anything about *how* that coherence is established. One way to think about this is to assume that tachyons exist inside the pion compton wavelength, but are *confined* by some mechanism connected to the strong interactions (eg colored quarks). If the time interval is large enough, and the quantum mechanical coherence of specific degrees of freedom is maintained by sufficient control over background signals, we know empirically that this region of coherence allowed by the uncertainty principle can extend to distances of 20 kilometers, as evidenced by tests of the Bell inequalities near Geneva. Such technological achievement of macroscopic coherence for *specific* degrees of freedom in a quantum mechanical system has unexpected consequences, as is further discussed in "Decoherence, Determinism and Chaos Revisited", in *Law and Prediction in the Light of Chaos Research*, P. Weingartner and G. Schurz, eds., Springer Lecture Notes in Physics, 1996 pp 152-188; see also SLAC-PUB-6397 (Jan. 1994), and Ch.15.

Still other points that already occur in this paper, which were only vaguely understood by me at the time (or impossible for me to grasp because not yet over the experimental horizon) are that the "3" at the first level of the hierarchy adumbers the $W^{\pm}, Z^0$ system, the "7" can be looked at as a first approximation to the proton-pion mass ratio, and that both weak-electromagnetic unification and a starting point for a new way to approach "quantum gravity" were already rattling around in my subconscious mind. But to understand why I think I can say this, the reader will have to follow the historical development represented by the papers presented in this volume.

International Journal of Theoretical Physics, Vol. 18, No. 7, 1979

On the Physical Interpretation and the Mathematical Structure of the Combinatorial Hierarchy[1]

Ted Bastin

12, Bove Town, Glastonbury, Somerset, England

H. Pierre Noyes

Stanford Linear Accelerator Center, Stanford University, Stanford, California 94305

John Amson

The Mathematical Institute, The University of St. Andrews, Scotland

and

Clive W. Kilmister

Department of Mathematics, King's College, University of London, England

Received March 5, 1979

The combinatorial hierarchy model for basic particle processes is based on elementary entities; any representation they may have is discrete and two-valued. We call them *Schnurs* to suggest their most fundamental aspect as concatenating strings. Consider a definite small number of them. Consider an elementary creation act as a result of which two different Schnurs generate a new Schnur which is again different. We speak of this process as a "discrimination." By this process and by this process alone can the complexity of the universe be explored. By concatenations of this process we create more complex entities which are themselves Schnurs at a new level of complexity. Everything plays a dual role in which something comes in from the outside to interact, and also serves as a synopsis or concatenation of such a process. We thus incorporate the observation metaphysic at the start, rejecting Bohr's reduction to the haptic language of common sense and classical physics. Since discriminations

[1]Work supported by the Department of Energy under contract number EY-76-C-03-051.

445

occur sequentially, our model is consistent with a "fixed past–uncertain future" philosophy of physics. We demonstrate that this model generates four hierarchical levels of rapidly increasing complexity. Concrete interpretation of the four levels of the hierarchy (with cardinals $3, 7, 127, 2^{127} - 1 \approx 10^{38}$) associates the three levels which map up and down with the three absolute conservation laws (charge, baryon number, lepton number) and the spin dichotomy. The first level represents $+$, $-$, and $\pm$ unit charge. The second has the quantum numbers of a baryon–antibaryon pair and associated charged meson (e.g., $n\bar{n}, p\bar{n}, p\bar{p}, n\bar{p}, \pi^+, \pi^0, \pi^-$). The third level associates this pair, now including four spin states as well as four charge states, with a neutral lepton–antilepton pair ($e\bar{e}$ or $\nu\bar{\nu}$), each pair in four spin states (total, 64 states)—three charged spinless, three charged spin-1, and a neutral spin-1 mesons (15 states), and a neutral vector boson associated with the leptons; this gives $3 + 15 + 3 \times 15 = 63$ possible boson states, so a total correct count of $63 + 64 = 127$ states. Something like $SU_2 \times SU_3$ and other indications of quark quantum numbers can occur as substructures at the fourth (unstable) level. Breaking into the (Bose) hierarchy by structures with the quantum numbers of a fermion, if this is an electron, allows us to understand Parker-Rhodes' calculation of $m_p/m_e = 1836.1515$ in terms of our interpretation of the hierarchy. A slight extension gives us the usual static approximation to the binding energy of the hydrogen atom, $\alpha^2 m_e c^2$. We also show that the cosmological implications of the theory are in accord with current experience. We conclude that we have made a promising beginning in the physical interpretation of a theory which could eventually encompass all branches of physics.

1. INTRODUCTION: GENERAL PRINCIPLES OF THE COMBINATORIAL HIERARCHY

In this section we are concerned with the basic principles of our combinatorial model of basic physical interactions. This theory was presented at two successive conferences on "Quantum Theory and the Structures of Time and Space" at Tutzing (Bastin, 1976b). We shall compare and contrast our own principles with the central position in those conferences as a convenient and brief way to present the relation of our theory to the basic principles of the quantum theory, since we may regard the central position established at Tutzing as the most coherent existing attempt to establish foundational principles for current quantum theory.

The combinatorial hierarchy model was originally developed (Bastin, 1966) as an attempt to base physics on a single binary process called "discrimination." Sets of "columns" containing only the existence symbols $0, 1$ closed under this operation are then viewed as new entities, and the process is repeated. In this way we generate a hierarchy of four levels of rapidly increasing complexity. Although the explicit representation of this hierarchy is not unique, the scheme itself is, as we demonstrate in this paper. Tentative contact with experiment can be made by specific interpretation of the representations, and structural features familiar in the study of elementary particle physics emerge, including some well-known numerical results. This theory is essentially intended as a conceptual underpinning of the existing formalism of the quantum theory.

The central idea at the Tutzing conferences was a theory of *Ur*'s—basic, discrete, two-valued entities. The claim made by the *Ur* theorists (see particularly von Weizsäcker's 1978 paper) has been that if finitism is firmly and clearly enough embraced, then something very like the usual quantum theoretical formalism can be sustained as a consistent theory and the paradoxes and other perplexities avoided.

A different position has been maintained by Finkelstein (1969, 1977, 1979), who accepts the finitist part of the *Ur* program but considers that further innovation in basic principles is necessary. He adopts a process philosophy, thinking that the elementary discrete constituents of nature must have a principle of concatenation, and that this principle, whatever it may be, must tell us a good deal about the interrelations of the classical and the quantum worlds.

Our theory accords with Finkelstein's demand for innovation beyond the finitist assumption; we adopt the general direction of his "process," or sequential concatenating conjecture. We present a definite model within the class specified by his conjecture, and can claim experimental backing for our model. Our model is distinct from quantum mechanics; it might become equivalent to the latter under special conditions. Some results that would normally be thought to be dependent upon quantum mechanics as a complete theory appear in our model at a more general stage than that at which we make contact with the special case of quantum mechanics. We discuss below how some recent work of Finkelstein's (1969, 1977, 1979) might allow such contact to be made.

The historical origins of the quantum theory concerned the experimental discovery of discreteness and an attempt to explain it using a continuum conceptual framework (we may consider that the Planck radiation formula was a striking experimental ratification of a theoretically arbitrary mathematical imposition of discreteness). Early quantum theory hardly claimed to be explanatory; the modern form of the theory has usually been seen as a successful reconciliation of the continuous and the discrete, and therefore as a satisfactory explanation of the latter. However, in view of the continuing unease with the conceptual foundations of the theory, it seems as appropriate today as it ever was to enquire (a) wherein the explanation lay, and (b) how successful it was. It is sensible to carry on our enquiry in the context of any of the traditional *Gedankenexperimente* (two-slit experiments, photon-splitting experiments, photon correlation experiments such as have been imagined by a sequence of theorists going back to Einstein, Podolsky, and Rosen).

As everybody knows, quantum theory has maintained that there is a distinct class of things in the universe called measurements or observations and that different rules apply to these from those that apply to interactions in which the acquisition of knowledge is not involved. In one way or another use is made of this principle to justify the importation into the formalism of a discrete principle. As everybody also knows, this principle has never produced peace of mind, even though the great thinkers of the quantum theory have concentrated their attention upon it. Consider, for example, the recent essay by Wheeler and Patton (1977). We shall refer to these arguments as the "observation metaphysic."

In Bohr's attempt to achieve an understanding of the observation metaphysic, an absolutely central part was played by his (Bohr's) insistence that all theoretical formulations had to be interpreted through the massively consistent and pervasive language which was at once classical physics and the common sense world. Bohr though it inconceivable that any underpinning or revision of this language using conceptual entities less evident to the senses was conceivable, practicable, or desirable. Indeed, his philosophy made a virtue of the necessity of this position.

In the *Ur* theory this position of Bohr's has been abandoned, though it would not be true to say that the "observer metaphysic" has gone with it. What has happened is that as a result of their finitist presupposition the *Ur* theorists have been able to present the conventional quantum theoretical view of measurement as a merely technical development free from its paradoxical characteristics, at the expense of a profound innovation in the application of probability to the quantum picture. The actual alternatives at any quantum process are finite, and the continuum of states out of which the measurement process picks one are in a different category, being "possibilities."

We, too, postulate entities that would be disallowed by Bohr's form of operationalism. We are equally concerned to find a comprehensible and still profound replacement for the "observation metaphysic," and claim to find it in the *individual* process.

Let us imagine a universe containing elementary entities which we may think of as our counterparts of the *Ur*s. To avoid confusion we will amend the terminology and call them *Schnurs* (German for "string")—a term that appropriately suggests computing concepts, in a way that represents their most fundamental aspect of concatenating strings. The Schnurs are discrete, and any representation they may have is two-valued. Consider a definite small number of them. Consider an elementary creation act as a result of which two different Schnurs generate a new Schnur, which is again different. We speak of this process as "discrimination." By this process, and by concatenations of this process, alone can the complexity of the universe be explored. It is also necessary that a record of these discriminations and resulting creations be kept as a part of the structure defined by the Schnurs; otherwise there is no sense in saying that they have, or have not, been carried out. Hence we consider a new lot of Schnurs, which consist of concatenations of creation processes preserving the discriminate structure explored by the original Schnurs. The members of the new class are themselves constituents of the universe and are also free to take part in the creation or discrimination process, and to map up to higher or down to lower levels. This last requirement is the stage at which the necessity becomes clear for a reflexive or recursive aspect to our model, which in current quantum theory takes the form of the "observation metaphysics." The construction of a hierarchy of new levels of Schnurs is necessary to obtain an approximation to a physical continuum; by means of it we can ultimately speak of a physical entity in a background of other physical entities in accordance with the requirements of common sense. However, it makes no sense to speak of the individual entities except in terms of the part they play in the construction. Everything plays a dual role, as a constituent in a developing process, where something comes in from outside to interact, and as a synopsis or concatenation of such a process where the external interaction becomes subsumed in one new entity.

How can a thing be both aspects at once? We do not think we are able at present to say clearly how it can, and we must let our model, which incorporates this duality, lead us forward without having a complete insight, as earlier theorists had to do in quantum theory. However, we are in a better position than current quantum theory, for we can adopt a strictly process view and insist that we always view the process from one viewpoint—albeit a viewpoint that can, and must, change. Then we are freed from conceptual confusion, and we progress by considering stability conditions under which the limitations of our way of approaching the inescapable duality are compensated. Indeed, we find in the stability of the hierarchy levels a profound condition under which we can be sure of a sort of automatic self-consistency which reflects itself in the properties of quantum objects, and which is the basis of our interpretation of our model.

We do not think it impossible that a mathematical way of thinking will emerge in which the dual function can be comprehended without the device of considering the structure of the universe from one point at which the decision making is occurring. One might revert to a more classical or synoptic mathematics. However, we do not think we can do it at present [though Parker-Rhodes (1978), whose work has played such an important part in our model, and who feels uncomfortable with a process philosophy, is trying to formulate something very similar in terms of a "mathematics of

indistinguishables," which transcends the process aspect]. We would conjecture that if such a conceptual framework ever is discovered, its proper
field of application would be wider than physics, and that the restricted
process view would probably be adequate for physics.

Our view of space–time is *constructive* in the sense that there is one set
of principles that gets us from the Schnurs to whatever approximation to
the continuum of space we decide we need. Our scheme is also constructive in the sense that we require that any mathematical constructions that
are needed to specify the attributes of any physical things, including the
space continuum, shall also be so derived. In this sense the *Ur* theory is not
constructive, and we have found our vital objection to it in this lack of
constructivity. This use of the term "constructive" is stringent. We are,
however, using it as in its *locus classicus*, Brouwer's theory of mathematical
intuition (which also stimulated the development of intuitionist logic).

Brouwer's basic concept is that of the free choice sequence. The
formal need for the free choice sequence is to construct the continuum
adequately. For Brouwer, the constructions of mathematics have no absolute quality but are creations of the intellect, whose validity is relative to
the state of mathematical understanding at a given epoch. They play a part
in guiding the development of the free choice sequences. So do other
considerations that we should normally regard as contingent. (An example
of Brouwer's was to make the development of a free choice sequence
depend upon whether, at the particular time in question, four successive
sevens were known to occur in the expansion of π). It would be possible
(and Brouwer was quite open to this suggestion) to regard the totality of
considerations that could influence free choice sequences as including the
contingent behavior of physical systems, in which case the similarity of the
processes in our constructive model and the basic entities with which
Brouwer constructed his universe would be quite close.

It would be fascinating to pursue this connection with Brouwer's
thought, but this cannot be the place. We introduce it at all here only
because it may be felt by some readers that our theory requires a mathematical ontology which is just wrong ; it may reassure the readers to
know that something very like what we propose has been authoritatively
put forward for analogous reasons in the literature of the foundations of
mathematics. The connection is also relevant to our present discussion,
because Brouwer's constructivism has no separate world of mathematical
entities; the difficulty we encounter with the *Ur* theorists is that they allow
themselves the use of continuous mathematical constructions where we felt
that a constructive development should include mathematical entities used
in the theory.

When one has a model for elementary processes one has to reconcile it with the macroscopic awareness of the world as an extended manifold of space and time. This is a large undertaking, which is usually not very explicitly faced. The traditional argument of a correspondence limit is only a small part of the problem for it presupposes that the problem has already been solved for the microscopic entities. Traditionally physicists rely on macroscopic experience to have universal application, and face the resulting confusions piecemeal. The position of the *Ur* theorists is not dissimilar for, as we have seen, they allow themselves to introduce continuum group theory, which then imports the principle of interpretation of extended spaces. We have left ourselves no such loophole, and the problem remains to be tackled.

Finkelstein's process approach also has to face this problem. Two arguments of his are relevant to it: (a) He has shown (Finkelstein, 1969) that the left–right moves of a dichotomous variable on a two-dimensional checkerboard generate, in the limit as the step size goes to zero, the full forward light cone of the Minkowski $(3 + 1)$-space. (b) Given any partial ordering relation, one can, by a theorem due to Galois, construct a lattice logic. If the lattice logic is that of bra and ket, then a theorem of Birkhoff's allows the construction of Hilbert space from the lattice (Finkelstein, 1979). It is not clear to us that this can meet the whole problem from our point of view; his approach might still end up with the commitment to macroscopic experience that we are trying to avoid. We hope that his treatment will turn out to be relevant to our problem. Certainly the lattice-theoretic result could be very significant in establishing a connection with Hilbert space.

We turn now to another difference between our Schnur theory and the *Ur* theory. This concerns the question whether we locate the reflexive character in the individual *Ur* processes or in statistical assemblages of them. We hold the former view, the *Ur* theorists the latter. The tradition is on our side, even though one is stretching a point in arguing as we have done that traditional quantum theory fails crucially at the point where it has to appeal to an observation metaphysic to introduce the reflexive character of quantum processes and yet claim support from that quarter. Still, the traditional argument that the essential character of quantum processes has to be defined for individual processes is very strong. One is accustomed to having to refute various facile approaches to the foundations of quantum theory by pointing out that the characteristic quantum-observation effect is individual and therefore cannot depend upon a statistical effect. For example, in the photon-splitting experiment, the incident beam can be attenuated to such a degree that the incident photons would have to be treated individually, and therefore could not interfere. Yet interference does take place. This piece of experimental evidence provides a very sharp refutation of any view whose attribution of simple atomic properties to the photons is subject to the restriction that one may consider only statistical distributions of these; von Weizsäcker's distinction between possibility and probability in conjunction with his principle of the finite alternatives allowed by the *Ur*'s is used to explain why the *Ur* theory is not in this class [J. H. M. Whiteman (1971) introduced a concept that he called *potentiality* to achieve a similar end.] However, this matter is crucial and one feels that the detailed mechanics that makes a statistical effect appear as an individual one should be presented. We think that our model, in which the effect is individual, has a crucial advantage, and that this advantage is a direct consequence of our constructive approach.

In all other respects, we find ourselves in complete agreement with the analysis of the use of probability that von Weizsäcker (1978) has undertaken. Probability is closely related to the concept of time in the quantum physics context. The concept of time that is commonplace in modern philosophical writing, and which owes more to Hume than to any other thinker, seems to be in conflict with a good deal of the thinking of physicists. Starting with Galileo, the time of physicists is based primarily upon the analogy between time "displacement" and displacement in space. Our model has developed partly from discussion that was designed to show that in a discrete approach one might have the advantage of adopting the Humean point of view without outrage to physical theory. Then one could take the past simply as the fixed domain and the future as the domain of uncertainty and of probabilistic inference. This point of view can be tagged "Fixed Past, Uncertain Future" (Noyes, 1975, 1976, 1977).

It is obviously tempting to identify the duality of function of our elementary discriminators or Schnurs with the duality of description in complementarity. Certainly the two are connected, but the connection is not simple, as must be clear from the foregoing discussion of the differences between our view and current quantum theory. Bohr's view of complementary descriptions seems to be very much a special form of a more general philosophy and to have had its special form dictated by the special form in which quantum physics has developed. It is probably safe to say that if one could state the general philosophy without such special reference, it would contain the reflexive or recursive character with which our discussion has been concerned. However, Bohr's philosophy has proved notoriously difficult to state in this bare form in spite of the best efforts of fifty years. We conclude this section by stating what we feel to be the reason for this recalcitrance.

In a discrete or finite theory it is not too perplexing to introduce a reflexive philosophy by using a recursive mathematical model, which is what we do. The really perplexing difficulties seem to appear if we associate this reflexive character with an observation imagined against an objectively existing background, as is done in so-called "measurement theory." Two incompatible principles are being appealed to. One principle requires entities in the universe to be constructed using the observation process; the other takes a realist view of them. Not surprisingly, no reconciliation of the resulting perplexities is achieved by studies at a technical level where fundamental principles tend to be assumed rather than discussed.

One question has been avoided till now. In our model the elementary entities have a dual function. One of the dual aspects is analogous to that of an observing system. Do we imagine that this aspect of its dual role would correspond to the quantum theoretical "observation," and if so how would we react to those writers on quantum theory who wish to see something irreducably mentalist in the observation? In reply, we would first observe that we are not compelled to answer this question before we can use our model. We have a model for interactions which are elementary (Ur) in the sense that all we know is built up from them, and we have an interpretation for the model in terms of scattering processes. This interpretation does not have to be the only one. We have tacitly assumed that the conditions of high energy are favorable for exhibiting the simplicity of the model and hence the scattering situation. However, under other conditions the interacting entities might even be living organisms with consciousness. The model should still apply. What we absolutely are not either compelled or allowed to say is that the phenomenon of consciousness as a separable ingredient is necessary for the interaction.

2. CONSTRUCTION OF THE HIERARCHY

In this section we develop the specific formalism by which we are implementing the program discussed above, using a very explicit representation of the abstract hierarchical structure. The mathematical structure itself is developed in group theoretic language in the Appendix. Our basic elements are the existence symbols 0 and 1, and our basic mathematical operation is symmetric difference or addition modulo 2: $(0+0=0, 1+0=1, 0+1=1, 1+1=0)$. The symbols are grouped as ordered sets ("columns") of height n $(n=1,2,3,...)$. The comparison between two such columns is called "discrimination." Each column x, whose height we can indicate by writing $(x)_n$ for x, has elements ("discriminators") x_i ($i = 1,\dots,n$); thus $x = (x_i)_n$. A column with every element $x_i = 0$ is called a "null" column. The basic binary operation of discrimination between two columns x,y of equal height is defined by

$$D_n(x,y) = x + y = (x_i + y_i)_n \tag{2.1}$$

The concept of such discriminators is abstracted from the more familiar idea of discrete quantum numbers, while the discrimination operation itself can be viewed, as we will discuss in another section, as an abstract model of a general scattering ("production") process in which the result of scattering two different systems is a third system that differs from either. Our mathematical model thus describes chains of atomic or elementary processes. Our policy for presenting the theory is first to establish a correspondence between the mathematical model that describes these chains of processes and the familiar structure of quantum numbers. In this way we can first view the mathematical model as providing a classification scheme. The basic dynamics of our theory is represented during the construction of this classification scheme by the concept of *discriminate closure*. We introduce this concept by the following argument.

Starting with columns of a given height, we imagine new columns formed by concatenating a sequence of them. Entities corresponding to the new columns are said to constitute a new level in the hierarchy. There is no difference between the new and the old in logical type; the only difference is that the boundary between the observing system and that which is observed has changed. The great conceptual and mathematical difficulties of such an idea can be handled in one special case, which is therefore of great importance. This case is that in which the entities at the new level represent all combinatorially possible concatenations of entities at the previous level, starting with a given set. Hence we get a *discriminately closed subset*.

A "discriminately closed subset" or DCsS consists of one or more nonnull columns, such that discrimination between any two distinct columns in the set yields a member of the set. Assume that we start from a basis of j linearly independent columns, that is, columns for which no sum of two or more different columns is null. Then there will be $2^j - 1$ distinct discriminately closed subsets. Symbolizing a DCsS by { }, a basis of two columns a,b gives the three DCsSs $\{a\}$, $\{b\}$, $\{a,b,a+b\}$; a basis of three columns a,b,c gives the seven DCsSs $\{a\}$, $\{b\}$, $\{c\}$, $\{a,b,a+b\}$, $\{b,c,b+c\}$, $\{c,a,c+a\}$, $\{a,b,c,a+b,b+c,c+a,a+b+c\}$. Proof of the general result is immediate either by noting that the number of DCsSs is simply the number of ways we can combine j things $1,2,\dots,j$ at a time, or by induction. The first step in constructing the hierarchy is then to consider the $2^j - 1$ DCsSs so formed as the basic entities of a new level.

The reason for seeking a constructive process of hierarchical nature that yields levels of rapidly increasing (in our case exponentiating) complexity is again abstracted from experience. We have detailed in the first section the reasons why we start from an elementary process (discrimination) which already implicitly contains the "observation metaphysic." There we also explained why, in our view, we adopt a constructive, process-oriented approach. The further requirement that the hierarchy so generated terminate is a basic requirement if we are to retain the principle of finitism. We defer the discussion of the reflexive character of the scheme until it is further developed. That the *combinatorial hierarchy* obtained by starting with columns of height $n = 2$ yields levels of interesting physical structure and sufficient complexity, and terminates at the appropriate level, has been shown previously (Bastin, 1966). We summarize the construction here.

We have seen that, given j linearly independent columns, we can always construct $2^j - 1$ DCsSs at that level. For them to form the basis of a new level, however, they must themselves be representable by linearly independent entities that contain the same information about discriminate closure as the sets themselves. For this purpose we introduce multiplication modulo 2 and matrices because linear operators preserve discrimination. We look for $2^j - 1$ matrices which (a) map each column in one of the subsets onto itself and onto no other column; (b) map only the null column onto the null column, and hence are nonsingular; and (c) are linearly independent. Provided this can be done, and the original basis consists of columns of height n, then the matrices themselves can be rearranged as columns (e.g., by putting one row on top of another by some consistent rule), and will then provide a linearly independent basis of $2^j - 1$ columns of height n^2. Such mapping matrices are easy to find for $n = 2$ (see below). Explicit examples have been found for $n = 3$, 4, and 16 (Noyes, 1978) proving the existence of the hierarchy. A formal existence proof has also been provided (Kilmister, 1978) based on unpublished work (Amson, 1976).

The use of matrix algebra could be misunderstood as implicitly incorporating into the scheme the basic assumptions of linear algebra. In fact, matrix algebra using the symbols 0, 1, discrimination, and multiplication mod 2 is the natural extension of the discrimination idea to incorporate mappings. This can be seen in more formal terms by following the group theoretic discussion given in the Appendix.

We can now present the general situation. We have seen that if at some level l there are $j(l)$ linearly independent columns of height $n(l)$, we can construct immediately $d(l) = 2^{j(l)} - 1$ DCsSs. Provided these can be mapped according to the restrictions given above, they form the basis for a new level with $j(l+1) = d(l)$ and $n(l+1) = n^2(l)$. The process will terminate if $n^2(l) < 2^{j(l)} - 1$ since at level l there are only $n^2(l)$ linearly independent matrices available; clearly this will always happen for some finite n. The situation for $n(1) = j(1) = N$, i.e., when the vectors at the lowest level which span the space are used as the basis, is exhibited in Table I. Thus, perhaps surprisingly considering the simplicity of the assumptions, the hierarchy with more than two levels turns out to be unique. See the Appendix for a more detailed discussion.

Although the cardinal numbers given by the hierarchy are unique, the specific representations used in the construction are not. It is important to understand this clearly because it is a complication in making any *simple* interpretation of the discriminators as representing the presence or absence of particular conventional quantum numbers in an isolated system. This ambiguity is present at the lowest level since for the two basis columns we have three choices: $a = \binom{1}{0}$, $b = \binom{0}{1}$; $a' = \binom{1}{0}$, $b' = \binom{1}{1}$; $a'' = \binom{1}{1}$, $b'' = \binom{0}{1}$. Corresponding to these three possible choices of basis, there are three different sets of mapping matrices. When, as here, the number of independent columns is equal to the height of the columns ($n = j$), the maximal discriminately closed set (MDCS) contains all the nonnull vectors in the space [here it is $\{\binom{1}{0}, \binom{0}{1}, \binom{1}{1}\}$] independent of the choice of basis; further, the only possible mapping matrix for the MDCS is then the unit matrix. For the first basis, the mapping matrices for $\{a\}$ and $\{b\}$ are $\binom{1\,1}{0\,1}$ and

TABLE I. The Possible Hierarchies Starting from $n(1) = j(1) = N$

l		1	2	3	4	Hierarchy terminates because
$N = 2$	$n(l)$	2	4	16	256	
	$j(l)$	2	3	7	127	$(256)^2 < 2^{127} - 1$
$d(l) = 2^{j(l)} - 1$		3	7	127	$2^{127} - 1 \approx 10^{38}$	
$N = 3$	$n(l)$	3	9			
	$j(l)$	3	7			$9^2 < 127$
	$d(l)$	7	127			
$N = 4$	$n(l)$	4	16			
	$j(l)$	4	15			$16^2 < 2^{15} - 1$
	$d(l)$	15	$2^{15} - 1$			
$N > 4$						$n^2(1) < 2^{j(1)} - 1$

$\binom{10}{11}$, respectively. For the second $a = a'$, so that matrix is the same but the mapping matrix for $\{b'\}$ is $\binom{01}{10}$; for the third we note that $a'' = b'$ and $b'' = b$. Rearranging the matrices as columns then give three different possible bases for the second level of the hierarchy, namely, with the rule

$$\binom{AC}{DB} \rightarrow \begin{bmatrix} A \\ B \\ C \\ D \end{bmatrix}$$

$$a_2 = \begin{bmatrix} 1 \\ 1 \\ 0 \\ 0 \end{bmatrix}, \quad b_2 = \begin{bmatrix} 1 \\ 1 \\ 1 \\ 0 \end{bmatrix}, \quad c_2 = \begin{bmatrix} 1 \\ 1 \\ 0 \\ 1 \end{bmatrix}; \quad a_2' = \begin{bmatrix} 1 \\ 1 \\ 0 \\ 0 \end{bmatrix}, \quad b_2' = \begin{bmatrix} 1 \\ 1 \\ 1 \\ 0 \end{bmatrix}, \quad c_2' = \begin{bmatrix} 0 \\ 0 \\ 1 \\ 1 \end{bmatrix};$$

$$a_2'' = \begin{bmatrix} 1 \\ 1 \\ 0 \\ 0 \end{bmatrix}, \quad b_2'' = \begin{bmatrix} 0 \\ 0 \\ 1 \\ 1 \end{bmatrix}, \quad c_2'' = \begin{bmatrix} 1 \\ 1 \\ 0 \\ 1 \end{bmatrix} \qquad (2.2)$$

In addition to this ambiguity, there is the further problem that we could have used any other rule for converting the matrices into column vectors, provided only the same rule is used for all three matrices. Thus the ordering of the rows has no significance, and *within* a level the properties of the system under discrimination are unaltered by a permutation of rows in the basis. An important structural property which does emerge, however, is that instead of the basis of three unit columns such as (1000), (0100), (0010), or any linearly independent set constructable on such a basis, at least two of the columns in the basis always contain two ones in the same two rows. This property guarantees that the MDCS (up to a permutation of rows) at the second level will always be

$$\left\{ \begin{bmatrix} 1 \\ 1 \\ 0 \\ 0 \end{bmatrix}, \begin{bmatrix} 1 \\ 1 \\ 0 \\ 1 \end{bmatrix}, \begin{bmatrix} 1 \\ 1 \\ 1 \\ 0 \end{bmatrix}, \begin{bmatrix} 1 \\ 1 \\ 1 \\ 1 \end{bmatrix}, \begin{bmatrix} 0 \\ 0 \\ 0 \\ 1 \end{bmatrix}, \begin{bmatrix} 0 \\ 0 \\ 1 \\ 0 \end{bmatrix}, \begin{bmatrix} 0 \\ 0 \\ 1 \\ 1 \end{bmatrix} \right\} \qquad (2.3)$$

Note that the first two rows may always be written as $\binom{1}{1}$ or $\binom{0}{0}$. We shall find this fact significant as a clue to physical interpretation. (Note that "rows" always refers to places in a column even though columns may be printed vertically or horizontally for purely typographical reasons.)

When it comes to constructing mapping matrices for the second level, we cannot use the unit matrix to represent the MDCS given in equation

(2.3) because it maps all 15 possible nonnull columns of height 4 onto themselves, and not just the required seven. The eight columns that must be excluded are of the form (10xy) or (01xy). A nonsingular matrix that has none of these as eigenvectors, but all the columns of equation (2.3), is exhibited in equation (2.4):

$$a\begin{bmatrix} 0 & 1 & 0 & 0 \\ 1 & 0 & 0 & 0 \\ 0 & 0 & 1 & 0 \\ 0 & 0 & 0 & 1 \end{bmatrix}\begin{bmatrix} x \\ x \\ y \\ z \end{bmatrix} = \begin{bmatrix} x \\ x \\ y \\ z \end{bmatrix}, \qquad \begin{pmatrix} a \\ b,c,d \\ e,f,g \end{pmatrix}\begin{bmatrix} x \\ \bar{x} \\ y \\ z \end{bmatrix} \neq \begin{bmatrix} x \\ \bar{x} \\ y \\ z \end{bmatrix},$$

$$\bar{x} = 1 + x, \qquad \begin{vmatrix} a \\ b,c,d \\ e,f,g \end{vmatrix} \neq 0$$

$$b\begin{bmatrix} 0 & 1 & 0 & 0 \\ 1 & 0 & 0 & 1 \\ 0 & 0 & 1 & 1 \\ 0 & 0 & 0 & 1 \end{bmatrix}\left\{\begin{bmatrix} 1 \\ 1 \\ 0 \\ 0 \end{bmatrix}\begin{bmatrix} 0 \\ 0 \\ 1 \\ 0 \end{bmatrix}\begin{bmatrix} 1 \\ 1 \\ 1 \\ 0 \end{bmatrix}\right\} c\begin{bmatrix} 0 & 1 & 0 & 0 \\ 1 & 0 & 1 & 0 \\ 0 & 0 & 1 & 0 \\ 0 & 0 & 1 & 1 \end{bmatrix}\left\{\begin{bmatrix} 1 \\ 1 \\ 0 \\ 0 \end{bmatrix}\begin{bmatrix} 0 \\ 0 \\ 0 \\ 1 \end{bmatrix}\begin{bmatrix} 1 \\ 1 \\ 0 \\ 1 \end{bmatrix}\right\}$$

$$d\begin{bmatrix} 1 & 1 & 0 & 0 \\ 1 & 0 & 0 & 0 \\ 0 & 0 & 1 & 0 \\ 1 & 0 & 0 & 1 \end{bmatrix}\left\{\begin{bmatrix} 0 \\ 0 \\ 1 \\ 0 \end{bmatrix}\begin{bmatrix} 0 \\ 0 \\ 0 \\ 1 \end{bmatrix}\begin{bmatrix} 0 \\ 0 \\ 1 \\ 1 \end{bmatrix}\right\} e\begin{bmatrix} 0 & 1 & 1 & 1 \\ 1 & 0 & 0 & 1 \\ 0 & 0 & 1 & 1 \\ 0 & 0 & 0 & 1 \end{bmatrix}\left\{\begin{bmatrix} 1 \\ 1 \\ 0 \\ 0 \end{bmatrix}\right\}$$

$$f\begin{bmatrix} 0 & 1 & 0 & 0 \\ 1 & 1 & 0 & 1 \\ 0 & 0 & 1 & 1 \\ 0 & 1 & 0 & 1 \end{bmatrix}\left\{\begin{bmatrix} 0 \\ 0 \\ 1 \\ 0 \end{bmatrix}\right\} g\begin{bmatrix} 0 & 1 & 0 & 0 \\ 1 & 1 & 1 & 0 \\ 1 & 1 & 1 & 0 \\ 0 & 1 & 1 & 1 \end{bmatrix}\left\{\begin{bmatrix} 0 \\ 0 \\ 0 \\ 1 \end{bmatrix}\right\} \qquad (2.4)$$

Choosing as a basis the columns (1100), (0010), (0001) we also exhibit six specific mapping matrices which have as eigenvectors only the columns in the six remaining DCsSs. This representation is not unique, since we find that of the 35 possible choices of three columns as a basis, omitting those that are not linearly independent or that are equivalent to others under a permutation of rows, there are 15 alternative choices. However, all of them have more than four descriptors in the three columns, so the choice exhibited is in that sense the simplest.

In order for these seven mapping matrices to form a basis for constructing the $2^7 - 1 = 127$ DCsSs of level III, they must be linearly independent. The linear independence is exhibited explicitly in equation

(2.5), after using a particular rule for rearranging the matrices as columns:

$$
\begin{bmatrix} I & A & K & L \\ B & M & G & E \\ O & P & C & F \\ J & N & H & D \end{bmatrix} \rightarrow
\begin{matrix}
 & a & a+b & a+c & a+d & b+e & b+f & b+c+f+g \\
A & 1 & 0 & 0 & 0 & 0 & 0 & 0 \\
B & 1 & 0 & 0 & 0 & 0 & 0 & 0 \\
C & 1 & 0 & 0 & 0 & 0 & 0 & 0 \\
D & 1 & 0 & 0 & 0 & 0 & 0 & 0 \\
E & 0 & 1 & 0 & 0 & 0 & 0 & 0 \\
F & 0 & 1 & 0 & 0 & 0 & 0 & 0 \\
G & 0 & 0 & 1 & 0 & 0 & 0 & 0 \\
H & 0 & 0 & 1 & 0 & 0 & 0 & 0 \\
I & 0 & 0 & 0 & 1 & 0 & 0 & 0 \\
J & 0 & 0 & 0 & 1 & 0 & 0 & 0 \\
K & 0 & 0 & 0 & 0 & 1 & 0 & 0 \\
L & 0 & 0 & 0 & 0 & 1 & 0 & 0 \\
M & 0 & 0 & 0 & 0 & 0 & 1 & 0 \\
N & 0 & 0 & 0 & 0 & 0 & 1 & 0 \\
O & 0 & 0 & 0 & 0 & 0 & 0 & 1 \\
P & 0 & 0 & 0 & 0 & 0 & 0 & 1
\end{matrix}
\qquad (2.5)
$$

The explicit choice of mapping matrices, which again is not unique, was again made in such a way as to get the simplest possible basis for level III in which all 16 rows are occupied.

To construct level IV we first find a basic matrix that has any of the 127 vectors that can be constructed from the seven given in equation (2.5), and none of the remaining 128 nonnull columns of height 16 that are not of this form, as eigenvectors. One possibility is exhibited in equation (2.6):

$$
\begin{bmatrix}
0 & 1 & 0 & 0 & 0 & 0 & 0 & 0 & 0 & 0 & 0 & 0 & 0 & 0 & 0 & 0 \\
1 & 0 & 0 & 0 & 0 & 0 & 0 & 0 & 0 & 0 & 0 & 0 & 0 & 0 & 0 & 0 \\
0 & 0 & 0 & 1 & 0 & 0 & 0 & 0 & 0 & 0 & 0 & 0 & 0 & 0 & 0 & 0 \\
0 & 0 & 1 & 0 & 0 & 0 & 0 & 0 & 0 & 0 & 0 & 0 & 0 & 0 & 0 & 0 \\
0 & 1 & 1 & 0 & 0 & 1 & 0 & 0 & 0 & 0 & 0 & 0 & 0 & 0 & 0 & 0 \\
0 & 0 & 0 & 0 & 1 & 0 & 0 & 0 & 0 & 0 & 0 & 0 & 0 & 0 & 0 & 0 \\
0 & 0 & 0 & 0 & 0 & 0 & 0 & 1 & 0 & 0 & 0 & 0 & 0 & 0 & 0 & 0 \\
0 & 0 & 0 & 0 & 0 & 0 & 1 & 0 & 0 & 0 & 0 & 0 & 0 & 0 & 0 & 0 \\
0 & 0 & 0 & 0 & 0 & 0 & 0 & 0 & 0 & 1 & 0 & 0 & 0 & 0 & 0 & 0 \\
0 & 0 & 0 & 0 & 0 & 0 & 0 & 0 & 1 & 0 & 0 & 0 & 0 & 0 & 0 & 0 \\
0 & 0 & 0 & 0 & 0 & 0 & 0 & 0 & 0 & 0 & 0 & 1 & 0 & 0 & 0 & 0 \\
0 & 0 & 0 & 0 & 0 & 0 & 0 & 0 & 0 & 0 & 1 & 0 & 0 & 0 & 0 & 0 \\
0 & 0 & 0 & 0 & 0 & 0 & 0 & 0 & 0 & 0 & 0 & 0 & 0 & 1 & 0 & 0 \\
0 & 0 & 0 & 0 & 0 & 0 & 0 & 0 & 0 & 0 & 0 & 0 & 1 & 0 & 0 & 0 \\
0 & 0 & 0 & 0 & 0 & 0 & 0 & 0 & 0 & 0 & 0 & 0 & 0 & 0 & 0 & 1 \\
0 & 0 & 0 & 0 & 0 & 0 & 0 & 0 & 0 & 0 & 0 & 0 & 0 & 0 & 1 & 0
\end{bmatrix}
\begin{bmatrix} a \\ a \\ a \\ a \\ b \\ b \\ c \\ c \\ d \\ d \\ e \\ e \\ f \\ f \\ g \\ g \end{bmatrix}
=
\begin{bmatrix} a \\ a \\ a \\ a \\ b \\ b \\ c \\ c \\ d \\ d \\ e \\ e \\ f \\ f \\ g \\ g \end{bmatrix}
\qquad (2.6)
$$

One then forms the 127 DCsSs, and finds nonsingular mapping matrices for each of them. This is done by leaving the first six rows of this basic matrix untouched—which guarantees that none of the unwanted columns from the 128 are brought back as eigenvectors—and adding ones one at a time to the remaining structure in such a way as to restrict the eigenvector set. Care must be used not to make the matrix singular and to maintain linear independence. The procedure is straightforward, if somewhat tedious, so the explicit result will not be given here. This empirical procedure thus proves the existence of all four levels of the hierarchy.

3. LEVELS 0, I, II, AND III: BARYONS, MESONS, LEPTONS, AND PHOTONS

In this section we attempt to correlate the mathematical structure developed above with some facts known from elementary particle physics. Because any physical process requires development of the hierarchy through the levels successively, the significant physical magnitude is not the cardinal of each level separately, but rather their cumulative sum, which gives the sequence $3, 10, 137, 137 + 2^{127} - 1 \approx 10^{38}$. Obviously these numbers could be interpreted immediately as the inverse of the superstrong, strong, electromagnetic, and gravitational coupling constants and suggest that in some sense the cumulative levels refer to systems of bosons with increasingly refined definitions of their possible interactions. One way to make this more specific would be to assume that the various systems at each cumulative level all have equal *a priori* probability, and that the probability of "coupling into" any one of them by the characteristic described at that level is therefore the inverse of the corresponding number. We will give this vague idea of coupling more specific content shortly. Further, the fact that the first three levels can be mapped up or down freely, but that any attempt to construct a linearly independent representation of the fourth level with $2^{127} - 1$ DCsSs must fail after $(256)^2$ linearly independent matrices have been selected, suggests that the destabilization of particle systems due to weak decay processes with coupling constant $10^{-5} m_p$ might also emerge from the scheme since $1/(256)^2$ has approximately this value (Bastin, 1966). This requires us to assume that the unit of mass in the scheme is the proton mass, but this is already clear from the initial sequence, since $\sim 10^{-38}$ *is* the gravitational coupling between two protons; the gravitational coupling constant between two electrons is 10^{-44}. Thus we can hope to derive the ratio of the electron mass to the proton mass once the scheme is sufficiently developed.

We are now in a position to state our policy toward the general question of the physical interpretation of the hierarchy so as to be consistent with the identification that has already been made of the basic scheme of cardinal numbers with dimensionless constants. This policy has two aspects. First we have the task of identifying the quantum numbers with configurations in the hierarchy, and secondly, we have to introduce fields corresponding to the quantum numbers, and show that we should expect these fields to have the characteristics that we find in nature. The second task will be coterminous with that of defining an extended space for the particles to be "in." Only the first task is confronted in this section.

We wish to identify places in columns with quantum numbers, and we wish to regard associations of quantum numbers in given columns as *systems* which, under conditions of stability that have yet to be established, will carry over unchanged into stable or unstable particles. For the moment we call them "systems." These ends require us to solve the following problems:

(1) How to get an initial distinguishing characteristic of a column which is available for taking the first step in interpretation in the sense that it cannot be eliminated by choosing a different basis.

(2) How to interpret the interrelations of columns in a set (including of course a DCS) at one level.

(3) How to relate the interpretation of a column at one level with that of columns of different lengths and hence different quantum numbers at another level.

The first step in the solution of these problems is to define *conservation* in respect of a set of properties to each of which a quantum number is conventionally assigned. These are the eight properties (1) of having z component of spin up, (2) of having z component of spin down, (3) of having charge $+$, (4) of having charge $-$, (5) of being a lepton, (6) of being an antilepton, (7) of being a baryon, (8) of being an antibaryon.

It makes things clearer to begin by speaking of properties and only later of the dichotomous variables that can correspond to quantum numbers. The latter require two rows to represent them.

Definition. A quantum number will be said to be *conserved* if the algebraic difference between the number of ones in the corresponding pair of rows of properties is constant *at each step* in the generation process.

The choice of the foregoing definitions (in particular that of the conserved quantities and of their relation to descriptors) embodies a lot of detailed argument whose correctness must be judged by the coherence of the resulting scheme. Moreover the choice of quantum numbers assumes the emergence of discrete quantities through the history of the quantum theory so that the theory is now at a stage that makes it ripe for combinatorialization. Thus in particular the use of the z component of spin as the appropriate quantum number for combinatorialization is obscured by the spatial idea of spin, but it is becoming more perspicuous as the "helicity state"; we stick to the earlier term.

Our definition of conservation introduces two novelties in principle. One is that of forming the algebraic sum of a set; the other is that of a primitive notion of simultaneity. The two are related since in forming the sum one is making an assertion about what is true collectively of the set *at each step*. The change is considerable since one abandons the principle of individual access in enumeration; although this change is already implicit in the hierarchy construction itself, it is appropriate to introduce it here with the motivation of the very fundamental idea of conservation. It is also at this point that we see the root of an idea of sequential delay that will take us from a purely sequential theory to one with a more conventional space and time. However, having recorded this starting point, we will not attempt to develop it further now.

We notice in the above account that all the structures that are going to be interpretable (dichotomous variables) require representation by *two* rows, so that systems of one row are not given a meaning. This principle already exists as a matter of logical necessity in the hierarchy construction —a correspondence that indicates satisfactory coherence in the theory as a whole. (The level of single elements cannot generate a hierarchy.) It follows that at level II the 4-columns are properly regarded as a pair of pairs.

We now return to the set of problems posed above. To handle the first problem—that of initial interpretation—we first draw attention to the "doubled discriminators" of Section 2, which we have shown must exist in the mapping construction. We note first that this asymmetry is already enough to refute the criticism that since one can always take a minimal basis using only columns of the form

and since row position is arbitrary, no interpretation that depends on relative position of ones can be significant. In fact we cannot always take such a minimal basis; we are therefore justified in beginning our interpretation with a nonminimal basis, and in particular with one in which we have the doubled discriminators which we have shown to be necessary.

The doubled discriminators enable us to develop a notation for putting together two systems, each of which is described by a dichotomous variable; we shall use spin as our first example (Bastin, 1976a). In Table II

TABLE II. Triplet–Singlet System from Two Dichotomic Vectors

Conventional notation				Hierarchy notation			
$S=1$			$S=0$	$S=1$			$S=0$
$S_z=1$	0	-1	$S_z=0$	$S_z=1$	0	-1	$S_z=0$
$\begin{pmatrix}1\\0\\1\\0\end{pmatrix}$	$\dfrac{1}{2^{1/2}}\begin{pmatrix}1\\1\\1\\1\end{pmatrix}$	$\begin{pmatrix}0\\1\\0\\1\end{pmatrix}$	$\dfrac{1}{2^{1/2}}\begin{pmatrix}1\\-1\\-1\\1\end{pmatrix}$	$\begin{pmatrix}1\\0\\1\\0\end{pmatrix}$	$\begin{pmatrix}1\\1\\1\\1\end{pmatrix}$	$\begin{pmatrix}0\\1\\0\\1\end{pmatrix}$	$\begin{pmatrix}0\\0\\0\\0\end{pmatrix}$

we assign the first two rows to indicate spin up or spin down of one system, and the second two rows to refer to spin up or spin down for the other system. The resulting singlet or triplet states are represented in Table II both in the conventional notation using algebraic vectors and in the hierarchy notation using only existence symbols. We see that the descriptive content of the two notations is identical so far as distinguishing the four possible singlet or triplet states goes. We also note that the singlet state is the null column; we can only give meaning to such a state in a richer system with more rows containing nonnull descriptors.

The spin-z state refers to a *single* system of spin 1/2 (which can be up or down) for which the algebraic notation is $\begin{pmatrix}1\\0\end{pmatrix}$ or $\begin{pmatrix}0\\1\end{pmatrix}$. A singlet/triplet system is either a spin-0 system with one state and a spin-1 system with three states, or the composition of two spin-1/2 to give the same result. The conventional notation for the result is given in Table II in comparison with our notation.

We have shown the above identification to be possible and consistent with the idea of conservation; we have not shown it to follow necessarily from the existence of doubled descriptors. The latter demonstration requires a new physical principle. In addition to the association of two existence symbols to make one dicotomous variable, we have encountered the association of two identical existence symbols in two rows to make an effective single existence symbol—a development that was forced by the necessary occurrence of "doubled descriptors." If we were to exploit, at level II, the full possibilities of the increased scope in our descriptive language offered by treating each row as independent, we would get $16-1=15$ possible systems. The mapping construction allows only 7 of these, consisting of one doubled existence symbol and one dichotomic variable. We see this more clearly if we enumerate four cases:

(I) Triple existence symbol plus single existence symbol. This is isomorphic to the basis for level I and gives nothing new.

(II) Triplet system exhibited in Table II. Since we are considering a system of four rows, the singlet possibility effectively represents nothing and is excluded. Further, we see that triplet system is simply a doubled representation of level I and again gives us nothing new.

(III) The doubled existence symbols together with the two-row dichotomous variables already exhibited in (2.3) is the unique MDCS forced by the hierarchy construction.

(IV) The maximum set, obtained by treating each descriptor independently, is excluded by the hierarchy construction. We are thus limited to case III, which exhibits the necessity of the interpretation.

We have used conservation as the basic interpretive principle. We have yet to display this in the context of the sequential dynamics of step-by-step discrimination which is implied in speaking of conservation. We approach this problem for the first three levels of the hierarchy by interpreting sequences of discriminations as the flow of quantum numbers through sequentially ordered "Feynman diagrams." As we will see below, the direction of the sequence has to be established *external* to the hierarchy as part of our construction of a finite representation of "space–time" which could, sometimes, approach conventional space–time in a large number regime.

The basic postulate by which we convert the symmetric discrimination operation $x + y = z = y + x$ into a partial ordering is to assume that when the discrimination occurs between two identical columns, i.e., when $x + x = 0$, there is some externally established criterion, which eventually is to be established recursively, by which the two x's are assigned to *different* sets. Our justification for this assumption is our equally basic postulate that the nonnull descriptors in a column refer to *conserved* quantum numbers. This is clearly impossible in the case at hand if both columns are on an identical footing, since then the symmetric operation would destroy quantum numbers. Abstracting from the empirical structure of elementary particle physics, we assume that $x + x = 0$ refers to a particle and an antiparticle which, so far as quantum numbers go, can indeed annihilate each other if they have opposite charge, opposite baryon number, opposite lepton number, and equal but opposite helicity ("z component of spin"), and *no* other distinguishing characteristics. This idea has yet to be worked out in deductive mathematical terms. Here we work it out, level by level through the first three levels of the hierarchy, using a diagrammatic technique abstracted from the familiar rules for Feynman diagrams.

Consider first a "universe" consisting only of identical columns x. This we call "level 0" of the hierarchy, since it clearly can be modeled by sets of "columns of height one" consisting of sets of the existence symbol 1, or the null 0. Notationally we represent the basic discrimination $1 + 1 = 0$ by Figure 1a, where the first 1 stands for a "particle" represented by the solid line and the second 1 stands for an "antiparticle" represented by the

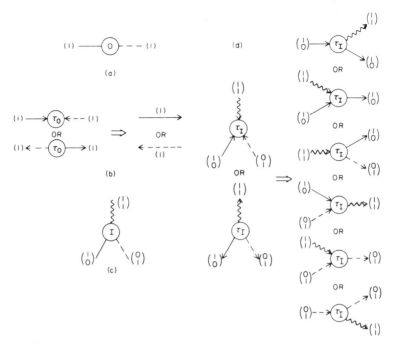

Fig. 1. (a) Level-0 discrimination interpreted as a particle–antiparticle diagram. (b) Level-0 sequentially ordered discrimination as particle or antiparticle. (c) Unique level-I DCsS as a particle–antiparticle–quantum vertex. (d) Sequential ordering of the level-I vertex.

dashed line. To make this into an ordered relation we assume each line has a "direction" relative to some externally established sequence of discriminations, which we indicate by placing an arrowhead on the line. The interpretation that conserves the (single type of) quantum number is then that both lines are "incoming" or "outgoing" as in the left side of Figure 1b. We now adopt the Feynman rule that a particle moving "forward" is the same as an antiparticle moving "backward." Since there is only one type in our level-0 "universe" we reverse one arrow. Then the discrimination has no effect and the universe consists only of particles moving forward (or of antiparticles moving backward). This is truly a Parmenidean universe in which there are no scatterings and nothing happens. In a broader context with higher columns, our (now ordered) discrimination makes a partial ordering selecting sets of "identical" columns with an "orthogonality" relation. So far as we can see, if this were all we had, we would have precisely the model discussed by Finkelstein (1977, 1979)—an "emission" followed by an "admission" which taken together form a detection which, when null, corresponds to the failure to detect—a partial ordering in which the "particles" continue to "move" undisturbed. But our structure is more complicated, as we will see shortly. An alternative to introducing an ordering relation might be to develop a metalanguage in which we can retain the nonordered discrimination operation; as we understand it, this is what Parker-Rhodes (1978) has done in his theory of indistinguishables, which allows the discussion of "twins" that are individually indistinguishable, and that cannot be ordered, but which allows the assignment of cardinal numbers to sets composed of them. We believe the route followed here is more consistent with our basic process philosophy.

We now proceed to level I, where we have three nonnull columns that can be symbolized as follows: $\left(\begin{smallmatrix}1\\0\end{smallmatrix}\right)$ by ———, $\left(\begin{smallmatrix}0\\1\end{smallmatrix}\right)$ by ----, and $\left(\begin{smallmatrix}1\\1\end{smallmatrix}\right)$ by $\sim\!\sim\!\sim$. We can symbolize the unique MDCS by the discrimination diagram given in Figure 1c. As a nonordered discrimination diagram this is to be interpreted as representing the fact that discrimination between any two of the columns yields the third. As a Feynman diagram with all three lines either incoming or outgoing (Figure 1d) there is still no internal way to assign order. If, however, we assign a sequential direction externally, and use the Feynman rule, we obtain six possibilities also given in Figure 1d. There is now a structural difference compared to "level 0," because we now need a rule to say what happens to the two rows when we reverse the direction of the arrow. We see that to conserve quantum numbers we must interchange the two rows.

Now physical interpretation becomes possible. Row one represents one dichotomic variable, or conserved quantum number, whose presence or absence is indicated by the exitence symbols 1 or 0, respectively. The second row represents a second distinct dichotomic variable. In order that both quantum numbers be conserved, they must be conjugate in the sense that reversing sequence interchanges rows. The simplest choice for interpretation, following Eddington's insight that the basic quantization is that of charge, is that the two quantum numbers are simply positive and negative unit electric charge. Then our rule that reversal of sequence must be coupled to interchange of rows translates to the usual Feynman rule that a positive particle moving forward "in time" is equivalent to a negative particle (antiparticle) moving backward "in time." Note that in contrast to $\left(\begin{smallmatrix}1\\0\end{smallmatrix}\right)$ and $\left(\begin{smallmatrix}0\\1\end{smallmatrix}\right)$ the column $\left(\begin{smallmatrix}1\\1\end{smallmatrix}\right)$ is self-conjugate, and we are free to assign it either direction until we have sufficient *external* structural information to specify that direction in another way.

High-energy physics allows us to provide an experimental model isomorphic to this lowest level of the hierarchy—a hydrogen bubble chamber in a magnetic field with a beam of antiprotons incident. Protons curve one way and antiprotons the other way, distinguishing the two quantum numbers. Annihilation produces electrically neutral quanta that leave no tracks in the chamber but whose presence can be inferred by the appearance of proton–antiproton pairs that can be spacially correlated with kinks in the tracks. A more detailed working out of this operational definition of quantized particles has been given elsewhere (Noyes, 1957). Such experiments provide direct empirical evidence for the quantization and conservation of unit electric charge. Note that the direction of the tracks must be inferred from external information on which side of the chamber the beam enters, or internally by the relation between density of bubbles along the track and velocity or energy. Relativistic kinematics allows the specific case when particle, antiparticle, and neutral quantum all have the same mass to be made into a model in which the three bind ("bootstrap") to form a single particle of the same mass and charge as one of the three, as has been discussed elsewhere (Noyes, 1979).

By such external considerations, we can talk about the ordered vertices of level I, symbolized in the figure by $\textcircled{\tau_I}$ since they will eventually become time ordered scattering vertices, and the nonordered $\textcircled{I}$, which represents the unique DCsS of level I. But it is easy to see that if we start with an arbitrary statistical assemblage of all three possible columns and all "directions," on the average nothing will happen. There will be the (unobservable and ignorable) discriminations of "level 0," and level I vertices with as many incoming as outgoing lines. Charge is conserved in the microscopic processes, and hence for the system as a whole. Any asymmetries would have to be established externally. This "universe" is still Parmenidean so far as observable consequences go.

When we go to level II the situation changes. If we represent columns $(11xy)$ by $=\!=$, and columns $(00xy)$ by $\curvearrowright\curvearrowright$ the seven DCsSs of Equation (2.4) can be pictured by the seven discrimination diagrams given in Figure 2. To convert these to ordered diagrams that conserve quantum numbers, we see that rows 3 and 4 are isomorphic to level I, and that we must interchange these two rows when we reverse the direction of an arrow. But rows 1 and 2 are self-conjugate and act within this group of seven columns like a single new dichotomic variable which is either present or absent. But now we have eight additional columns $(10xy)$ and $(01xy)$ *outside* the hierarchy. Under our basic statistical assumption that initially all columns have equal probability, and that all rows are to be interpreted in terms of conserved quantum numbers, we see that we have added not one but two new dichotomic variables. Further, they also can form DCsSs, as we can see for example in Figure 3, ignoring for the moment the arrows. If, as we did within the hierarchy, we assigned all lines as incoming (or outgoing) the quantum number in the first row would be annihilated, contrary to our basic interpretive postulate. Hence, for these new vertices we *must*, in order

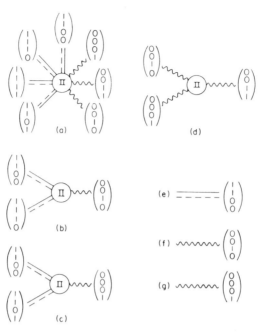

Fig. 2. (a) Unique level-II MDCS as a discrimination vertex within the hierarchy. (b)–(g) Six DCsS of level II in a particular representation as discrimination vertices within the hierarchy.

to conserve quantum numbers, assume that we have one incoming and two outgoing arrows. Then for any sequence of processes involving all 15 nonnull columns, *and* ordered vertices when $(10xy)$ or $(01xy)$ are involved, all four dichotomic quantum numbers will be conserved, provided (in our specific representation) we interchange both row 1 with row 2 *and* row 3 with row 4 when we change the direction of an arrow.

How are we to interpret this situation physically? We claim that the structural characteristic of the hierarchy—which, as proved in the last

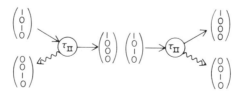

Fig. 3. Two level-II vertices outside the hierarchy sequentially ordered to conserve quantum numbers; the two shown can be interpreted as the emission or absorption of a charged quantum by a baryon.

section, necessarily doubles one descriptor in the basis used for constructing level II—implies two different types of dichotomic quantum numbers
at level II. Since one pair is isomorphic to level I, we retain the identification of this with electric charge. The second we propose to interpret
as baryon number in row 1 and antibaryon number in row 2. The
seven columns within the hierarchy can then be interpreted as four
baryon–antibaryon pairs $B^+ \bar{B}^0, B^+ B^-, B^0 B^-, B^0 \bar{B}^0$, and three quanta
$Q^+, Q^\pm, Q^-$. We also have the start of a sequential dynamics because of
the necessity of ordered vertices once we deal with single baryons. If we
identify these with nucleons and antinucleons, and the quanta with
charged and neutral mesons, we have a crude model for nuclear physics.
We will not develop this here, as we have a more specific calculation closer
to reality to present in level III. Note also that, again by use of the bubble
chamber, it is possible to distinguish baryons from mesons since, empirically, the number of baryons minus the number of antibaryons is conserved, while the number of mesons is not; charge is, of course, still
conserved. A little thought should convince the reader that our rules
guarantee this contact with experiment.

We wish to emphasize here the structural features of the hierarchy
which allow this comparison with experiment. Suppose we ignored the
doubling of one of the descriptors in going from level I to level II. Then we
could model the $2^3 - 1 = 7$ DCsSs with columns of height 3, e.g., the basis
$(100), (010), (001)$ which would put all three rows on the same footing.
Quantum numbers could be conserved, and the DCsSs mapped by 3×3
matrices, which would then give $2^7 - 1 = 127$ DCsSs of columns of height 9.
But these cannot be mapped by the $9 \times 9 = 81$ linearly independent
matrices available; this hierarchy terminates too quickly. Another alternative would be to use the $2^4 - 1 = 15$ DCsSs constructed from columns of
height 4, e.g., using the symmetric basis $(1000), (0100), (0010), (0001)$, which
again makes all rows indistinguishable. These again can be mapped and
provide a basis for $2^{15} - 1$ DCsSs of columns of height 16; again these
cannot be mapped by the $16 \times 16 = 256$ linearly independent matrices
available, so this also terminates too quickly. Only the asymmetric basis
obtained from the mapping of level I allows the continuation to both level
III and level IV. Further, as we have seen, this asymmetric basis, by
distinguishing between columns inside and outside the hierarchy allows us,
for the first time, to introduce meaningful sequence along with conservation. Thus, discrimination, conservation, and the existence of DCsSs that
can be mapped at a single level are even conjointly not enough. We *must*
use the unique hierarchy construction to get a rich enough physics without
additional postulates. When we do, we are rewarded by finding a structure
that can be interpreted as exhibiting the asymmetry between baryons and
mesons that lies at the core of all nuclear theory.

Going now to level III, we have seen in equation (2.5) that we have a representation of the seven basis columns with one quadrupled and six doubled descriptors. Here we are not on so firm ground in interpretation, because this representation is no longer unique, and we have to argue instead that it is the simplest and most symmetric representation we can construct. But we believe, though we have not proved, that *any* level-III representation will have a quadrupled descriptor. Guided by the hypothesis that the third level, being stable, should contain quantum numbers corresponding to the absolute conservation laws of charge, baryon number, lepton number, and helicity ("*z* component of spin"), and our successful handling of the doubled descriptor as baryon–antibaryon quantum numbers at level II, we assume that the quadrupled quantum number represents a baryon–antibaryon pair in conjunction with a lepton–antilepton pair. Then, following the scheme given in Table II, two of the doubled descriptors represent the four spin states obtained by putting together a spin-1/2 baryon with a spin-1/2 antibaryon to form a singlet–triplet system, two of the doubled descriptors correspond to putting together (the same) baryon antibaryon pair to form a singlet–triplet isospin system, and the last two doubled descriptors to the singlet–triplet spin system obtained from spin-1/2 lepton and a spin-1/2 antilepton. Explicitly, the 16 column of equation (2.6) is then $(B, \bar{B}, l, \bar{l}, s_B^+, s_B^-, s_{\bar{B}}^+, s_{\bar{B}}^-, i_B^+, i_B^-, i_{\bar{B}}^+, i_{\bar{B}}^-, s_l^+, s_l^-, s_{\bar{l}}^+, s_{\bar{l}}^-)$. Anyone familiar with Feynman rules will see immediately that if we interchange rows pairwise when we change the direction of an arrow, we have the usual rule that spins, particle–antiparticle designation, and charge reverse under time reversal, and that we can conserve quantum numbers in the same way we did at lower levels.

The physical interpretation of the individual states in the hierarchy is now straightforward. When we have (1111...) we have 16 spin states×4 isospin states or 64 in all. Note that we now have to talk about conservation of the "*z* component of isospin," which is equivalent to charge conservation in this context. Note also that we are referring to helicity rather than "spin" in a 3-space or 4-space sense. This is not to carry any implications about "rotations" until we have constructed some discrete approximation to "space–time," which we have yet to do. The (0000...) columns are also easy to interpret. Three of them carry isospin without spin, like pions; nine of them carry both spin and isospin, like the three spin × three charge states of the ρ mesons; three have isospin zero in three spin states like the ω meson. All of these 15 mesons come from rows associated with the baryons. The remaining three spin states associated with the leptons we identify with the two helicity states of the photon (γ) and the Coulomb field. The mesons can be put together with the γ to form $3 \times 15 = 45$ states. Thus the total number of states is $64 + 15 + 3 + 3 \times 15 = 127$ as required.

The presence of the γ is particularly interesting. Empirically the mesons all have finite mass and dimensionally speaking explore "distances" of the order of 10^{-13} cm, where many people agree our ordinary ideas of space–time are suspect. But the γ, being massless, has effects of infinite range, and in the large-number limit goes over, via the correspondence principle, to the classical electromagnetic field. One might think that, prior to the development of an explicit dynamics, we should not be able to get quantitative results from the theory at this stage. But the presence of the γ and the Coulomb field in our interpretation allows us to discuss, at least heuristically, a remarkable calculation.

The calculation was originally achieved by A. F. Parker-Rhodes (1978), who justifies his physical interpretation of the hierarchy, and of more extended structures, on the basis of his theory of indistinguishables. Unfortunately, this theory requires considerable logical development for consistent presentation since objects that can be counted as two when together, but that are truly indistinguishable when separate (called "twins"), cannot be grouped in ordered sets; they can, however, be grouped in such a way as to define a unique cardinal for the group or "sort." Thus a "sort theory" dealing with this possibility has to be developed, based on the three parity relations "identical," "distinguishable," and "twins"—together with their negations. This requires a semantic theory, using two-valued logic, for discussion of the object theory, and an implication language, again using two-valued logic, for the statement and proof of theorems. However important the theory of indistinguishables may be, Parker-Rhodes' ideas of interpretation are inconsistent with those developed in this paper, and we give his deductions in an amended form. We expect that before very long a consistent presentation on our own principles will have been reached, but the form we give below is to some extent a compromise with conventional thinking. Our excuse for (in a sense) premature publication is the astonishing accuracy of the result. We believe that the presentation we give here is believable in terms that are closer to ordinary quantum mechanical usage—once one is willing to make the conceptual leap that allows the discussion of quantum ideas *prior* to any mention of space–time.

We have seen that the three stable levels of the hierarchy can be viewed as systems carrying the quantum numbers of baryon–antibaryon pairs and lepton–antilepton pairs and the associated bosons. Since comparison between any two such systems leads to a third, and all three levels map up or down, it seems appropriate to think of the hierarchy as containing all 137 possibilities with equal *a priori* probability. But to discover the actual structure, we must somehow "break into" this closed system, which necessarily requires a column that is not one of the members of the hierarchy. The example we pick is the electron.

Using the specific choice of row designations already introduced, i.e., $(B, \bar{B}, l, \bar{l}, s_B^+, s_B^-, s_{\bar{B}}^+, s_{\bar{B}}^-, i_B^+, i_B^-, i_{\bar{B}}^+, i_{\bar{B}}^-, s_l^+, s_l^-, s_{\bar{l}}^+, s_{\bar{l}}^-)$, an electron with "spin up" is (0010 0000 0101 1000) and with "spin down" is (0010 0000 0101 0100).

In order to couple this column into the hierarchy, we have to introduce some new sort of vertex that does conserve quantum numbers; just how does not have to be specified for our current purpose. Presumably this can be done in the same way that we introduced an ordered meson–baryon vertex at level II. The only member of the 137 columns in the hierarchy that does not change the electron spin or charge, or refer to irrelevant quantum numbers, is the Coulomb case. So we assume that the electron couples to this with a probability of 1/137. This member of the hierarchy then communicates with all the others in a random fashion, eventually ending up again with the Coulomb case and back to the electron. In this respect we view the hierarchy as resembling something like the "vacuum fluctuations" of quantum field theory. The reason that this can lead to a result is that the electron cannot coincide with those members of the hierarchy that contain electron–positron pairs while this process is taking place, thanks to the exclusion principle. Particularly since we have as yet not made use of the exclusion principle, an assumption more in keeping with our basic statistical approach (which has the same effect on the calculation) is that the statistical uncertainties in the concept of "length" at nuclear dimensions do not allow us to discuss Coulomb energy separations for lengths smaller than some distance d. Thus the process necessarily involves some space–time separation or interval between the electron and the hierarchy, which we will estimate statistically. Further, since we have no reference frame to refer this distance to, the resulting charge distribution relative to this space–time interval must also be distributed statistically, subject only to charge conservation. The calculation we present is of the ratio of the square of this statistically smeared-out charge to the statistically estimated distance of separation, equated, as is often assumed, to the electron rest energy $m_e c^2$. Schematically, the process we are computing is shown in Figure 4.

Our first step is to take out the dimensional factors and thus reduce the statistical part of the calculation to dimensionless form. The square of the charge is e^2; it is smeared out into two (or more) parts over some

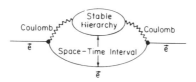

Fig. 4. Schematic representation of the electron self-energy.

distance r. We introduce a random variable x to represent the charge in one part, and, in order to conserve charge between two parts, write the square of the charge as $e^2x(1-x)$. As we have already argued, the coupling we should use at this stage in the development of the theory is $1/137$, not the empirical value of the fine-structure constant α, so $e^2 = \hbar c / 137$.

Because of the statistical uncertainty in the concept of length at nuclear dimensions, or because of the exclusion principle, there is some distance of closest approach d, which acts as a cutoff in the distance r. Since the only stable mass other than the m_e we are computing is the proton mass m_p, and proton–antiproton pairs occur in the levels of the hierarchy, it seems reasonable to take this shortest distance we can define to be the Compton wavelength of a proton–antiproton pair $d = h/2m_pc$; our second random variable y is then defined by $r = yd$, with $y \geqslant 1$. We like the idea of introducing Planck's constant into the theory as a basic measure of the breakdown of the concept of macroscopic length.

The random variable x represents the charge in a system with three degrees of freedom smeared out statistically and interacting with the remaining charge $1-x$. If we could cut the charge into two pieces, like a hunk of butter, x would vary between 0 and 1. But in our interpretation the hierarchy contains pieces with both positive charge $(p\bar{n}, \pi^+, \rho^+, \ldots)$ and negative charge $(\bar{p}n, \pi^-, \rho^-, \ldots)$ as well as neutral and internally neutralized systems, all of which communicate with each other in the stabilization process. Hence, if we look at all the possibilities, and maintain overall charge conservation, x can have any value between $-\infty$ and $+\infty$. Once we have gone beyond the first separation, we have no way of knowing whether the Coulomb energy we are evaluating is attractive (unlike charges) or repulsive (like charges) outside of the interval $0 < x < 1$. Statistically the positive and negative effects outside this interval must cancel. This statement is not obvious, as has been pointed out to us by F. Levin. To explain it, we note first that if Figure 4 represented a single process the charge would have to follow the electron line and there would be no charge smearing. But in fact we are computing a statistical average of such processes in which we assign the charge to two pieces according to the random variable x as ex and $e(1-x)$. The probability of this separation taking place at one vertex is proportional to the dipole $e^2x(1-x)$. Once we have two smeared out charge distributions further smearing will come from the virtual appearance of charged particle-antiparticle pairs; thanks to our diagrammatic rules, charge is conserved at each vertex in such processes and the overall electric neutrality guarantees that for the first distribution the charge remains ex or $e(1-x)$. Thus although this further smearing can lead to regions with any positive or negative value for the charge, these effects cancel outside the interval $0 \leq x \leq 1$. Further, after the initial smearing, the effective squared charge of each piece is e^2x^2 or $e^2(1-x)^2$, a fact we will need below. As we can see from Figure 4, in order to reform the electron from these smeared out distributions, we need a second vertex. By microscopic time reversal invariance, which is guaranteed by our equal *a-priori* probabilities, the probability of this closure is again proportional to the dipole $e^2x(1-x)$. We conclude that the overall weighting factor $P(x(1-x))$ to be used in computing $[e^2x(1-x)]$ is proportional to $[x(1-x)]^2$, and is to be normed on the interval $0 \le x \le 1$.

Putting this together, we see that

$$m_e c^2 = \langle q^2 \rangle \left\langle \frac{1}{r} \right\rangle = \frac{\hbar c}{137} \langle x(1-x) \rangle \frac{2m_p c}{h} \left\langle \frac{1}{y} \right\rangle$$

or

$$\frac{m_p}{m_e} = \frac{137\pi}{\langle x(1-x) \rangle \langle 1/y \rangle} \tag{3.1}$$

To calculate the expectation value of $1/y$ we need some probability weighting factor $P(1/y)$. We have seen above that the hierarchy has three distinct levels with different interpretations, each carrying charge, so we assume that the distribution of charge in the statistical system has three degrees of freedom, each of which brings in its own random $1/y$. Thus we assume $P(1/y) = (1/y) \cdot (1/y) \cdot (1/y)$ and find that

$$\left\langle \frac{1}{y} \right\rangle = \int_1^\infty \left(\frac{1}{y} \right) P \left(\frac{1}{y} \right) \frac{dy}{y^2} \Big/ \int_1^\infty P \left(\frac{1}{y} \right) \frac{dy}{y^2} = \frac{4}{5} \tag{3.2}$$

If the charge splitting x had only one degree of freedom, the expectation value of $x(1-x)$ using the weighting $P(x(1-x)) = x^2(1-x)^2$ would be

$$K_1 = \langle x(1-x) \rangle_1 = \int_0^1 x(1-x) P(x(1-x)) \, dx \Big/ \int_0^1 P(x(1-x)) \, dx = \frac{3}{14}$$

$$\tag{3.3}$$

Actually, as already noted, we have three degrees of freedom coming from the three levels of the hierarchy. Once the distribution has separated into x and $1-x$ the effective squared charge of each piece is x^2 or $(1-x)^2$, so we can write the recursion relation

$$K_n = \int_0^1 \left[x^3(1-x)^3 + K_{n-1} x^2(1-x)^4 \right] \Big/ \int_0^1 x^2(1-x)^2 \, dx$$

$$= \int_0^1 \left[x^3(1-x)^3 + K_{n-1} x^4(1-x)^2 \right] \Big/ \int_0^1 x^2(1-x)^2 \, dx$$

$$= \frac{3}{14} + \frac{2}{7} K_{n-1} = \frac{3}{14} \sum_{i=0}^{n-1} \left(\frac{2}{7} \right)^i \tag{3.4}$$

Putting this back into formula, using K_3, because of the three degrees of freedom of the hierarchy, we have

$$\frac{m_p}{m_e} = \frac{137\pi}{\frac{3}{14} \times \left[1 + \left(\frac{2}{7} \right) + \left(\frac{2}{7} \right)^2 \right] \times \frac{4}{5}} = 1836.151497 \cdots \tag{3.5}$$

as compared with the latest empirical result 1836.15152 ± 0.00070 (Barash-Schmidt et al., 1978).

Clearly, in presenting our calculation in this way, we have leaped ahead of what we are justified in doing as an explicit dynamical calculation. But the calculation illustrates one way in which two algebraic quantities can be introduced into the theory in the form of the square of one divided by the other. The specific interpretation is compelling because of the high quality of the numerical result; the critical integer 3 which enters both the charge distribution and the separation as three degrees of freedom is, we are confident, correctly identified as the three levels of the hierarchy. That we should be able to interpret this calculation within our framework is evident. This fact alone puts us in a strong position.

The quality of the result makes it important to discuss corrections which might destroy it. To begin with, we have used the value 137 for $1/\alpha$ rather than the empirical value. As discussed below, because of coupling to level IV, we can anticipate corrections to $1/\alpha$ of order $1/256^2$, which is of the correct order of magnitude. The second correction we can anticipate is in the cutoff parameter d. Our first estimate is almost certainly approximately correct, but does not account for the fact that electrons in the hierarchy are sometimes present and sometimes absent. Hence, we can anticipate a correction to d of order $m_e/2m_p$ as well as in the calculation of the correction to $1/\alpha$. Thus we anticipate something like the empirical result for $1/\alpha$ and must hope that the correction to d will almost exactly compensate for it in our formula. Looked at this way, the calculation can be viewed as a guide to *how* to construct the dynamics, rather than as a prediction of our theory. It has already proved of great value in setting up the classification scheme given in the last section, and in obtaining the kinematic bootstrap (Noyes, 1979) at level I.

Since the language we use for justifying the calculation when exhibited pictorially as in Figure 4 makes the stable hierarchy look like a photon, we can try to extend this analogy. To begin with, if we look at coupling into the hierarchy through transverse photons, these will flip the spin of the electron. But again, for a specified spin of the electron, this can happen in only 1 of the 137 possible cases, so the coupling constant is the

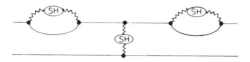

Fig. 5. Single-"photon" exchange between electron and proton.

same as we used in the Coulomb calculation (and including this in our "self-energy" calculation does not alter the result), which is encouraging. So consider an electron and a proton which exchange a "photon" so described. Making the static, nonrelativistic assumption that the mass of the proton does not change with velocity and that its motion does not affect the energy of the system, the additional effect we must consider is that the electron must acquire its own mass both before and after the exchange by the process already considered. This leads to the diagram given in Figure 5.

If the "photon" exchanged in the figure carries any momentum, the diagram cannot represent the whole story, since there will also be the emission of "bremsstrahlung" in the final state. So we consider the diagram only for the case when both electron and proton are at rest, but as far apart as we like. This is to be interpreted as an electron and proton bound in the ground state of hydrogen, and contrasted with a free electron and proton with the Coulomb effect shielded out. The second case then is the one already considered except that an inert proton has been added, and the first can be calculated as before, provided we multiply the coupling by the two additional powers of α shown in Figure 5; the statistical calculation remains unaltered. We conclude that the binding energy of the ground state of hydrogen is given by $\alpha^2 m_e c^2 = m_e e^4/\hbar^2$, which is indeed the correct result, in the static case. To obtain the center-of-mass correction we must allow for the motion of the proton, which requires more dynamics than we have developed. Further, to get the excited states, we must be able to describe unstable systems that decay via photon emission, for which we are as yet unready.

We summarize the results of this section by pointing out that we already have in hand the basic ingredients from which atomic theory could be built—stable electrons in correct mass ratio to the proton, photons, and time-ordered Feynman vertices, together with a hint as to how the statistical smearing out of the (not time ordered internally) three levels of the hierarchy's DCsS can stabilize dynamical systems with finite "self-energies." Calculation of genuine dynamical processes such as $\pi^0 \rightarrow 2\gamma$ will provide a critical test of whether we are on the right track. We also have the basic ingredients for an approximate nuclear physics—nucleons and antinucleons coupled to pions and the vector mesons. What is still missing are processes involving beta decay and neutrinos. For these we must go on to level IV.

4. LEVEL IV: WEAK INTERACTIONS AND COSMOLOGY

Returning to the basic mapping matrix from which level IV can be constructed as given in equation (2.6), we see that it will lead, rearranged as a column, to an 18-fold descriptor. The specific mapping actually constructed leads in addition to 126 single descriptors, so it leaves 112 of the 256 available rows unaccounted for. We are confident that mappings that fill all rows can be constructed, and that the multiple descriptor can be $20, 22, \ldots$ in other representations, still using our preferred basis. This enormous ambiguity is actually what we expect from elementary particle experiments at high energy. For a while it was thought that there were only three "quark" quantum numbers (up, down, and strange) and two types of leptons (muon and electron with associated neutrinos). But now the "charm" quantum number has been observed, the "upsilon," the heavy lepton called "tau," and most experimental physicists believe that is unlikely to be the end of the story. The important point for us is that none of these new quantum numbers (in contrast to charge, baryon number, lepton number, and helicity) are reliably known to be exactly conserved. Many (e.g., strangeness) are violated in weak decays. But this is what we expect from our combinatorial hierarchy. From our basis of 127 linearly independent columns, we can construct $2^{127} - 1$ DCsSs. But only $(256)^2$ linearly independent matrices are available to map, and hence stabilize, them. Thus, even if we happen to start with a particular DCsS, because of our basic statistical assumption that all possible columns are randomly available, after roughly $(256)^2$ discriminations we can expect this set to encounter some column from outside. This will destabilize the system and lead to a "weak decay." Our rough estimate of the coupling constant as $1/(256)^2$ is close to the "weak decay constant" of $10^{-5}m_p$, where, as we saw in our last section, we are constrained to use m_p as our unit of mass. Thus, qualitatively, the scheme predicts weak decays, as was already noted in the first presentation of the hierarchy (Bastin, 1966). We will not even attempt to sketch how this dynamics might work in this paper.

However, we can proceed a little way with the implications of our structure as a classification scheme. The number 18 that occurs in the simplest possible construction of a mapping is already suggestive of quark–antiquark pairs with three colors and three flavors. The 18-fold descriptor then corresponds to conservation of baryon (or quark) number for this system. Doubling 18 of the single descriptors will give us 36 spin states. We can use $9 \times 18 = 162$ rows to assign color and flavor, leaving $256 - 18 - 36 - 162 = 40$ rows to describe various types of leptons. Working out the details of all this will be a fascinating puzzle. We will certainly want to be intimately in touch with high-energy experiments and current quantum chromodynamics and lepton theories in order to obtain empirical guidance. But, since we can count on destabilization of these "partially conserved quantum numbers," we know that, qualitatively at least, we are dealing with the right structure. We believe it will be more profitable to tackle this problem after we have worked out a firmer hold on atomic and nuclear physics at level III than to plunge into it now.

Before leaving the subject of "weak interactions" we note that we may have been too hasty in placing all of them in level IV. Just as we were able to interpret two columns at level III as a charged lepton (electron) with two spin states, we could leave off the charge descriptors and interpret the resulting columns as an electron-type neutrino. Then, if we can find a way to couple this to the baryons—which we have not yet succeeded in doing —we might be able to include ordinary beta decay at level III. This would not only complete our picture of low-energy nuclear physics, but also could lead to a Weinberg–Salam-type of weak-electromagnetic unification with the same coupling constant of $1/137$. The difficulty will be to show that the coupling to baryons generates a sufficiently large mass for the "W boson" so that the fact that it has yet to be observed experimentally can be accounted for. Then only the more exotic leptons, like the quark quantum numbers, would come in at level IV. We suspect this is the correct route to follow. The check will be whether the extension to level IV gives the quantitatively correct modification of α in accord with experiment. The correction will clearly be of order $1/(256)^2$, which is the right order of magnitude.

But this "high-energy physics" aspect of level IV only deals with the lower levels of its potential complexity—the $2^{127} - 1 \approx 10^{38}$ DCsSs, each of which is a distinct and discriminable entity. Just as we interpret $1/137$ as an approximation to α, we interpret 10^{-38} as an estimate of the gravitational coupling constant between two protons—protons rather than electrons, since we have already accounted for the rest mass of the electron in terms of this unit. At this point a more conventional argument, adapted from a remark of Dyson's (1952), becomes relevant. If we try to count N_e charged particle–antiparticle pairs within a volume whose radius is their compton wavelength, their electrostatic energy is

$$N_e e^2 / (\hbar/2mc) = N_e (e^2/\hbar c) 2mc^2 \tag{4.1}$$

We interpret this result as saying that if we try to determine the number N_e

for a system with more than 137 pairs by electromagnetic means, we are unable to do so because the energy has become so large that additional pairs could be present, and the counting breaks down. Hence, $N_e = 137$ is the maximum meaningful number of charged particle pairs we can discuss electromagnetically in such a volume (Noyes, 1974).

Extending the argument to gravitation, we see that, since

$$N_G Gm_p^2 / (\hbar / m_p c) = N_G (Gm_p^2 / \hbar c) m_p c^2 \qquad (4.2)$$

the maximum number of gravitating protons we can discuss within the compton wavelength of any one of them is $N_G \cong 10^{38}$. In this case, the gravitational field at the surface is so intense that light cannot escape, so this system forms a Laplacian "black hole" (Laplace, 1795). Hence, just as failure of the "fourth level" of the hierarchy to possess linearly independent mappings gives us an estimate of instability to weak decay, the upper limit $2^{127} - 1 \approx 10^{38}$ represents a gravitational instability for systems with large numbers of particles.

Since we have $\sim 10^{38}$ discriminate entities in the scheme, we are logically justified in starting our discussion with the $(10^{38})^2$ possible discriminations between them. For stability, these systems should contain lepton number and baryon number $(10^{38})^2$, although we cannot as yet prove such a conjecture. Given it, the initial discriminations will create all sorts of ephemeral forms of the type already discussed, and a historical system of loci that provides an initial space–time mesh. Once the decays and scattering have proceeded a while, these will settle down to protons, electrons, photons, hydrogen atoms,... and we have started the "big bang." The radiation soon breaks away from the matter, and provides a unique discrete approximation to a space–time framework, locally defined in terms of the cosmic background radiation. Since this "black body spectrum" can be measured locally, it provides us both a cosmic time scale from the temperature, and an absolute frame for measuring particle velocities. Our hope is that we can use this idea to define space–time frameworks more easily connected to laboratory observation than abstract definitions. In particular, since our W boson–photon coupling is discrete, and defined at proto-space–time loci, we should be able to use our dynamic scheme to explain what we mean by a local discrete coordinate system for physical measurement. Only when this task is complete can we tackle the question of what we might mean by a "wave function," and how we are to relate our particular formalism to the successful results obtained by conventional quantum mechanics.

5. CONCLUSION

In this paper we have sketched a physical interpretation of the combinatorial hierarchy, which, if the program can be carried through, should provide a finitist conceptual frame for that fundamental revision of physics which we seek. Our philosophical reasons for adapting this approach are discussed in detail in the opening section. Here we stress that the contact with experiment already established in this paper, together with the indications of structural contact with the classification schemes used in elementary particle physics, and conceptual contact with the fundamental ideas underlying current cosmology, make it clear that no field of physics need be omitted in this synthesis. The original coincidence between the cardinals of the hierarchy and the inverse boson field coupling constants allows us to believe that we have indeed unified strong, electromagnetic and gravitational phenomena in one framework. The weak decay instability is also indicated. Our proposed classification scheme brings in the absolute conservation laws at the correct level, and points toward a weak-electromagnetic unification at that or the next level. Structural contact exists between SU_2, SU_3, and SU_6 (quark) classifications, including an appropriate three-color–three-flavor option flexible enough to allow for new flavors and new heavy leptons. The cosmology should yield the conserved quantum numbers of the universe, some sort of "big bang," and hence the cosmic background radiation as a unique reference system. Since this background is not time reversal invariant, it might even lead ultimately to the explanation of the $K_L - K_S$ decay. So far as we see, no major area of physics has been omitted as potentially outside the reach of a scheme of this structure.

APPENDIX: MATHEMATICAL STRUCTURE OF THE HIERARCHY

This appendix contains a short formal account of the essential mathematical features of a "discrimination system." Throughout, S will denote a nonempty finite set, and $C = \{0, 1\}$ will denote either the cyclic group of order 2 (with addition mod 2, or equivalently, Boolean Exclusive-Or as group operation) or the field of two elements (with addition as before, and integer multiplication, or equivalently, Boolean And as field multiplication operation); the context makes clear which usage is intended. The empty set is $\varnothing$; $\mathbb{N} = \{1, 2, 3, \ldots\}$, $\mathbb{N}_0 = \{0, 1, 2, 3, \ldots\}$; $|X|$ is the cardinality of a set X.

A.1. Discrimination System

A.1.1. Definition. A *discrimination system of type* N ($N \in \mathbb{N}$) is a group S isomorphic to the Abelian group $C^N = C \oplus \cdots \oplus C$, direct sum of N copies of C, together with additional structure as detailed later; its order is $|S| = 2^N$. Thus $x \in S$ iff $x = (x_1, \ldots, x_N)$ ($x_i = 0, 1$), the group operation (written $+$) on S is termwise addition mod 2 (or equivalently, Boolean Exclusive-Or applied to strings of length N), and the group neutral is $e = (0, \ldots, 0)$.

A.1.2. Fact. Besides being symmetric (Abelian) and associative, the group operation $+$ on S is also *discriminative*:

$$(\forall x, y, z \in S) \qquad x + x = e \quad \text{and} \quad y \neq z \Rightarrow y + z \neq e$$

i.e., $+$ can "discriminate" between a pair of equal elements and a pair of unequal elements; the (unique) group neutral e is called the "(discrimination) neutral for S."

A.1.3. Theorem. Any set S equipped with a binary operation that is symmetric, associative, and discriminative (with respect to a unique discrimination neutral e) is isomorphic to an Abelian group C^N for some $N \in \mathbb{N}$; the discrimination neutral e is then (identified with) the neutral element in the group C^N.

A.1.4. Remarks. (1) The Boolean dual $u = (1, \ldots, 1)$ of the discrimination neutral $e = (0, \ldots, 0)$ in a discrimination system S is referred to as the "antineutral" for S. (2) The group $C^N, +$ can always be given a multiplication $*$ so that $C^N, +, *$ becomes a field F of prime power 2^N isomorphic to the Galois field $GF(2^N)$; its neutral-free part $C^N \setminus \{e\}$ forms a multiplicative cyclic group of order $2^N - 1$ with identity element which can be chosen to be the antineutral $u = (1, \ldots, 1)$.

A.2. Discriminately Closed Subsets

A.2.0. Remark. To avoid repetition, the abbreviation "d" or "d-" will be used for the words "discrimination, discriminate, discriminately" as appropriate throughout this and later sections.

A.2.1. Definition. Let S be a d-system with neutral e and let $T \subseteq S$ be a subset. Then T is a *dc-subset* (discriminately closed subset) (alias, subset T is *dc*) iff (a) T is *neutral-free* (i.e., $T \subseteq S \setminus \{e\}$), (b) the *e-join* $T \cup \{e\}$ of T is a subgroup in S.

A.2.2. Facts. (1) Conditions (i) and (ii) are equivalent: (i) T is a *dc*-subset of S; (ii) $(\forall x, y \in T)$ $x \neq y \Leftrightarrow x + y \in T$. (2) $S \setminus \{e\}$ is a *dc*-subset of S. (3) $\varnothing$ is a dc-subset of S. (4) Singleton $\{x\}$ is a dc-subset of $S \Leftrightarrow x \neq e$. (5) T is a subgroup of $S \Leftrightarrow T \setminus \{e\}$ is a dc-subset of S. (6) $R \cap T$ is a dc-subset of S whenever R and T are dc-subsets of S.

A.2.3. Definition. The *d-closure* T^{dc} of a neutral-free subset $T \subseteq S$ is the smallest dc-subset of S containing T; T^{dc} is *d-generated* by T. The *d-union* $R \cup {}^{dc}T$ of two neutral-free subsets $R, T \subseteq S$ is the dc-subset $(R \cup T)^{dc}$. The latter definition extends in an obvious way to families $(T_i)_{i \in I}$ of neutral-free subsets of S.

A.2.4. Facts. Let R, T be dc-subsets of S; then (1) T is $dc \Leftrightarrow T = T^{dc}$. (2) $(T^{dc})^{dc} = T^{dc}$, hence T^{dc} is dc. (3) $R \subseteq T \Rightarrow R^{dc} \subseteq T^{dc}$. (4) $R^{dc} \cup T^{dc} \subseteq (R \cup T)^{dc}$. (5) $(R^{dc} \cup T^{dc})^{dc} \subseteq (R \cup T)^{dc}$. (6) $R^{dc} \cup {}^{dc}T^{dc} = (R^{dc} \cup T^{dc})^{dc} = (R \cup T)^{dc} = R \cup {}^{dc}T$.

A.2.5. Fact. If $\langle T \rangle$ denotes the subgroup generated in S by a subset $T \subseteq S$ and T is neutral-free then $T^{dc} = \langle T \rangle \setminus \{e\}$ and $\langle T \rangle = T^{dc} \cup \{e\}$.

A.2.6. Definition. A subset T of a d-system S is a *d-subsystem* of S iff T is a subgroup of the group S and $T \neq \{e\}$.

A.2.7. Fact. If T is a d-subsystem of a d-system S of type N then T is isomorphic to a nontrivial subgroup $C^M \subseteq C^N$ with $1 \leqslant M \leqslant N$, and $1 < |T| = 2^M \leqslant |S| = 2^N$, and the neutral elements of T and S coincide.

A.2.8. Definition. The *d-complement* of a dc-subset T in a d-system S is the unique dc-subset R such that the subgroup $R \cup \{e\}$ is the direct complement of the subgroup $T \cup \{e\}$ in the group S.
Notation: $R = S \ominus^{dc} T$ and $S = R \oplus^{dc} T$.

Remark. Since $R = S \ominus^{dc} T \Leftrightarrow T = S \ominus^{dc} R$, so s is said to be *d-decomposed* by the *d-complementary* subsets R and T.

A.2.9. Fact. R and T are d-complements in $S \Leftrightarrow R \cap T = \varnothing$ and $R \cup {}^{dc}T = S \setminus \{e\}$.

A.2.10. Definition. Subset T is *d-independent* in S iff T is neutral-free and $(\forall t \in T)$ $\{t\}^{dc} \cap (T \setminus \{t\})^{dc} = \varnothing$; otherwise T is *d-dependent* in S.

Remark. A neutral-free subset T is d-independent in d-system $S \Leftrightarrow T$ is an independent subset in the group S. The definition extends in an obvious way to families $(T_i)_{i \in I}$ of neutral-free subsets of S, thus $(T_i)_{i \in I}$ is a d-independent family iff $(\forall k \in I)$ $T_k^{dc} \cap [\cup_{i \in I, i \neq k} T_i]^{dc} = \varnothing$.

A.2.11. Fact. These four conditions are equivalent: (a) Neutral-free family $(T_i)_{i \in I}$ is d-independent in d-system S. (b) Subgroup family $(\langle T_i \rangle)_{i \in I}$ is independent in group S. (c) Subgroup $\langle \cup_{i \in I} T_i \rangle =$

group direct sum $\oplus_{i \in I} \langle T_i \rangle$. (d) Each $x \in \cup_{i \in I}^{dc} T_i$ has a unique representation $x = \oplus_{i \in I} x_i$ with each component $x_i \in T_i^{dc} \cup \{e\} = \langle T_i \rangle$ and $x_j \neq e$ for at least one index $j \in I$.

A.3. Discriminate Morphisms

A.3.1. Definition. A morphism $f : S \to T$ between two d-systems S and T is a *d-morphism* iff f is an injective group homomorphism between the groups S and T. (Thus every automorphism $f : S \to S$ is a d-morphism on the d-system S.)

A.3.2. Remark. An automorphism $f : S \to S$ on a d-system S leaves fixed the neutral e $[f(e) = e]$ and permutes some or all of the members of the maximal neutral-free subset $S \setminus \{e\}$ in S.

A.3.3. Facts. Let $f : S \to T$ be a d-morphism; then (1) $R \subseteq S \Rightarrow f(\langle R \rangle) = \langle f(r) \rangle$; (2) $R \subseteq S \setminus \{e\} \Rightarrow f(R^{dc}) = [f(r)]^{dc}$; (3) $R = R^{dc} \Leftrightarrow f(R) = f(R)^{dc}$; (4) $(S_i)_{i \in I}$ is independent (or, respectively, d-independent) in $S \Leftrightarrow [f(S_i)]_{i \in I}$ is independent (or, respectively, d-independent) in T.

A.3.4. Remark. Recall the following: Let G be a group with neutral e; let $E(G)$ be the group of endomorphisms of G with respect to the composition operation $\circ$, $[(f \circ h)(g) = f(h(g)) \; \forall g \in G]$; let $A(G) \subset E(G)$ be the subgroup of automorphisms of G. Introducing a second group operation $(+)$ on $E(G)$, namely, pointwise addition $[(f + h)(g) = f(g) + h(g) \; \forall g \in G]$, makes $E(G), +$ Abelian when G is Abelian, and $E(G), +, \circ$ becomes a ring with neutral endomorphism $\mathbf{e}(g) = e$ $(\forall g \in G)$ and identity automorphism $\mathbf{u}(g) = g$ $(\forall g \in G)$. The case where G is a d-system is of special interest.

A.3.5. Theorem. Let S be a d-system of type N with neutral e. Let $E(S)$ be the Abelian group of endomorphisms of S under pointwise addition, with neutral endomorphism $\mathbf{e}$. Then $E(S)$ is a d-system of type $M = N^2$ and order $2^{(N^2)}$, with neutral $\mathbf{e}$ and antineutral $\mathbf{u}$. [Hence pointwise addition is discriminative on $E(S)$.]

A.3.6. Theorem. The ring $E(S), +, \circ$ of endomorphisms of a d-system S of type N is isomorphic to the ring of square $N \times N$-matrices over C.

A.3.7. Theorem. Let $A(S) \subset E(S)$ be the subset of automorphisms of S; then $A(S) \subseteq E(S) \setminus \{\mathbf{e}\}$, i.e., is a neutral-free subset in the d-system $E(S)$, but is not a dc-subset in $E(S)$. Indeed, $A(S)^{dc} = E(S) \setminus \{\mathbf{e}\}$.

Corollary. Every nonneutral endomorphism of S is a finite sum of distinct automorphisms of S.

A.3.8. Remarks. Let S be a d-system of type $N = t$; then $|S| = 2^t$, $|E(S)| = 2^{(2^t)}$, $|A(S)| = (2^t - 1)(2^t - 2)(2^t - 4) \cdots (2^t - 2^{t-1})$; let $r_t = |A(S)|/|E(S)|$, then $r_1 = 0.5$ and r_t decreases monotonically, and $\lim_{t \to \infty} r_t = 0.288788\ldots$ (i.e., for large t about 29% of all endomorphisms on S are automorphisms). For example:

t	1	2	3	4	5	16
$|S|$	2	4	8	16	32	65,536
$|A(S)|$	1	6	168	20,160	9,999,360	$3.34\ldots(10^{76})$
$|E(S)|$	2	16	512	65,536	33,554,432	$2^{256} = 1.1579\ldots(10^{77})$

A.3.9. Remark. Recall the following: If $\mathbf{a} \in A(S)$ then we may define an equivalence relation $[\mathbf{a}]$ on S by $x[\mathbf{a}]y \Leftrightarrow x = \mathbf{a}^k y$ for some $k \in \mathbb{N}_0$. Let $s \in \mathbb{N}$ be the least integer such that $\mathbf{a}^s = \mathbf{u}$ (the identity automorphism on S). Then an equivalence class mod$[\mathbf{a}]$ is called an $\mathbf{a}$-cycle of size r where r is its cardinality. Each $\mathbf{a}$-cycle is an orbit of the subgroup $\langle \mathbf{a} \rangle = \{\mathbf{u}, \mathbf{a}, \mathbf{a}^2, \ldots, \mathbf{a}^{s-1}\}$ in $A(S), \circ$, and vice versa. Its size $r = \min\{0 < k \leqslant s | x = \mathbf{a}^k x\} = \text{index} |\langle \mathbf{a} \rangle : A_x|$ where A_x is the stabilizer subgroup $\{\mathbf{b} \in A(S) | x = \mathbf{b}x\}$ of an arbitrary element x in the $\mathbf{a}$-cycle.

A.3.10. Definition. An automorphism $\mathbf{a} \in A(S)$ is *maximal* iff there exists an $\mathbf{a}$-cycle equal to $S \setminus \{e\}$ (i.e., of maximum size $r = 2^N - 1$); $\mathbf{a}$ is *minimal* iff there exists an $\mathbf{a}$-cycle equal to $\{x\}$ for some $x \neq e$ (i.e., a nonneutral $\mathbf{a}$-cycle of minimum size $r = 1$).

A.3.11. Theorem. Each d-system S of type N has (a) at least $2^N - 2$ distinct maximal automorphisms, and (b) at least $2^N - 1$ distinct minimal automorphisms.

A.3.12. Remark. The proof of the above theorem requires the construction of some not immediately obvious automorphisms:

(a) Let F be the (Galois) field associated with S [Remark A.1.4(2)], $F_0 = F \setminus \{e\}$ its neutral-free multiplicative group, cyclic of order $n = 2^N - 1$. Choose any one of the $n - 1$ generators $b \in F_0$ (thus $b^n = \mathbf{u} \neq b$, and $\mathbf{u}, b, b^2, \ldots, b^{n-1}$ are all the distinct elements of F_0). Let $\mathbf{a} : S \to S, \mathbf{a}(x) = bx$ (if $x \neq e$), $\mathbf{a}(e) = e$; then $\mathbf{a} \in A(S)$ and for a fixed element $x \in S \setminus \{e\}$ the images $\mathbf{a}^k(x) = b^k x$ $(k = 1, \ldots, n)$ are all distinct and exhaust $S \setminus \{e\}$. Thus each such $\mathbf{a}$ is a maximal automorphism on S, one for each of the $2^N - 2$ generators $b \in F_0$.

(b) Given a dc-subset T in S, let R be its d-complement (Definition A.2.8). Define $f = \mathbf{u}' \oplus \mathbf{m}' : S \to S$ by $f(x) = \mathbf{u}'(x)$, $(\forall x \in T \setminus \{e\})$, where $\mathbf{u}'$ is the identity automorphism on the d-subsystem $T \setminus \{e\}$, and $f(y) = \mathbf{m}'(y)$, $(\forall y \in R \setminus \{e\})$, where $\mathbf{m}'$ is a maximal automorphism [by (a) above] on the

d-subsystem $R \setminus \{e\}$. Then $\forall x = t \oplus r \in S = (T \setminus \{e\}) \oplus (R \setminus \{e\})$, we have $f(x) = (u' \oplus m')(t \oplus r) = u'(t) + m'(r)$. Thus $f \in A(S)$ and f fixes the dc-subset T. Taking $T = \{x\}$ $(x \neq e)$ makes f minimal, and T is one of its cycles of size 1 (the only nonneutral one).

A.3.13. Definition. The *df-set* (discrimination fixed set) of an automorphism $\mathbf{a} \in A(S)$ is the subset $DF(\mathbf{a}) = \{x \in S \setminus \{e\} \mid x = \mathbf{a}(x)\}$.

Remark. Each $DF(\mathbf{a})$ is a dc-subset; it may be empty [e.g., if $\mathbf{a}(x) = x$ only if $x = e$, i.e., if $\mathbf{a}$ "unfixes" every nonneutral member of S].

A.3.14. Theorem. Each dc-subset T in a d-system S is the df-set of some automorphism $\mathbf{a} \in A(S)$.

A.4. Discrimination Hierarchies

A.4.1. Definition. A d-system S of type N *determines* iteratively a sequence $E^0, E^1, E^2, \ldots, E^m, \ldots$ of d-systems where $E^0 = S$ and $(\forall m \in \mathbb{N})$ $E^m = E(E^{m-1})$ is the group of endomorphisms of the d-system E^{m-1} under pointwise addition. Then E^m is the *d-system of level m determined by the base d-system $E^0 = S$*; it is of type $t(m)$ and order $|E^m| = 2^{t(m)}$; A^m denotes the subset of automorphisms in E^m; the neutral (or, respectively, identity) morphisms in E^m are denoted by $\mathbf{e}^m$ (respectively, $\mathbf{u}^m$).

Remark. Where need be, the previous single-level notations such as $S, e, u, E(S), A(S), \mathbf{e}, \mathbf{u}$ may now be replaced by $E^0, \mathbf{e}^0, \mathbf{u}^0, E^1, A^1, \mathbf{e}^1, \mathbf{u}^1$.

A.4.2. Fact. Since $t(0) = N$ and $t(m) = t(m-1)^2$, so $t(m) = N^{(2^m)}$ and $|E^m| = 2^{t(m)}$ $(\forall m \in \mathbb{N}_0)$.

A.4.3. Remark. Plainly, each E^m is isomorphic to E^0 if E^0 is of type $N = 1$. More generally, if $N \geq 2$, we have an injective mapping f^m of E^m strictly into E^{m+1} given $(\forall m \in \mathbb{N}_0)$ by $f^m(\mathbf{e}^m) = \mathbf{e}^{m+1}$ and $f^m(x) = \mathbf{a}_x \in E^{m+1}$ $(\forall x \neq \mathbf{e}^m$ in $E^m)$ where $\mathbf{a}_x$ is an automorphism with singleton dc-subset $\{x\}$ as its df-set (Definition A.3.13). This mapping can be extended, in certain circumstances, to one that maps many more df-sets in one level injectively into an independent set of automorphisms in the next level, in a way now to be made precise.

A.4.4. Construction. Let $N \geq 2$; let $m \in \mathbb{N}_0$; let $K^m \subset E^m$ be an independent subset with $|K^m| = k(m)$ $[\leq t(m)]$ (Fact A.4.2). Let $K \subseteq K^m$ be any one of the $2^{k(m)} - 1$ nonempty subsets of K^m; and define $W_K = \{\mathbf{a} \in A^{m+1} \mid K^{dc} = DF(\mathbf{a})\}$. (Note that distinct subsets K of K^m have distinct d-closures K^{dc} because K^m is an independent set of elements.) Simple examples with $N = 3$, $m = 1$, show that we may have $|W_K| > 1$. In such a case, choose precisely one automorphism $\mathbf{a}_K$, say, in W_K and let C^{m+1} be the set of all such choices; thus

$$C^{m+1} = \left\{ \mathbf{a}_K \in A^{m+1} \mid K \subseteq K^m, K \neq \varnothing, DF(\mathbf{a}_K) = K^{dc} \right\}$$

$[= C^{m+1}(K^m)$, if we need to refer to the particular set K^m in use]. Hence $C^{m+1} \subset A^{m+1} \subset E^{m+1}$, and $|C^{m+1}| = 2^{k(m)} - 1$ $[= c(m+1)$, say, by way of definition]. To initialize this construction we define C^0 to be a basis (i.e., a maximal independent subset) for E^0 ($= S$), so that $c(0) = |S| = N$.

A.4.5. Question. Is it possible to construct iteratively the sequence $K^0 = C^0, K^1 = C^1(K^0), \ldots, K^{m+1} = C^{m+1}(K^m), \ldots$ with each K^m an independent subset (as required by Construction A.4.4)? An obviously necessary condition for this to be possible is this inequality:

$$k(m+1) = c(m+1) = 2^{k(m)} - 1 \leqslant t(m)^2 \qquad [*]$$

since $K^{m+1} \subset E^{m+1}$. A sufficient condition will be given below (A.4.8).

A.4.6. Definition. Let $S = E^0$ be a base d-system of type $N \geqslant 2$.

(A) For $m \in \mathbb{N}_0$, E^m is *d-injectable* into E^{m+1} *via* an independent subset $K^m \subset E^m$ iff there exists at least one choice of automorphisms for the set $C^{m+1}(K^m)$ which makes the latter set independent in E^{m+1}.

(B) A finite sequence $(E^0, E^1, \ldots, E^H)$ of d-systems determined by E^0 is a *d-hierarchy (discrimination hierarchy)* of *height* $H + 1$ iff these three conditions hold: (1) $H \geqslant 1$. (2) Independent subset K^0 is maximal in E^0. (3) For each $m = 0, 1, \ldots, H - 1$, but not for $m = H$, E^m is d-injectable into E^{m+1} via the independent subset $K^m \subset E^m$, where $K^m = C^m(K^{m-1})$ for each $m = 1, \ldots, H - 1$.

(C) A d-hierarchy is *trivial* if $H \leqslant 1$, otherwise *nontrivial*.

A.4.7. Remark. The sequence of independent subsets $(K^0, \ldots, K^H)$ in a d-hierarchy $(E^0, \ldots, E^H)$ is a "discriminate spine"; examples show that it need not be unique. If for some m the choice of C^{m+1} is not unique, each possible choice of C^{m+1} gives rise to a different "branch" of the spine.

A.4.8. Main Theorem. A necessary and sufficient condition for the existence of a nontrivial discrimination hierarchy is that the base discrimination system $S = E^0$ be of type $N = 2$; and then the discrimination hierarchy is of height $H + 1 = 4$.

A.4.9. Remarks. (1) The necessity follows from the condition $[*]$ in (A.4.5): Since $H \geqslant 2$ and E^m is d-injectable into E^{m+1} for $m = 0, 1, \ldots$, $H - 1$, condition $[*]$ holds in particular for $m = 0, 1$. Thus, $k(0) = N$ and $2^N - 1 \leqslant t(0)^2 = N^2$ so that $2 \leqslant N \leqslant 4$; and $k(1) = 2^{k(0)} - 1 = 2^N - 1$, hence

106

$2^{k(1)} - 1 = 2^{(2^N - 1)} - 1 \leqslant t(1)^2 = [t(0)^2]^2 = N^4$ so that $N = 2$ as asserted. But then [*] can be satisfied for $m = 0$, 1 or 2 but not for $m \geqslant 3$; in particular E^3 is not then d-injectable into E^4 so that $2 \leqslant H \leqslant 3$; i.e., the discrimination hierarchy must have height 3 or 4. (2) A theoretical proof of the existence of a discrimination hierarchy with $N = 2$ and height $H + 1 = 4$ has been provided by C. W. Kilmister (1978) and will be reported elsewhere; an empirical representation using matrices over the field $\{0, 1\}$ has been constructed by H. P. Noyes and described in the main body of the paper to which this Appendix is attached.

A.4.10. Remark. The connections between the notations used in this Appendix and those used in the main text (e.g., in Table I) are as follows:

Table I	Appendix
Index of level = 1, 2, 3, 4	$m = l - 1 = 0, 1, 2, 3$
"Dimension" of level = $n(l)$	"Type" of level = $t(m) = n(l - 1)$
Number of independent columns is $j(l)$	Number of independent elements in subset K^m is $k(m) = j(l - 1)$
Number of discriminately closed subsets used in level is $d(l) = 2^{j(l)} - 1$	Number of automorphisms chosen for the subset C^{m+1}, corresponding one-to-one with dc-subsets used in previous level is $c(m + 1) = 2^{k(m)} - 1 = d(l)$

ACKNOWLEDGMENTS

We are deeply indebted to A. F. Parker-Rhodes, who first formulated the necessary condition for the existence of a discrimination hierarchy, for permission to present the calculation of m_p/m_e drawn from his unpublished manuscript, "The Theory of Indistinguishables," the more particularly since his interpretation of the hierarchy as it occurs in his theory, and his justification for the calculation itself, differ in significant ways from our own views. The crucial notion of a discriminately closed subset arose out of discussions with Ted Bastin. The characterization of a discrimination system (Section 1), together with the proofs of many of the main results quoted here, are due to C. W. Kilmister. The verification of the conjecture that a discrimination hierarchy of height 4 existed owes a great deal to the patience and persistence of C. W. Kilmister and H. Pierre Noyes. One of us (H.P.N.) wishes to thank the Research Institute for Theoretical Physics of the University of Helsinki and K. V. Laurikainen, for generous hospitality, support, and active collaboration during the month after the 1978 Tutzing Conference, in which the existence proof for the hierarchy and one connection with SU_3 were hammered out. T.B. and H.P.N. are both indebted to the Max-Planck-Institut zur Erforschung der Lebensbedingungen der Wissenschaftlich-Technischen Welt for travel support and living expenses at the 1978 Tutzing Conference, where a preliminary version of this work was presented.

REFERENCES

Amson, J. (1976). "Discrimination Systems," 20 pp. (unpublished).

Barasch-Schmidt, N. et al. (1978). "Particle Properties Data Booklet April 1978," p. 2, from *Physics Letters*, **75B**, 1–250.

Bastin, T. (1966). "On the Origin of the Scale Constants of Physics," *Studia Philosophica Gandensia*, **4**, 77–101.

Bastin, T. (1976a). "A Combinatorial Model for Scattering," Report to the Science Research Council (U.K.), 57 pp. (unpublished).

Bastin, T. (1976b). "An Operational Model for Particle Scattering Using a Discrete Approach," Report to the Conference on "Quantum Theory and the Structures of Time and Space 2," Tutzing, 6 pp. (unpublished).

Dyson, F. J. (1952). "Divergence of Perturbation Theory in Quantum Electrodynamics," *Physical Review*, **85**, 631–632.

Finkelstein, D. (1969). "Space–Time Code," *Physical Review*, **184**, 1261–1271.

Finkelstein, D. (1979). "Holistic Methods," submitted to *International Journal of Theoretical Physics*.

Kilmister, C. W. (1978). private communication.

Laplace, P. S. (1795). *Exposition du Monde*, Vol. II, p. 305, Paris.

Noyes, H. P. (1957). "The Physical Description of Elementary Particles," *American Scientist*, Vol. 45, 431-448.

Noyes, H. P. (1974). "Non-Locality in Particle Physics," 55 pp., SLAC-PUB-1405.

Noyes, H. P. (1975). "Fixed Past and Uncertain Future: A Single-time Covariant Quantum Particle Mechanics," *Foundations of Physics*, **5**, 37–43 (Erratum **6**, 125, 1976).

Noyes, H. P. (1976). "A Democritean Phenomenology for Quantum Scattering Theory," *Foundations of Physics*, **6**, 83–100.

Noyes, H. P. (1976). "A Democritean Approach to Elementary Particle Physics," from Proceedings of the Summer Institute on Particle Physics, Martha Zipf, ed., pp. 239–259, SLAC Report No. 198 (issued as a separate document as SLAC-PUB-1956, 1977).

Noyes, H. P. (1978). Private communication.

Noyes, H. P. (1979). "The Lowest Level of the Combinatorial Hierarchy as a Particle Antiparticle Quantum Bootstrap," SLAC-PUB-2277.

Parker-Rhodes, A. F. (1978). "The Theory of Indistinguishables," 208 pp., (unpublished).

Weizsäcker, C. F. von. (1978). "Temporal Logic and a Reconstruction of Quantum Theory," 88 pp., presented at the Conference on "Quantum Theory and the Structures of Time and Space 3," Tutzing (unpublished).

Wheeler, J. A., and Patton, C. M. (1977). "Is Physics Legislated by Cosmogony?" pp. 19–35, in *The Encyclopedia of Ignorance*, R. Duncan and M. Weston-Smith, eds., Pergamon, Oxford.

Whiteman, J. H. M. (1971). "The Phenomenology of Observations and Explanation in Quantum Theory," in *Quantum Theory and Beyond*, T. Bastin, ed., pp. 71–84, Cambridge.

Comment on "*On the Physical Interpretation and Mathematical Structure of the Combinatorial Hierarchy*"
Ted Bastin (1999)

Pierre Noyes invites me to make comments on how I see the place of the 'combinatorial hierarchy' in physical theory and what its use may be. I suppose my comments go together with his "Comment on PITCH" which refers to Chapter 2. What I see is quite a different picture from that painted by Pierre and represents the point of view of myself and Clive Kilmister in the first place: many others have had a hand in it though they might not take on the main responsibility.

In Pierre's' "Introduction" to this volume he mentions our alternative view, but declines to present it, particularly its up-to-date developments, because he says "...it still leaves untouched the mystery of why the Parker-Rhodes algorithm can -let alone should- lead to a replacement for the conventional understanding of the foundations of physics." A sufficient reason for Pierre's mystery which has to be solved at the outset is the problem of how calculated numbers can fit the real world. From the conventional point of view, including that of Pierre, calculated numbers and empirically determined numbers are things of quite different kind and both cannot be the results of measurements, since there can never be a pre-established harmony. The one sentence solution is that you must take them as the paradigm case of measurement itself ('paradigm' in the dictionary -not the trendy- sense) and develop the more general, normal, view of measurement subsequently. We have always made this quite clear: indeed it is rather obvious, and there is no case for Pierre's maintaining the existence of a problem when the solution is staring him in the face - little though it may fit in with his principles. For such a change to be carried through, philosophical innovation is inevitable. Such change was envisaged before the arrival on the scene of Frederick Parker-Rhodes. Pierre avoids going into this by starting with Parker-Rhodes 'algorithm'. In fact Frederick, on his own clearly stated claim at the time, produced his astounding work by formalizing the algebraic ideas of hierarchy structure that were then being used, using binary algebra instead of the continuum for the first time.

That this change in philosophical orientation (though only the tip of the iceberg was manifest at that stage) should have called into play the mathematics which culminated in Parker-Rhodes 'algorithm', could well be seen as a remarkable endorsement of empirical method. Facts compelling a change in thinking.

I cannot develop this philosophical change and its attendant mathematics here, but I can point to the profound difficulties which attend current theory which show that however difficult the philosophical changes may be, they are not being undertaken frivolously. In the early seventies I made several extended trips as Pierre's guest at SLAC. Part of the agenda was for Pierre to teach me the principles of high energy physics and his approach to it. I learned a certain amount of course, but the further we went the more I found it impossible to see the whole subject as a consistent series of arguments. It wasn't just that there were obscurities and gaps in the understanding: you would expect that. There seemed to be a built-in confusion at work. As near as I could pin this down, it was due to switches from classical realism to the combinatorial principles appropriate to the high energy situation without any principles to guide which set of principles one was to follow or when it was legitimate to switch. To cope with the new ideas one would expect the classical concepts to have been replaced by combinatorially based ideas with careful prescription of the steps to take to get to the continuum thinking. In particular the notions of mass, charge and spin were spoken of faut de mieux with the assumption that their definition in classical mechanics would always be legitimate and applicable. Of course this criticism can be made of the foundations of quantum theory itself. But there the theology which forces the two irreconcilables into a marriage which never was, and whose most colourful sobriquet is the 'collapse of the wave-function', is done in one step. In high energy work it is disseminated throughout. Moreover, discretizing the mechanics gets you nowhere if you still rely on the intuition of unrevised classical concepts.

The change in our view of measurement and the evident need for a rebuilding of the classical dynamical concepts are but the two sides of one coin. This complementarity becomes evident when one looks carefully at the hierarchy construction. The mathematics does not work unless one allows that the constructive stages are sequential. Without making the change from a static mathematical model the numbers do not come out right and it is then wrong to claim them as experimental successes. This sequential principle is so all-pervading that in current work we speak of using a process theory.

Comment on *"On the Physical Interpretation and Mathematical Structure of the Combinatorial Hierarchy"*
Clive Kilmister (1999)

I am delighted to have had Pierre Noyes' invitation to add comments on " Physical Interpretations ...", the more so as it caused me to reread it and I found with some surprise more in it than I remembered. I want to use this opportunity to state how PITCH*, some of which Pierre probably disagrees with.

The emphasis in the introduction to PITCH is on the two primitive ideas of *Schnurs* and process. In fact it seems now that we can get by with only one. As I see Pierre's later developments, he has gone for strings, and the idea of process has taken a back seat. I have preferred to go for process. The process is one of generating labels for entities which have come into consideration and in the simplest case these labels are equivalent to strings. This makes some changes. For example, in such a process it is not possible to restrict the choices of matrices at the second level change to be a linearly independent set — the process simply constructs matrices. At this level the set will most often, but not always, be linearly independent. When one takes account of this, the figure 137 becomes [1] 137.3051...

Other later developments are interestingly foreshadowed in section 3 of PITCH, where it is suggested that when $x + x = 0$ "there is some externally established criterion, which is eventually to be established recursively by which the two x's are assigned to different sets." If the process is one of labeling and if, as the paper already emphasises, all it can do is to discriminate between two entities to see if they are the same or not then this act of discrimination need not be commutative, as the paper assumes. For if a has already been labeled and an entity b is discriminated against a and proves to be the same, then the label b has to be discarded in favor of a. If a is compared to b the opposite is true. Of course such a difference can sometimes be

---

*The title of the initial draft for this work was "Physical Interpretation of the Combinatorial Hierarchy" which in ANPA circles was abbreviated to PICH. When the existence proof of the combinatorial hierarchy became available, and John Amson's appendix was added, we changed the title to "Physical Interpretation and Mathematical Structure of the Combinatorial Hierarchy". In our own thinking, this allowed us to believe that THE Combinatorial Hierarchy we were discussing had both existence and uniqueness and we changed the acronym to PITCH to emphasize the "THE". (note added by hpn)

ignored, so that the new system is restricted by the need to collapse onto the old one. In this way the numerical improvements are not lost (and may in fact be improved).

Finally I take up Pierre's continuing puzzlement about how a process model of physics can possibly give rise to numerical values of constants measured in the laboratory. He is surely right that no start from an orthodox position will ever be able to justify such results. The conclusion I would draw is that a different starting point is needed. Ted Bastin and I have nailed our colours to one such mast: the few scale constants, as we call them, must be taken as logically prior. The problem is then turned on its head; it is the more normal view of measurement that needs to be developed. This is surely a major task but we hope not an impossible one.

References

[1] T.Bastin and C.W.Kilmister, *Combinatorial Physics*, World Scientific, Singapore, 1995, p.109.

Comment on *"On the Physical Interpretation and Mathematical Structure of the Combinatorial Hierarchy"*
Pierre Noyes (1999)

This is the first paper on the combinatorial hierarchy to be published in the mainstream literature. It was presented by me (with Bastin in attendance as a backup) at the 1978 Tutzing Conference organized by C.F.von Weizsacker. Bastin presented a paper at the 1976 Tutzing conference which contained some new work by John Amson, but the written version did not meet the deadline for the proceedings. The paper reproduced above missed the 1978 proceedings because we were never informed when the deadline was! Fortunately David Finkelstein agreed to publish this expanded version in the *International Journal of Theoretical Physics*.

Two major advances are recorded here. The first is the Parker-Rhodes calculation of the proton-electron mass ratio. He had submitted his version of this calculation to *Nature* without trying to relate it to what the rest of the group were doing. It was rejected, but was presented in a more general context and eventually published in his book, *The Theory of Indistinguishables*[2]. The version given here is primarily my attempt to recast his calculation in a language more familiar to physicists. I don't think I made any mistakes, but I do not find the line of reasoning I present particularly compelling. We will return to this problem, and in particular to David McGoveran's way of arriving at the same algebraic formula (published in *Proc. ANPA 12*, pp 15-16 and reprinted here as *The Electron-Proton Mass Ratio* by David McGoveran in Ch. 6).

The second major advance was a proof of the *existence*, in a mathematical sense, of the combinatorial hierarchy. This problem had fallen between stools, primarily because there were so many representations of the hierarchy beyond level 2, and so many possible mapping matrices, that it seemed "obvious" that many representations must exist. Fortunately K.V.Laurikainen had invited me to spend some time in Finland after the Tutzing Conference, which gave me time to explore the problem systematically, in a very pedestrian way, and to actually construct specific representations — which in themselves convey no insight. As mentioned in the paper, a much more elegant proof was eventually constructed by Clive Kilmister, but not in time to include in the published article. This was in the spirit of the mathematical appendix by John Amson, which was the first systematic, published treatment of the relation between *discriminate closure* and the combinatorial hierarchy.

A minor point is that in the paper I tried to relate the operations used in the hierarchy construction to a discrete version of Feynman diagrams. Although much progress along that line has been made subsequently, a definitive formulation still eludes me. The very speculative excursion into cosmology at the end of section IV, as we will see, developed into a definite prediction about the density of matter in the universe a decade before the observational evidence was good enough to test it. As Ch. 16 shows, this prediction has been triumphantly confirmed in the year 2000.

Addendum: I am most grateful to Ted Bastin and Clive Kilmister for the comments on our differences which immediately precede my own comments on PITCH. They make clear that, rather than starting with the Parker-Rhodes formula as a puzzle to be explained, as I do, they feel that a *process* point of view must be adopted from the outset, and the existence of the structure constants treated as logically prior to the development of the "normal view of measurement", to use Clive's phrase. This allows me to state my difficulty somewhat differently: I have never understood what either of them mean by "process" in such a way as to be able to apply it to the study of physical problems. Clive admits that the difficult task of using their approach to develop a view of measurement that allows comparison with standard empirical results has yet to be achieved. Until that "major task" is at least under way and preliminary results presented, I also fear I will be unable to make that connection on my own despite a quarter century of effort.

References

[1] T.Bastin and C.W.Kilmister, *Combinatorial Physics*, World Scientific, Singapore, 1995, p.109.

[2] A.F.Parker-Rhodes, *The Theory of Indistinguishables*, Reidel, Dordrecht, 1981.

Comment on *"On the Physical Interpretation and Mathematical Structure of the Combinatorial Hierarchy"*
John Amson (2000)

My contribution to the paper "On the Physical Interpretation and the Mathematical Structure of the Combinatorial Hierarchy" consisted largely of the Appendix (Mathematical Structure of the Hierarchy). This appendix was based on work I did during the period 1964-1978 in consort with Ted Bastin, Frederick Parker-Rhodes and Clive Kilmister. The formal nature of the crucial idea of "Discriminate Closure" and "Discriminately Closed Subsets" was first identified late one evening in the summer of '65 in a brain-storming session between me, Ted Bastin and Margaret Masterman (the then director of the Cambridge Language Research Unit whose Information Structures Unit was supporting our investigations at that time).

Curiously, Frederick, the discoverer of the basic idea of the Combinatorial Hierarchy, was not involved in that session. I think this was because of the rather different mathematical ways in which Frederick and I thought of the Hierarchy. The background mathematics for Frederick was always in terms of Linear (Matrix) Operators and their Eigenvectors whilst for me it was in terms of those Operators and their Invariant Subspaces. The differences between these two views are mathematically slight in contrast to the way they affected our reasoning processes. I do not think Frederick's emphasis on eigenvectors could have led him to conjure up the idea of a Discriminately Closed Subset in itself, despite the idea being intrinsically latent in his work (see especially his seminal paper, subsequently edited by me and posthumously published as "Hierarchies of Descriptive Levels in Physical Theory", International Journal of General Systems, Vol.27, Nos.1-3, 1998, 57-80, and my comments to his section 2.2).

To keep the published size of the Appendix down, almost all proofs were omitted, and in hindsight insufficient attention was drawn to the important fact that a subset of a vector space was a Discriminately Closed Subset if and only if the subset in question when enlarged by the inclusion of the null vector was a vector subspace (see my "A.2.1. Definition"). Of course, all the vector space structures for the Combinatorial Hierarchy were always over the 'discrimination' field J2 of two elements 0,1. Because of this, in Frederick's eigenvector approach there is only one candidate for a nonzero eigenvalue, namely "1", and hence any eigenvector is necessarily mapped by a hierarchy matrix into itself ($Mx = 1x = x$). By contrast, for me, a vector that was mapped by a matrix to itself was "fixed" by that matrix and any other kind of vector was "un-fixed". But a subset of non-null vectors which contained not only all of its fixed and unfixed vectors and their images, but no others, had to be of special interest: these became our Discriminately Closed Subsets.

Also because of this inherent restriction to vectors spaces over the field of two elements, it follows that any "ray" of points through the origin (the null-vector) in any vector space in our hierarchy contains but a single non-null vector. Now it is well-known that to every vector space of finite dimension N, say, there corresponds another important mathematical structure, a Projective Geometry, of dimension $N-1$, in which the "points" are the "rays" of the original vector space, and that the sub-geometries are just the vector subspaces of the original vector space with the null-vector deleted (!); and vice versa. In the case of our hierarchy vector spaces this means that our Discriminately Closed Subsets are simply the sub-geometries of the corresponding projective geometries. In other words, we hadn't actually discovered anything 'mathematically' new when we hit on the idea of Discriminately Closure and Discriminately Closed Subsets; we simply explored and developed these ideas quite innocently from scratch, motivated by our need to create a formal mathematical setting for Frederick's Combinatorial Hierarchy. But it would also mean that a wealth of previously known combinatorial facts about projective sub-geometries could later be taken over 'en bloc' and translated into information in the context of our Combinatorial Hierarchy, as I subsequently showed in my later paper "Discrimination Systems and Projective Geometries", Proceedings of ANPA 9, 1987, 158-189. However, that voluminous quantity of information was not yet knowingly available to us at the time my Appendix was being written. Our Combinatorial Hierarchy turned out to be a far richer mathematical structure than we envisaged at the time.

One other combinatorial aspect that still lies dormant in my Appendix, and which could still prove a valuable gold-mine of information about the Combinatorial Hierarchy, is adumbrated in my "A.4.7 Remark" about a "discriminate spine". Paraphrased, this says that the way in which Frederick and any one else actually "constructs" an instance of the Combinatorial Hierarchy (as Pierre heroically did in his 'Finland' demonstration of the "existence" of the Combinatorial Hierarchy) is subject to "choice" when passing up from one Level to the next: the discriminate spine is the route-map for our given construction. This leaves open at least these questions: (1) how many such route-maps are there ? (2) could the combinatorial number of possible route choices at each level-change be significant in a quantum-physical context ?; (3) could the existence of these alternative route-maps affect the deductions which might be drawn from any actual implementation of the Manthey-Noyes "Program Universe" computer algorithm ? (4) if all route-maps turned out to be 'equivalent' in some precise sense (as I conjecture will be the case) then what Combinatorial Hierarchical importance should we attach to the membership of individual elements in each equivalence class of route-maps ?

John Amson

5 Shore, ANSTRUTHER, Fife KY10 3DY, Scotland

Tel and Fax: +44(0)1-333-310-087

e-mail: john.amson@which.net

ON THE CONSTRUCTION OF RELATIVISTIC
QUANTUM THEORY: A Progress Report*

H. PIERRE NOYES

Stanford Linear Accelerator Center
Stanford University, Stanford, California, 94305

ABSTRACT

We construct the particulate states of quantum physics using a recursive computer program that incorporates non-determinism by means of locally arbitrary choices. Quantum numbers and coupling constants arise from the construction via the *unique 4-level* combinatorial hierarchy. The construction defines indivisible quantum events with the requisite supraluminal correlations, yet does not allow supraluminal communication. Measurement criteria incorporate $c, \hbar$ and m_p or (*not* " and") G, connected to laboratory events via finite particle number scattering theory and the counter paradigm. The resulting theory is discrete throughout, contains no infinities, and, as far as we have developed it, is in agreement with quantum mechanical and cosmological fact.

Submitted to *Proceedings of the 7th Annual International Meeting*
ALTERNATIVE NATURAL PHILOSOPHY ASSOCIATION
King's College, Cambridge, England, 28 June - 1 July, 1985

---

* Work supported by the Department of Energy, contract $DE-AC03-76SF00515$.

1. INTRODUCTION

Although a successful challenge to the experimental predictions of quantum mechanics has yet to be mounted, and subtle features such as the supraluminal correlations without supraluminal signaling implied by Aspect's[1] and other EPR-Bohm type experiments have been demonstrated, for some physicists a conceptual unease continues to persist. I present here my current attempt to meet this problem.

What framework do I accept for physics? I believe that most practicing physicists would agree with me that physics is an empirical science. Physics in historical practice has rested on quantitative measurement, or at least on "operational procedures" which lead to "replicable" results. Here I will insist on the stricter standard that the results of experiment be reduced to "counting", taking due account of the (again specified) expected range of uncertainty. I cannot accept the concept "infinite" (or "infinitesimal") as valid in physics; for us finite beings (physicists or no) this great renunciation is (in my opinion) the first step toward acquiring knowledge. Quantum mechanics brought this issue to the fore for physicists; it had been raised for chemists by Dalton and Prout long before, and for philosophers and mathematicians by Democritus and Zeno in antiquity.

For me, any formulation of quantum mechanics as we know it must contain the idea that quantum events are unique and indivisible. In contrast, classical physics is *scale invariant*; any arbitrarily chosen standards of mass, length and time- or any three experimentally independent combinations of those unit standards will suffice. There is in it no place for unique events; the "microscopic" laws are "time reversal invariant". Events (which in classical physics are always in principle decomposable into "microscopic" substructure) acquire what uniqueness they possess due to the imposition of boundary conditions by the analysis of the physicist, or their embedding in a large-number "statistical" background, or their relation to a "cosmological time", or ... Modern physics removes the scale invariance of classical physics by recognizing a limiting velocity, a quantized unit of action (and angular momentum!), and quantized masses. Experimentally the only stable (lifetime $> 10^{33}$ years) "elementary" mass values are those of the proton m_p and the electron m_e in the ratio $\frac{m_p}{m_e} = 1836.1515 \pm 0.0005$, again stable within the stated standard deviation. (According to current standard cosmologies the stable nuclei that could have at least as long "lifetimes" (eg. He^4) are only about 10^{16} years old, or less; whether protons have an "age" is less clear.)

In my approach, I adopt from classical physics the dynamical definition of mass ratios from Newton's third law as articulated by Mach, but generalised to recognise the limiting velocity by using mass invariance $(E^2 - (p \cdot p)c^2 = m^2 c^4)$ and 3-momentum conservation $(\Sigma p_{initial} = \Sigma p_{final})$. Experimental contradiction of this assumption would be of immediate practical interest for those interested in the exploration of the solar system and beyond! The so far undefined energy (E) in this equation is a global quantity. As Wick saw in the late 30's [2] the easiest way to make a compelling argument for Yukawa's finite range meson theory of nuclear forces [2] is to insist on (relativistic) 3-momentum conservation, but allow energy fluctuations consistent with Heisenberg's energy-time uncertainty principle and Einstein's mass-energy equivalence. This is also a basic principle underlying Heisenberg's and Chew's S-Matrix theory.

There is already a well known conceptual puzzle at this stage in our discussion. If we fasten on macroscopic (gravitational) rather than microscopic (particulate) phenomena as basic, the fundamental mass unit we would choose would be the "Planck mass" $M_{Planck} = \sqrt{\hbar c / G} \simeq \sqrt{1.7 \times 10^{38}} m_p$ rather than the proton mass. Contemporary physics meets this problem by using the "equivalence principle". The postulated equivalence of microscopic ("inertial" or 3-momentum conserving) mass ratios and macroscopic ("gravitational") mass ratios allows gravitation, and (perhaps) all other "interactions" along with it, to be "geometricised". But this need not be the only route to "supergravity", or whatever catch phrase becomes current when this paper appears. In my opinion, one of the strengths of the approach to physics developed here is that our theory can have only one type of mass, and that the first approximations to both M_{Planck}/m_p and m_p/m_e are calculated.

Once one accepts quantum events as basic, and the limiting velocity as well established experimentally, the "supraluminal correlations" [1] predicted - but for some people not explained - by quantum mechanics also cry out for conceptual clarity and a deeper insight. I do not believe that this can be achieved by first developing the full technical apparatus of quantum mechanics and then presenting these startling results as a deductive consequence. Etter, McGoveran, Manthey,

Gefwert and I[4] believe that the issue *can* indeed be clarified by invoking a minimal set of postulates that do not depend on the idea of space-time, let alone quantum mechanics. What follows in this paper is consistent with that point of view, and with earlier papers.

Another aspect of contemporary physics that I would like to see emerge at an early stage is that our "universe" start out as simply as possible and evolve by a finite number of steps through recognisable stages into the complex situation that we encounter as we now explore it. When I started on this research I was at least open to the possibility that the universe we are exploring is "indefinitely extensible" in both "space" and "time". That we find great simplicity as we retrodict the past on the cosmological scale, could (as Bastin has often emphasised) simply be a consequence of impoverished data, – i.e. of the successive disappearance of relevant observable points of reference as we extend our horizons. I do not think this problem arises in acute form while retrodicting the last 15 billion years. I have been greatly impressed during the course of my own professional career by the convergence of seemingly disparate and very detailed measurements to a reasonably consistent "time scale" of that length. The past was *different* from the present in the probable range and type of configurations that occurred, but there is no indication as yet that the elementary *possibilities* were significantly different (except, possibly, during the *very* early stages). In the current paper, the very early stages of the evolution are *simpler* and not just different. For those who are more comfortable with a universe that has no beginning and no end John Amson provides a nested hierarchy which can be explored (past ↔ future; small ↔ large) so far as information is available (Appendix VI by John Amson entitled *"Bi-Orobourous"* - *a Recursive Hierarchy Construction*). So far as I can see, the consequences when this point of view is articulated in the current practice of physics are likely to be practically indistinguishable from those of the approach developed here.

To the best of my knowledge we can retrodict the universe backward in time for only about fifteen billion years. There is an "event horizon" and a preferred coordinate system defined by the radiation that broke away from the matter when the cosmic fireball became optically thin; within the event horizon there are particles whose baryon and lepton number add up to approximately the square of $2^{127} + 136$. Our construction yields all of these observed features as stable consequences of the construction independent of the details. That the theory developed here has a starting point and achieves evolving complexity possessing dynamic and heritable stability in the presence of a "random" background of quantum events, is for me a satisfactory result. The "universe" we construct in this paper will go on increasing in complexity in the future, and hence contains indefinitely extensible possibilities. This theory has a fixed past, an event horizon, and yet and indefinitely extendable (though uncertain) future. It may be that I have found what I was looking for, but I can assure the reader that the steps along the way were taken for more immediate reasons, - so far as I am aware.

In this current attempt to meet these basic requirements when reconstructing quantum theory I have made use of many ideas and techniques conceived and developed by other people[5,6] . In the series of papers on "Concept of Order"[7,8] Bastin and Kilmister presented reasons why distinct "events" should relate to a basic algebraic structure connected to "3+1 space". By 1966 this research, to which Amson, Bastin, Kilmister, Parker-Rhodes and Pask had all contributed, had led[9] to the closed 4-level *combinatorial hierarchy* with the cumulative cardinals 3, 10, 137, $2^{127} + 136 \simeq 1.7 \times 10^{38}$.

The work on the combinatorial hierarchy did not face directly the statistical aspect of quantum mechanics, which I have already indicated I see as fundamental. I therefore start my technical discussion in Chapter 2 by calling on more recent work by Manthey and McGoveran to spell out what current computer practice means by "non-determinism" and "arbitrary choice". As the names of Amson and Pask will indicate, the earlier work had also made use of concepts used in computer science, but before the nondeterminism born of asynchronous communication over a shared memory had come to the fore. I turned that way because Gefwert demanded that anything that laid claim the description "constructive physics" had to be computable. Fortunately the expert I first turned to was Manthey; the result was PROGRAM UNIVERSE.

Our use of a computer simulation to model the theory is sometimes misunderstood. I do not think of the universe as a "big computer in the sky". What the coding does for us is to keep us honest; if we can show that the program is indeed computable, then we have protected ourselves from making all sorts of logical errors. A computer simulation is a specific type of "model". If it succeeds, all that we can say is that, within current experimental error, we have succeeded in isolating those aspects of experience which act in a manner isomorphic to the action of our model. When the program fails, then we will have isolated a situation from which we might learn some new physics, or possibly something that goes beyond physics. There can well be things in heaven and earth that are not contained in this philosophy. I trust it is clear that I am not a reductionist or a mechanist. Materialism is a separate issue, which will not be discussed here.

PROGRAM UNIVERSE, a peculiarly simple algorithm, *automatically* develops some representation of the hierarchy, necessarily specifies unique, correlated, global events and provides address ensembles for these events labeled by the fixed elements (eventually connected to quantum numbers, masses and coupling constants) provided by the combinatorial hierarchy. The technical details are given in Chapter 2 where we provide a specific construction of the four level hierarchy and the address ensembles; Mike Manthey's coding for this construction is given as Appendix IV.

In order to meet our objective of constructing a quantum mechanical physical theory, we must somehow relate the structure we now have in hand to measurements of mass, length and time in the ordinary sense. We do this in Chapter 3 by noting that quantum events "fire counters" and allow velocities, momenta and energy to be measured by well known techniques to an accuracy only limited by available budget (or space and time available to conduct meaningful experiments). As in Heisenberg's and Chew's S-Matrix philosophy, momentum measurements, and the momentum space formulation of quantum mechanics are a strategically useful place to connect theory to experiment. We make this more than usually explicit by starting from a "counter paradigm" which relates the bit string universe to laboratory measurement. We find that the relativistic version of the "wave-particle dualism" emerges with little effort. We also discover that some fundamental cosmological observations find a ready explanation within this simple framework, independent of the details by means of which it is articulated.

The next step, spelled out in Chapter 4, is to construct from the strings and events provided by PROGRAM UNIVERSE a relativistic quantum scattering theory which, via the counter paradigm, conserves quantum numbers and 3-momenta in a manner consistent with laboratory experience. The basic idea in this scattering theory is to use Faddeev-Yakubovsky equations for the dynamics rather than a Hamiltonian, or Lagrangian or analytic S-Matrix formulation.

Probably the most significant step taken since ANPA 7 is the derivation of the "propagator" for the scattering theory directly from the bit string universe via the counter paradigm. We give a more detailed explanation here than was possible in our conference report[10]

In Chapter 5 we return to the four level hierarchy labels and connect them to the conserved quantum numbers in the standard model for quarks and leptons. Briefly, level 1 describes the simplest neutrinos, level 2 describes electrons, positrons and the associated electromagnetic quanta, while level 3 gives us two flavors of quarks and the associated gluons related to each other in a color octet. This pattern will repeat until the possibilities close off at the Planck mass, but the coupling between successive generations will be weak because of the combinatorial explosion in possibilities. Thus we can anticipate that the Kobiyashi-Maskawa mixing angles will indeed be small. The count of quantum numbers is correct, and the quantitative or qualitative results so far achieved produce no glaring contradictions

For completeness we repeat in Chapter 6 the Parker-Rhodes calculation of m_p/m_e as it looks from the present context. The question of whether the result will be stable when we compute the correction to the fine structure constant and the "recoil corrections" is still unanswered.

We will discuss in the concluding section what it might mean if this qualitative-quantitative success persists up to a point where a definite quantitative conflict with experiment counter-indicates the acceptance of the theory.

2. GENERATING AND DISCRIMINATING BASIS STATES; EVENTS

The first problem we must face in constructing our theory is where the "random" aspect of quantum mechanics enters. As one will see in Appendix I, Kilmister allows his generation and discrimination operations to be interleaved in an order which is not explicitly specified. The route we follow below, which depends on the explicit use of a "pseudo-random number generator" in the computer program, is in my opinion a specific articulation of Kilmister's more general discussion. In Appendix II Bastin discusses, among other things, the idea of "inexact matching" which he and Kilmister have explored; this might also end up in something that could be shown to be equivalent to my approach, but has as yet not been articulated far enough to settle the issue.

The method of actually writing down a computer program forced Manthey and me to tackle the "randomness" issue head on. For Manthey, the non-determinism born of asynchronous communication over a shared memory – one basic problem in concurrent programming – is viewed as at least analogous to the non-determinism encountered in quantum mechanics[11,12] . Thus the basic coding for the routine which returns either a zero or a one with "equal prior probability" (and whose output is symbolized below by "R") as given by Mike Manthey(in Appendix IV) starts from two memory locations which flip a bit backward and forward on one time interval; one bit is read whenever the (asynchronous) operation of the main program calls up this routine. However, when Manthey and I had occasion to need this routine for an EPR computer simulation we are working on, he fell back on the standard (pseudo)-"random number generator" available on his local computer. I turn to another expert for discussion of this issue.

The term McGoveran prefers to use when talking about what is often called "randomness" is "arbitrary choice". By this he means "not due to any finite, locally specifiable algorithm"; of course in standard practice, the *local* operation of the computer is deterministic (if it is working properly), so this means calling on some number generated in a larger system not under local control. In the same termininology, he would "define" *random* as "not due to any finite, local *or* global algorithm". Since we have no operational way to meet this requirement, the concept of "random" is, strictly speaking, meaningless in our context;

we must content ourselves with the currently available pseudo-random number generators for our simulations. McGoveran's basic thinking on this was spelled out recently[13] in response to a query from Kilmister. I quote:

"I think that we must insist on computability to the detriment of randomness for a number of reasons, each of which I have previously discussed. This position does no harm to the power of the model since, as proved by Shannon (1965), – an infinite state machine with a random element can be replaced by an infinite state machine (infinite being "constructively infinite"), and as I demonstrated (1984 ANPA West Proceedings), there will always be a method for constructing certain repeating binary inputs which a given finite state machine with finite memory can not distinguish from 'random' binary inputs.

"In some sense, randomness is a local phenomenon. So long as a 'generator' is available which has significantly more states than the 'detector', there will be a possibility of generating strings which are random from the detector's point-of-view. Similarly, given a string which passes all computable tests for randomness of a fixed complexity (i.e. by a finite state machine with m possible states), it will be possible to construct a finite state machine with $n >> m$ possible states which produces that string. In algebraic terms: there exists a computable function g (pseudo-random number generator) for each finite class of computable functions f_i such that, whenever the range R_g of g is sufficiently large compared to the union of the domains of f_i (call this D), it is impossible to prove that G is computable based on the f_i. In pictures:

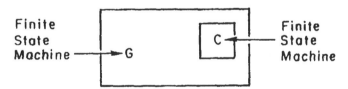

[C sees some outputs of G as random; G sees no outputs of C as random. The computational cost of detecting the orderliness of all G's output is too great for C.]

"Because we have neither the means for specifying what we mean by randomness, nor for detecting it in a finite system, and we can be certain that a means exists for constructing ANY sequence, I have insisted that we have no need for the concept of randomness, replacing it with 'arbitrary' We implicitly recognize an as-yet-unspecified 'finite computable function' as the source of 'arbitrary' strings.

"So long as we are blind to the nature of a deterministic system, the effect is the same *locally* as having "random choice". Furthermore, true randomness *implies* infinities (an infinite state machine is required for generating random output – i.e. undecidable output). I think we need consistency here and so deny randomness in favor of parsimony."

Now that we have spelled out how, in practice we can select either of our two symbols 0, 1 with what is close enough in practice to "equal prior probability" in the frequency theory sense of probability, we can understand the basic arbitrary choice from which the algorithm called PROGRAM UNIVERSE starts. But the generation of the strings in this universe and the discrimination between them require considerably more background if the algorithm is to be followed.

The basic elements with which the hierarchy work started are ordered strings of the symbols 0, 1 of the form $S^a(N_U) = (....., b_n^a, ...)_{N_U}^a$ where N_U is the integer specifying the number of symbols in ("length" of) the string, n is the usual integer ordering parameter $n \in 1, 2, 3, ..., N_U$, and $b_i^a \in 0, 1$. We will discover below that when our construction has proceeded far enough we can specify the label a in terms of the sequence of symbols in the N_L positions $n \in 1, 2, 3, ..., N_L < N_U$. For those who wish the integers themselves to be constructed, one can follow Gefwert's approach in terms of primitive recursive functions[14,15] , or follow Kilmister's foundational discussion in Appendix I where in a sense they come to us along with the hierarchy itself. If we define the null string $0_N = (0, 0, ..., 0)_N$, the operation $\oplus$ which tells us whether two strings are the same or different (and hence *discriminates* between them) gives $S^a \oplus S^a = 0_{N_U}$ when they are the same and has two equivalent definitions:

$$(..., b_n^a +_2 b_n^b, ...)_{N_U} \equiv S^a \oplus S^b \equiv (..., (b_n^a - b_n^b)^2, ...)_{N_U} \qquad (2.1)$$

whether they are the same *or* different. For the first definition the operation $+_2$ is addition (mod 2), or symmetric difference, or exclusive "or", the symbols are bits and the operation is the standard XOR of computer practice. For the second definition the symbols are integers and we can define operations such as

$k^a(N) = \Sigma_{n=1}^N b_n^a$ which gives us the number of "1" 's in the string. This fruitful ambiguity was first noted by Kilmister and myself; we refer to *either* operation as *discrimination* in order to preserve the generalisation that goes beyond XOR. The anti-null string is symbolised by $1_N \equiv (1, 1, ..., 1)_N$, allowing us to define the "bar operation" $\bar{S}^a(N) \equiv 1_N \oplus S^a(N)$ which interchanges "0"'s and "1" 's in a string.

By 1980 Kilmister[16] realised that the discrimination operation by itself would not suffice for the theory, since it gives us no clue as to how the strings arise in the first place. He therefore introduced a generation operation by modifying a construction of the integers used by Conway (originally due to von Neumann), and found out how to go on to arrive at the discrimination operation using this approach. The final(?) version of his approach (which was sketched out at ANPA 7 and completed since) is given as Kilmister's Appendix I. In this way, or by using discriminate closure and the matrix mapping due to Parker-Rhodes[9][17] or the set-theoretic derivation due to John Amson[18] one arrives at the *unique*, 4-level *combinatorial hierarchy* with the cumulative cardinals $3, 10, 137, 2^{127}+136 \simeq 1.7 \times 10^{38}$.

As Kilmister and I soon realised, once one has introduced the generation operation, there is nothing to stop it from generating additional elements even when the full hierarchy has closed off. In terms of the bit-string representations used in my work, this means that the first bits in the string can be put into 1-1 correspondence with *any* representation of the hierarchy, and that as we go on cranking out new elements of still greater length there will come to be many strings with the same label. Kilmister and I called the portion of the string beyond the *label* the *address* and thus arrived at the idea of *labeled address ensembles*. These come to play the role in our theory of *quantum state vectors*, but there are subtle differences from the conventional quantum mechanics which we will discuss at the appropriate point.

When Christoffer Gefwert heard of our work, he saw that constructive mathematics could offer the appropriate philosophical framework in which to achieve consistency, and suggested to me that if we were indeed trying to create a "constructive physics", it would have to be expressible as a computer program. This encouraged me to re-establish contact with Michael Manthey and led to the first version of PROGRAM UNIVERSE[19] . Since we did not see any simple way to code up Kilmister's generation-discrimination construction, we decided to generate strings in the simplest way we could think of. What we now have is simply described. If there are SU strings in a universe of length N_U, it is allowed to evolve in only two ways. Two strings are picked arbitrarily and discriminated. If the resulting string is not already in the universe, it is adjoined; the number of strings goes up by one. If the string is already in the universe, an arbitrary bit is selected for each string and concatenated with that string; the length of the strings goes up by one. This second operation is called TICK.

We generate the strings according to the flow chart:

Figure 1.The flow chart for Program Universe.

PROGRAM UNIVERSE

NO. STRINGS = SU $R \Rightarrow 0,1$ (FLIP BIT)

LENGTH = N_U PICK := SOME $U[i]$ $p = 1/SU$

ELEMENT $U[i]$ TICK $U := U \parallel R$

$i \in 1,2,\ldots,SU$ $\bar{S} = 1_N \oplus S$

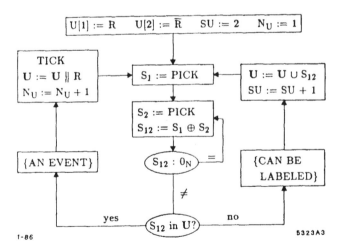

1-86   5323A3

For those who prefer explicit coding, this is provided by Manthey in Appendix IV. The program is initiated by the arbitrary choice of two distinct bits: $R :=$ 0 or 1, $R = 1 \oplus R$. Entering at $PICK$, we take $S_1 := PICK; S_2 := PICK; S_{12} := S_1 \oplus S_2$. If $S_{12} = 0_{N_U}$, we recurse to picking S_2 until we pass this test. [A still simpler alternative, which occurred to me in writing this paper, would be to allow the null string to occur as one of the elements in the universe. So far as I can see, this would not affect the running of the main program, this change might require a little care, and perhaps some change in the coding when we turn below to the extraction of the hierarchy from the results of a run of the program.] The program then asks if S_{12} is already in the universe. If it is not it is adjoined, $\mathbf{U} := \mathbf{U} \cup S_{12}, SU := SU + 1$, and the program returns to $PICK$. If S_{12} is already in the universe, we go to our third, and last, arbitrary operation called $TICK$. This simply adjoins a bit (via R), arbitrarily chosen for each string, to the growing end of each string, $\mathbf{U} := \mathbf{U} \| R$, $N := N + 1$, and the program returns to $PICK$; here "$\|$" denotes string concatenation.

What may not be obvious is that TICK results either from a *3-event* which guarantees that at string length N_U the universe contains three strings constrained by $S^a \oplus S^b \oplus S^c = 0_{N_U}$ or a *4-event* constrained by $S^a \oplus S^b \oplus S^c \oplus S^d = 0_{N_U}$ that these are the only ways events happen in the bit string universe is illustrated in Figure 2.

Figure 2. How events happen in *Program Universe*.

3-EVENTS

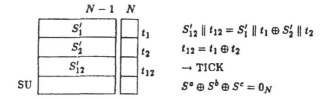

$$S'_{12} \parallel t_{12} = S'_1 \parallel t_1 \oplus S'_2 \parallel t_2$$
$$t_{12} = t_1 \oplus t_2$$
$$\rightarrow \text{TICK}$$
$$S^a \oplus S^b \oplus S^c = 0_N$$

4-EVENTS

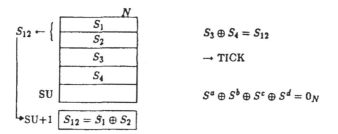

$$S_3 \oplus S_4 = S_{12}$$
$$\rightarrow \text{TICK}$$
$$S^a \oplus S^b \oplus S^c \oplus S^d = 0_N$$

EACH TICK "RECORDS" A UNIQUE EVENT "SOMEWHERE"
IN THE UNIVERSE

1-86
5323A2

In the case of 3-events the universe just before $(N-1)^{st}$ TICK contained three strings constrained by $S_1' \oplus S_2' \oplus S_{12}' = 0_{N-1}$ which were replaced by $S_1 = S_1' \| t_1$, $S_2' \| t_2$ and $S_{12}' \| t_{12}$ respectively as a consequence of that TICK. Before the Nth TICK S_1 and S_2 are picked and S_{12} generated by discrimination. Clearly, if $t_{12} = t_1 \oplus t_2$ then S_{12} is *already* in the universe, and the program will proceed to carry through the Nth tick. However it can also happen that when two strings are picked the S_{12} generated by discrimination is *not* already in the universe, and hence will be adjoined to it. Eventually however (if the program does not encounter some circumstance that produces a 3-event first) it will pick two strings S_3 and S_4 (which could even be a second pick of S_1 and S_2) such that $S_3 \oplus S_4 = S_1 \oplus S_2$; clearly this will then lead to the Nth TICK as a 4-event.

In the original version of PROGRAM UNIVERSE, I was hung up with the idea that only 4-events should occur, because energy *and* 3-momentum cannot in general be conserved in 3-events (a fact familiar to particle scattering data analysts). I therefore went to some elaboration to insure this, and only later stripped down the program to the present form. Once I had done this, James Lindesay then saw that 3-events could also occur by the mechanism just described. As we will see below both are needed in the scattering theory, so this fact turned out to be extremely fortunate. This is only one of many instances in the course of this research where the attempt to arrive at simpler formulations has had profound consequences.

It is important to keep in mind both here and in what follows that the actual structure of the memory and the specific strings in it generated by our computer simulation are *not* to be thought of as modeling "real" elements in the world. We are not allowed to access them directly, even conceptually, when it comes to interpretation. The string length, whether a specific event is a 3-event or a 4-event and how many other combinations of strings satisfy the event constraint at that TICK are hidden from us. We can only talk about them as structural constraints and in terms of statistical arguments. Contact with experiment can only be made *indirectly* via the counter paradigm. This "simulates" in another sense what actually goes on in the laboratory. We can never know "what it is" that initiates the chain of happenings which leads to the firing of a counter. All we can do is to use the connections provided by theory and experiment by means of some more or less successful type of analogical thinking to refine and improve the *statistical* behavior of our counters, or more sophisticated detectors.

In order to see that this program *also* leads to some representation of the combinatorial hierarchy and to the label-address schema, we must first discuss the idea of discriminate closure, originally due to John Amson. Given two distinct (*linearly independent* or l.i.) non-null strings a, b, the set $\{a, b, a \oplus b\}$ *closes* under discrimination. Observing that the singleton sets $\{a\}$, $\{b\}$ are closed, we see that two l.i. strings generate three *discriminately closed subsets* (DCsS's). Given a third l.i. string c, we can generate $\{c\}$, $\{b, c, b \oplus c\}$, $\{c, a, c \oplus a\}$, and $\{a, b, c, a \oplus b, b \oplus c, c \oplus a, a \oplus b \oplus c\}$ as well. In fact, given j l.i. strings, we can generate $2^j - 1$ DCsS's because this is the number of ways we can choose j distinct things one, two,... up to j at a time. This allows us to construct the combinatorial hierarchy[9] by generating the sequence $(2 \Rightarrow 2^2 - 1 = 3), (3 \Rightarrow 2^3 - 1 = 7), (7 \Rightarrow 2^7 - 1 = 127), (127 \Rightarrow 2^{127} - 1 \simeq 1.7 \times 10^{38})$ provided that we can find some "stop rule" that terminates the construction.

The original stop rule was due to Parker-Rhodes. He saw that if the DCsS's at one level, treated as sets of vectors, could be mapped by non-singular (so as not to map onto zero) square matrices having uniquely those vectors as eigenvectors, and if these mapping matrices were themselves linearly independent, they could be rearranged as vectors and used as a basis for the next level. In this way the first sequence is mapped by the second sequence $(2 \Rightarrow 2^2 = 4)$, $(4 \Rightarrow 4^2 = 16)$, $(16 \Rightarrow 16^2 = 256)$, $(256 \Rightarrow 256^2)$. The process terminates because there are only $256^2 = 65,536 = 6.5536 \times 10^4$ l.i. matrices available to map the fourth level, which are many too few to map the $2^{127} - 1 = 1.7016... \times 10^{38}$ DCsS's of that level. This (unique) hierarchy is exhibited in Table 1.

Table 1

The combinatorial hierarchy

hierarchy level	ℓ	$B(\ell+1)=H(\ell)$	$H(\ell)=2^{B(\ell)}-1$	$M(\ell+1)=[M(\ell)]^2$	$C(\ell)=\Sigma_{j=1}^{\ell}H(j)$
	(0)	-	2	(2)	-
	1	2	3	4	3
	2	3	7	16	10
	3	7	127	256	137
	4	127	$2^{127}-1$	$(256)^2$	$2^{127}-1+137$

Level 5 cannot be constructed because $M(4) < H(4)$

Although this argument proves the *necessity* of the termination (which is no mystery in the sense that an exponential sequence must cross a power sequence at some finite term), it did not establish the existence of the hierarchy. This was first done by me by creating explicit constructions of the mapping matrices[17] and later more elegantly by Kilmister[20] . That the termination, and indeed the combinatorial hierarchy itself, is a much more fundamental object that the apparently *ad hoc* mapping procedure which first led to it can be seen either by Kilmister's latest derivation as included here or by the very different way Parker-Rhodes now gets it out of his *Theory of Indistinguishables*[21] ; a useful discussion of that theory is provided by him IN Appendix III.

The method Manthey and I use to construct the hierarchy is much simpler; in fact some might call it "simple-minded". We claim that all we have to do is to demonstrate explicitly (i.e. by providing the coding) that any run of PROGRAM UNIVERSE contains (if we enter the program at appropriate points during the sequence) all we need to extract some representation of the hierarchy and the label address scheme from the computer memory *without* affecting the running of the program. The obvious intervention point exists where a new string is generated, as indicated on the flow chart (Figure 1) by the box [CAN BE LABELED]. The subtlety here is that if we assign the label i to the string $U[i]$ as a *pointer* to the spot in memory where that string is stored, this pointer can be left unaltered from then on. It is of course simply the integer value of $SU+1$ at the "time" in the simulation [sequential step in the execution of that run of the program] when that memory slot was first needed. Of course we must take care in setting up the memory that *all* memory slots are of length $N_{max} > N_U$, i.e. can accommodate the longest string we can encounter during the (necessarily finite) time our budget will allow us to run the program. Then, each time we TICK, the bits which were present at that point in the sequential execution of the program when the slot $[i]$ was first assigned will remain unaltered; only the growing head of the string will change. Thus if the strings $i, j, k....$ labeled by these slots are linearly independent at the time when the latest one is assigned, they will remain linearly independent from then on.

Once this is understood the coding Manthey and I give for our labeling routine should be easy to follow. We take the first two linearly independent strings and call these the basis vectors for *level 1*. The next vector which is linearly independent of these two starts the basis array for *level 2*, which closes when we have 3 bases vectors linearly independent of each other and of the basis for level 1, and so on until we have found exactly $2+3+7+127$ linearly independent strings. The string length when this happens is then the *label length* N_L; it remains fixed from then on. During this part of the construction we may have encountered strings which were *not* linearly independent of the others, which up to now we could safely ignore. Now we make one *mamouth* search through the memory and assign each of these strings to one of the four levels of the hierarchy; it is easy to see that this assignment (if made sequentially passing through level 1 to level 4) has to be unique. From now on when we generate a new string, we look at the first N_L bits and see if they correspond to any label already in memory. If so (since the address part of the string *must* differ) we assign the address to the *address ensemble* carrying that label. If the new string also has a new label, we simply find (by upward sequential search as before) what level of the hierarchy it belongs to and start a new labeled address ensemble. Because of discriminate closure, we must eventually generate $2^{127} + 136$ distinct labels, organised in the four levels of the hierarchy. Once this happens, the label set cannot change, and the parameters i for these labels will retain an *invariant* significance no matter how long we continue to TICK. We emphasize once more that *what* specific representation of the hierarchy we generate in this way is irrelevant.

Each event results in a TICK, which increases the complexity of the universe in an irreversible way. Our theory has an ordering parameter (N_U) which is conceptually closer to the "time" in general relativistic cosmologies than to the "reversible" time of special relativity. The arbitrary elements in the algorithm that generates events preclude *unique* "retrodiction", while the finite complexity parameters (SU, N_U) prevent a combinatorial explosion in *statistical* retrodiction. In this sense we have a *fixed* – though only partially retrodictable – *past* and a necessarily *unknown future* of finite, but arbitrarily increasing, complexity. Only structural characteristics of the system, rather than the bit strings used in computer simulations of pieces of our theory, are available for epistemological correlations with experience.

What was *not* realised when this program was created was that this simple algorithm provides us with precisely the minimal elements needed to construct a finite particle number scattering theory. The increase in the number of strings in the universe by the creation of novel strings from discrimination is our replacement for the "particle creation" of quantum field theory. It is not the same, because it is both finite and irreversible; it also changes the "state space". The creation of novel strings by increasing the string length (TICK) implies an "exclusion principle"; if a string (state) already exists, the attempt to fill it leads to an "event", and a universe of increased complexity. Note that the string length N_U is simply the number of events that have occurred since the start up of the universe; this order parameter is irreversible and monotonically increasing like the cosmological "time" of conventional theories. Our events are unique, indivisible and global, in the computer sense; consequently events cannot be localized, and will be "supraluminally" correlated.

3. THE COUNTER PARADIGM; THE COSMIC FRAME

To make contact with physics we must now relate our bit string universe to the laboratory measurement of mass, length and time or three independent dimensional standards which can be related to these measurements. Laboratory practice in elementary particle physics is to use "counter" experiments or their equivalent for velocity measurement, momentum conservation for mass ratio measurement, and to find some phenomenon that brings in Planck's constant for the third connection (charge via $e^2/\hbar c$, Compton scattering, deBroglie wave interference, black body radiation, photo-effect,...). The inter-relationships between these measurements provide tight standards of self-consistency, and numerical values for the fundamental constants which in the end are more important than the comparisons with the standard meter, kilogram and second. Thus all we need do in principle is to make contact with three aspects of our theory in such a way that these connections follow.

As Heisenberg realised long ago, one of the easiest ways to make contact with macroscopic laboratory physics is through particulate momentum measurement, for example the firing of two counters a distance L apart with a time interval T, and identifying the "particle" which naively speaking "fires the counters" by measuring its mass (eg by momentum conservation in a scattering from a particle of known mass). Since the counters can, in principle, be placed as far apart as we like the velocity $V = L/T$ can be measured to as high precision as our budget allows. Empirically all such velocities are less than or indistinguishable from the limiting velocity c, and the momentum P and energy E are related to the mass m by $P = m\beta c/\sqrt{1-\beta^2}$ and $E = mc^2/\sqrt{1-\beta^2}$ (or $E^2 - p^2 c^2 = m^2 c^4$) where $\beta = V/c$. Thus if the basic quantum mechanics used is written in momentum space, and all physical quantities can be computed from the momentum space scattering theory, then contact with laboratory measurement is about as direct as possible. This is sometimes called the S-Matrix philosophy, and is adopted here. From this point of view, the representation of quantum mechanics in space-time is then obtained by Fourier transformation, and has only a formal significance, particularly for short distances where direct measurement with rods and clocks is impossible. Hence if we can show that our bit string universe supports a momentum space scattering theory of the same structure as conventional relativistic quantum mechanics (or at least in close enough correspondence to that structure so as not to be in conflict with experiment), our interpretive job has been done for us by Heisenberg and Chew. We develop this scattering theory in the next chapter, but still find it instructive to go as far as we can in interpretation without invoking that formal apparatus.

The means used to connect the bit string universe to the practice of particle physics is to assume that

any **elementary event,** *under circumstances which it is the task of the experimental physicist to investigate, can lead to the firing of a counter.*

The typical laboratory situation we envisage is that in which one of a beam of particles of some known type (which eventually we will have to connect to some label a in the bit string universe) enters and fires a macroscopic counter, and at time T later a counter a distance L from the first which is sensitive to the same type of particles also fires. Ignoring the practical details which will occur to the experimentalist, and the many sophisticated steps he will have to take to convince his colleagues that neither firing was "spurious", we follow conventional practice and say that this sequence of happenings means that a particle of type a has been shown to have a velocity $v = L/T$, and until something else happens will (if it carries a conserved quantum number such as charge or baryon number) continue to have that velocity in the geometrical direction defined by the first two counters. This assumption can be checked by adding counters down stream and checking that indeed (within uncertainties of measurement and corrected for energy loss in the counters) the expected velocity is again measured. We call this the "counter paradigm".

The first step in connecting the counter paradigm to the bit string universe is to assume that the first firing is connected to some unique event involving label a and that N TICK's later there was a second event involving the same label connected to the second firing. Further we assume that for some relevant portion (to be spelled out in detail later) of the address ensemble with this label the average number of ones added by these TICK's was $< k^a(N) > = < \Sigma_{i=1}^{N} b_i^a >$ allowing us to define a parameter $\beta^a = \frac{2<k^a(N)>-N}{N}$. Since $-1 \leq \beta^a \leq +1$ we identify it with a velocity measured in units of the limiting velocity c, and connect it to the experiment by requiring that $\beta^a = v/c = L/cT$. Following Stein[22] we interpret this ensemble of strings of length N as a biased random walk in which a 1 represents a step in the positive and a 0 a step in the negative direction.

Since we now know how to relate sub-ensembles of bit strings to velocities in laboratory events, the question naturally arises as to what coordinate system the full ensembles generated by PROGRAM UNIVERSE refer to. Fortunately this is an easy question to answer. We now know that the solar system is moving at approximately 600 km/sec with respect to the coordinate system in which the $2.7^\circ K$ background radiation is at rest; we also have measured the direction of this motion with respect to the distant galaxies. But the statistical method by which the strings are generated guarantees that on average they will have as many zeros as ones, defining uniquely a zero velocity frame with respect to which non-null velocities have significance. Clearly this must be identified with the empirically known "cosmic" zero velocity frame. Further there are two strings, 1_{N_U} and 0_{N_U}, which describe two states in which corresponding labels, 1_{N_L} and 0_{N_L}, have had the limiting velocity in opposite directions from the start. Thus we have an event horizon, to which we cannot assign any further content even after we have constructed our version of 3+1 "space"; the event horizon must be *isotropic*. Of course within that event horizon we could still be receiving signals from the remnants of collections of events which can be expected to be isotropic only in a statistical sense. We find it very satisfactory that these observed cosmological features emerge so readily from our interpretation of the model.

Now that we have confidence that the address strings do indeed specify discrete velocity states in general and not just in the laboratory, we next note that once the hierarchy has closed off at level 4, the set of available labels is *fixed* and simply keeps on reproducing itself in subsequent events. Thus labels have an *invariant* significance no matter how many subsequent TICK's occur, and can be used to identify both quantum numbers and elementary particle masses. Of course it will then become the task of the theory to compute the ratios of these masses to m_p (or to M_{Planck}). The problem is to make this assignment in such a way as to guarantee both quantum number conservation and 3-momentum conservation between *connected* events. Just how to do this is not obvious, and I have made several false starts on the problem, from each of which I learned something. The key turned out to lie in the parallel development of a finite particle number relativistic quantum scattering theory[23-37] which I hope will one day be considered as a candidate to replace both quantum field theory and S-Matrix theory as the theory of choice for practical problems in relativistic quantum mechanics. That, of course, lies in some *very* uncertain future. Fortunately the development has proceeded far enough to give the essential clues as to how to connect the bit string universe to at least one version of relativistic quantum mechanics.

We now spell out in more detail precisely how the counter paradigm is used to connect the firing of two laboratory counters as described above to two events in the bit string universe. These two events will involve some label L^a of length N_L. We assume that the address string A^a is of length $N_A = N_U - N_L$ when the first firing occurs and of length $N_A + N$ when the second firing occurs. The laboratory velocity $V = \beta c$ is then to be computed from the bit string model by $\beta^a = (2k^a/N) - 1$ where for a single string $k^a = \Sigma_{n=N_A+1}^{N_A+N} b_n^a$. As we have already discussed, we are not allowed to access the computer memory directly, so our knowledge is not this precise. The macroscopic size of the counters ΔL and finite time resolution ΔT necessarily require us to consider discuss all strings in the bit string universe in some range $\beta \pm \Delta\beta/2$ where $\Delta\beta = (L+\Delta L)/c(T - \Delta T) - L/cT$. We will see in the next chapter that this "wave packet" description is essential for the calculation of the "propagator" in the scattering theory.

Before I fastened on the counter paradigm as the correct point of contact between the theory and experiment, I tried to make use of Stein's[22] "derivation" of the Lorentz transformation and the uncertainty principle. He assumed that the basic "objects" underling what we call particles are ensembles of biased random walks of N steps of length ℓ with a probability p of taking a step in the positive direction and $q = 1 - p$ in the negative direction, and hence the probability distribution $N!/p!q!$ for the most probable position of the peak. To relate this to the velocity of the "particle" take $p = \frac{1}{2}(1 + \beta)$ and $q = \frac{1}{2}(1 - \beta)$, where βc is indeed the velocity of the most probable position. From the fact that the standard deviation from the peak is $\sqrt{Npq} = \sqrt{\frac{N}{4}(1 - \beta^2)}$ Stein then arrives at the Lorentz transformation, and by taking $\ell = h/mc$ gets the uncertainty principle as well.

Once I had the counter paradigm in mind, I took over Stein's "random walk" idea by assuming that the 1's in the $N_A + 1 \to N_A + N$ portion of the address strings represented steps in the positive (first firing to second) direction between the counters and the 0's steps in the opposite direction. The definition of β remains the same, and by taking the step length as $\ell = hc/E$ the velocity of the most probable position and the momentum are correctly related to the energy. Further the velocity of each step is the deBroglie "phase velocity". If we make up "wave packets" from these discrete "velocity states", it is easy to show that the most probable position still moves with the mean velocity and that the "coherence length" which determines interference phenomena based on these periodicities is indeed[38,39] the deBroglie wavelength $\lambda = h/p$. Our discrete theory therefore relates momentum measurement to interference phenomena and the "wave-particle dualism" in much the same way that it is done by following the S-Matrix philosophy.

We now have $\hbar, c$, and m/m_p related to measurement in a precise way. In the next chapter we complete this part of the argument by showing that we can indeed construct a scattering theory with 3-momentum and quantum number conservation in events using the strings of program universe. But our identification of address strings with velocity states already allows a number of cosmological connections between our theory and experimental fact to be made independent of the technical details of the scattering theory. As was spelled out above, we have the cosmological event horizon and its isotropy, and the identification of the coordinate system in which the theory is constructed with that coordinate system in which the $2.7^\circ K$ cosmic background radiation is at rest.

4. SCATTERING THEORY; CONSERVATION LAWS

We must now proceed to show that the events discussed above can be interpreted as supporting conservation laws that will be preserved by all relevant TICK-connected happenings. This will be done by invoking a new multi-particle relativistic quantum mechanical scattering theory[23-37]. The basic idea in this scattering theory is to use Faddeev-Yakubovsky equations for the dynamics rather than a Hamiltonian, or Lagrangian or analytic S-Matrix formulation. The basic input to the *linear* integral equations is then a two-particle scattering amplitude with one or more spectators. Because neither particle from the scattering pair is allowed to scatter again with its partner until something else has happened, there can be no "self-energy-loops" or infinities such as occur in field theory. Because the equations are linear, the solutions are unique, in contrast to the non-linear ambiguities that occur in the analytic S-Matrix theory. Because of the algebraic structure of the equations probability flux is conserved for those degrees of freedom which are included.

The basic theory allows any finite number of distinguishable particles. Fortunately we will not have to explore the combinatorial explosion that results in the standard Faddeev-Yakubovsky theory when one tries to go from N to $N+1$ with $N > 4$ because elementary events in PROGRAM UNIVERSE can involve at most four distinct strings. The 4-process has two cases: (3,1) three particles can coalesce to one (or one dissociate to three) with the fourth particle as a spectator; (2,2) two particles can scatter, and the scattering of the second pair can be the spectator. The 3-process allows one pair to scatter with the third particle as a spectator; adding a spectator to this process will lead to one of the two previous possibilities on the first iteration.

The Faddeev (3-particle) theory has three input processes: $a + b \leftrightarrow a + b$, c spectator; $b + c \leftrightarrow b + c$, a spectator; $c + a \leftrightarrow c + a$, b spectator. But when "crossing" is considered[40] the dynamics have to describe as well the anti-particles $\bar{a}, \bar{b}, \bar{c}$ with no change in the dynamical degrees of freedom. In quantum field theory or S-Matrix theory, any particulate state with velocity v and quantum number(s) $Q_{(s)}$ must enter the theory in such a way that no prediction of the theory is altered by changing the (conventional and arbitrary) choice of sign of the quantum numbers and choice of reference direction for velocities to $-v, -Q_{(s)}$ and inverting the coordinates (parity operation); the *relative* sign between velocities and quantum numbers *is* significant. Since for any labeled address A^a, $A^{\bar{a}} = 1_{N_U - N_L} \oplus A^a$, $\beta^{\bar{a}} = -\beta^a$, all we need do to insure this rule is to require that for any quantum number we define using the (unique) label string for label a ($a \in 1, 2, ..., 2^{127} + 136$)) $Q^{\bar{a}} = -Q^a$; all rules used below meet this requirement. Then any 3-event can be viewed as a two particle amplitude

$$\begin{matrix} a \to \\ b \to \end{matrix} \leftarrow \bar{a} \oplus \bar{b} \equiv a \oplus b \to \begin{matrix} \to a \\ \to b \end{matrix}; \ c = a \oplus b \ spectator \qquad (4.1)$$

or the velocity reversed equivalent; note that we cannot distinguish this locally from any cyclic or anti-cyclic permutation on a, b, c. In terms of the scattering theory we have developed[23-37] the basic scattering process starts from a collision between a particle and an anti-particle with opposite velocities, which is isomorphic to the bit-string 3-event described by Eqn. (4.1) if we look at is as $a + b \to \leftarrow \bar{a} + \bar{b}$. Because the distinction between the symbols 0 and 1 does not depend on which is which (a point brought home forcefully by John Amson's discussion of the *Bi-Orobourous* included here, the masses of particle and antiparticle must be the same. Consequently the basic process has zero total momentum, which is consistent with the assumption made above that the construction refers to the zero momentum frame. The extension to 4 events, taking proper account of the two cases, is immediate:

$$(3,1): \ a \oplus b \oplus c \to d \equiv \bar{d} \leftarrow \bar{a} \oplus \bar{b} \oplus \bar{c}; \ d = a \oplus b \oplus c \ spectator \qquad (4.2)$$

$$(2,2): \ a \oplus b \to (ab) \equiv (\bar{a}\bar{b}) \leftarrow \bar{a} \oplus \bar{b}; \ (cd) = a \oplus b \ spectator \qquad (4.3)$$

Since our basic process is what is called in high energy physics "anelastic" (2 in, 2 out but not necessarily the same two), it would appear that there are only two degrees of freedom – energy and scattering angle or the manifestly covariant Mandelstam variables. Actually this is true so far as the coupled integral equations go, but the coupling between the three Faddeev channels necessarily brings in a third dynamical degree of freedom. If we take these three degrees of freedom to be the magnitudes of the three momenta p_a, p_b, p_c, 3-momentum conservation guarantees that the vector triangle formed by them closes, so the magnitudes fix the internal angles. One vector in the plane of the triangle then can be used to relate the scattering triangle to space-fixed (laboratory) axes, providing 3 kinematic degrees of freedom. Since 3-momentum is conserved the plane of the triangle is fixed, as is the *total* 3-momentum in any arbitrary laboratory frame; total 3-momentum provides 3 more kinematic degrees of freedom. Since the particles are "on mass shell" ($E^2 - p^2 = m^2$ with $c = 1$), 9 of the 12 degrees of freedom are needed only to relate the fundamental dynamics to the manifestly covariant description in terms of the 4-vectors $\vec{k}_a, \vec{k}_b, \vec{k}_c$. A similar analysis shows how the Faddeev-Yakubovsky dynamics used in the 4-particle equations in the zero momentum frame, again under the assumption of 3-momentum conservation, suffices to provide all that is needed for the interpretation of the results in terms of standard relativistic kinematics.

Now that we know where we are headed, we can try to connect this theory up to the events in the bit string universe. The scattering theory uses single particle basis states with energy, momentum, mass and velocity connected (with $\beta^2 = (V/c)^2$ and $c = 1$) by

$$E^2 - p^2 = m^2 \geq 0; \ 0 \leq \beta^2 = p^2/E^2 \leq 1 \qquad (4.4)$$

Calling these states $|m_a, \beta_a > \equiv |a >$, the single particle mass $\mathbf{M}^a$, velocity $\mathbf{B}^a$, momentum $\mathbf{P}^a$ and energy $\mathbf{E}^a$ operators have these states as eigenvectors:

$$\mathbf{M}^a|a> = m_a|a>; \mathbf{B}^a|a> = \beta_a|a> = \frac{p_a}{\sqrt{m_a^2 + p_a^2}}|a> \qquad (4.5)$$

$$\mathbf{P}^a|a> = \frac{m_a\beta_a}{\sqrt{1-\beta_a^2}}|a>; \mathbf{E}^a|a> = \frac{m_a}{\sqrt{1-\beta_a^2}}|a> \qquad (4.6)$$

So far the connection to the bit strings of specified address length thought of as states is immediate if we make the identification $S^a(N_U) = L^a(N_L)\|A^a(N_U - N_L) = |a>$. All values of the parameters compatible with the constraints expressed in Eqn. (4.4) are allowed in the conventional scattering theory. Bit string dynamics is more specific. Only the discrete velocity eigenvalues $\beta_a = \frac{2\Sigma_{n=N_1+1}^{N_2} b_n^a}{N_2 - N_1} - 1$ are allowed. Here N_1 and N_2 are the string lengths of the universe when the two events of interest in defining the velocity state space occurred; of course $N_L \leq N_1 < N_2 \leq N_U$.

There is an interesting convergence between the basis states the bit string universe generates automatically and the "light cone quantization" states which Pauli and Brodsky[41] find peculiarly appropriate to simplify the quantum field theory problem. They introduce a finite momentum cutoff Λ and discretize the problem by using a finite quantization length L (the old trick of periodic boundary conditions). In the Introduction to their second paper Pauli and Brodsky call the parameters L and Λ "artificial", which indeed they are in their context of trying to discretize a "continuum" theory; for us two related finite parameters are *necessary*. Our theory has a finite momentum cutoff: the smallest finite mass particle recoiling from the rest of the universe. The maximum invariant energy we can discuss in our theory is, so far as we can see now, $M_U c^2 = (2^{127}+136)M_{Planck}c^2 = (2^{127}+136)^2 m_p c^2$. Working out the connection to the maximum finite values for N we can discuss consistently in our framework would get us into a discussion the issues raised by Amson's *Bi-Ourobourous*, so we defer it to ANPA 8 or later. We see no likelihood of finding *direct* experimental confirmation of our finite philosophy by exploring that limit experimentally. Both for the Pauli-Brodsky approach and for ours the momentum cutoff is set by the computing budget rather than more fundamental considerations.

Fortunately the minimum resolution we can achieve sets practical limits that are simpler to discuss, and which are directly related to the states Pauli and Brodsky use. They relate their invariant 4-momentum M to their "harmonic resolution" K by requiring, as we do, zero center-of-mass momentum (cf. p. 1999, Ref 41). Their unit of length is $\lambda_C = h/Mc$ which is the same as the step length ℓ in our random walk. Hence their harmonic resolution $K = L/\lambda_C = L/\ell = N$, that is the number of steps taken in the random walk, or the number of bits in the relevant portion of the address strings. As they say "One must conclude, that the wave function of a particle in one space and one time dimension depends on the .... value of the harmonic resolution K". This should make it clear that we can map our results onto theirs or visa versa, and find out the equivalent of their Lagrangian, creation and destruction operators, etc. in our context - or visa versa. The details remain a problem for future research.

There is a difficulty in their approach in going to 3+1 space since one needs two basic operators in addition to the invariant four momentum and the harmonic resolution. But, as Pauli and Brodsky assure me the obvious high energy particle physics choice of $M, P_{\parallel}, P_{\perp}, L_z$ works very well. Our problem is different in that there is nothing in the definition of "event" which insures that 3-momentum will be conserved, a fact which Kilmister pointed out rather forcefully at ANPA 7. Hence it is not obvious how to put these single particle states together to describe a 3-event or a 4-event. Actually this is a difficulty in *any* quantum theory, not just ours. The quantum framework is in fact more general than the 3-momentum conservation which (so far) is always observed. Non-relativistic quantum mechanics meets this problem by requiring that any interaction used either conserve 3-momentum, or be an approximation in a system where some large mass is allowed to take up arbitrary amounts of momentum. In quantum field theory the problem is met by assuming certain symmetries in the space of description *and* the allowed interactions, which lead to 3-momentum conservation for *observable processes*. "Vacuum fluctuations" (or disconnected graphs) which violate various conservation laws can still occur; they correspond to the events in our theory which we also wish to exclude. If one takes the symmetries as more fundamental, then momentum conservation can be "derived"; however, I would claim that the symmetries were introduced in the first place in order to insure this result. From a logical point of view momentum conservation is an added postulate.

S-Matrix theory starts from physically observable processes, and hence imposes momentum conservation from the start. The finite particle number scattering theory I am modeling simply requires 3-momentum conservation for all driving terms in the integral equations, and the structure of the equations guarantees that this propagates through the solutions. I claim we have at least as much right to restrict the interpretation of the bit string theory to those connected events which conserve 3-momentum when we discuss physical predictions as does any other quantum theory.

Actually the recent work by McGoveran and Etter[4] puts us on still firmer ground in making this restriction. The basic fact about a discrete topology is that distance cannot be defined until ordering relations, which define attributes of the resulting partially ordered sets, are imposed on the initially *indistinguishable* finite elements. Once this is done, the "distance" depends on the number discrete ways in which the information content of two different collections differs with respect to each attribute. Thus the metric, and the rate of information transfer, is attribute-dependent. Consequently there will be various "limiting velocities", the one which refers to *all* attributes being the minimum of these maximum allowed velocities. In the physical case, this is clearly the velocity of light and is the maximum rate at which information (i.e. anything with physical efficacy in producing change) can be transferred. However correlations (or in computer terminology *synchronization*) can occur supraluminally. This is our basic explanation of the EPR effect. With regard to the point under discussion, since 3-momentum conservation is one of the known attributes of physical effects, we are clearly justified in requiring this of the events that enter our scattering theory. Our bit string universe is then richer that the physical portion we discussion this paper, – a point worth pursuing in the future.

A second difficulty which emerges is that even though we restrict ourselves to (eg for 3-events) those strings for which

$$|p_a - p_b| \leq p_c \leq p_a + p_b; \ a, b, c \ cyclic \qquad (4.8)$$

we will not have all the richness of Euclidean geometry. We can of course define our angles in the triangle implied by 3-momentum conservation [which will close if Eqn. (4.8) is imposed] by $p_c^2 = p_a^2 + p_b^2 + 2p_a p_b cos\theta_{ab}$, but the digitization of the momenta (via the digitization of the velocities) will allow only certain angles to occur. This, of course, is familiar in the old "vector model" for quantum mechanical angular momentum; it is sometimes still called "space quantization". One loose end that still needs to be tied up is the connection of angular momentum quantization in units of $\hbar$ to the h we have already introduced via our random walk. We obviously cannot introduce Planck's constant twice. In a metric space restricted to commensurable lengths Phythagoras' Theorem does not always hold. McGoveran has pointed out that when we try to close triangles in a discrete space the restriction to integer values is one way that non-commutativity can enter a discrete topology. So all of this should work out in the long run. If it doesn't we are in serious trouble.

We nail down the 3-momentum conservation law by allowing only those labeled address ensembles for which it holds to provide dynamical connection between TICK - separated events The next step is to show that there are conservation laws arising from the labels which can stay in step with the kinematics. This is considerably easier. A 3-event requires that $L^a \oplus L^b \oplus L^c = 0_{N_L} \equiv 0_L$ and hence that $L^a \oplus L^b \oplus c = 1_{N_L} \equiv 1_L$ cyclic on a, b, c, where $L^a = 1_L \oplus L^a$. If we define the quantum number operators for some attribute x by $Q_x^a|a> = q_x^a|a>$ and require that

$$\mathbf{Q}_x|0_L> = 0 = \mathbf{Q}_x|1_L> \tag{4.9}$$

and that $q_a = -q_{\bar{a}}$, quantum number conservation in 3-events follows immediately. Further, velocities and particle-antiparticle status reverse together, as in usual in the Feynman rules. We defer the discussion of "spin" to the next chapter.

Probably the most significant step taken since ANPA 7 is the derivation of the "propagator" for the scattering theory directly from the bit string universe via the counter paradigm. The breakthrough was achieved last fall in collaboration with Mike Manthey, who got me out of the rut of a "binomial theorem" connection between TICK-separated events I had failed to make work. In the scattering theory, the connection between events is provided by the "propagator" $\frac{1}{E'-E-i0^{\pm}}$. Here +(-) refer to "incoming" ("outgoing") boundary conditions, and are all that remains in the "stationary state" scattering formalism to record the "time dependence" of the wave function in the Schroedinger representation. The unitarity of the S-Matrix $\mathbf{S} = 1 + i\mathbf{T}$, that is $\mathbf{S}^\dagger\mathbf{S} = 1$ or the corresponding restriction on the scattering amplitude T, is then all that is needed to insure flux conservation, detailed balance and time reversal invariance in the conventional formalism. This is the formal expression of the Wick-Yukawa mechanism, which attributes quantum dynamics to the "off-shell" scatterings at short distance which conserve 3-momentum but allow the energy fluctuations consistent with the Heisenberg energy-time uncertainty principle and the Einstein mass-energy relation. In words, the propagator is the probability amplitude for having the energy E' in an intermediate state in the scattering process when one starts from energy E for the incoming state. Since only the value at the singularity survives in the end (i.e. "asymptotically", or to use more physical language, in connections between numbers that can be measured in the laboratory), the normalization of this singularity can be fixed by the unitarity (flux conservation) requirement, and need not concern us. The scattering equations are simply the sum over all the possibilities allowed by the conservation laws with this weighting.

To obtain the statistical connection between events, we start from our counter paradigm, and note that because of the macroscopic size of laboratory counters, there will always be some uncertainty $\Delta\beta$ in measured velocities, reflected in our integers k_a by $\Delta k = \frac{1}{2}N\Delta\beta$. A measurement which gives a value of β outside this interval will have to be interpreted as a result of some scattering that occurred among the TICK's that separate the event (firing of the exit counter in the counter telescope that measures the initial value of $\beta = \beta_0$ to accuracy $\Delta\beta$) which defines the problem and the event which terminates the "free particle propagation"; we must exclude such *observable* scatterings from consideration. What we are interested in is the probability distribution of finding two values k, k' within this allowed interval, and how this correlated probability changes as we tick away. If $k = k'$ it is clear that when we start both lie in the interval of integral length $2\Delta k$ about the central value $k_0 = \frac{N}{2}(1 + \beta_0)$. When $k \neq k'$ the interval in which both can lie will be smaller, and will be given by

$$[(k + \Delta k) - (k' - \Delta k)] = 2\Delta k - (k' - k) \tag{4.10}$$

when $k' > k$ or by $2\Delta k + (k' - k)$ in the other case. Consequently the correlated probability of encountering both k and k' in the "window" defined by the velocity resolution, normalized to unity when they are the same, is $f(k, k') = \frac{2\Delta k \mp (k' - k)}{2\Delta k \pm (k' - k)}$, where the positive sign corresponds to $k' > k$. The correlated probability of finding two values k_T, k_T' after T ticks in an event with the same labels and same normalization is $\frac{f(k_T,k_T')}{f(k,k')}$. This is 1 if $k' = k$ and $k_T' = k_T$. However, when $k' \neq k$, a little algebra allows us to write this ratio as

$$\frac{1 \pm \frac{2(\Delta k - \Delta k_T)}{(k'-k)} + \frac{4\Delta k \Delta k_T}{(k'-k)^2}}{1 \mp \frac{2(\Delta k - \Delta k_T)}{(k'-k)} + \frac{4\Delta k \Delta k_T}{(k'-k)^2}} \qquad (4.11)$$

If the second measurement has the same velocity resolution $\Delta\beta$ as the first, since $T > 0$ we have that $\Delta k_T < \Delta k$. Thus, if we start with some specified spread of events corresponding to laboratory boundary conditions, and tick away, the fraction of connected events we need consider diminishes in the manner illustrated in Figure 3.

150

Figure 3. The connection between the address strings in tick-separated events resulting from an initial uncertainty in velocity measurement.

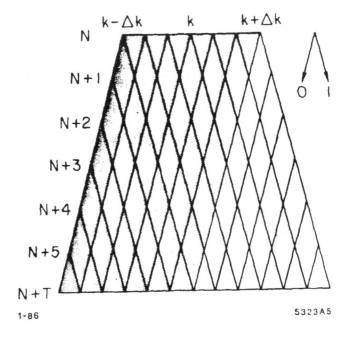

1-86

5323A5

Consequently if we ask for the correlated probability of finding the value β' starting from the value β we have proved that in the sharp resolution limit this is 1 if $\beta = \beta'$ and ± 0 otherwise. That is we have shown that in our theory a free particle propagates with constant velocity with overwhelming probability – our version of Newton's first law.

Were it not for the $\pm$, the propagator would simply be a δ-function, and since we are requiring 3-momentum conservation the theory would reduce to relativistic "point particle" scattering kinematics. But the limit we have derived approaches 0 with a sign that depends on which velocity is greater, which in turn depends on the choice of positive direction in our laboratory coordinate system, and hence in terms of the general description on whether the state is incoming or outgoing. In order to preserve this critical distinction in the limit, instead of something proportional to a δ-function we must write the propagator as

$$P(\beta, \beta') = \frac{lim}{\eta \to 0^{\pm}} \left[\frac{-i\eta}{\beta' - \beta - i\eta} \right] \tag{4.12}$$

where the limit is to be taken *after* summing over the allowed possibilities. Thus we find that the complexity of the wave function, and the propagator needed for scattering theory, can actually be derived from our interpretation of the bit strings. As already noted, the actual normalization of the propagator depends on the normalization of states, so we can use the conventional choice $\frac{1}{(E' - E \mp i0^+)}$ just as well.

What I like best about this derivation is that the macroscopic dimensions of the counters enter explicitly into the structure we need, just as "wave packets" have to be brought in for careful discussion of fundamental problems in standard quantum theory. It is also very satisfactory that the dichotomic choices at the lattice connections arising from TICK are strongly reminiscent of Finkelstein's "space-time-code" checkerboard. Scattering theory is one way to connect the imaginary time dependence in the Schroedinger equation of the conventional treatment to the discrete time scale we have to use to describe our time evolution. For consistency, this must also connect to the complex representation of angular momentum, and non-commutativity. As mentioned above McGoveran has some profound ideas here that are crying out to be explored.

Now that we have the propagator for a free particle, it is easy to write down the basic two-particle scattering operators as poles in the invariant two-particle 4-momentum which occur when the two particles coalesce to form a "bound state" of the mass appropriate to the resulting label and clothe this with 3-momentum conservation. We now have derived all the ingredients needed for the scattering theory. Since we have on hand a preliminary description of the theory[37] we repeat here the portion relevant to this paper.

Fortunately the "zero range scattering theory" developed in a non-relativistic context[28] allows scattering amplitudes to be inserted in Faddeev equations without specifying their relation to the non-invariant concept of "potential energy distribution". The model then reduces to the *kinematic* requirement that the "elementary" (or input) two-particle amplitude for meson-nucleon scattering have a pole when the invariant four-momentum of this pair is equal to the nucleon mass. As has been noted many times[25] the use of Faddeev dynamics guarantees unitarity without ever producing the self-energy infinities caused by the quantum field theory formalism. Clearly our general philosophical framework is that of S-Matrix theory, although we part company from the usual approaches to that theory by restricting ourselves to finite particle sectors. The second critical physical input is that 3-momentum be conserved in each elementary scattering. All particles are "on-shell"; only the energy of the system as a whole is allowed to fluctuate within the limits provided by the uncertainty principle. Again this is hardly new; Wick used this idea long ago[2] to provide physical insight into Yukawa's[3] meson theory. Putting this together with the requirement that observable probabilities be conserved specifies a minimal theory, as we now show.

Although the two-nucleon one meson system described by four-vectors has twelve degrees of freedom, our mass shell requirement $(\vec{k})^2 = \vec{k} \cdot \vec{k} = \epsilon_m^2 - \underset{\sim}{k} \cdot \underset{\sim}{k} = m^2$ reduces these to 9, and total 3-momentum conservation to 6. We restrict the Faddeev treatment (which would include the kinematic equivalent of particle "creation" and "destruction") by assuming that we start and end with a "bound pair" plus a free particle, and hence need only consider the residues of the double poles in the Faddeev amplitudes. Under these circumstances, 3-momentum conservation fixes the scattering plane in the *external* (and then laboratory) frame and reduces the *dynamical* (internal) degrees of freedom to 3. The remaining 3 simply allow the result of solving our dynamical equations to be related to external, and via the *total* 3-momentum to laboratory, coordinates. In general there will be nine "elastic and rearrangement" amplitudes (for example if we have a nucleon and an anti-nucleon, there will be a pole at the mass of the meson), but our "confined quantum" assumption[30,31] reduces these to four. Finally, the δ-function on spectator momentum reduces the 3 degrees of freedom to two dynamical degrees of freedom for each Faddeev channel (of course care must be exercised because the Faddeev description is "overcomplete"); we take these to be the magnitude of the momentum and the scattering angle, as in nonrelativistic potential scattering, or a single vector variable $\tilde{p}$ with the understanding that the azimuthal angle (or magnetic quantum number) is an "ignorable coordinate".

There is a further non-trivial kinematic fact which simplifies our result. We use the Goldberger-Watson[42] propagator $R_0^{-1}(z) = \epsilon_1 + \epsilon_2 + \epsilon_\mu - z$ where $\epsilon_i = \sqrt{p_i^2 + m_i^2}$, $i \in 1, 2$ and $\epsilon_\mu = \sqrt{q^2 + \mu^2}$. Since we are in the zero momentum frame, this is related to the invariant $S = (\vec{k}_1 + \vec{k}_2 + \vec{k}_\mu)^2 = (\epsilon_1 + \epsilon_2 + \epsilon_\mu)^2$ by $R_0(z) = (\sqrt{S} - z)^{-1}$ Here p_1, p_2, q refer to the "internal" coordinates where all three particles are "free". But the "external" coordinates refer to a particle of mass m_a and "bound state" of mass μ_a, with the invariant $s_a = (\epsilon_a + \epsilon_{\mu_a})^2$ or $\epsilon_a = \frac{1}{2}\sqrt{s_a} + \frac{m_a^2 - \mu_a^2}{2\sqrt{s_a}}$ because $p_a^2 = \epsilon_a^2 - m_a^2 = \epsilon_{\mu_a}^2 - \mu_a^2$. The model requires the driving terms to have a pole at $S_{i\mu} = (\vec{k}_i + \vec{k}_\mu)^2 = m_i^2 = (\epsilon_i + \epsilon_\mu)^2 - p_j^2$ where we have used the fact that $p_1 + p_2 + q = 0$. Hence (for equal mass nucleons) $S_{i\mu} - m_i^2 = (\sqrt{S} - \epsilon_j)^2 - \epsilon_j^2 = \sqrt{S}(\sqrt{S} - 2\epsilon_j)$, and the pole also occurs at $S = 4\epsilon_j^2$. Finally, we note that on shell, $S = s_i = s_j = 4\epsilon_j^2$ and $p^2 = (p^0)^2$, so the pole also occurs when the two momenta are equal. This allows us to write the driving terms as

$$\frac{g^2 \delta^3(p - p_0)}{p^2 - (p^0)^2 - i\eta} \tag{4.13}$$

[In this treatment we are using the continuum approximation]

Now that our space and the operators in it are defined, we can start from the Faddeev decomposition of the three body transition operator

$$\mathbf{T}^{(3)} = \Sigma_{i,j \in \alpha, \beta, \gamma} \mathbf{M}_{ij} \tag{4.14}$$

where the Faddeev operators $\mathbf{M}_{ij}$ are defined by the operator equations

$$-[\mathbf{M}_{\alpha\beta} - \mathbf{t}_\alpha \delta_{\alpha\beta}] = \mathbf{t}_\alpha \mathbf{R}_0 [\mathbf{M}_{\beta\beta} + \mathbf{M}_{\gamma\beta}] = [\mathbf{M}_{\alpha\alpha} + \mathbf{M}_{\alpha\gamma}] \mathbf{R}_0 \mathbf{t}_\beta \tag{4.15}$$

154

The δ-function in the driving terms reduces the corresponding integral equations immediately to coupled equations in two variables. Further, since we are concerned here only with the (2,2) sector, and hence with the residues of the double poles, which in a non-relativistic context would be called "elastic and rearrangement amplitudes", we can define

$$M_{\alpha\beta} - t_\alpha \delta_{\alpha\beta} = \frac{g_a}{s_a - \mu_{bc}^2} H_{\alpha\beta}(\underset{\sim}{p}_a, \underset{\sim}{p}_b^0; z) \frac{g_b}{s_b^0 - \mu_{ca}^2} \qquad (4.16)$$

For the 3-nucleon paper we are relying on here, we assumed two nucleons and one meson with no direct nucleon-nucleon scattering, and called the four surviving amplitudes K_{ij}.

The final result for the nucleon-nucleon amplitude in this (scalar) model is then that

$$T(\underset{\sim}{p},\underset{\sim}{p}') = K_{11}(\underset{\sim}{p},\underset{\sim}{p}') + K_{12}(\underset{\sim}{p},-\underset{\sim}{p}') + K_{21}(-\underset{\sim}{p},\underset{\sim}{p}') + K_{22}(-\underset{\sim}{p},-\underset{\sim}{p}') \qquad (4.17)$$

where

$$K_{ij}(\underset{\sim}{p}_i,\underset{\sim}{p}_j) - V_{ik}(\underset{\sim}{p}_i,\underset{\sim}{p}_j) = \int d^3 p_k' \frac{V_{ik}(\underset{\sim}{p}_i,\underset{\sim}{p}_k')K_{kj}(\underset{\sim}{p}_k',\underset{\sim}{p}_j)}{(p_k')^2 - p_j^2 - i\eta} \qquad (4.18)$$

and

$$V_{ij} = -(1 - \delta_{ij}) \frac{g^2}{\epsilon_\mu^{ij'}(\epsilon_\mu^{ij'} - \epsilon_{\mu_i} + \epsilon_j')} \qquad (4.19)$$

with $\epsilon_\mu^{ij'} = \sqrt{(\underset{\sim}{p}_i + \underset{\sim}{p}_j')^2 + \mu^2}$.

If the "bound state" is required to contain exactly one particle and one meson, three particle unitarity fixes a unique constant value for the coupling constant[27]. However, as has been discussed in connection with the "reduced width" (also the residue of a "bound state" pole) in the non-relativistic theory[43] it is possible to treat the residue as a measure of how much of the state is "composite" and how much "elementary"; the density matrix derivation given in the reference is due to Lindesay. In the case at hand, since the K_{ij} satisfy coupled channel Lippmann-Schwinger equations, their unitarity and that of the T constructed from them is immediate, and is independent of the value of g^2, making this, as well as the meson mass available as adjustable parameters for use in low energy phenomenology. In fact the equations in the non-relativistic region correspond to an ordinary and exchange "Yukawa potential" or for negligible meson mass and $g^2 = e^2$ to the usual coulomb potential. Thus we finally have made contact with both Rutherford Scattering and the Schroedinger equation for the hydrogen atom starting from bit strings!

5. THE STANDARD MODEL OF QUARKS AND LEPTONS; COSMOLOGY

We saw in the last section that our quantum numbers are to be defined in such a way that they reverse sign under the "bar" operation $\bar{S}^a = 1_U \oplus S^a = 1_L \| 1_A \oplus L^a \| A^a$, as do the velocities in the address part of the string. Hence for each string we can single out one quantum number which defines the relative sign between velocities and quantum numbers, and hence defines a "direction" in the space of quantum numbers which is correlated with the directions in ordinary space. This obviously is "helicity" which can be directed either along or against the direction of particle motion. Putting this together with the 3-momentum conservation we have already assured, the fact that this does not reverse sign when the coordinates are reflected makes this a "pseudo-vector" or "spin", and we must assume that it is to be measured in units of $\frac{1}{2}\hbar$ if we are to make contact with well known experimental facts. As already noted, one remaining foundational problem is to connect up the unit with the "orbital angular momentum" from our definitions of 3-momentum and the lengths that occur in our random walks (deBroglie phase and group wavelengths using h rather than $\hbar$ as the unit with these dimensions). In what follows we will assume that this can be done without encountering difficult problems.

Once we have identified the necessity for one quantum number in each label being interpretable as "spin", or more precisely "helicity", (including of course the possibility of the value 0 for some strings), the interpretation of Level 1 is essentially forced on us. The dichotomous spin state with no other structure is the "two component neutrino" familiar since the parity non-conserving theory of weak interactions was created by Lee and Yang, and demonstrated experimentally by Wu. A simple way to represent this is, for a two-bit representation (b_1, b_2), is to take $h_z = (b_1 - b_2)\frac{1}{2}\hbar$, as is shown in Table 2a.

Table 2
Conserved Quantum Numbers

2a. Level 1.

String (b_1, b_2)	$q_0 = b_1 - b_2$
$\begin{pmatrix} 1\ 0 \end{pmatrix}$	+1
$\begin{pmatrix} 0\ 1 \end{pmatrix}$	−1
$\begin{pmatrix} 1\ 1 \end{pmatrix}$	0
$\begin{pmatrix} 0\ 0 \end{pmatrix}$	0

2b. Levels 2 and 3.

String $(b_1 b_2 b_3 b_4)$	q_1 $b_1 - b_2 + b_3 - b_4$	q_2 $b_1 + b_2 - b_3 - b_4$	q_3 $b_1 - b_2 - b_3 + b_4$
$\begin{pmatrix} 1\ 1\ 1\ 0 \end{pmatrix}$	+1	+1	−1
$\begin{pmatrix} 0\ 0\ 0\ 1 \end{pmatrix}$	−1	−1	+1
$\begin{pmatrix} 1\ 1\ 0\ 1 \end{pmatrix}$	−1	+1	−1
$\begin{pmatrix} 0\ 0\ 1\ 0 \end{pmatrix}$	+1	−1	+1
$\begin{pmatrix} 1\ 1\ 0\ 0 \end{pmatrix}$	0	+2	0
$\begin{pmatrix} 0\ 0\ 1\ 1 \end{pmatrix}$	0	−2	0
$\begin{pmatrix} 1\ 1\ 1\ 1 \end{pmatrix}$	0	0	0
$\begin{pmatrix} 0\ 0\ 0\ 0 \end{pmatrix}$	0	0	0
$\cdots$			
$\begin{pmatrix} 0\ 1\ 1\ 1 \end{pmatrix}$	−1	−1	+1
$\begin{pmatrix} 1\ 0\ 0\ 0 \end{pmatrix}$	+1	+1	−1
$\begin{pmatrix} 1\ 0\ 1\ 1 \end{pmatrix}$	+1	−1	−1
$\begin{pmatrix} 0\ 1\ 0\ 0 \end{pmatrix}$	−1	+1	+1
$\begin{pmatrix} 1\ 0\ 1\ 0 \end{pmatrix}$	+2	0	0
$\begin{pmatrix} 0\ 1\ 0\ 1 \end{pmatrix}$	−2	0	0
$\begin{pmatrix} 1\ 0\ 0\ 1 \end{pmatrix}$	0	0	+2
$\begin{pmatrix} 0\ 1\ 1\ 0 \end{pmatrix}$	0	0	−2

Then, if we adopt the usual convention that the electron neutrino is "left handed" and has negative helicity relative to the positive direction of motion, we have (for massless neutrinos) $\underset{\sim}{\nu_e} = (01)\|1_A$ and $\underset{\sim}{\bar{\nu}_e} = (10)\|1_A$. There are only two states because (thanks to invariance under the bar operation) for intermediate states a neutrino moving the positive direction is indistinguishable from an anti-neutrino moving in the negative direction. Only in the laboratory, where we can use macroscopic "rigid bodies" to establish directions, can we measure both parameters.

Having an obvious interpretation of the basis states for Level 1, the interpretation of Level 2 is almost as straightforward. We use the representation (b_1, b_2, b_3, b_4), which allows three quantum numbers which meet our restrictions to be defined: $q_1 = b_1 - b_2 + b_3 - b_4$, $q_2 = b_1 + b_2 - b_3 - b_4$, $q_3 = b_1 - b_2 + b_3 - b_4$. These are exhibited explicitly in Table 2b. Since Level 2 only has three linearly independent basis vectors, we require $b_1 = b_2$, which arises naturally from the mapping matrix construction of the hierarchy, as we have discussed in detail in previous work. Under this restriction $q_1 = -q_3$, so there are only two independent quantum numbers. The obvious choice is to identify q_1 as lepton number (or electric charge) and q_2 as helicity in units of $\frac{1}{2}\hbar$, which leads to the particle identifications in Table 3. The graphical representation of these numbers given in Figure 4. may be more informative. We defer discussion of the "Coulomb interaction" called C until we have made the Level 3 assignments.

For level 3 we use one four-bit string allowing all 16 possibilities concatenated with a second for-bit string resembling level 2 and hence having only 8 possibilities. The first has four basis vectors and the second three, making up the required 7. Together we have 128 possibilities, or if we subtract the null string, the usual hierarchy 127. Assuming for the moment that the second 4-bit string is (1111) or (0000) – which we will see shortly is a QCD (quantum chromodynamics) color singlet – we have in fact the 16 states which can be formed from two distinguishable fermions and antifermions. These are clearly the nucleons with the associated pseudoscalar and vector (since a fermion-antifermion pair has odd parity) mesons. The identifications are spelled out in Table 3.

Table 3

Particle identifications for Levels 2 and 3

String	Level 2	Level 3 (color singlet)
$(1\ 1\ 1\ 0)$	$e^+_{+\frac{1}{2}}$	$n_{+\frac{1}{2}}$
$(0\ 0\ 0\ 1)$	$e^-_{-\frac{1}{2}}$	$\bar{n}_{-\frac{1}{2}}$
$(1\ 1\ 0\ 1)$	$e^-_{+\frac{1}{2}}$	$\bar{n}_{+\frac{1}{2}}$
$(0\ 0\ 1\ 0)$	$e^+_{-\frac{1}{2}}$	$n_{-\frac{1}{2}}$
$(1\ 1\ 0\ 0)$	γ_{+1}	ρ^0_{+1}, ω_{+i}
$(0\ 0\ 1\ 1)$	γ_{-1}	ρ^0_{-1}, ω_{-1}
$(1\ 1\ 1\ 1)$	C	$\pi^0, \rho^0_0, \omega_0$
$(0\ 0\ 0\ 0)$	(C)	$(\pi^0, \rho^0_0, \omega_0)$
$(0\ 1\ 1\ 1)$		$\bar{p}_{-\frac{1}{2}}$
$(1\ 0\ 0\ 0)$		$p_{+\frac{1}{2}}$
$(1\ 0\ 1\ 1)$		$p_{-\frac{1}{2}}$
$(0\ 1\ 0\ 0)$		$\bar{p}_{+\frac{1}{2}}$
$(1\ 0\ 1\ 0)$		d_0
$(0\ 1\ 0\ 1)$		$\bar{d}_0$
$(1\ 0\ 0\ 1)$		π^+, ρ^+_0
$(0\ 1\ 1\ 0)$		π^-, ρ^-_0

Figure 4. Level 2 quantum numbers represented as strings.

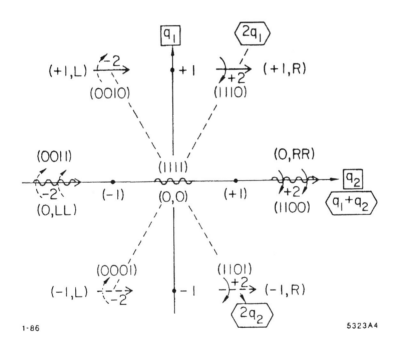

1-86   5323A4

To make this into quantum chromodynamics, we need only note that the level 2 quantum numbers also define an SU_3 octet, as is shown in Table 4 in terms of I, U and V -spin; again Figure 4 illustrates the relationships.

Table 4

The SU3 octet for "I,U,V spin"

	$(b_{11}b_{12}b_{13}b_{14})$	$2I_z$	$2U_z$	$2V_z = 2(I_z + U_z)]$
STRING:	1110	+1	+1	+2
	0010	-1	+2	+1
	1100	+2	-1	+1
	1111	0	0	0
	0000	0	0	0
	0011	-2	+1	-1
	1101	+1	-2	-1
	0001	-1	-1	-2

$2I_z = b_{11} + b_{12} - b_{13} - b_{14}$

$2U_z = -2b_{11} + b_{12} + 2b_{13} - b_{14}$

$2V_z = -b_{11} + 2b_{12} + b_{13} - 2b_{14}$

For color we could take red = (0001), anti-red = (1110); yellow = (0010), anti-yellow =(1101); blue = (1100), anti-blue = (0011). Then three colors or three anti-colors give the color singlet (1111), as do the appropriate combinations of color and anti-color. The three basis strings so constructed concatenated with the four already discussed give us two distinguishable colored quark and the associated gluons. Since $a \oplus a \oplus a \equiv a$, three colored quarks (or anti-quarks) add to give a color singlet and yield the spin and helicity states of a nucleons and anti-nucleons as we have just shown. Speculatively, since the scattering theory employed allows three states of the same mass to combine to single state of that mass, we can take both the quark and the nucleon mass to be the same; this would mean that quark structure would only appear at the 3 Gev level, which is desirable if nuclear physics is to continue to use mesons and nucleons as a first approximation.

Clearly we now have the quantum numbers for the first generation of quarks and leptons familiar from the standard model. Because of the closure properties of the hierarchy it is obvious that we will get higher generations simply by duplicating the structure we already have as many times as we need to get to $2^{127} + 136$ quantum states. We see that level 4 gives us a combinatorial explosion of higher generations with the same structure, but only weakly coupled because of the large number of combinatorial possibilities.

This is all very satisfactory until we ask (a) how to interpret the level one closure $(11)0_A$ (or $(00)1_A$) and (b) how to extend this interpretation to label strings of length L, which our PROGRAM UNIVERSE construction forces us to do. This problem has not been faced in previous discussions, and the conclusions reached for purposed of this report are frankly speculative. The problem is to get the coupling *between* levels and generations right. The speculative idea starts with the conjecture that the label $1_L = (11)_{L/2}$ is simply the universal Newtonian gravitational interaction which couples to any pair of labels with probability $[2^{127} + 136]^{-1}$ But then, so far as level 1 labels go, it is indistinguishable from (11). To go on, the analagous level 1 - level 2 cross level coupling would be the unit helicity Z^0 with unit helicity extensions to the $W\pm$. The 1-2-3 cross level coupling would be (as before) the coulomb interaction, with care taken so that the neutrinos carry no charge. As we have noted before, our theory is analagous to doing QED in the "coulomb gauge", so the spin-flip $\gamma_{\pm|}$ which come along are down in probability by $1/137$. Including these must correct our first approximation $\alpha = 1/137$ toward the *observed* value, but all strong as well as weak interactions will enter the calculation of the correction. Thus the mixing between generations cannot be ignored *a priori*. Conventional theories are now struggling with the problem of how best to combine weak-electromagnetic unification with the standard model, the generation structure, and gravitation in some sort of "super-unification" scheme. So the problem we hit in our own theory lies close to the cutting edge of conventional physics, as promised.

The cosmological implications of our theory are also interesting. We have already noted that our first approximation gives us Newtonian gravitation, so a "flat space" cosmology can also be anticipated. Our "big bang" – like that in an early version of Parker-Rhodes' *Theory of Indistinguishables* starts out "cold" in that we have to generate the labels first and only begin to develop "heat" after the basis vectors close and we begin to accumulate addressed label ensembles. Since the initial scatterings can take place in $\simeq (1.7 \times 10^{38})^2$ ways, and baryon number and lepton number appear to be very well conserved in our scheme, this initial condition gives approximately the baryon number and lepton number of the universe within the (rather broad - but all "flat space") observational limits. Since the initial address strings are short, they correspond to very high velocities and the resulting temperature will be extremely high. Even though we start "cold" we get a cosmic fireball early on. Once the average temperature falls (due to the expanding event horizon) down to the Tev range now being explored by particle accelerators our cosmology will develop much like others. The question which lies open is whether our rather unusual boundary condition will have consequences at variance with more conventional models in such a way as to lead to feasible observational tests. Only the uncertain future can decide.

6. THE MASS SCALE

[This section is quoted from SLAC-PUB-3566, "A discrete foundation for Physics and Experience (January, 1985).]

What is still missing in our fundamental theory are the mass ratios of the particles relative to our standard m_p identified by $\hbar c / G m_p^2 = 2^{127} + 136$. Here we adapt a calculation of Parker-Rhodes[44] based on his alternative, but closely related, approach to the problem of constructing a fundamental theory. He confronts the problem of *indistinguishability*, which in modern science goes back at least to Gibbs, but poses the problem in the logical (static) framework of how we can make sense of the idea that there are *two* (or more) things which are indistinguishable other than by the *cardinal* number for the assemblage *without* introducing either "space" or "time" as primitive notions. Clearly his starting point is distinct from the constructive program, and the "fixed past - uncertain future" implicit in our growing universe with randomly selected bit strings.

We have seen above that, for a system at rest in the coordinate system defined internally by $< \beta > = 0$ or externally by zero velocity with respect to the background radiation, the minimal fundamental length is $h/m_p c$, inside which length we have no way of giving experimental meaning to the concept of length without external coupling [2]. We have also seen that our scattering theory has, for zero mass coulomb photons, a macroscopic limit in Rutherford scattering, a non-relativistic limit in Bohr's theory of the Hydrogen atom, a continuum approximation in deBroglie's wave theory provided by continuum interpolation using Fourier analysis, and hence the usual formalism for the macroscopic e^2/r "potential" up to $O(1/137)$ spin-dependent corrections or relativistic corrections of the same order (either of which corrections — relativistic spin(Dirac) or relativistic motion (Sommerfeld) — account quantitatively for the empirical hydrogen fine-structure to that order). We have also seen that our momentum-space S-matrix theory has (within our digital restrictions) the usual properties of rotational and Lorentz invariance in $3 + 1$ momentum-energy space, and hence by our interpretive paradigms in 3-space.

We therefore can assert that outside a radius of $h/2m_p c$, the energy associated with the (minimally three) partons connected to an electron, the electrostatic energy of an electron can be calculated statistically from $< e^2/r >$ with three degrees of freedom and $r \geq (h/2m_p c)y, y \geq 1$. Since the conservation laws we have already established require charge conservation, the electrostatic energy must be calculated from the charge separation outside this radius with charges ex and $e(1-x)$, so $<e^2> = e^2 <x(1-x)>$. At first glance x can have any value, but in any statistical calculation the charge conservation we have already established requires that these cancel outside of the interval $0 \leq x \leq 1$. We have seen that the leptons are massless until they are coupled to hadrons at level 3 of the hierarchy (with, as the first approximation, $e^2/\hbar c = 1/137$). Hence, in this approximation, we can equate $m_e c^2$ with $<e^2/r>$, and arrive at the first Parker-Rhodes formula

$$m_p/m_e = \frac{137\pi}{<x(1-x)><1/y>}; \ 0 \leq x \leq 1; \ 0 \leq (1/y) \leq 1$$

From here on in, the only point to discuss is the weighting factors used in calculating the expectation values, since we now have from our S-matrix theory the same number of degrees of freedom (three) as Parker-Rhodes arrives at by a different argument based on the *Theory of Indistinguishables*. For the $(1/y)$ weighting factor this is almost trivial; our carefully constructed derivation of the Coulomb law and the symmetries of 3-space imply that $P(1/y) = 1/y$. For $x(1-x)$ the two-vertex structure of our S-matrix theory requires one such factor at each vertex in any statistical calculation: $P(x(1-x)) = x^2(1-x)^2$. The calculation for three degrees of freedom is then straightforward, and has been published several times[5,6,10,17,38,39]. [45-47] . The result is $<1/y> = 4/5$, $<x(1-x)> = (3/14)[1+(2/7)+(2/7)^2]$, leading immediately to the second Parker-Rhodes formula

$$m_p/m_e = 137\pi/[(3/14)[1+(2/7)+(2/7)^2](4/5)] = 1836.151497...$$

in comparison with the experimental value of 1836.1515 ± 0.0005. Although this result has been published and presented many times, we know of no published challenge to the calculation.

The success of this calculation encourages us to believe that the seven basis vectors of level 3 will lead to a first approximation for $m_p/m_\pi \approx 7$ with corrections of order $1/7$, but this has yet to be demonstrated.

7. CONCLUSIONS

We have tried to show in this paper that a complete reconstruction of relativistic quantum mechanics, elementary particle physics and cosmology can be based on a simple computer algorithm organized to exploit the closure properties of the combinatorial hierarchy and known basic principles of modern physics. Some people found the initial success in 1966 already impressive; others called it numerology. By now there are a number of quantitative and qualitative successes to our credit and no known failures. In the language of high energy physics, it may not yet be convincing as an "experiment", but it is beginning to look like the basis for a reasonable "research proposal".

The question remains – first asked me at Joensuu – what it will mean if the results are close to experiment, but can be shown to be in *quantitative* disagreement in a way that no fundamental remedy seems likely to fix. We will assume that a reasonable amount of effort has gone into minor tinkering with the fundamentals – they are so rigid that this gives very little scope. I expect this to be the case some day, hopefully soon. That I anticipate *close* agreement with experiment once the scattering calculations are carried out rests on the overdetermination Chew has already shown to exist between the structure of scattering theory, unitarity and "crossing". The theory has "bootstrap" properties, a finite and convergent equivalent of "renormalization", and once we get the quantum number assignments nailed down, will*have to* (if Chew is correct come close to experimental results. Our advantage over his theory is that our equations are *linear* allowing unique answers for any finite number of particulate degrees of freedom. So what if we are wrong? I would claim that *any* theory must contain the minimal elements we have used, and so must *also* fail unless additional basic principles are added. So a "failure", if we have done our job properly will either point to an are in which to look for *new* physics (eg. extension of the present three (MLT) dimensional concepts [to 7??]) or even a way to look *beyond* physics. This is, as I see it, the most one should ask of any model or theory or philosophy.

ACKNOWLEDGEMENTS

In this paper I have reported on a program aimed at reconstructing quantum theory "from the ground up" which started with work by Bastin and Kilmister in the 1950's[7,8], reached an intermediate success with the discovery of the combinatorial hierarchy by Amson, Bastin, Kilmister and Parker-Rhodes in the 1960's[9], and to which I started contributing in the 1970's[17]. For a survey of the historical development one must follow "preprint" literature and conference proceedings [5,6,10,38,39,46−48], [48] . My own recent work on this problem, carried out primarily in collaboration with David McGoveran, Michael Manthey, and Christoffer Gefwert, including the developments that have occurred since ANPA 7, forms the body of this paper. None of it would have taken place without those collaborators, and all of the ANPA members have contributed in one way or another, but I have decided on this occasion to present my own views without qualification, and make no attempt at achieving consensus. I trust this will lead to lively discussion and criticism at ANPA 8.

REFERENCES

1. A.Aspect and P. Grangier, "Tests of Bell's Inequalities...", in *Symposium on the Foundations of Modern Physics*, P. Lahti and P. Mittelstaedt, eds., World Scientific, Singapore, 1985, pp.51-71, and references therein.

2. G.C.Wick, *Nature, 142*, 993 (1938).

3. H.Yukawa, *Proc.Phys.Math.Soc., Japan, 17*, 48 (1935).

4. H.Pierre Noyes, David McGoveran, Tom Etter, Michael J. Manthey and Christoffer Gefwert, "Minimal Postulates for a Discrete Relativistic Quantum Physics"; ABSTRACT: "Starting from a minimum number of postulates, which do not refer to supposed physical properties of space and/or time, we show that any parametrized change in the distinguishable number of states for a specific attribute satisfying the postulates has the properties of the relativistic doppler shift and the relativistic composition law for velocities; additionally, such changes satisfy the Lorentz transformation. Each of these parameterized changes has an upper limit which may be characterized as a velocity. For a finite set of attributes, we show that the minimum of these maximum limiting velocities dominates relationships between the attributes. This yields a theory which is a discrete, finite version of the accepted properties of relationships in physical spacetime. However, this space is multiply-connected and allows for correlations exceeding the minimal limiting velocity. Hence this construction is relevant to the supraluminal correlations demonstrated in Aspect's and other EPR-Bohm type experiments." Submitted to the 2nd *INTERNATIONAL SYMPOSIUM ON FOUNDATIONS OF QUANTUM MECHANICS -In the Light of New Technology*, Tokyo, September 1-4, 1986.

5. For a summary of the early history cf. H. Pierre Noyes, Christoffer Gefwert, and Michael J. Manthey, "Toward a Constructive Physics" SLAC-PUB-3116 (rev. September, 1983); a short version of the paper based on this material was presented at the 7th *Congress of Logic, Methodology and Philosophy of Science*, Salzburg, Austria, July 11-16, 1983 (see next reference).

6. H.P.Noyes, Christoffer Gefwert, and Michael J. Manthey, "Program for a Constructive Physics", in *Foundations of Physics*, P. Weingartner and G. Dorn,eds, Hölder-Pichler-Tempsky, Vienna, Austria, 1986.

7. E.W.Bastin and C.W.Kilmister, *Proc. Roy. Soc.*, **A 212**, 559 (1952).

8. –*Proc.Camb.Phil.Soc.*, **50**, 254, 278 (1954), **51**, 454 (1955);**53**, 462 (1957); **55**, 66 (1959).

9. T. Bastin, *Studia Philosophica Gandensia*, **4**: 77, (1966).

10. H.Pierre Noyes, Christoffer Gefwert and Michael J. Manthey, "A Research Program with No Measurement Problem", *Proc. New York Academy of Sciences*, "New Techniques and Ideas in Quantum Measurement Theory", D. Greenberger, ed., New York City, 21-24 January, 1986; available as SLAC-PUB-3734

11. M.J.Manthey and B.M.E.Moret, "The Computational Metaphor and Quantum Physics", *Communications of the ACM*, **26**, 137-145, 1983.

12. M.J.Manthey, "Non-Determinism Can Be Causal", Tech. Report No. CS(83)-7, Univ. of New Mexico, Albuquerque, NM 87131.

13. David McGoveran, *ANPA Newsletter* **6**, No.1, 6-7 (1986).

14. Christoffer Gefwert, *On The Logical Form of Primitive Recursive Functions*, SLAC-PUB-3334 (May 1984), to be submitted to the *Journal of Philosophical Logic*.

15. –, *On The Logical Form of Mathematical Language*, SLAC-PUB-3344 (May 1984), to be submitted to the *Journal of Philosophical Logic*.

16. C.W.Kilmister, Appendix II.1 "Brouwerian Foundations for the Hierarchy" (based on a paper presented at ANPA 2, King's College, Cambridge, 1980) and Appendix II.2 "On Generation and Discrimination" (based on a paper presented at ANPA 4, King's College Cambridge, 1982) in H. Pierre Noyes, Christoffer Gefwert, and Michael J. Manthey, "Toward a Constructive Physics" SLAC-PUB-3116 (rev. September, 1983).

17. T.Bastin, H.P.Noyes, J.Amson and C.W.Kilmister, "On the Physical Interpretation and Mathematical Structure of the Combinatorial Hierarchy", published in *The International Journal of Theoretical Physics*, **18**, 455-488 (1979).

18. J.Amson, Appendix to Ref. 17.

19. M.J.Manthey, Appendix IV. in SLAC-PUB-3116 (rev. September, 1983).

20. C.W.Kilmister, **Appendix II.3** "HIERARCHY CONSTRUCTION(second version)" in SLAC-PUB-3116 (Rev. September, 1983).

21. A.F.Parker-Rhodes, *The Theory of Indistinguishables*, Synthese Library, *150*, Reidel, Dordrecht, 1981.

22. Irving Stein, seminar at Stanford 1978, and papers at ANPA 2,3, King's College, Cambridge, 1980,'81.

23. H.P.Noyes, "Three Body Forces", in *Few Body Problems*, I.Slaus, et.al. eds, North Holland, Amsterdam, 1972, p.122.

24. P.Noyes, J.Bell, and T.E.Phipps, Jr., *Fixed Past and Uncertain Future; an exchange of correspondence*, SLAC-PUB-1351, December 1973 (unpublished),p.30.

25. H.P.Noyes, *Czech.J.Phys.*,**B 24**, 1205 (1974).

26. H.P.Noyes, *Foundations of Physics, 5*, 37 (1975) [Erratum *6*, 125]; *6*, 83 (1976).

27. J.V.Lindesay, PhD Thesis, Stanford, 1981, available as SLAC Report No. 243.

28. H.P.Noyes, Phys.Rev *C 26*, 1858 (1982).

29. H.P.Noyes and J.V.Lindesay,*Australian J. Physics, 36,* 601 (1983).

30. H.P.Noyes and G.Pastrana, "A Covariant Theory with a Confined Quantum", in *Few Body Problems in Physics*, B. Zeitnitz, ed., North Holland, Amsterdam, 1983, p. 655.

31. J.V.Lindesay and A.Markevich, "A Minimal Unitary(Covariant) Scattering Theory", in *Few Body Problems in Physics*, B. Zeitnitz, ed., North Holland, Amsterdam, 1983, p. 321.

32. H.P.Noyes and M.K.Orlowski, *Nuovo Cimento* **81 A**, 617 (1984).

33. J.V.Lindesay, A.J.Markevich, H.P.Noyes and G.Pastrana, "A Self- Consistent, Poincaré-Invariant, and Unitary Three-Particle Scattering Theory", *Phys.Rev.*D`33`, 2339-49 (1986).

34. A.J.Markevich, "Angular Momentum and Spin within a Self-Consistent, Poincaré-Invariant, and Unitary Three-Particle Scattering Theory", *Phys.Rev.*] **33**, 2350-56 (1986).

35. H.P.Noyes and G.Pastrana, "A Covariant Approach to Mesonic Degrees of Freedom in Light Nuclei", presented at the *International Symposium on Mesons and 'Light Nuclei*, Bechyně Castle, Czechoslovakia, May 27- June 1, 1985.

36. H.P.Noyes, "Comment in Reply", SLAC-PUB-3916, April 1986 (unpublished).

37. H.P.Noyes "A Minimal Relativistic Model for the Three Nucleon System" in *Three Body Forces in the Three Nucleon System*, B.Berman, ed, Springer-Verlag (in press), available as SLAC-PUB-3973, June, 1986.

38. H.P.Noyes, "A Finite Particle Number Approach to Quantum Physics", SLAC-PUB-2906 (April, 1982, unpublished).

39. H.P.Noyes, "A Finite Particle Number Approach to Physics", in *The Wave-Particle Dualism* S.Diner, et.al., (eds), Reidel, Dordrecht, 1984, pp. 537-556.

40. G.Pastrana, PhD thesis, Stanford (in preparation).

41. H-C. Pauli and S.J.Brodsky, *Phys. Rev.* D **32**, 1993, 2001 (1985)

42. M.L.Goldberger and G.M.Watson, *Collision Theory*, Wiley, New York, 1964.

43. Ref.43, pp. 1869-70.

44. Ref. 21, pp 182-185.

45. H.P.Noyes, M.J.Manthey and C.Gefwert, "Constructing a Bit String Universe: a Progress Report" in *Quantum Concepts in Space and Time*, R. Penrose and C.J.Isham, eds, Oxford University Press, pp. 260-273 (in press).

46. H.P.Noyes, "A Discrete Foundation for Physics and Experience", SLAC-PUB-3566, January 1985 (presented at the Esalen Mind-Physics Conference, 4-8 February 1985).

47. H.P.Noyes, C.Gefwert and M.J. Manthey, "The Measurement Problem in Program Universe", *Symposium on the Foundations of Modern Physics*, P. Lahti and P. Mittelstaedt, eds., World Scientific, Singapore, 1985, pp.511-525.

48. H.P.Noyes, "Non-Locality in Particle Physics", SLAC-PUB-1405 (1974).

Comment on *"On the Construction of Relativistic Quantum Theory: A Progress Report"*
H. Pierre Noyes (2000)

Ch.2 had already gone through a couple of drafts and discussions by the time that the founding meeting of the Alternative Natural Philosophy Association was held at Clive Kilmister's red Tiles cottage in Sussex in the fall of 1979. Clive's cottage is not far from the White Hart Inn where Tom Paine and some like-minded revolutionaries met a couple of centuries ago. For a while we hoped that this accident was a good omen for our own attempted revolution in the foundations of physics. Bastin, Kilmister, Noyes and Parker-Rhodes were present; we were in touch with Amson by phone. We approved the final draft of PITCH, agreed on the name and structure of the Alternative Natural Philosophy Association (ANPA), and then went on to scientific discussion. So this meeting became "ANPA 1". All subsequent meetings have been held annually in Cambridge. This chapter shows some of the progress made during the first seven years of the organization. Together with the appendices (see note below) it formed part of the unpublished proceedings of ANPA 7.

Section 1 struggles with the foundational ideas we were trying to develop into a theory. Since my own preferred assumptions have, by now, come to diverge from this attempt at consensus, I give a recent formulation here. My three basic principles are: (1) measurement can determine whether two quantities are the same or different; (2) measurement can tell something from nothing; (3) the events which physics explores by measurement, and which can be modeled by binary addition and multiplication, have the same structure as events which leave historical, paleontological, cosmological,... records and hence can be modeled by a growing universe of bit-strings.

Section 2 started our from an early ANPA presentation of Clive Kilmister's which was exploring the view (since abandoned by Clive — see comments by Bastin and Kilmister on Ch. 2) that in addition to the fundamental operation of discrimination the ANPA research program would probably also have to introduce an equally fundamental operation which would generate bits (and bit-strings) in some sort of arbitrary sequence of interleaved generations and discriminations). Mike Manthey and I seized on this idea and tried to model it in the simplest way we could think of. This was the origin of *program universe* As Ch.16 shows, this simplistic approach produced cosmo-logical results in agreement with current observations long before those observations were made, perhaps the most important *predictive* success the research program has achieved to date. At a more fundamental level, we particularly direct the reader to McGoveran's discussion (pp 8-10) of why there is a conflict between the concepts of "computable" and "random", and how this can be resolved. This also explains why he uses the term "arbitrary" when some would say "random", a practice we have tried to adopt.

The remaining sections record the introduction of other significant novelties into the ANPA program. The interested reader can use the following list to guide him to the appropriate section: Section 3: the counter paradigm (p.22). Section 4: scattering theory, conservation laws, derivation of Newton's first law, identification of poles in propagators with bound states and coupling constants. Section 5: a first attempt to sketch out the bit-string version of the standard model of quarks and leptons (see Ch. 16 for the latest attempt). Section 6: the mass scale.

Note:

The original preprint had six appendices, which were six papers presented at ANPA 7. No currently available "Proceedings" exist for any ANPA conference prior to ANPA 9; the original preprint is an incomplete *Proc. ANPA 7*. Missing from this reprinting are the six appendices mentioned below. These can be obtained as pp 54-132 of SLAC-PUB-4008, June 1986 from the SLAC website: www.slac.stanford.edu/spires/hep, then "find author noyes, h.p." and search for this item. The six appendices are:

I. A FINAL FOUNDATIONAL DISCUSSION?, Clive W. Kilmister

"My aim is to derive the Parker-Rhodes construction ... from the acknowledged properties of quantum mechanics."

II. A SYSTEMS-THEORETIC CRITIQUE of PROCESS-HIERARCHY MOD-ELS, Ted Bastin

"*Of hierarchies, why some be abolished and some retained.* (Edward VI Prayer Book)"

III. AGNOSIA *A Philosophical Apologia for INDISTINGUISHABLES*, A.F. Parker-Rhodes

A decade or so later, when I revived this essay, Fredrick remarked "It wears well, doesn't it?" I suggest that this is a good starting point for anyone interested in deciding whether or not to take on studying his book *The Theory of Indistinguishables*.

IV. PROGRAM UNIVERSE, Michael J. Manthey

This is an actual program in Pascal. This program has never been run, and I am told there are a couple of "obvious" glitches in it, but that these are easily correctable. It should be obvious that running time is at least exponential, so any one who wants to do the computer experiment should do his own coding, and think hard about why he or she wants to run it for more than three levels.

V. Some Fundamental Characteristics of a Discrete Geometry, David McGoveran

Some of this material is covered in Ch.4.

VI. "BI-OROBOUROS" — A RECURSIVE HIERARCHY CONSTRUCTION, John Amson

This construction shows that the duality between "0" and "1" implies that, once the Parker-Rhodes hierarchy has been completed, its dual also exists. Hence the complete hierarchy and its dual could serve as the "0" and "1" to start off a new construction. From this point of view, complexity of structure can increase indefinitely, and exploration of underlying structure can extend to lower and lower levels indefinitely.

FOUNDATIONS OF A DISCRETE PHYSICS*

David McGoveran† and Pierre Noyes

Stanford Linear Accelerator Center,
Stanford University, Stanford, California 94309

ABSTRACT

Starting from the principles of finiteness, discreteness, finite computability and absolute nonuniqueness, we develop the ordering operator calculus, a strictly constructive mathematical system having the empirical properties required by quantum mechanical and special relativistic phenomena. We show how to construct discrete distance functions, and both rectangular and spherical coordinate systems (with a discrete version of "π"). The richest discrete space constructible without a preferred axis and preserving translational and rotational invariance is shown to be a discrete 3-space with the usual symmetries. We introduce a local ordering parameter with local (proper) time-like properties and universal ordering parameters with global (cosmological) time-like properties. Constructed "attribute velocities" connect ensembles with attributes that are invariant as the appropriate time-like parameter increases. For each such attribute, we show how to construct attribute velocities which must satisfy the "relativistic Doppler shift" and the "relativistic velocity composition law," as well as the Lorentz transformations. By construction, these velocities have finite maximum and minimum values. In the space of all attributes, the minimum of these maximum velocities will predominate in all multiple attribute computations, and hence can be identified as a fundamental limiting velocity. General commutation relations are constructed which under the physical interpretation are shown to reduce to the usual quantum mechanical commutation relations.

1. INTRODUCTION

The purpose of this paper is to present a self-contained mathematical foundation for the modeling of diverse phenomena—in particular, physical phenomena—and to demonstrate its utility.

Contributed to the 9th Annual International Meeting of the Alternative Natural Philosophy Association, Cambridge, England, September 23–28, 1987. To appear as Chapter 2 in DISCRETE AND COMBINATORIAL PHYSICS: Proc. of ANPA 9, H. P. Noyes, ed.; published by ANPA WEST, 25 Buena Vista Way, Mill Valley, CA 94941

---

*Work supported by the Department of Energy, contract DE–AC03–76SF00515.

† Permanent address: Alternative Technologies, 150 Felker Street, Suite E, Santa Cruz, CA 95060, USA.

Twentieth century foundational mathematics is caught on the horns of several dilemmas. Perhaps the most difficult of these dilemmas is also the most ancient: the separation of description and process or, as more usually encountered, the separation of mind and body. This dilemma manifests itself in the split-mind with which the practitioner of mathematics must operate. On the one hand, we perform finite computations by prescribed methods; on the other, we must keep forever in mind that these are artificial limitations of space, time, energy and symbolism—as is evident in the ever present use of ellipses and the infinity symbol. The description ignores the process of describing.

Somehow the student of mathematics must simply accept the fact that we never quite complete (and in principle cannot complete) many tasks of either description or describing, but must extrapolate. Such acts of faith are deeply embedded in the foundations. Of course, one should not be too concerned that counterfactual paradoxes arise as a result of following the faith with fervor or that one can prove that a mathematical system, if moderately powerful, cannot be both consistent and complete[1]. One must simply accept. At once, the student must pretend that the system is faithful (generates trustworthy results) and unfaithful (is either inconsistent or incomplete).

Twentieth century foundational physics is caught on the horns of a similar dilemma. The practitioner of laboratory physics appeals to the theoretician to completely describe his practice in an objective manner. Again, on the one hand, we perform finite measurements and computations by prescribed methods, while on the other hand we are asked to accept the fact that these are artificial limitations of space, time, energy and symbolism. Again the description ignores the process of describing. Dirac[2], seems to have been acutely aware of this separation of practice and formalism in dealing with the physical interpretation of discrete eigenvalues versus a range of eigenvalues:

"An eigenstate of ξ belonging to an eigenvalue ξ' lying in a range is a state which cannot strictly be realized in practice, since it would need an infinite amount of precision to get ξ to equal exactly ξ' . . . . Thus an eigenstate belonging to an eigenvalue in a range is a mathematical idealization of what can be attained in practice. All the same such eigenstates play a useful role in the theory and one could not very well do without them. Science contains many examples of theoretical concepts which are limits of things met with in practice and are useful for the precise formulation of laws of nature, although they are not realizable experimentally, and this is just one more of them. It may be that the infinite length of the ket vectors corresponding to these eigenstates is connected with their unrealizability, and that all realizable states correspond to ket vectors that can be normalized and that form a Hilbert space."

Neither the general nor the special theories of relativity readily admit of quantization. These theories are formulated within the space-time continuum using differential geometry. In conflict with this, quantum events are unique, discrete, irreversible, nonlocal, and yet indivisible. Conventional quantum theory tries to embed them in a space-time continuum, which is the source of many conceptual difficulties such as the "collapse of the wave function," the EPR "paradox" and the infinities of second quantized field theory. The properties of quantum events are more fundamental mathematically and conceptually than the properties of an abstract continuum.

One cannot construct a basis which is adequate for this thinking and for the description of phenomena with a language which is dependent on an embedding of discrete structures in a continuous one. We will develop terminology afresh, without the taint of a continuum (and infinities). Our point of view is more process-oriented than just descriptive: it must be possible to generate the structures and the properties which we explore.

L. E. J. Brouwer and others have attempted to constructivize mathematics since 1907 based upon severe and successful criticisms of classical mathematics. As noted by Bishop[3], "Many mathematicians familiar with Brouwer's objections to classical mathematics concede their validity but remain unconvinced that there is any satisfactory alternative." These are valid criticisms, but so are similar criticisms of various constructive attempts, which fail to recognize any of what we feel are some of the more essential aspects of the practice of mathematics. In particular, mathematics which is not process-oriented, context-sensitive, finite, discrete and constructive (primarily in Bishop's sense)[4] is of little use in practice, since the Universe in and about which mathematics is to be used is all of these things. The Universe is only knowable as a complete, consistent system: there exist no black holes arising from undecidability, halting problems, incompleteness or magic of any kind. It is not knowable or understandable in terms of its parts alone. We are strict, constructive, *systems* mechanists.

While we contend that the mathematical foundation presented here will indeed prove useful outside of physics (and we have reason to believe it will), the focus of this paper is restricted to demonstrating the utility of the mathematics for physics.

In order to construct a discrete basis for physics, we limit ourselves from the start to a finite number of symbols (e.g., 0, 1) and to an order parameter defined in terms of primitive recursion. In ordinary language, this allows us to count up to (or down from) some finite integer N which we specify in advance. No construction will be allowed to exceed this integer without additional articulation of the extant theory. This additional articulation will be consistent with and guided by our approach. These restrictions allow us to d–map our construction onto any "operational" description of physics in a sense even more strict than Bridgman's "pointer readings" and the finite specification of what operations are needed to make "pointer readings" are allowed only if we can reduce the operations to "counting." That this apparently impoverished starting point leads to interesting physics will be demonstrated in what follows. In particular, we achieve a fresh understanding of a number of the best established physical facts.

The context-sensitive process of ordering is fundamental in this: simple but subtle notions of ordering, carefully formalized, result in a rich mathematical structure. If one insists on finiteness, discreteness and a strong constructive approach, the power of the system is surprisingly undiminished from that of continuum mathematics[5]. For example, where others have claimed that a finite, discrete topology was indefinable, we assert that the proper notion of open set defined within the formalism is in fact more constructive than the usual definition from point set topology or Intuitionistic Zermelo–Frankel (IZF) set theory, and clearly avoids the paradoxes generated by the usual continuum-oriented definition of open neighborhood or open set.

In this paper, five principles will be introduced which should not strain the reader's credulity: finiteness, discreteness, finite computability, absolute nonuniqueness and strict constructionism. Then, after presenting eight key concepts (indistinguishables, d–sorts, ordering operators, d–sets, open d–sets, d–subsets, parameterization, dimension or basis and attributes) within the context of a larger development, the following consequences will be constructed: the 3+1 dimensional structure of space-time, a combinatoric construction of π, identification of the speed of light constant, the Lorentz Transformations, the relativistic Doppler shift, the relativistic composition law for d-velocities, the uncertainty principle, superluminal correlations without supraluminal communication, a combinatoric construction of the exponentiation operator, the commutation relations for linear and angular momentum, the de Broglie relations, the relativistic mass change, identification of Planck's constant and momentum conserving events.

1.1 PRINCIPLES

We will develop a theory which, both in terms of the constructs and operations defined on those constructs, possesses the properties expressed in the following five principles.

Principle I: The theory possesses the property of strict finiteness.

By finiteness, we mean that no infinities or infinitesimals are allowed in the theory. By infinities, we mean an x such that x is larger than any finite y in the system. By an infinitesimal, we mean an x such that x is smaller than any finite y in the system and is not identical to 0. In particular, no x in the system can be arbitrarily large or small. Furthermore, and in keeping with strict finiteness, we require finite definability of any derived (constructed) system, subsystem or attribute of a system.

Principle II: The theory possesses the property of discreteness.

By discreteness, we mean that the depth of partitioning by recursive descent (as by Dedekind cuts) or construction by recursive ascent (as in the construction of the transfinites) is bounded in advance from outside the theory. This absolute bound on the practice is a pragmatic constraint. Over the course of any effort, a particular bound will evolve by refinement.*

★ As an example, consider any practice which is realizable on a physical computer. The bound is fixed in advance by the amount of accessible storage. It is our point of view that altering this bound constitutes an alteration of the system (computer plus algorithm) which cannot be understood or modeled within the system. Thus, a system which allows for changes to this bound is ill-defined. If the behavior of a program written to run on a computer having a certain amount of memory is in any way dependent on the amount of memory available, then it is clear that changing the amount of memory available requires the programmer to reevaluate the program for unplanned behavior. If the programmer is wise, this is taken into account by coding "system parameters" into the program such that the system alterations will be "automatically" handled.

Principle III: The theory possesses the property of finite computability.

By finite computability, we mean that the theory is constructive in the following strict sense. It must always be possible to specify any procedure or argument used in the theory as an algorithm having a finite number of finitely definable steps and consuming a finite amount of memory. Such a theory is Turing computable, but only theories which are both Turing computable and which use a finite tape are finite computable. Those which use countably infinite tapes or countably infinite algorithms are excluded by this principle.

Principle IV: The theory possesses the property of absolute nonuniqueness.

Simply put, we assume indistinguishability and uniformity unless we have constructively stated otherwise. By absolute nonuniqueness, we mean that no property which serves to single out or distinguish a construct within the theory from any other construct within the theory may be used in the construction in the absence of an explicitly stated computational mechanism. That is, we will invoke a finite algorithm within the theory whenever a property is to be used in a construction and will otherwise be required to deal with the absence of that property (i.e., by probabilistic means). Any two differently labeled, but otherwise indistinguishable, constructs must be treated as interchangeable in the absence of such an algorithm. Thus, the only *a priori* property that is acceptable is recognition of a lack of information as evidenced by indistinguishability.[†]

Principle V: The formalism used in the theory is strictly constructive.

Following Bishop[6], and in addition to the preceding Principles I–IV, we will argue by constructive means. As such, proof by contradiction will be considered to be justified, since we are restricted by Principle I to finite situations. The only way in which we may show that an object exists is to give a finite means for constructing it. Bishop would say "finding it," but we do not accept the idea of *a priori* existence of nonfundamental objects. Complex (derived) objects are constructed, not found.

A property P is called definable in the system, if, for every object x constructively shown to exist, x has a property P or it does not. This is different from saying that it has the property P or else it has the property "NOT P." If this cannot be said, then the property P is not constructively defined or even definable within the finite system. Within these constraints on the allowed subject matter, we will deny arguments by the principle of omniscience and of limited omniscience, except (again contrary to the position of Bishop) where the latter may be supported by a finite search. Because our theory is finitary, we embrace the Law of the Excluded Middle (as would Bishop).

We call this position **strict constructionist** because we understand it to be more restrictive than the constructive positions of both Bishop[7] and Beeson[8], which are among the more restrictive statements of the position, and clearly more so than Brouwer.

---

† As we will see, this very general principle is at the heart of most invariance principles, including the assumption of equal *a priori* probabilities, isotropy, homogeneity and relativity.

2. MATHEMATICAL FOUNDATIONS I

In this chapter and the next two, we develop a strict constructive mathematical system which we refer to as the **ordering operator calculus**. This system will be shown to have sufficient power to be a foundation for, or simply replace, significant aspects of conventional mathematics including set theory, lattice theory, differential topology, real and complex analysis and differential geometry.

2.1 PRELIMINARY CONCEPTS

Several concepts will be taken as fundamental in the development of our theory. These concepts are well-known to computer scientists and are rigorously defined by them. Nonetheless, we will provide definitions which limit the scope and applicability of the terms, since our usage will in general be more restrictive. It is especially important for the reader to keep in mind that we do *not* import the additional theoretical framework which is normally accepted within computer science and discrete mathematics.

Recursively Definable

By **recursively definable**, we shall mean simply that an abstract term is definable with a finite number of steps from simpler terms and values.

Computable

By **computable**, we shall mean that an effective procedure has been given by which an abstract construct can be constructed in a finite number of steps and with finite resources. We shall use the term *recursive* in a manner similar to that used in recursive function theory, in that it includes both recursive and iterative algorithms and is not restricted to mean a "recursive procedure call" in the computer programming sense.

Computational Complexity

By the **computational cost** $C(O)$ of an abstract, finite, discrete algorithm O, we shall mean a measure of the time cost and the space cost of the algorithm. Each of these is usually expressed as a procedure, which shows how to compute from the cardinality and/or ordinality of the domain upon which the algorithm operates (usually called the *size* of the problem), and yields a measure of the **computational time cost** $C_t(O)$ or the **computational space cost** $C_s(O)$ of the algorithm in time-like units (e.g., CPU cycles or algorithmic steps) or space-like units (e.g., bits), respectively. Note that, for us, these costs include the cost of running and storing the algorithm itself.

It is considered normal to express the **computational complexity** measure in terms of the dominant term of the appropriate polynomial, logarithmic, exponential or combinatorial expression; we will consider this to be shorthand for the exact expression. An algorithmic procedure $g(n)$ will be said to be of **computational complexity**

$$O[f(n)]$$

read "of order $f(n)$," if there exists a rational constant c such that

$$g(n) \leq cf(n)$$

for all n.[*] By the **total computational cost** of an abstract, finite, discrete algorithm O, we shall mean the result of a procedure which computes for each pair of inputs $C_t(O)$ and $C_s(O)$ a finite number $C(O)$ in a finite number of steps. Such a procedure (which in classical mathematics is representable by a polynomial expression) is said to represent a **computational metric**.

Representational Resources

By **representational resources** of an abstract, finite, discrete system, we shall mean the maximum of the spatial complexities of those algorithms which may be expressed within the system, without appeal to either spatial or time resources outside the system.

2.2 THE CONCEPT OF ORDER REVISITED

Ensembles

Consider a collection of mathematical (in the sense that physical properties are neither implied, nor are they denied) objects about which we have no knowledge, other than their quantity (cardinality), together with a collection of (mathematical) operators for selecting some of those objects. We call this collection of objects an **ensemble**, because it differs from the usual set-theoretic notion of a collection in ways which we now explain.

Ordering Operators

The notions of distinguishability and indistinguishability of such objects are relative. Without a stated computational mechanism, we are required to assume indistinguishability in keeping with Principle IV.[†] When asked if two objects are distinguishable, one must respond with a question, "distinguishable with what algorithm?". If presented with such an algorithm, we can think of that algorithm as inducing a property on the objects on which the algorithm operates; then the question of distinguishability becomes, "distinguishable with regard to such and such a property." Indeed, whether the objects are "truly" indistinguishables or not in the sense of

---

[*] It is usually permissible that a finite number of values of n violate the inequality. We do not allow this.

[†] This is not an ontological statement.

Parker–Rhodes[9] is irrelevant: our inability to directly access the objects makes the properties of the computational mechanism used on the objects the essential knowledge in building our theory.

We choose a single means of establishing structure in our formalism, namely, the generalized concept of ordering relation called an **ordering operator.** These computational mechanisms are defined as having the following properties:

1) they are only defined on a finite ensemble (a domain);

2) the ensemble must have fixed cardinality N;

3) they take as single input a label;

4) each label carries an embedded unique inaccessible sequence number;

5) they operate on the ensemble or some portion of it;

6) they generate as output one or more labels;

6) the labels successively generated are not necessarily unique;

7) the labels so generated constitute a finite ensemble;

8) the mechanism has a stop rule,

9) the details of the mechanism, including the stop rule, are not inferable.

Note that without either the identification of the ensemble and the input or recognition of the output, there is no knowledge that the operator has been used. By recursively applying this mechanism, we generate an ordered sequence of labels. Clearly, the ordering operator counts as a generating function in the sense used by Kilmister[10], although it does not require the same mathematical foundation and has additional computational power. Since we lack knowledge about the nature of the indistinguishables, we need to specify a few more characteristics of the mechanism of ordering operators. Having done this, ordering operators then also serve an essential function in our axiomatic system as general rules of inference, since they determine precisely what can be constructively exhibited or evaluated.

By **indistinguishables,** we mean that, given the ordering operator mechanism, the objects in an ensemble come in two forms which we now define. By **identicals** we mean that there exists no algorithm constructed within the formalism which serves to distinguish two objects. Thus, identicals is[‡] what one gets when an ordering operator operates twice on the "same" object. By **twins,** we mean that the algorithm used to manipulate the objects does not distinguish them, but that there exists some algorithm constructed within the formalism which does distinguish them. Thus, twins are what one obtains when the ordering operator operates on two objects, but does not distinguish between them in its output; that is, the objects seem to us to be the same within the context of the specific ordering operator. Thus, two objects are indistinguishables only for a specific alogrithm.

---

‡ The grammatical "agreement" as used here is intentional.

The output which results from using the ordering operators in either of these first two cases is two indistinguishable but sequence ordered object descriptions which we will call **labels** for short.§ Thus, the ensemble of objects has no inherent ordinality as far as we are able to know.

Above, we said that the ordering operator operates on the ensemble. Specifically, we mean that the ordering operator picks a finite number of indistinguishables, given a label as input. If the operator picks more than one object, successive recursions of the operator via input of a label generate one sequence numbered label per object in the subcollection, until the subcollection has been exhausted. The sequence numbers "stick onto" the objects, and their significance can only be recognized by the ordering operator that generates them (it maintains the equivalent of a symbol table which allows it to look up the sequence number(s) associated with a label and vice versa); thus, other ordering operators simply ignore the sequence numbers if operating on the same ensemble. The subcollection is then returned to the ensemble. Further input of a label returns one to the initial situation.

Note that this mechanism allows the operator to generate both total and partial orderings of the labels. The ordering operator algorithm has a stop rule (it halts in a well-defined manner) and will not allow, without repetition, recursive generation of more than a fixed and finite number of labels. The process is defined with (a) the operator and (b) a unique starting label. For some label input, the number of labels output by recursively feeding the output label into the input (i.e., **recursive generation**) is a maximum. The maximal label output of the operator and the ordering operator, itself, are **mutually defining**. Thus, given a finite ordering of labels on a given ensemble, we define an operator, and vice versa. Finally, the complexity of the operator mechanism (i.e., the algorithm) is too great to be represented by the labels alone. We would also have to know the intrinsic nature of the ensemble, but this can only be investigated with (other) ordering operators.

Suppose that a particular ordering operator O on a specific ensemble C (its "domain" in this instance) is given as input a specific label L_0. Let the resulting output of O be the label L_1. On input of L_1, O generates L_2, etc., up to some finite number of labels N. These labels need not be unique; however, each corresponds to some object in the ensemble C. Suppose that this correspondence is such that at least one label has been generated for each object in the ensemble C. If we keep a record of the labels so generated, we can be certain that rerunning the generation will produce an identical record, given the same ensemble C and the same initial label L_0. Indeed, if we begin with L_1, O produces an identical record with exception of the missing entry for L_0. If we begin with L_2, O produces an identical record with missing entries for L_0 and L_1.

If several successive entries in the record are equivalent except for the sequence in which they were generated, we cannot know whether the objects in the underlying ensemble to which these labels correspond are twins (indistinguishable but distinct) or

---

§ We suggest the use of **tags** where the term labels would be otherwise confusing as, for example, in Noyes[11] where label refers to a particular kind of label in our sense of the term.

identicals (indistinct). This can only be known by detailed knowledge of the objects in the ensemble and the algorithm by which the operator works. One might argue that in reproducing a record starting from equivalent records, that the recursion would not terminate since the labels are equivalent. However, note that conveying both the label and its record entry sequence is required for entry if the record is to be reproduced. The use of the notation L_n is not accidental. Two pieces of information are conveyed; the label and a number representing the sequence in which it was produced.

Define an operator $O^{\#}$ associated with O that behaves as follows: if the sequence number is left out, then $O^{\#}$ selects a default sequence number for the particular label consistent with the possible sequence numbers with which it might be produced. Thus, the operator may generate labels in two modes: with the sequence number or without it. When the sequence number is excluded, the recursive use of the ordering operator is similar to a sampling algorithm, subject to the constraint of an ordering relation. This ordering relation is not, in general, transitive. In this case, it is possible for $O^{\#}$ to generate L_M multiple times, given only label L for input. The output of $O^{\#}$ must then be ordered on the output sequence numbers to recover the ordering relation which $O^{\#}$ mutually defines. Unless we refer to sampling with repetition allowed, we will mean that $O^{\#}$ has as input and output both the sequence number and the label. However, we will ordinarily refer only to the input and output labels, the sequence being assumed and, since we then do not distinguish between O and $O^{\#}$, we will simply use O notationally.

Note that without the underlying objects, the algorithm for the ordering operator cannot in principle be defined, since the nature of the algorithm will depend upon the nature of the ensemble. Furthermore, depending upon which ensemble an ordering operator operates on, the statistical distribution of labels so generated (with or without repetition) is determined by the intrinsic character of the objects in the ensemble; namely, the cardinality of the ensemble and whether there exist indistinguishables or not, and how many. At best, having run through the operator once, we may use the output "sequence"[*] as a "look-up" table for additional runs, but only in the sense of checking off what has been generated so far and, in the case of a partial ordering, what is left.

2.3 CLASSIFICATIONS OF FUNDAMENTAL OBJECTS AND OPERATORS

The elementary unstructured object of our mathematics is taken to be a d–sort. We define a d–sort as any ensemble of n indistinguishable objects having cardinality n, of which it is NOT asserted that every pair of members is either identical or distinct[12]. A perfect d–sort is a d–sort for which every pair of members is either identical or twins (indistinguishable). We allow the members of a d–sort to be labeled by an

---

[*] We do not mean to imply that the output is "sequential" or totally ordered by the use of this term.

ordering operator, noting however that such internal labeling is inaccessible except via the ordering operator which performs the labeling.[†]

In order to formally define the relationship between ordering operators and the objects of our formalism, we have need of some additional definitions.

An **ordering relation** $\leq$ is a binary relation (a relation on two arguments). From time to time, we will take the liberty of writing $y \geq x$ in place of $x \leq y$. We shall mean by the symbol for **equivalence** $=$, a binary relation such that the two arguments are either identicals or twins; that is, they are the members of a perfect d–sort. A **successor function** $'$ is an ordering relation such that P1 and P2 are satisfied:

P1: Given arbitrary objects a, b, b' and x in a d–sort S such that if $a \leq x \leq b$, $a \leq b$, $a \leq b'$, and either $a = x$ or $x = b$, then $b = b'$ (uniqueness).

P2: There exists an e in d–sort S such that $e \leq x$ for all x in S (infimum).

A **recursive enumeration** E is an ordering operator which provides or recovers a label for each member of a d–sort. It is, therefore, an effective procedure for listing the members of a d–sort, with repetition allowed. More formally, a recursive enumeration is a rule with successor function ($'$) such that, given a label for object x in a d–sort S as input, the recursive enumeration generates a label for object x' in S, not necessarily distinct from x.

Theorem 1:   Neither the enumeration nor the successor function on a given d–sort are unique.

A **partial ordering relation** $\leq$ is a binary relation between two members of a d–sort S which, for all arguments x, y, or z in S, satisfies the following conditions:

P3: For all x, $x \leq x$ (reflexive).

P4: If $x \leq y$ and $y \leq x$, then $x = y$ (antisymmetric).

P5: If $x \leq y$ and $y \leq z$, then $x \leq z$ (transitive).

A **parameterization** is a partial ordering relation induced on a d–sort by an ordering operator O such that, given a label x for a member x of S as argument, the parameterization generates a label x' for the successor of x, called x'. That is, the partial ordering relation satisfies P1 and P2, as well as P3, P4 and P5. We may refer to a ordering operator O as a parameterization if O is used to induce a parameterization.

A **total ordering relation** is a partial ordering relation which satisfies P6.

P6: Given arbitrary x and y in d–sort S, either $x \leq y$ or $y \leq x$.

---

† Parker–Rhodes was insistent that indistinguishables could not be labeled at all. In this respect, we suspect that Parker–Rhodes would have likened d–sorts to a kind of multiset[13]. With a suitable mapping we could identify the notion of an ordering operator with Parker–Rhodes' functor in which case the ordering operator would produce ordinal sorts. However, this would relegate our theory to the domain of sort theory, an extra degree of ontological freedom and notational complexity which we cannot afford.

We shall deem it convenient, at times, to speak of a particular type of recursive enumeration. In particular, we will want a recursive enumeration without repetition, and in which the binary relation is a partial ordering relation.

A **rule of correspondence** is a total ordering relation induced by an ordering operator O on a d–sort of cardinality 2.

A member x of a d–sort S of cardinality 2, with partial ordering, is called a **supremum or sup** if, for member x and arbitrary y in S, $y \leq x$ and it is not the case that $x = y$. Similarly, a member z of S is called an **infimum or inf** if, for members y and z, $z \leq y$ and it is not the case that $y = z$.

Theorem 2:   Every total ordering relation induced on a finite d–sort defines a supremum and an infimum.

Theorem 3:   A rule of correspondence defines a supremum and an infimum.

A member a of a d–sort S with ordering operator O inducing an ordering relation $\leq$ on S is an **upper bound** if there exists a member b such that for the d–sort S' consisting of a and b, with ordering operator O' inducing the ordering relation $\leq$, a is the sup of S'. Similarly, a member b of a d–sort S with ordering operator O inducing an ordering relation $\leq$ is a **lower bound** if there exists a member a such that for the d–sort S' consisting of a and b, with ordering operator O' inducing an ordering relation $\leq$, a is the inf of S'.

Note that we have used partial ordering in the definitions of sup and inf so that we may create d-chains of d–sorts with the same partial ordering. Defining the concepts of upper and lower bound in this way insures transitivity across d–sorts; then these concepts take on the usual lattice theoretic definitions and the uniqueness of the sup and inf (in a given d–sort of cardinality n with specified partial ordering) is assured[14].

2.4 CONSTRUCTED OBJECTS: FROM d–SORTS TO COORDINATES

A d–**set** is a d–sort with ordinality m imposed by one or more recursive enumerations. The ordinality m of the d–set is just the cardinality of the d–sort of labels given by the recursive enumeration. Note that there may be more than one such recursive enumeration associated with the d–sort.

We may now classify the generations of an ordering operator in terms of the cardinality and the ordinality of the labels it generates. When these are the same, the output is a d–set and the O is said to be a **total ordering operator**. When these are not the same, the output is a d–sort and O is then said to be a **partial ordering operator**. Like sets in set theory, d–sets have no members that cannot be counted uniquely, while d–sorts have members that can not be counted uniquely. Unlike sets, a d–set is only defined with respect to one or more *specific* ordering operators. It, like all other objects in our system, is *constructed*.

We say an enumeration E is **monotonic increasing** if it gives labels to elements in the order of the recursive enumeration of the d–set on which it is defined. Similarly, we say E is **monotonic decreasing** if it gives labels to elements in reverse of the order of the recursive enumeration of the d–set. An enumeration E will be said to be **nonmonotonic** if it cannot be shown to be either monotonic increasing or monotonic decreasing constructively.

We are now in a position to define an important concept of topology; the **notion** corresponding to an open set. Note that our definition in no way appeals to the notions of continuity (in the usual sense of the word) or infinitesimals.[*]

The **boundary** of a d-set S defined with ordering operators O_i generating ordering relations R_i, consists of those elements of the underlying d-sort which, when operated on by any of the O_i, generates a label which is either a sup or inf of the R_i.

A d-set S' is said to be an **open** d-set with respect to a d-set S defined with ordering operators O_i, if S' is just S without the elements which would generate the boundary of S. Thus, for the defining ordering operators of the d-set S, none of the defining ordering operators of the d-set S' on an element of the underlying d-sort of S' generates a label which is either a supremum or an infimum of the defining ordering operators for S. It will generally be the case that d-sets are formed from multiple ordering operators. The extension of the definition to more complex d-sets having multiple ordering operators is straightforward.

Clearly, from the definition of ordering operator (i.e. an ordering operator and its productions are mutually defining), S' is, itself, a closed d-set, but *for a different ensemble of ordering operators*. This makes clear the importance of the notion that a d-set is a d-set by virtue of the defining ordering operators.

Note also that this eliminates the possibility of d-sets with deleted points. The transitivity of an ordering relation is, itself, defined constructively. Thus, the transitivity of any ordering relation is broken if a point is "deleted" and new ordering relations are induced, resulting in a new d-set. By defining open d-set as above, we have insured that there is no means of specifying a *classical boundary* for the a set independent of the construction of the set, as is done with a classical (infinite or continuum) set. Thus, every enumeration of the elements of a d-set would have to be an infinite enumeration, either by allowing (infinite and therefore not constructively definable) repetition (in the case of d-sorts) or without repetition (i.e., only if the d-sort is itself infinite and, in which case the d-set cannot be constructed as the ordering operator, is finitely definable only when all the members of the d-sort can be finitely specified). Thus, for d-sorts of sufficiently large cardinality, open d-sets are not distinguishable from the open sets of the classical definition.[†]

The fundamental concept of local topology is now within reach; i.e., an open neighborhood. A **d-subsort** (or **d-subset**) S' of a given d-sort (or d-set) S is itself a d-sort (d-set) which is defined with the same ordering operators, and for which x is an element of S' if and only if x is an element of S. We say that the S **contains**

---

[*] In addition, the definition is purely constructive and recovers the "classical" definition for sorts of sufficiently large cardinality n. We shall define what we mean by "sufficiently large" in a future paper in which we will discuss measurement—both abstract and physical—using the terminology developed here. Briefly, one measures the number of partitions n of a model by a d-map G to a d-sort of cardinality n'. Then, if $n > n'$ for all n' chosen, n is "sufficiently large," as no finiteness will be detectable independent of the measurement.

[†] This will be more obvious after we provide a constructive definition of a smooth recursive enumeration, below.

S'. Note that this precludes the possibility of "supersets" of d–sets being equivocated with d–subsets contained in the d–sets and, thus, the "set of all sets" of Russell's Paradox. For arbitrary elements of a d–sort S, an **open neighborhood of** x is any open d–sort containing an open d–sort containing x.

A **closed** d–**sort** (d–**set**) is a d–sort with defining ordering operators O_i, such that at least one element of the d–sort is either a supremum or an infimum for at least one of the O_i.

A d–**map** G on a d–sort S is an ensemble of rules of correspondence defined on S (i.e., inducing a d–sort of cardinality 2), such that there exists a d-subsort of S with members x_i all of which are the infimums of the rules of correspondence, and there exists a d-subsort of S with members all of which are the supremums of the rules of correspondence. We refer to one of these d-subsorts as the **domain** of the d–map G and to the other as the **range** of G. The range is said to be a d-subsort of some d–sort called the **image** of G. The domain and range have ordinality 1.

The **union** of two d–sets S and S' is a d–set S'', whose members consist of the members of either or both of S and S'. Similarly, the **intersection** of two d–sets S and S' is a d–set S'', whose members consist of the members of S which are also members of S'. The **symmetric** d–**set difference** of two d–sets S and S' is a d–set S'', whose members are either members of S but not of S', or members of S' but not of S.

A d–map M on a d–sort S is said to be **one-to-many**, if and only if the cardinality of the domain is less than the cardinality of the range. Such a d–map is called an **operation**. A d–map M on a d–sort S is said to be **many-to-one**, if and only if the cardinality of the domain is greater than the cardinality of the range and one-to-one, if and only if the cardinality of the domain is equal to the cardinality of the range. Such a d–map is called a **function f()**. A d–map M on a d–sort S is said to be **onto**, if and only if every element of S is either a supremum or an infimum of the rules of correspondence. If a d–map M on S is both many-to-one and onto, it is called an **isomorphism**. M is said to be order-preserving or **isotone** if, given an ordering operator on two elements x_1 and x_2 in the domain of M, there exist corresponding elements y_1 and y_2 in the range of M, which are also valid arguments of the induced ordering relation in the following sense: given that x_1 corresponds to y_1 and x_2 corresponds to y_2, it follows that if $x_1 \geq x_2$ then $y_1 \geq y_2$ and similarly, if $y_1 \geq y_2$ then $x_1 \geq x_2$.

Theorem 4:   An isomorphism is isotone.

Argument:

If we could not make this choice, there would exist some order on the d-subsorts, and the d–map would not exist. Note that such an ordering relation on the range or domain provide structure and thus increase the ordinality of the d-subsort.

$$\text{QED}$$

A **bisection** of a d–sort S is any one-to-one d–map defined on S and having both domain and range in S, in that a one-to-one d–map divides the d–sort into a range and a domain.

A **partitioning** is an ensemble of d-maps on a d-sort S such that no element is in the domain or range of any other d-map. Thus, a partitioning of the d-set S is a selection of disjoint d-subsets from the ensemble of d-subsets of S, which are disjoint union S (their union is equivalent to S). Note that a partitioning provides a natural means of "dividing" a d-set into parts, each d-sort distinguishable from the other.

We designate by $\{\}$ a d-set S and by $\{|\}$ the bisection of S. For convenience, we label the d-sorts thus created by a bisection of S, L and R. We call the label for $\{\}$ the **identity label**. The bisection $\{L|R\}$ is the bisection of S into d-sorts L and R. We have defined bisection in such a way that it is invertible. Thus, we may speak of the inverse process from time to time and call this **adjoining**. Note that for any d-set of ordinality 2, bisection yields 2 d-sorts of ordinality 1, namely L and R. Thus, adjoining L to R yields a d-set of ordinality 2. Similarly, adjoining L to S or S to R yields a d-set of ordinality 3. Note that adjoining L to itself (or R to itself) yields a d-sort of cardinality 1. We call this recursive adjoining, beginning with a single d-sort, the **von Neumann recursion**.

We leave the details of this recursion to the reader, but note that it differs from the original recursive process defined by von Neumann and the more recent explication by Conway[15], only in that we do not define a cardinal 0, nor do we require the (∞) real number line, since we are merely generating labels. Note that neither 0 nor ∞ are defined for us, since we cannot show how to construct (or even find in Bishop's terminology) either of them constructively and finitely. For us, the infimum is the successor of 0 and ∞ is the successor of the supremum of a recursive enumeration. Clearly, these are not constructable within the context of the specific recursive enumeration nor are they unique for the collection of recursive enumerations in the system, even though they are ordered with respect to its properly generated labels.

By a **number**, we mean a label given to an element x of a d-set via a monotonic recursive enumeration. We define a **primary enumeration** as follows: first establish a unique label to represent identicals (i.e., the identity element); then, a recursive enumeration which labels the initial element with that of the identity element, and which, on recursion, generates labels for which there exists an isomorphism with the recursively generated elements of the von Neumann recursion is called a primary enumeration. A primary enumeration generates the integers up to the cardinality of the d-sort.

We call a ternary relation $+$ **addition** if, operating on a d-set S of arithmetic elements (i.e., *numbers*), for arbitrarily chosen elements x, y, z, there exist elements of S x' and constant element e_1, such that the following relations hold:

$$x + (y + z) = (x + y) + z$$

$$x + e_1 = x$$

$$x + x' = e_1$$

$$x + y = y + x \quad .$$

188

We call a ternary relation × multiplication on a d-set S with addition if, for arbitrarily chosen elements of S, w, x, y and z, there exists elements of S w' and e_2, such that the following relations hold:

$$w \times (x + y) = (w \times x) + (w \times y)$$

$$(w + z) \times x = (w \times x) + (z \times x)$$

$$(w \times z) \times x = w \times (z \times x)$$

$$w \times x = x \times w$$

$$w \times e_2 = w$$

$$w \times w' = e_2$$

Note that addition and multiplication are both intended to be relations manifested by the ordering operators and could be defined much like the reverse-polish notation calculator which has a single (arithmetic label) display and accepts one input (label) at a time. For operators, closure is not defined explicitly. The existence of a unique starting label and a stop rule guarantees that, for some input labels, the operator will simply stop and perhaps generate a special label. This is quite similar to an "overflow" or "underflow" condition in a physical calculator. In practice, arithmetic closure simply guarantees that a calculator is well-behaved and does not suddenly generate a symbol which is not a number. Our operators have this deterministic element built in, and so there is no need of a closure property.

A **reparameterization** is an ordering relation induced on a d-set by an ordering operator, such that, given a label for the element x of the d-set as an argument, the reparameterization gives a label for the successor of x; namely, x'. A reparameterization is not a primary enumeration.

A **segment** [x,y] for x and y in d-set P, is the d-set of all elements z which satisfy $x \leq z \leq y$. A partially ordered d-set is said to be **locally finite** if every segment is finite. Clearly, all partially ordered d-sets within the ordering operator calculus are locally finite.

Enumerations on d-sets may be divided into two classes. **Normal enumerations** are those for which, though not necessarily monotonic, there exists an isomorphism to the von Neumann recursion and which begin with the identity label. **Subclass enumerations** are those for which there exists an isomorphism to the von Neumann recursion, and which (up to redundancy) establish the identity label as the final label* of the d-set.

Theorem 5: A subclass enumeration is a reparameterization.

A **fractional enumeration** is a subclass enumeration in which the labeling of each element is given in comparison to the final label of the d-set. The labeling thus generated follows a recursion relation induced by the inverse algorithm for multiplication. A monotonic subclass enumeration on a d-set of nonfinite cardinality would define the **real number line segment on [0,1]**. Since this is strictly not defineable within the formalism, we define the **discrete real number line segment [0,1]** to be a monotonic subclass enumeration on a d-set S of finite cardinality. We call the cardinality of S the **precision** of the segment. A reparameterization of the discrete real number line segment [0,1], with final label n and initial label m, defines the **discrete real number line segment** $[m, n]$.,[*]

Random versus Arbitrary

As noted in the introduction, where information regarding the construction of a property is not available, we shall be required to deal with the property by probabilistic means. In order to do this, we must introduce a concept of randomness which is constructive and finite. We are now in a position to do so.

Kolmogorov[17] and Chaiten[18] have defined the measure of randomness of a string in terms of the length of its shortest description, an inherent property of individual strings. Namely, if the space complexity of the algorithm is greater than the length of the string it produces, then the string is random. Unfortunately, this definition is not acceptable for three reasons: (a) it allows for infinitely long strings and infinitely complex algorithms, (b) it is nonconstructive (i.e., it does not tell how to construct a random string) and (c) the set of Kolmogorov-random strings is nonrecursive. A number of extensions have been considered, but none give an effective procedure for writing pseudorandom generators.

Suppose that the algorithm for an particular computation O is not known. Select a computational metric. Let the computational cost $C(O)$ of representing the algorithm for O be greater than the **representational resources** n within the finite discrete system S to be constructed. Under certain conditions which we now determine, the algorithm may not be discovered or even constructed within S.

Theorem 6:
An algorithm O with computational cost $C(O)$ is indistinguishable from a "true" random number generator within a discrete, finite system S with representational resources $R(S)$ whenever

$$R(S) < C(O) + log_2 C(O) \tag{1}$$

(where the operations $<$, $+$ and log_2 have their usual meanings). Call O an **arbitrary binary number generator**.

---

[*] Note that if the reparameterization is a one-to-many d-map with range of cardinality greater than the cardinality of the system N, and if monotonic decreasing, we obtain $(m, n]$, and if monotonic increasing, $[m, n)$, and the adjoin of these segments is (m, n). Furthermore, this provides a formal definition of the hierarchical nature of the real numbers. As pointed out elsewhere[16] the equivocation of this hierarchy of classes is the source of a number of apparent paradoxes.

190

Argument:

Consider (1) a system composed of a Universal Turing machine with a finite memory, and (2) a binary number generator G. Such a system is incapable of deciding whether or not the number generator produces repeating binary strings of length n whenever the memory is smaller than an amount m equal to $n + \log_2 n$.

Suppose that the Turing machine takes as input a particular substring of length n output by G, and we wish it to determine whether or not the number generator G is producing this substring repeatedly. Select as a computational metric the computational space cost C_s, without regard to the computational time cost C_t. The Turing machine must consume an amount of memory equal to n in order to store the string; then, the computational space cost C_s for any computation on the substring, including direct comparison with a second input substring, is at least as great as C_s, for a count of the number of symbols n in the substring ($log_2 n$). Thus, $n + log_2 n$ sets a lower bound on the computational space cost $C_s(O)$ for any algorithm which may be selected to make the decision.

It follows that the system cannot decide whether or not the target string has been produced if it has memory less than $n + log_2 n$. But this means that the system cannot distinguish between number generators which produce repeating strings and random numbers. Clearly, the symbols in the repeating strings will occur with equal probability, as required for a random distribution. However, since the system cannot detect that a given string is repeating, it cannot detect that some string of cyclicity n is repeating. Thus, for systems with less than $n + log_2 n$ memory, a generator producing repeating strings of minimal cyclicity n is indistinguishable from a generator producing random numbers.

QED

This theorem means that we may actually construct ordering operators which are "perfect" pseudorandom generators, in our terms more properly called "arbitrariness generators." Thus, ordering operators can be constructed from existing ordering operators, and not all ordering operators need have *a priori* existence. According to the Theorem, such a situation will give rise to a nondeterminism born of computational complexity and representational impoverishment:[*] we cannot predict the output of the ordering operator, because we could not even express the complete algorithm if it were "known."

Given this situation, it is possible to understand the ordering operator foundations as arising from a complex, though finite system of space complexity $n + log_2 n$ greater than the space complexity n of the finite system in which we are working. The complexity of the *a priori* ordering operators is greater than the space complexity n of the known system. Note that this does not introduce an infinite regress, since we need postulate an extension of the known system only once to account for conditions of "randomness" and indeterminacy. The notion of *truly random* can have no meaning within the theory.

---

[*] The variables involved in an ordering operator's algorithm may rightly be called von Neumann hidden variables. This does not mean hidden variables in the usual "quantum mechanical" sense, since we do not have a Hilbert space. It is interesting to note that von Neumann had similar ideas when he referred to systems with "partial knowledge."

d–Spaces

A d–set S' is said to be a **permutation** of the d–set S, if the only difference between them is the partial ordering relation. Consider a d–set S partitioned into n mutually disjoint d–subsets.[†] These d–subsets need not be formed by equipartitioning of S, although this is what will usually be meant. In general, however, we will denote cardinality of each of the n partitions by $m_1, m_2, m_3, \ldots m_r, \ldots m_n$, respectively. For each d–subset r, by definition, there exists an ordering operator which generates m_r distinct labels. Call an ordered d–set of the labels, one from each of the n partitions, an n–**tuple** or d–**point**. Form a d–set R from all the possible ordered n–tuples of labels. We call such a d–set R a d–**space**.

A d–space S on which addition holds for the d–points of S and for which there exist elements (defined via a primary enumeration) between which multiplication holds, is called a **vector d–space**. The d–points of a vector d–space are called d–**vectors**. By a d–**basis**, we mean an ensemble of n totally ordered d–sets.

A d–**curve** on a d–space is a d–set of d–points for which there exists at least one basis d–set which can be mapped 1–1 onto the d–curve, and for which there exists at least one total ordering on the d–basis, which is preserved by the d–map. A **smooth d–map** is defined here to mean that there exists a partial ordering over the domain and a partial ordering over the range of the d–map, such that the nth enumeration in the partially-ordered domain maps to the nth enumeration in the partially-ordered range (i.e., the d–map is isotone but not necessarily 1–1), and for every reparameterization of the domain there exists a reparameterization of the range which is isotone.

The **derivate** of a recursive enumeration $f(n)$ is the number x, where

$$P7: \quad x = \frac{f(n) - f(n+h)}{h} \; .$$

Thus, x is just the divided difference $[n, n+h]$ of $f()$. Because the primary enumeration generates the integers, h is just 1 if $f(n)$ is a primary enumeration. Then x is also the **forward difference**. Indeed, most of the results of the calculus of finite differences may now be taken over intact[19][‡].

A recursive enumeration $f(n)$ (or rule of correspondence or a d–map), is **locally differentiable** over some d–sort S if the corresponding primary enumeration exists; then there exists a number x such that $f(n) + x = f(n+1)$. Although there may exist a recursive enumeration on a d–sort, the primary enumeration need not exist if, for example, the d–sort is only partially ordered. A smooth d–map, which is locally differentiable for all the labels in the domain of the d–map, corresponds to the classical notion of a continuous function.

---

† We could just as well begin with n mutually disjoint d–sets S_n, and form a new d–set S which is the union of the disjoint d–sets; however, this would require care in specifying the ordering operator on which S is defined.

‡ Note that for d–set with sufficiently large cardinality m, one may reparameterize P7 to read $P7'$: $f(n) + x\frac{1}{m} = f(n+\frac{1}{m})$, which reduces to the classical definition (L'Hopital's Rule) in the limit of large m, though without appeal to infinitesimals.

192

The series formed by summing a recursive enumeration $f(n)$ for successive values of n is said to **converge** if the recursive enumeration is monotonic decreasing, and **diverges** if the recursive enumeration is monotonic increasing.

If the domain of a function, with range defined on the discrete real number segment [m,n], depends on the parameterization chosen for the range, then we call it a **real-valued function**; otherwise we call it a **scalar function**.

For recursive functions with multiple arguments, we define the **partial derivate** $f_i'(x)$ as

$$f_i'(x_1, x_2, ... x_n) = f(x_1, ..., x_i + 1, ..., x_n) - f(x_1, ..., x_i, ..., x_n) \quad .$$

With this definition, we recover the inverse function theorem for inverse suitably defined over d–sets. Further, the determinant (computed in the usual manner) will go to the infimum of the d–set, if the x_i are dependent. If the x_i are independent, then we are guaranteed that there exist linearly independent recursive functions

$$f_i(x_i)$$

such that some linear combination of the f_i yield f.

A **chart** is an 2–tuple consisting of a neighborhood N and a d–map from N to some d–space R^N, whose N disjoint d–subsets are each defined on a discrete real number line segment. If it is possible to construct a system of charts in such a way that each d–point of a d–space M is in at least one neighborhood, we call this system an **atlas**. A d–space M with an atlas A is called a d–**manifold**. Each d–map on a manifold M associates with each element of M an n–tuple of the space called the **coordinate** of the element under this d–map. A manifold can therefore also be understood as a d–set of d–points (N–tuples) where, for each d–point of the d–set there exists an open neighborhood which has a smooth one-to-one d–map onto an open d–set of R^N for some N.

A **coordinate** d–**basis** x^i parameterized on the generations t of an ordering operator O^i of a d–space S, is a basis such that the d–sets of the basis have no element (indeed, no d–point) in common (they are mutually disjoint), other than a uniquely and arbitrarily identified d–point called the **origin**.

Each d–space is characterized by a unique number n, which is the maximum number of disjoint d–subsets of S of equal cardinality, such that the union of the d–points formed from these disjoint d–subsets is indistinguishable from the d–space S. This number is called the d–**dimension of S**. For a coordinate d–basis of d–dimension n, we may refer to one of the disjoint d–subsets as a d–coordinate of S.

Having defined a vector d–space, we can assume the usual definition for linear combination, linear independence, maximal linear independent set (d–set), basis, dimension, components, metric functions, inner product, etc. We may also define the usual continuum notions, as long as we adhere strictly to the conditions for d–sets. We can now include eigenvalues and eigenvectors as usual, except that a range of eigenvectors is not allowed. Keep in mind that each recursive function may have multiple arguments. Thus, the d–vector at d–point P of M is not just a real number, except in the case when the number of arguments is one, and even then it may have a sign. It is truly a discrete vector with n components.

For any coordinate system x^i in an open neighborhood of a d–point on a vector d–space, the coordinates define a coordinate d–basis x^i (since there are n linearly independent d–vectors in the tangent d–space, these being the vectors formed from first derivates at the d–point of the underlying d–space vectors). Good coordinates are those where the x^i are linearly independent—this is just the condition on them to provide a 1–1 d–map to some neighborhood of the d–point in M onto a region in R^N.

Notice that we have nowhere restricted the definitions of the elements of a given d–sort or d–set: the structure is always extrinsic, thus supplying a local topology. This also means that we can use our constructs as fundamental elements in the definitions we have just given, thus generating a further layer of recursion. In this way, we will be able to define hierarchical structures.

An **inner product n()** is a recursive function on a vector d–space V, which satisfies the following if x is a d–vector of V and a is any (discrete) real number, and $|a|$ is the value independent of sign:

P8: $n(ax) = |a| \times n(x)$.

An inner product n is a **distance function** or *norm* in a vector d–space V, if it satisfies the following, where x and y are vectors of V and 0 is the inf of the d–set of all such vectors:

P9: $n(x) \geq 0$ and $n(x) = 0$ iff $x = 0$.

P10: $n(x + y) \leq n(x) + n(y)$.

Note that the relations of $+$ and $\times$ need not be the usual addition and multiplication. For example, $\times$ can be multiplication modulo 2.

By treating d–sets in R^N, we are in no difficulty, as long as we remember that defining the members of the image d–set must be recursive. Our recursion serves to maintain the class of elements in the d–set (insuring the existence of what is usually called, and which we will continue to refer to as, an **equivalence class**). In practice, one may use the standard notation and properties of continuum mathematics as a kind of shorthand (effectively making a Dedekind cut to obtain the real number line segment). If this is done, we must remind ourselves that in so doing, we have changed class membership (e.g., 1/2 is not in the same class as 0 and 1): we have effected a reparameterization. Since our definition of d–set is dependent upon the ordering operator which generates it, this means we must reconstruct any relationships between the d–set and any other d–sets.

In a sense, the d–set of d–points S in R^N is the union f of the image d–sets for all classes in the domain. Thus, we may define a distance function on S. We may also define a distance function on a d–set (a non–Hausdorff space), but it will be "multivalued" in the sense that the ordering between elements need not produce a single chain; thus, there may be more than one "path" between elements. If we take the distance function such that the number of elements traversed is minimal, then at best we must assume that elements in the string defining the distance (i.e., the minimal simple chain between any two elements) may in fact be twins under the equivalence class defined by the distance function.[*]

A **bilinear and symmetric** inner product n satisfies P11:

P11: $n(x + y)^2 + n(x - y)^2 = 2[n(x)]^2 + 2[n(y)]^2$.

If we say that a curve passes through a d-point P of a manifold M, it follows from the definitions that M is a recursively enumerable d-set and the curve is then a monotonic recursive function f on M. Thus, we say that the d-points of M (or objects of M) form an equivalence class ordered under f.

A derivate of f at P is then the **motion** along f at P (how fast is the ordering parameter increasing and in what direction $+$ or $-$). Furthermore, for monotonic f, there are $n!$ distinct orderings of M without redundancy, if M has n objects and a sufficiently large number of reparameterizations for a given f (i.e., adding $1, 2, \ldots n$ to f).

If a is an element of a d-set A and b is an element of d-set B, then a and b are comparable if and only if, for some ordering operator O, $a \leq b$ or $b \leq a$.

We say that a **covers** b when the segment $[a, b]$ has two elements.

Theorem 7: If M consists of objects which are not in the same equivalence class, then we order them such that there exists an f for each class in M; then we may cover M by a suitable choice of

$$x^i(P_i)$$

with $f(x_1, \ldots x_i, \ldots x_m)$ for m classes in M. Thus, we can establish a basis for M.

There are, of course, many such bases. A coordinate basis is, then, one for which M is covered and the f_i do not order the same P_i (i.e., they are maximal and (linearly) independent).

A d-**fiber** consists of the d-set of all the derivates for all the possible parameterizations at a d-point of the base manifold M. A **projection** d-**map** assigns each d-fiber to a d-point of M.

A **product space** $M \otimes N$ consists of all ordered pairs (a, b), with a in M and b in N.

A **vector field** is a rule which chooses precisely one tangent d-vector from the tangent d-space at each d-point and assigns this to the point. For every vector field, there exists a curve, just as every curve has a tangent vector at each point. A d-set of curves which cover a manifold is called a **congruence**.

It will be useful to step outside the theory from time to time in order to understand the relationship between the ordering operator calculus and other mathematical endeavors. Certain computational phenomena in the practice of standard mathematics as applied to laboratory physics may be accounted for in this way. For this purpose, we introduce two special terms. If a recursive function were to be defined on

---

[*] Clearly, we do not care if the d-maps are strictly recursive; they may be analytic in the usual sense, as long as we keep in mind the constraints on the space.

an infinite set, then it would be said to be **analytic**. The **analytic interpolation** of a recursive function defined on a d–set over the segment $[m, n]$, is just an analytic recursive function for which there exists a d–map between the reparameterization of some discrete real number line segment $[m, n]$ of the recursive function and some monotonic sequence belonging to the infinite set generated by the analytic recursive function.

Theorem 8:   A monotonic recursive function on a finite d–set of cardinality n has at most $n - 1$ derivates.

From time-to-time, we may say that some aspect of our construction is **global**, by which we mean that it is characteristic of or applicable to the entire d–space. Similarly, we may say that some aspect of our construction is **local**, if it is characteristic of or applicable only to some proper d–subspace.

d-Vector Functions

A **one-form** is a recursive function which generates a (discrete) real number for each d–vector on M and follows the usual linearity. The formation of this number is called the **contraction** of the one-form on the d–vector. A **metric** is a linear, symmetric function of two d–vectors (the "dot" product).

The recursive enumeration for a general d–set provides a parameterization for recursive functions defined on the d–set. Clearly, the parameter takes values from 1 to n over a d–subset of cardinality n. Note that the function deals, in general, not with cardinality but with ordinality, and this is arbitrary under the permutation group. The input and output are only symbols. Consider finite d–sets only. Interpretation as having cardinality n induces (via the function) an ordering on the d–set; thus, some structure is supplied. The function is intrinsically a mechanics of typography—how we can combine and use symbols is a recursive function.

In general, the notion of a recursive enumeration of a d–subset of the recursive d–set goes over to a parameterization under a d–map: $J \to R^1$. The parameterizations must cover the d–set. If they do so independently without repetition, then we have a **coordinate parameterization**.

Theorem 9:   Exterior differentiation (defined as usual) commutes with any differentiable mapping of the manifold.

It is interesting to note that the cohomology groups depend only on the topological structure of M, and *not* on its differentiability. That is to say, cohomology theory passes over from standard differential geometry to the present theory almost intact, as there is no dependence on the definition of differentiability. This is particularly important for applications in physics, where Gauss' and Stokes' theorems are of such great use.

A Note on Computing Numeric Roots

It will often be the case that d-functions are needed, which make use of various roots, such as the square root. For us, not all numbers have a "square" root, meaning two **equivalent** roots, and similar comments hold regarding higher-ordered roots, such as the "cube" root. Where such references are made in the remainder of this paper, we refer to the so-called "symmetric" root. Symmetric roots are defined as being a rational root of the number plus or minus some other rational number. Thus, in general, any square may be expressed as

$$(a - \epsilon) \times (a + \epsilon) = a^2 - \epsilon^2$$

such that ϵ is a rational fraction up to the precision of the computation. Note that this definition literally inverts the Pythagorean construction of the irrationals, but in a manner which requires no irrationals. This is, of course, just the operational definition which is taken in finite computation such as that using Newton's Method. One performs a recursive computation until the error (our ϵ) is sufficiently small. The nonequivalence of the two *root*'s in the ordering operator calculus holds a special significance: it suggests that the d–space is intrinsically noncommutative and that a commutative d–space is meaningful only if constructed of perfect squares, perfect cubes, etc.

2.5 CONSTRUCTED OBJECTS: PARTITION LATTICES

In this section, we provide the concepts necessary to make the appropriate connections to generating functions and the finite operator calculus, as well as incidence algebras and von Neumann's theory of games. These concepts will prove useful when we begin the process of interpretation of physical phenomena.

An **order ideal** in a partially ordered d–set P is a d–subset Z of P, which has the property that if x is an element of Z and $y \leq x$, then y is an element of Z.

The **product** $P \otimes Q$ of two ordered d–sets is (p, q), where p is an element of P and q is an element of Q endowed with order $(p, q) \geq (r, s)$ whenever $p \geq r$ and $q \geq s$. The **direct sum** or *disjoint union* $P \oplus Q$ consists of elements x and y with order $x \leq y$ if and only if

(i) x, y are elements of P and $x \leq y$ in P

or

(ii) x, y are elements of Q and $x \leq y$ in Q.

The **blocks** of a partition of a d–set S are the d–subsets of S making up the partition. A partition $\prod$ is a **refinement** of a partition $\sum$ if every block of $\prod$ is contained in a block of $\sum$. The inf or 0 of $\prod(S)$ is the partition whose blocks are the one element subsets of S, and the sup or 1 of $\prod(S)$ is the partition with one block. The **lattice of partitions** $\prod(S)$ of a d–set S is the d–set of partitions of S ordered by refinement.

Note that there is a natural correspondence between equivalence relations on a d–set S and partitions of S, since the equivalence classes of an equivalence relation form the blocks of a partition and, hence, there is an induced lattice structure on the family of equivalence relations of S.

2.6 CONSTRUCTED OBJECTS: COMBINATORIAL SYSTEMS

A **combinatorial system** consists of a unique initial label or *word* called the **axiom** of the system, and a finite d-set of strings called the **productions** of the system. Productions are the recursively generated words of a combinatorial system. The **alphabet** of the system is all the **symbols** or **letters** that occur in the axiom or productions of the system. A word of the system contains only the alphabet of the system.

Theorem 10:   For every combinatorial system there exists a combinatorial system with precisely a two letter alphabet, whose decision problem is recursively solvable if and only if that for the first system is also recursively solvable.

Theorem 11:   The d-set of integers generated by a combinatorial system is recursively enumerable.

Theorem 12:   If the d-set of integers generated by the combinatorial system is not recursive, the decision problem for the combinatorial system is unsolvable.

Any n-form field divides all d-vector basis into two classes: those for which it is, on contraction, $+$ and those $-$. This is called right- and left-handedness, respectively. If it is possible to be consistent in specifying handedness at each (not continuously here, since M is discrete) d-point P of the manifold, then M is said to be **orientable**. For every orientable manifold M, there exists an inverse to the derivate function called the **integral**.

3. MATHEMATICAL FOUNDATIONS II: ATTRIBUTE SPACE

By a **combinatorial attribute**, we mean a property of a d-sort that has been constructed by an ordering operator. In particular, for binary sequences or ensembles labeled by the generations of i ordering operators O_i, an attribute is any property which is recursively definable or computable on the ensembles of labels generated by the ordering operators. Let the sequence of labels output by each of the O_i be represented by a directed graph G_i (this is a Hasse diagram if there exist no cycles in the ordered labels produced by the ordering operator), consisting of labels as nodes and connecting arcs to represent the pairwise orderings between labels. Call the graph g_i which results from G_i by the removal of any number of arcs and/or nodes, a reduction of G_i. Consider the i collections of reductions R_i for each of the G_i. Form a new collection of reductions, consisting of no more than one reduction from each from of the R_i. If there exists an isomorphism between the all the pairs of reductions in such a collection, the reduction represents an attribute of the collection of ordering operators O_i.

Let the i ensembles of labels generated by the O_i be operated on by a new ordering operator O^*, so that the ensembles are (partially) ordered and labeled. Each generation of O^* can be classified according to whether or not there is a reduction corresponding to the underlying ensemble, which represents an instance of a particular attribute. We will call the instances of an attribute the **attribute states** of the attribute over the ordering operator O^*.

Since the output of any ordering operator O for each generation may be arbitrarily complex and have considerable internal structure (inaccessible unless constructed in the manner above), we will, henceforth, drop the notational distinction between O and O^*. The reader should, however, keep in mind the considerable structure which is implied when we refer to a combinatorial attribute or an attribute state.

A combinatorial attribute (or simply attribute where no confusion will result from the usage) is conceptually akin to a set-theoretic property, although mathematically distinct. First, attributes are constructive, whereas set-theoretic properties are not, being generally of an *a priori* nature and giving set theory that "tacked-on" look. Second, they are not "properties" of a set, but rather of a d–sort which has been constructed with an ordering operator. The definition of an attribute is thus much stronger than the definition of set-theoretic property, in that an attribute would certainly be a property but all set-theoretic properties are not attributes. Clearly, an ensemble has attributes as a set has properties, if one remembers that this similarity is metaphorical rather than mathematical.

As an example, consider the generation of the permutations of a discrete, finite, ordered collection S (noting that such a collection is a special case of a d–sort; in fact, it is a d–set). If the generator is specified via a recursive algorithm, has a unique starting ordering of the set and halts after generating all possible permutations generated in a specific order, the generator is a special (and useful) type of ordering operator. The notion of permutation so defined and used is an attribute of the ordered set P of all permutations of the original ordered collection S (i.e., it is possible to recursively define the permutations of a particular finite ordered collection and to recursively give their complete denotation), and any specific permutation P_i of the ordered collection S is an attribute state. Thus, a permutation is an attribute with respect to a reference ensemble (the starting ordering of S) and the ordering operator which generates permutations. Similarly, any specific subensemble is an attribute, with respect to the subensemble and the **identity ordering operator**—the ordering operator which, given a label as input, returns it as output. In this sense, we may search a d–set for d–subsets which are equivalent; i.e., those which have the same (identity) attribute.)

3.1 MULTIPLE ORDERING OPERATORS

Please note that more than one attribute (indeed, more than one ordering operator which generates permutations) may be defined on the ordered set. This is an essential characteristic of d–sorts which must not be overlooked. For example, given an ordered d–set of labels of cardinality N, there are $N!$ additional distinct permutations possible. There are then $(N! - 1)!$ possible ways in which the d–set of all permutations can be generated and, therefore, $(N! - 1)!$ permutation attributes definable starting from the ordered d–set of labels.

In any given construction, we must explicitly state what ordering operators generate the structure, as these provide the connectivity of the elements. We may then construct the ways in which two or more operators combine or interact with each other. Suppose a d–sort of labels L of cardinality N are independently generated by i ordering operators O_i. In order to treat the d–sorts generated by O_i as a single construction, it must be possible to demonstrate the constructive existence of a total ordering operator O' such that O' generates L. O' is said to be **decomposable** into the O_i if, to each generation of O', there corresponds one and only one O_i which generates the corresponding label. This is the first instance of the label being produced by this O_i, and all other O_i (except the O_i that generated the previous label of O' on its previous generation) generate the same label again. The O_i are said to be **serializable**.

Two or more ordering operators are said to be **intrinsically coupled**, if they generate at least one label or attribute state which is mutually indistinguishable (i.e., if the first operator cannot distinguish some label, (called the **coupling label**), generated by the second operator from a label which it generates and vice versa). Ordering operators which are serializable are **locally orthogonal**, since they are not intrinsically coupled. This does not mean that they are **globally orthogonal**, since they may be **extrinsically coupled** via a third ordering operator with which they are both intrinsically coupled.

The coupling of two ordering operators which are extrinsically coupled via a third ordering operator is said to be of **coupling degree** one; if via a third and a fourth, such that the output of the third is input to the fourth and the output of the fourth is indistinguishable from one of the labels of the first two, then the coupling is said to be of **coupling degree** two; the number of intervening ordering operators gives the degree of the coupling. The number of coupling labels gives the **coupling order**. Note that coupling is dependent on the specific ordering operators involved; two ordering operators may be coupled in multiple ways.

The coupling of two ordering operators O_1 and O_2 is characterized by a unique rational fraction called the **scale**, which is just the ratio of the cardinality N_1 of the labels, which may be produced by O_1 when coupled to O_2 to the cardinality N_2 of the labels, which may be produced by O_2 when coupled to O_1. The degree, order and scale, and the cardinalities and ordinalities of the ordering operators give all the information necessary to compute the probability (frequency) of one ordering operator interacting or **mixing** with another ordering operator. In a system of coupled ordering operators, the labels output by two ordering operators will be said to **superpose**.

We may relax the "set" restriction in our example: if a partial ordering is generated by an ordering operator, the collection (ensemble) is a d–sort with respect to that operator. In other words, distinguishability is meaningful only in terms of the ways (i.e., ordering operators) one has specified how to generate the ensemble. If the generator treats the order of "two" permutation states indifferently, then they are indistinguishable for that ordering operator, and we have no other means of determining distinguishability.*

---

* A further example may be helpful. In the d–space which is the positions on a chess board, the sequence of moves which any given chess piece takes during a game determines an ordering operator. The form or rule that specifies the legal moves that a piece can make specifies an attribute. If the ordering operator is parameterized on the attribute which defines a piece's legal moves, then each such legal position, generated in allowed sequence, is an attribute state.

200

It will be necessary to form quite complex attributes: attributes, then attributes of attributes, then attributes of attributes of attributes, etc. We will refer to these as **attributes of first order, second order, third order,** etc., respectively. From the definition of combinatorial attribute given above, it can be seen that the construction of attributes is potentially recursive. One can form a collection of ordering operators O_i^*, each of which generate the attribute states of an attribute; perhaps being distinct (although they need not be) in the order of generation of the attribute states. From the collection of outputs of the O_i^*, one may construct a new attribute and define a new ordering operator O^{**}, which generates the attribute states of this attribute of second order. This method of constructing higher-ordered attributes may be recursively continued, up to the point at which the only reduction possible is the graph consisting of a single node.[†]

3.2 ESTABLISHING A DISTANCE FUNCTION

Based upon the definitions of ordering operator, dimension, and coordinate d–basis, a one–dimensional d–space coordinate basis behaves as a totally ordered d–set. It is convenient to represent this d–set by a sequence of binary strings; i.e., a string in an alphabet two symbols, where the order of the symbols is dictated by the ordering operator. For example, given a string composed of n unique labels, one may use Huffman encoding[21] as a way of unambiguously giving a binary representation of the sequence of labels.

Now, define **attribute distance** for a specific attribute generated by an ordering operator O_i, as the measure dependent solely upon the number of distinguishable states s between two ensembles of labels which O_i may generate, normalized by the total number of states which O_i may generate, N_i. This is equivalent to the unique number of generations of O_i required to generate the first ensemble A from the second (called the *reference ensemble*) B, and results in a **distance function d()** on the closed interval of rational fractions $[0,1]$:

$$d(A, B) = s/N$$

By a reparameterization on an attribute (from the definition of reparameterization), we mean a mapping of the labels for the attribute states generated by some ordering operator O, to the tags for the attribute states generated by a second ordering operator O'. Via a reparameterization, then, we may remap $d()$ into $d'()$, defined on the closed interval of rational fractions $[-1,1]$. Since the cardinality of O may be

---

[†] Note that the mathematical objects of our construction are each defined relative to one or more operations. This forces an intrinsic connection between the usual static form of mathematics and the dynamics found in physics. It also precludes arbitrary identification of constructed entities. If our analysis of paradox[20] is complete, then many (we hope all) of the paradoxes which arise in logic, set theory and philosophy are not possible here.

smaller than that of O', indistinguishable attribute states for O may be induced in the mapping in order to properly map all the tags of $O'^{\ddagger\S}$ as the "maximum number [N—added for clarification] of distinguishable orientations between" two measured attribute values divided by the square root of N for normalization. This is just a measure of the distinguishability of two measurements, based on the number of values which an attribute being measured can take. Clearly, if the statistical distance between two values is zero, they are indistinguishable from an information theoretic point of view.

Wootters presents strong evidence that "statistical distance equals actual physical distance." The specific relationship is derived for the case of photon polarization measurements. We will demonstrate that this relationship is even more general than (apparently) assumed by Wootters and that the relationship serves as the basis for an extension of relativity and explains much in quantum mechanics.)

Suppose that we reparameterize $d(A, B)$. Represent a generation of O_i which decreases $d(A, B)$ as a 0 and one which increases $d(A, B)$ as a 1. The total number of 1's is simply the Hamming distance, and is defined on the interval $[0, N]$. By subtracting the number of 0's and then dividing by N, the result is a Hamming measure on the interval $[-1,1]$ (i.e., centered on 0, which has an ordinal but not a cardinal significance), and is independent of the number of generations of O_i. In general, we will find this to be a more useful form of the attribute distance function.

3.3 SYNCHRONIZATION

In order to perform operations on multiple ensembles with some ordering operator O, some means must be given for establishing a label in each as the common input to O. If the ensembles are not identical, then a d–map between the ensembles will be useful. In this section, we define a particularly useful mechanism for achieving such a mapping.

Pick three ensembles, A, B and C. Let the attribute distance between A and C be zero, but with $t_C > t_A$ by some ordering operator O local to A and C, with generations parameterized by t. Let there be an ordering operator O'' local to B. Furthermore, let the attribute distance between A and B be nonzero. We say that the ordering

---

‡ In the chess example, the attribute distance in terms of the ordering operator which generates a specific piece's moves is just the number of moves that the piece has made, divided by the total number of moves it will make in the game. Then the distance is always some rational fraction of the total distance the piece will travel in the context of the game. Note, however, that having all the sequences of moves for all the pieces in a game does not allow us to reconstruct the game; we must know how the moves are interleaved. This can be accomplished by specifying the ordering operator in terms of the game clock; that is, for each move of the game, each piece's ordering operator must generate some attribute state. For us, whenever a piece does not move, the attribute state generated is indistinguishable from the previous attribute state. Reparameterizing the ordering operators in this way normalizes all the attribute distances, thus providing a global distance function topology on the d–space of the chess board.

§ Note the similarity between the notion of attribute distance and statistical (read with the frequency interpretation) distance, as defined by Wootters[22].

202

operators O and O'', with generations parameterized by t and t'', respectively, are **synchronized** if condition (1) holds, and A and B are said to be **synchronous** if the O and O'' are synchronized and conditions (2) and (3) hold:

1) $t_B - t_A = t''_C - t_B$;

2) if A is synchronous with B, then B is synchronous with A;

3) if A is synchronous with B and B is synchronous with C, then A is synchronous with C.

In other words, (1) states that the ordering operators are synchronized if there exists a binary symmetric relation between t and t'' over the specified attribute. By reason of the nonuniqueness principle (Principle IV), the property of synchrony between *ensembles* must also be reflexive and transitive as in (2) and (3), respectively. This simply means that it is possible to define a new distance function defined by T across the ensembles, which is consistent with the distance functions defined by t and t''.

Henceforth, we drop the notational difference between synchronous t and t'', since these may be replaced by a single universal ordering operator with parameter T.

3.4 THE DIMENSIONALITY OF D–SPACES

We are now in a position to construct a unique global property of d–spaces which have a distance function that is coordinate independent. We will begin by examining how such a coordinate independent distance function can be established, using the concepts we have defined and constructed. We will then investigate a global property of the resulting d–space.

Under the condition that the cardinality of the d–space or d–subspace precludes explicit representation of the algorithm for an ordering operator O, we are clearly faced with a severe lack of knowledge, the algorithm for O is arbitrary and the output may, therefore, may be treated as random for our purposes (Theorem 6).

Consider the labels produced by O to be represented by bit strings [i.e., strings of 1's and 0's] or, if repetition is being allowed and only two labels are allowed, to be arbitrarily treated as 1 and 0. The sequences of 1's and 0's thus produced meet the conditions of Bernoulli trials (see Figure 1): being unable to specify the algorithm used by O forces us to see successive productions of O as independent. Only after the fact, may we label the resulting output of O as representing some specific, previously known, ordering relation.

Now, let there be r such arbitrary binary number generators, $O_1, O_2, \ldots O_r$, with string (d–set) outputs $S_1(n), S_2(n), \ldots S_r(n)$ up to some maximum, R. We will refer to n as the ordinality of each string, parameterized by a counter we will call t. In the absence of other information about $O_1, O_2, \ldots O_r$, we assume them to be independent operators and, indeed, cannot discover otherwise, due to the computational complexity of the operators and our relatively impoverished (representationally) system of

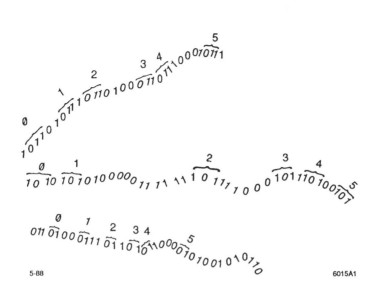

5-88   6015A1

Figure 1   Independent trials.

just two symbols.* Clearly, $S_1, S_2, \ldots S_r$ are synchronized on n via t, but this provides only trivial structural information about any relationships between the strings, given our criteria for "arbitrary" and the independence of $O_1, O_2, \ldots O_r$. $S_1, S_2, \ldots$ S_r constitute a coordinate d–basis, as long as we identify the initial outputs of $O_1, O_2, \ldots O_r$ as "identical."

Look for other means of synchronizing $S_1, S_2, \ldots S_r$; pick as an attribute any sequence of length M of binary symbols (i.e., a specific substring of $S_1, S_2, \ldots S_r$); interpret the first M symbols of $S_1, S_2, \ldots S_r$ as matching this substring; then examine the output of $O_1, O_2, \ldots O_r$ for further synchronized productions of this string. As long as the algorithm governing the ordering operator which generates the strings is of sufficiently great computational complexity, the occurrence of the substrings is arbitrary and simple statistics for concurrent, independent Bernoulli trials apply.

At $L = Mn$, (i.e., the position of the $n + 1^{th}$ string of length M), the probability that the number of occurrences of the substring is the same across multiple strings produced in this way is just

$$u_L = \frac{1}{2^{rL}} \sum_{k=0}^{L} \binom{L}{k}^r \tag{2}$$

where $r \in \{1, 2, \ldots R\}$.

⋆ Proving nonindependence would require a knowledge of the ordering operator's algorithm.

Now, interpret the normalized number of occurrences of the substring as defining a discrete distance function across the R–dimensional d–space. That the normalized number of occurrences of the substring constitute an attribute distance, and, therefore, a distance function, is trivial; the identity attribute for the given substring may be satisfied (the substring may occur) only finitely many times in a string of finite length. This defines the number of possible instances of the attribute. Take as a reference ensemble, the null string (the string of zero length, i.e., the empty string). The attribute distance follows from the definition, immediately.

Call a specific generation of an r–dimensional d–space with any distance function f, a **reference frame**. We see that Eq. (2) is just the probability that the distance function has a r–independent value, given a value of R and L (i.e., the distance function is length preserving).

Theorem 13:   The upper bound on the global d–dimensionality of a d–space of cardinality N with a discrete, finite and homogeneous distance function is 3 for sufficiently large N.

Argument:

Note that for $R > 3$, the terms of Eq. (2) are monotonic decreasing (i.e., Eq. (2) converges). That is, for sufficiently long strings, the probability of another synchronized occurrence of the specified substring must approach zero. For $R \leq 3$, however, the terms of Eq. (2) are typically monotonic increasing [i.e., Eq. (2) diverges]—there is always the probability of another occurrence of the specified substring. Hence, the possibility of an isotropic (Principle IV) distance function across more than three dimensions is unlikely, while it is certain if $R \leq 3$[23]. Clearly, the case of $R = 3$ contains the greatest representational power.*

Formal Argument:

Two or more attributes are mutually independent if the generating or defining ordering operators on the space of ensembles are mutually disjoint, with the exception of a single element; in which case, they define the dimensions of the space. An attribute admits a distance function if and only if there exists a total ordering on the ensembles which possess the attribute. Two or more (R) such distance functions are symmetric (do not introduce inhomogeneities into the space—Principle IV), if and only if they can be synchronized (this is equivalent to demanding that there is exist an R–way matching criteria between the productions of the R ordering operators,

---

* The concentration of synchronizable events for "short" attribute distances, even with $R \gg 3$ will be related to the big bang and to quantum fluctuations.

such that a match is found arbitrarily often in sufficiently long productions).[†]

However, this is not possible if $R > 3$. Let the productions of the ordering operators be mapped to binary strings. These strings may be treated as the results of Bernoulli trials. The probability of a specific occurrence at the nth trial is given by

$$u(n) = \frac{1}{2^{rn}} \left[\binom{n}{0}^r + \binom{n}{1}^r + \binom{n}{2}^r + \cdots + \binom{n}{n}^r \right] \tag{3}$$

for n trials, with $r = R$ equal to the number of dimensional metrics. The maximal term of the binomial distribution

$$\binom{n}{k} 2^{-n} \tag{4}$$

is of the order $\sqrt{2/\pi n)}$ and $< n^{-1/2}$. Therefore,

$$u(n) < n^{-(r-1)/2} 2^{-n} \left[\binom{n}{0} + \binom{n}{1} + \cdots + \binom{n}{n} \right] = n^{-(r-1)/2} \tag{5}$$

and so the sum of $u(n)$ converges for $R > 3$.

That it diverges otherwise is proved as follows:

Case 1: For $R = 2$, and from the normal approximation to the binomial distribution

$$u(n) = \binom{2n}{n} 2^{-2n} \simeq \frac{1}{(n\pi)^{1/2}}$$

and so $\sum u(n)$ diverges for $R = 2$. Note, however, that $u(n) \to 0$ as $n \to$ large N. Therefore, while synchronization is certain, it has a mean recurrence time on the order of $\sqrt{N}$, so that, in two dimensions, the synchronization is sparse.

---

† More importantly, note that these conditions are identical to those demanded by Einstein in deriving the Lorentz transformations. Namely, the demand that clocks be synchronizable is equivalent to demanding spatial homogeneity (i.e., that there is no preferred coordinate). In addition, the property of transitivity (from the definition of synchronization) implies that there exists, at least mathematically, a "universal" clock. This is ironic, in as much as special relativity is usually understood to have removed the Newtonian concept of Universal Time. In fact, Einstein did not remove the concept, but rather showed that this global time need not be accessible, as long as synchronization with transitivity was allowed. Under these conditions, local time is sufficient.

206

Case 2: For $R = 3$, and from the normal approximation to the binomial distribution, for sufficiently large n and

$$\frac{1}{2} n - n^{1/2} \leq k \leq \frac{1}{2} n + n^{1/2} \; ,$$

we have

$$\binom{n}{k} 2^{-n} > cn^{-1/2} \; ,$$

where c is a (small) positive constant. Therefore,

$$u(n) > 2n^{1/2} \left(c^3 n^{-3/2} \right) = \frac{2c^3}{n} \; ,$$

and so $\sum u(n)$ diverges for $R = 3$.

Thus, as a recurrent event, any given sequence will be shared between more than three runs only a finite number of times, and, hence, is unlikely; whereas, between three or fewer, the same sequence will be shared in position arbitrarily often for sufficiently large strings.

It follows immediately, that we cannot define a metric-homogeneous discrete space with more than three spatial dimensions. Any other space must introduce either asymmetries or inhomogeneities over the metric.

QED

It might be argued[24] that in a d–space of finite cardinality N, the theorem no longer applies. However, consider what happens if the global distance function is defined with $R = 4$; then there exist local distance functions defined on the three–dimensional d–subspaces.

Suppose that some relationship is to be defined between the local distance function and the global distance function. For large, finite N, this becomes impossible. A comparison of $u(n)$ for each distance function shows that, as N becomes large, "meter marks" for the three–dimensional d–subspace become relatively more frequent, whereas those for the four–dimensional space become less frequent. Thus, the d–map becomes impossible, unless the one–dimensional d–subspace grows more rapidly than the three–dimensional d–subspace; i.e., unless one dimension is different from the remaining three. However, by hypothesis and in keeping with Principle IV, this is not possible, since it makes the four–dimensional d–space inhomogenous.

Furthermore, there would then be a three–dimensional d–subspace, composed of the one–dimensional d–subspace and any two other dimensions, which would generate as rapidly as the four–dimensional d–space, and the difficulty of defining a relationship between the distance function on this d–subspace and the global distance function would be undiminished. Notice that this difficulty becomes apparent for relatively small runs (as soon as $n^{1/2}$ is significant), since the ratio of expectations for synchronization between a d–space and its largest d–subspace is bounded by $n^{-1/2}$.

Constructing a Coordinate System

It is important to understand how one constructs a coordinate system using Theorem 13 and the definitions that preceded it. We make explicit use of the notion of independence in order to construct an orthogonal basis, since independence is the essential constructive notion underlying orthogonality when a geometry (i.e., some notion of "angle") does not, as yet, exist. Having taken this step, we are then *required* to construct a norm which vanishes when the two arguments are orthogonal. This is, of course, trivial if the usual operations of addition and multiplication are available; but care must be taken, since we deny the need for the usual properties of closure and commutativity.

Having once identified an attribute *independently* in each of three binary strings generated as in the discussion preceding Theorem 13, computed the distance from an *arbitrarily identified* origin using the appropriate one of three distance functions (each need only be defined on one of the strings), and, finally, established synchronization across the three strings, the only quantities of interest in performing d–vector computations in this three–dimensional d–space are the "meter marks" established by synchronization. This synchronization establishes a new distance function uniquely defined in the ordering operator sense, which is independent of which of the three strings are involved in the computation.

Thus, if we now treat the three strings as generating a coordinate d–basis x, y and z, we may say that a d–vector of a certain "magnitude" has a particular "direction." In the simplest case, the direction is either "parallel" or "antiparallel" to x, y or z. In such a case, the norm which we use must give the magnitude of the d–vector, when the arguments to the norm are the d–vector and the appropriate unit d–vector in the "same" direction and the infimum of the distance function, if either of the other two unit d–vectors are used. This proscription on the construction of a norm results in a unique norm only in that all such norms will be "orthogonality" preserving.

Similar comments hold with regard to the construction of a "vector product." Great care must be taken not to assume any intrinsic notion of direction and connectivity of the d–space, such as that which is often imposed by the Pythagorean theorem (which is valid only in a Euclidean or flat–space and generally not valid in a discrete, finite space). Furthermore, great confusion and apparent contradictions result if one insists on using the distance function defined, in order to construct "meter marks" on a particular string as though it were global (i.e., useful for all three strings or identifiable with the distance functions defined by the process of synchronization).

Note that if the d–vectors are represented by binary strings, it is necessary that the independent attributes be represented consistently; thus, the attributes must be independent under the operation of discrimination (exclusive or). If the d–vectors in a d–space are represented in a manner consistent with "meter marks," we then have a means of forming the vector product. In the canonical form, the independent attribute substrings for a three dimensional d–space are just "001" (x), "010" (y) and "100" (z). These substrings form a complete representation and are independent under the operation of discrimination.

A *d*–vector **represented** by the binary string "001011110," then has an attribute distance in the x direction of 2, in the y direction of 2 and in the z direction of 1. Such a representation gives more information than just the direction and magnitudes—it contains a history of the generation of the *d*–vector. This explicit representation of the process nature of mathematical objects is an important characteristic of the ordering operator calculus. In order to use the usual notions of a vector space, including computation of components, this historical information must be obscured. Thus, one only considers the magnitudes and the directions, without the explicit binary representation of the *d*–vectors.

3.5 CHARACTERISTICS OF A DISCRETE GEOMETRY

Having developed a *d*–space with a coordinate independent distance function, we may now explore certain other symmetry relations on the *d*–space. In particular, we will find it useful to understand the *d*–space equivalents of the familiar orthogonal and rotational symmetries.

It is a central point of this section that a measure of the discrete cardinality N and of the curvature of a discrete geometry in a *d*–space is given by the precision with which two ratios are identical in value: the ratio of the area of the maximally-sided symmetric polygon, which may be constructed in the *d*–space to the area of a square $(\pi(N)_{areas})$ in the *d*–space, and the ratio of the perimeter of that same polygon to the perimeter of the square $(\pi(N)_{perimeters})$; see Figure 2.

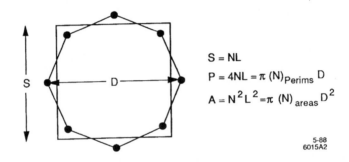

$$S = NL$$
$$P = 4NL = \pi (N)_{Perims} D$$
$$A = N^2 L^2 = \pi (N)_{areas} D^2$$

5-88
6015A2

Figure 2   Relation between $\pi(N)_{areas}$ *and* $\pi(N)_{parameters}$.

Indeed, the relationship between these values has global significance, and we shall have need of understanding that significance in later sections, as well as being able to explicitly use one or other of the ratios thus constructed.

In what follows we construct a square and a circle, and construct an algorithm for a rational fraction ratio which plays the role of π. We begin by constructing the equivalent of a square: an orthogonal, two–dimensional coordinate patch. The only elements allowed for construction are a finite (perhaps large) number of discrete elements (essentially indistinguishable mathematical objects), ordering operators, the ability to count and the ability to label the objects through an operator.

By **nearest** n **neighbor** of a label e in a sequence of **generations of an ordering operator** O, is meant any label n such that for labels a, b, e and n, $a : a = O(e)$ (read "label a such that a is generated on input of label e to ordering operator O') or $b : e = O(b)$; then for any ordering operator O' mutually disjoint from an ordering operator O, at least one of n=O'(e), n=O'(a), n=O'(b), e=O'(n), a=O'(n) or b=O'(n) holds. Clearly, a and b are nearest neighbors of e, as well. The process of identifying nearest neighbors simply defines a new binary ordering relation between a pair n and e, if there exists a third label a for which (possibly distinct) binary ordering relations exist between a, e and n.

Before proceeding with the constructions, a comment on notation: the ordering operators used will be total ordering operators. Those differing only in a subscript will denote a d-set of mutually disjoint ordering operators having domains of equal cardinality (and ordinality by definition). The symbols O and O' will be used for ordering operators whose output is to be taken as orthogonal: that is, the generations are mutually disjoint (distinguishable) except for a single generation of each which are indistinguishable. The generations of the O and O' will be notationally distinguished by x and y, respectively. A prefixed superscript of either 1 or -1 will denote the generations of the ordering operator as coming either before or after some specified and unique label, respectively. The subscripts associated with O and O' will be carried over to the respective generations.

A Discrete Coordinate Patch

Without reference to a particular geometry or distance function, a "square" can be defined as a closed d-set having the following properties:

a) two–dimensionality;

b) the edges or boundary consists of two d-sets of two mutually disjoint totally ordered d-subsets (four sides);

c) fixed center under interchange of the coordinate parameters;

d) it is possible to establish a distance function on the edges such that each of the totally ordered d-subsets is of equal length.

The criteria for two–dimensionality is satisfied by requiring two mutually disjoint ordering operators. The algorithm is as follows:

1) Select a label L_0; see Figure 3.

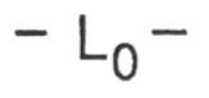

$$- L_0 -$$

5-88　　　　　　　　　　6015A3

Figure 3   A starting label L_0.

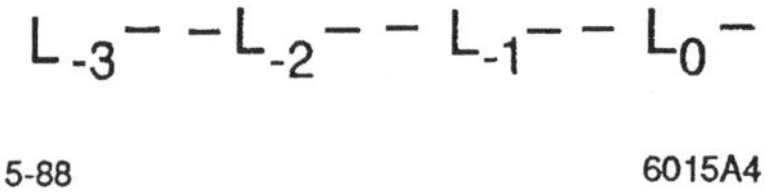

$$L_{-3} - - L_{-2} - - L_{-1} - - L_0 -$$

5-88   6015A4

Figure 4   The subchain of length $n = 4$, $^{-1}x_0$ with L_0 as supremum.

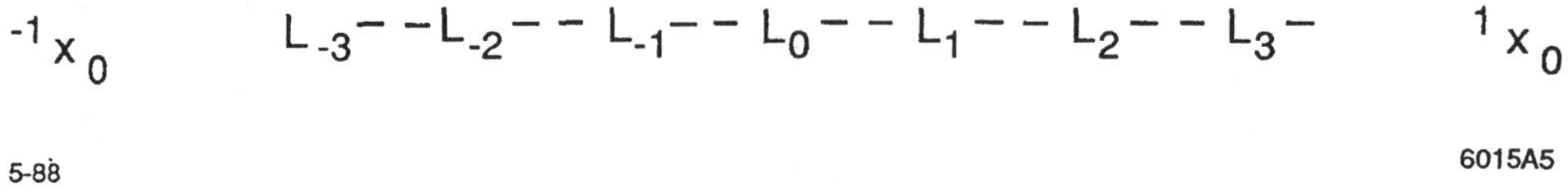

$^{-1}x_0$ $L_{-3} - - L_{-2} - - L_{-1} - - L_0 - - L_1 - - L_2 - - L_3 -$ $^1x_0$

5-88   6015A5

Figure 5   The subchain of length $n = 4$, $^1x_0$, with L_0 as infimum added.

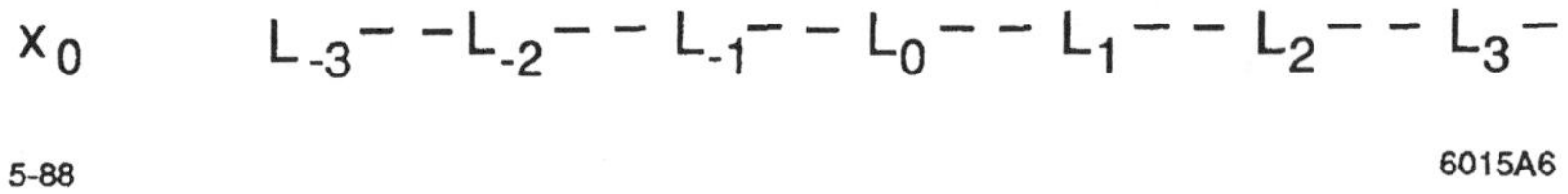

x_0 $L_{-3} - - L_{-2} - - L_{-1} - - L_0 - - L_1 - - L_2 - - L_3 -$

5-88   6015A6

Figure 6   The chain length of $n = 7$, x_0.

2) Establish a totally ordered d–set **chain** $^{-1}x_0$ of length n with L_0 as the supremum, using the ordering operator O_x; see Figure 4.

3) Establish a chain $^1x_0$ of length n, with L_0 as the infimum, using the ordering operator O_x; see Figure 5.

4) Call the union of $^{-1}x_0$ and $^1x_0$: x_0. Require that x_0 be totally ordered; see Figure 6.

5) For each label L_i of x_0, establish chains $^{-1}y_i$ and ^1y_i of length n, under the ordering operator O'_y, with the selected label of x_0, as either the supremum and infimum of the chain. Require that the y_i are disjoint, as are the pairs $(^{-1}y_i, {}^1y_i)$. This is a unique labeling or total ordering requirement on the entire construction (i.e., there must exist an ordering operator O'' such that the labels of the entire construction are totally ordered; see Figure 7.

6) Require that the nth label of the y_i form chains x_i, ordered by ordering operator O_{x_i}; see Figure 8.

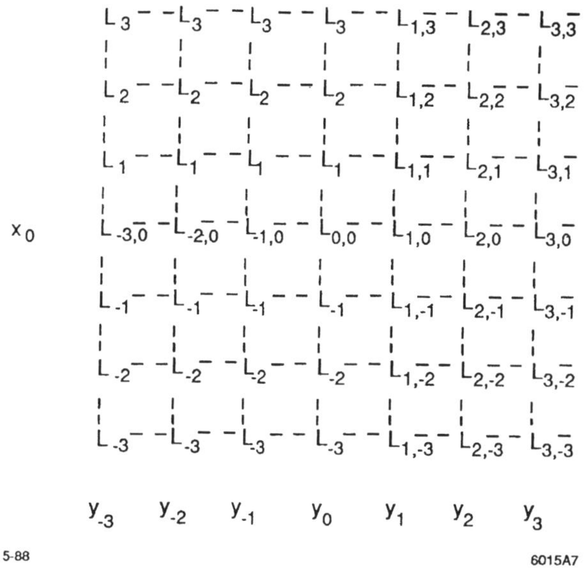

211

$$
\begin{array}{ccccccc}
L_3 & L_3 & L_3 & L_3 & L_{1,3} & L_{2,3} & L_{3,3} \\
L_2 & L_2 & L_2 & L_2 & L_{1,2} & L_{2,2} & L_{3,2} \\
L_1 & L_1 & L_1 & L_1 & L_{1,1} & L_{2,1} & L_{3,1} \\
L_{-3,0} & L_{-2,0} & L_{-1,0} & L_{0,0} & L_{1,0} & L_{2,0} & L_{3,0} \\
L_{-1} & L_{-1} & L_{-1} & L_{-1} & L_{1,-1} & L_{2,-1} & L_{3,-1} \\
L_{-2} & L_{-2} & L_{-2} & L_{-2} & L_{1,-2} & L_{2,-2} & L_{3,-2} \\
L_{-3} & L_{-3} & L_{-3} & L_{-3} & L_{1,-3} & L_{2,-3} & L_{3,-3}
\end{array}
$$

x_0 (row label at left of center row)

$$y_{-3} \quad y_{-2} \quad y_{-1} \quad y_0 \quad y_1 \quad y_2 \quad y_3$$

5-88   6015A7

Figure 7   *The chains y_i of length $n = 7$ added.*

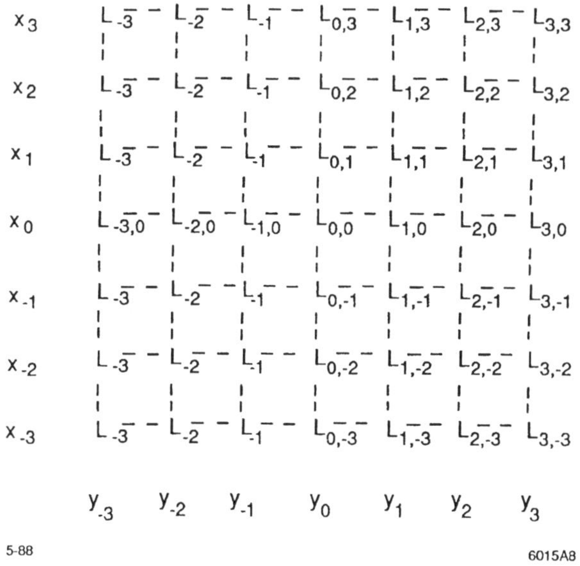

	y_{-3}	y_{-2}	y_{-1}	y_0	y_1	y_2	y_3
x_3	L_{-3}	L_{-2}	L_{-1}	$L_{0,3}$	$L_{1,3}$	$L_{2,3}$	$L_{3,3}$
x_2	L_{-3}	L_{-2}	L_{-1}	$L_{0,2}$	$L_{1,2}$	$L_{2,2}$	$L_{3,2}$
x_1	L_{-3}	L_{-2}	L_{-1}	$L_{0,1}$	$L_{1,1}$	$L_{2,1}$	$L_{3,1}$
x_0	$L_{-3,0}$	$L_{-2,0}$	$L_{-1,0}$	$L_{0,0}$	$L_{1,0}$	$L_{2,0}$	$L_{3,0}$
x_{-1}	L_{-3}	L_{-2}	L_{-1}	$L_{0,-1}$	$L_{1,-1}$	$L_{2,-1}$	$L_{3,-1}$
x_{-2}	L_{-3}	L_{-2}	L_{-1}	$L_{0,-2}$	$L_{1,-2}$	$L_{2,-2}$	$L_{3,-2}$
x_{-3}	L_{-3}	L_{-2}	L_{-1}	$L_{0,-3}$	$L_{1,-3}$	$L_{2,-3}$	$L_{3,-3}$

5-88   6015A8

Figure 8   *The chains x_i of length $n = 7$ added. Note that all labels are now subscripted twice, since they are identified as the production of two ordering operators.*

7) The resulting object satisfies the requirements; it is the discretum version of a two–dimensional (square) coordinate patch. In particular, the two–dimensionality of the construction is satisfied by the definition of mutually disjoint ordering operators: at most, one label in a chain resulting from one operator will be found in a chain resulting from the other. For the given construction: at most, two operators can be used; a third would result in a partial ordering, instead of a total ordering, of the labels of the construction, and this would then represent an object which is not connected or result in an object for which "multiple" labels are doubly labeled. Thus, the ordering operators "parameterize" the object.

A Discrete Circular Patch

We can now proceed to construct an object which behaves as a discretum version of the 2–sphere. A 2–sphere (again, without reference to distance functions) has the following properties:

a) two–dimensionality;

b) every edge (boundary) label is indistinguishable from every other, under interchange of the corresponding ordering operators;

c) existence of a unique label, which remains fixed in the construction, under interchange of any two ordering operators which generate it.

The constructive algorithm is as follows.

1) Select a (square) coordinate patch with center $L_{0,0}$ and all labels uniquely subscripted. Call this patch S_0; see Figure 9.

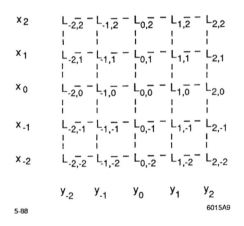

Figure 9   Select a patch, s_0.

2) Constrain the possible ordering operators (as before) to those operators which produce chains of length n and which generate from $L_{0,0}$ a nearest neighbor of $L_{0,0}$,

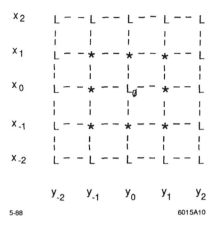

Figure 10   *The nearest neighbors of L_0 are shown as 'an asterisk (*)*.*

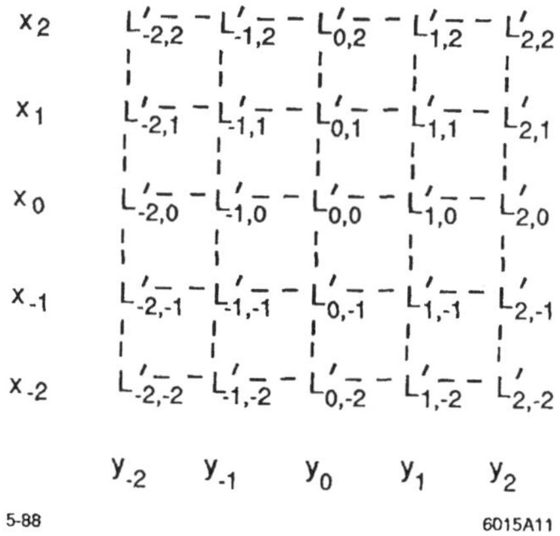

x_2 | $L'_{-2,2} - L'_{-1,2} - L'_{0,2} - L'_{1,2} - L'_{2,2}$
x_1 | $L'_{-2,1} - L'_{-1,1} - L'_{0,1} - L'_{1,1} - L'_{2,1}$
x_0 | $L'_{-2,0} - L'_{-1,0} - L'_{0,0} - L'_{1,0} - L'_{2,0}$
x_{-1} | $L'_{-2,-1} - L'_{-1,-1} - L'_{0,-1} - L'_{1,-1} - L'_{2,-1}$
x_{-2} | $L'_{-2,-2} - L'_{-1,-2} - L'_{0,-2} - L'_{1,-2} - L'_{2,-2}$

$y_{-2} \quad y_{-1} \quad y_0 \quad y_1 \quad y_2$

5-88   6015A11

Figure 11   *A new patch, P_i.*

then a nearest neighbor of this label, and so on. We refer to the operators which generate these labels as **radial permutations** of the coordinate patch; see Figure 10.

3) Starting from $L_{0,0}$, construct a coordinate patch with a new pair of ordering operators which are radial permutations of the coordinate patch; see Figure 11.

4) Map the labels of this patch S_i to patch S_0, and eliminate any labels which do not have at least i subscripts; see Figure 12.

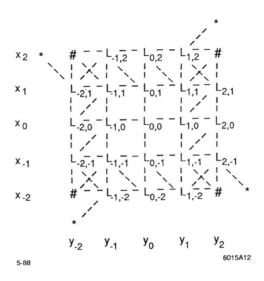

Figure 12   Mapping the new patch to the old.

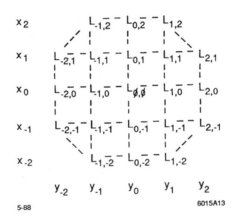

Figure 13   Elements remaining after all allowed radial permutations.

5) Repeat this process for all pairs of allowed radial permutations; see Figure 13.

The result is a discretum version of the circle, in that it has a fixed center ($L_{0,0}$) with radial symmetry (isomorphic to its radial permutations with identified center $L_{0,0}$). It has built-in bounds on "precision." The relation between the number of "sides" of the polygon formed by a set of cardinality n and the number of permutations is fixed: it gives a measure of the "size" of the circle.

Note that in Figures 3–13, the radial permutations which were not invoked would result in either the same labels of the construction being deleted, as here, or else would not maximally d–map the coordinant patch. The reader may readily demonstrate this. Also note that starting with a central label is a matter of technical convenience for the algorithms and may be circumvented.

$\pi(N)$

Given these two geometric objects, it is possible to define a ratio which plays the role of the ratio of the area of the circle to the area of the square patch from which it was formed. This number is obtained by counting the number of labels contained in the circle and the number of labels contained in the square and forming the ratio.

A second ratio is obtained from the ratio of the cardinality of the d–set of all radial permutations (obtainable by counting the labels on the perimeter of the circle) and the cardinality of the generations of one such radial permutation (e.g., the length n of the chain x_0).

In general, these ratios will be functions of the length n of the chain x_0. Furthermore, the values of the ratios will not, in general, be those obtained under Euclidean geometry. However, if one insists on isotropy, homogeneity and "density" (i.e., large n), it is easy to see that these values must be those obtained by the standard polygonal approximation to the circle. In particular, these ratios will be approximations to $\pi/4$ and π, with the appropriate precision. These constructions, and the results, are closely related to numerical and statistical "approximation" methods, as seen from within the traditional geometric paradigm. In fact, Archimedes came close to the construction used here (Measurement of the Circle). However, the definitions used here are completely constructive and general, matching the continuum definitions (which we prefer to think of as the "analytic interpolation") as desired.

CALCULATIONS

By Areas:

A (square patch) = 25

A (polygon) = 21

Ratio = π (areas)/4

   = A (polygon)/A (square)

   = 21/25 = 0.84

π (area) = 3.36

By Perimeters:

C (polygon) = 12

C (square patch) = 16

Ratio = π (lengths)/4

   = C (polygon)/C (square patch)

   = 12/16 = 0.75

π (perimeter) = 3.00

Indeed, if the cardinality of the d–space (N) is changing (evolving), then the two values of $\pi(N)$ will be changing, also. Furthermore, if the relevant discrete cardinality is related to a spatial volume, then, as this region becomes smaller, calculations involving $\pi(n)$ cannot be treated in a naive manner. Specifically, the multiple computational definitions of $\pi(n)$ must be disassociated, if the values are different (i.e., the ratio $\pi_{areas}/\pi_{circumference}$ will not be 1). The value of pi can no longer be taken as a constant, independent of spatial volume. Indeed, if the d–space is inhomogenous, the value will depend on the local inhomogeneities; it will have different values depending on the "density" of the local d–space. Even more important, if the d–space is discrete and finite, and if the values of $\pi(n)$ are not related to the local spatial volume via a cardinality of the local volume, it follows that the values of $\pi(n)$ used in calculations relate only to the cardinality of the d–space. In other words, $\pi(N)$ becomes a true global discrete topological constant, and local physical properties are then immediately dependent on the global properties.[*]

In the remainder of this paper we will use $\pi(N)$ to refer to the combinatoric computation of π, based on the ratios of perimeters for a d–space of cardinality N. We cannot use the ratio computed from area ratios, since we will not, in general, know the "curvature" of the d–space. Note that measuring the difference between the two ratios gives a means of locally measuring the flatness of the d–space. Similarly, the curvature can be measured by examining the ratio computed on the basis of "volumes."

Radian and Trigonometric Measures

Having constructed the largest coordinate patch and the corresponding inscribed "circle," we may now pick an orientation and specify a total ordering operator which generates the sequence of attribute states constituting the perimeter as labels. We then reparameterize the generations of this ordering operator into the interval of rational fractions $[0, 2\pi(N)]$. We call this parameterization the **radian measure** on a d–space of cardinality N. Similarly, we shall refer to the cardinality or length of the total ordering generated by one of the radial permutations used in constructed any circle, the **radius r** of the circle.

A radius r and a radian measure θ then correspond to that d–point which results from a translation in coordinate distance of attribute distance r from the origin, followed by θ generations of the reparameterized perimeter ordering operator. Since every pair r and θ correspond to a unique point on the perimeter, and $\pi(N)$ is constructed from the maximal coordinate d-patch, we may regard θ as a direction and define the trigonometric computations of θ in the usual manner using the norm function. In particular, take the cosine to be the unit normal projection on the x-axis and the sine to be the unit normal projection on the y-axis. Note that this does not assume the Pythagorean theorem, unless it is already entailed in the norm function.

3.6 PROPERTIES OF EVOLVING SYSTEMS: ATTRIBUTE VELOCITIES

Given a d–space, we require that there exist a total ordering operator on the space, so that a distance function (such as that produced by the Program Universe ordering operator) is possible. The universal ordering parameter T, on which the generation of this ordering operator is based, provides a local total ordering for the

---

[*] Applying this fact to physical phenomena, that π should then be of cosmological (global) significance is not surprising. Consider these results where the d–space is the physical Universe.

evolution of each ensemble, such that the local total orderings are isomorphic up to reparameterization. This in turn provides for synchrony.

We now define the **increment** I of an ensemble as the number of generations of some ordering operator[†] O needed to describe (establish local isomorphism with) the increases in attribute distance between an ensemble and a reference ensemble, with respect to T. This operator parameterizes the generation of the attribute states. Similarly, we define the **decrement** D of an ensemble as the number of generations t of the ordering operator O needed to describe the decreases in attribute distance between an ensemble and a reference ensemble, with respect to T. The **total size** S of an ensemble is defined as the arithmetic sum $+$ of the I and D. Use $[I, D]$ to denote an ensemble with increment I and decrement D and total size $I + D$. Note that the total size S is not generally the same as the maximum cardinality N since total size refers to increments and decrements of the ordering operator, and not to the cardinality of the d–sort of labels produced by the operator.

Attribute velocity v is defined as the mathematical rate of change in attribute distance of an ensemble, with respect to generations t of an ordering operator O, computed as the difference between I and D, divided by the total size S:

$$v = \frac{I - D}{S} \, . \tag{6}$$

The relative attribute velocity v' is just v computed relative to a third ensemble (reference), having attribute velocity u. The relative attribute velocity may be regarded as a discrete map which transforms an ensemble $[I, D]$ into an ensemble $[I', D']$, where I' and D' depend only on I, D and u, and where v' depends only on u and v. This is just a change in the reference ensemble. The **increment quotient** is defined as the ratio of I' to I,

$$q = \frac{I'}{I} \, . \tag{7}$$

The **attribute speed** of an ensemble is the magnitude of the attribute velocity (note that direction is given by arithmetic sign or the degenerate cosine in the one–dimensional case, a discrete version of the $x^1 - x^2$ cosine in the two–dimensional case, and a discrete version of the $x^1 - x^2$ and $x^2 - x^3$ cosines in the three–dimensional case). Finally, we define independent ensembles as those having all states generated with respect to an ordering operator O, distinguishable. We will discuss the impact of indistinguishable states in a later section. Having defined these terms, we may now prove a series of theorems regarding the properties of such ensembles.

Theorem 14: The increment and decrement are additive for independent ensembles when aggregated; that is, the number of distinguishable states and the number of

---

† In general, this is not the same ordering operator which generated the ensemble.

generations t relative to an ordering operator O required to describe I and D for independent ensembles is conserved.

$$[I, D] + [I', D'] = [I + I', D + D'] \qquad (8)$$

Argument:

As long as two states of an attribute are distinguishable over t, we are certain that a generation of O is required for each. It follows that the total number of generations T for independent ensembles (those having all states distinguishable) is given by the .arithmetic sum (total count) of the generations of O, for the increment and decrement of each. Indeed, the total size of the ensemble is just $S + S'$.

QED

Theorem 15: The attribute velocity v of an ensemble $[I, D]$ is a function of I and D, and nothing else.

Argument:

If ensembles A and B have the same attribute speed, then the aggregate ensemble $A + B$ must also have that attribute speed. Hence, v cannot depend on total size, but only on the ratio of I to D. Let $r = I/D$; then we can write

$$v = v(r) . \qquad (9)$$

QED

Theorem 16: v is an increasing function of the ratio r.

Argument:

Trivially, the case from the definitions.

QED

Theorem 17: If the values of I and D are reversed, then v is reversed;

$$v([D, I]) = -v([I, D]) . \qquad (10)$$

Argument:

Inverting I and D is equivalent to counting distinguishable states from above, as .compared to from below—i.e., if one counts from 0 to the maximum number of distinguishable states, one obtains the usual definition of I and D. If one counts from the maximum number of distinguishable states down to 0, consistency with the definition of additivity can be maintained if this is equivalent to a reparameterization resulting in a change of arithmetic sign.

QED

Theorem 18:   If neither I nor D is 0,

$$v\left(\frac{1}{r}\right) = -v(r) \ . \tag{11}$$

Argument:

Trivially, from Theorem 4 and the supposition.

QED

Theorem 19:   The attribute distance between any two ensembles has an upper and lower bound.

Argument:

Trivially, from the finitary principle (Principle I).

QED

Theorem 20:   The lower bound of v CANNOT BE ZERO for independent (i.e., distinguishable) ensembles.

Argument:

If the lower bound of v were zero, the ensembles would be attribute indistinguishable and hence not independent.

QED

Theorem 21:
There is a limit to v as D approaches 0, which we can take as 1 by appropriate reparameterization; i.e., $v([I, 0]) = 1$ and, hence, $v([0, D]) = -1$. We shall refer to this upper bound as v_{max}.

Theorem 22:   Etters Velocity Relationship,

$$v(r) = \frac{(r - 1)}{(r + 1)} \tag{12}$$

holds for attribute velocities.

Proof:

Consider a d–space with distance function as previously defined. Now, examine the region between synchronization (metric marks or ticks). In this region, as we have shown, there exists a value for the isotropic distance function. Let I be the total number of 0's and D the total number of 1's generated up to n generations of the ordering operator which defines the distance function (called the **metric ordering operator**); then the total attribute "displacement" in $I + D$ generations is just $I - D$. This gives an Etters velocity relationship of $I - D / I + D$ or, if $r = I/D$,

$$v(r) = \frac{(r - 1)}{(r + 1)} \ .$$

QED

220

4. MATHEMATICAL FOUNDATIONS III: COORDINATE TRANSFORMATIONS

In order to explore the invariant properties of a system, we must have a means of expressing not only the coordinate bases defined in the previous chapter, but also transformations between coordinate bases. Of particular interest are those coordinate bases which define a reference frame. In the present chapter, we develop a series of theorems regarding transformations between reference frames.

Theorem 23:   Suppose that synchronizable reference frames K, with coordinate bases x^i and k with coordinate bases y^i, i in $\{1,2,3\}$, are defined so that the origin of k has attribute velocity v in the direction x^1, with respect to the origin of K in the universal ordering parameter T; then the coordinates transform according to:

$$t' = \gamma \, \frac{t - vx^1}{v_{\text{max}}^2} \, , \tag{13}$$

$$y^1 = \gamma \, (x^1 - vt) \, , \tag{14}$$

$$y^2 = x^2 \, , \tag{15}$$

and

$$y^3 = x^3 \, , \tag{16}$$

where

$$\gamma = \frac{1}{\left[\frac{1-v^2}{v_{\text{max}}^2}\right]^{1/2}} \, .$$

Argument:

Select A, B and C synchronous with a distance function $d()$. Let $d(A,B)$ be the attribute distance between A and B and $d(B,C)$ be the attribute distance between B and C. Given $d(A,C) = 0$, as above; then, by symmetry (Principle IV), we require that $d(A,B) = d(B,C)$, so that for maximum attribute velocity v_{max}, we have

$$v_{\text{max}} = \frac{2d(A,B)}{t(A) - t(C)} \, . \tag{17}$$

Since $d(A,C) = 0$, note that A and C are indistinguishable, except by parameter t. Furthermore, with reference to a third ensemble with attribute velocity v,

$$t(B) - t(A) = \frac{d(A,B)}{[v_{\text{max}} - v]} \quad \text{and} \quad t(C) - t(B) = \frac{d(A,B)}{[v_{\text{max}} + v]} \, . \tag{18}$$

Now, suppose that we wish to compare the attribute distances d and d' and the operators t and t', with reference to third and fourth ensembles with attribute velocities 0 and v, respectively. Call these systems K and k. Furthermore, assume that there exist at least two independent attribute distances (generated from mutually disjoint ordering operators, except for a single element) for K and k; call these x^i and y^i, respectively. We seek one-to-one transformations (discrete maps) between these operator values. Given (in the absence of specific cause—i.e., an ordering operator) homogeneity (Principle IV) of the system K and k in the parameters, these transformations must be linear and homogenous.

Let $x^{1'} = x^1 - vt$; then k has a system of values $x^{1'}$ independent of t. Define t' as a function of $x^{1'}$, x^2, t. Let $d'(A, B)$ be the attribute distance between A and B, and $d'(B, C)$ be the attribute distance between B and C. Given $d'(A, C) = 0$ as above; then, by symmetry, we require that $d'(A, B) = d'(B, C)$, and

$$t'(B) = \frac{1}{2}[t'(A) - t'(B)] \,,$$

or

$$\frac{1}{2}\left[t'(0,t) + t'\left(0, t + \frac{x^{1'}}{v_{\max} - v} + \frac{x^{1'}}{v_{\max} + v}\right)\right] = t'\left(x^{1'}, t + \frac{x^{1'}}{v_{\max} - v}\right) \,. \qquad (19)$$

Let $x^{1'}$ be chosen small, and use an appropriate reparameterization, so that we may use the calculus of finite differences in solving for the proper transformations. Then, taking the *finite derivates* (not the derivatives),[25]

$$\frac{1}{2}\left(\frac{1}{v_{\max} - v} + \frac{1}{v_{\max} + v}\right)\frac{dt'}{dt} = \frac{dt'}{dx^{1'}} + \frac{1}{v_{\max} + v}\left(\frac{dt'}{dt}\right) \,, \qquad (20)$$

or

$$\frac{dt'}{dx^{1'}} + \frac{v}{[v_{\max}^2 - v^2]}\left(\frac{dt'}{dt}\right) = 0 \,, \qquad (21)$$

and

$$\frac{dt'}{dx^2} = 0 \,. \qquad (22)$$

Since t' is linear, and we can assume $t' = 0$ when $t = 0$, the solution is just

$$t' = a\left(t - \frac{v}{v_{\max}^2 - v^2}\, x^{1'}\right) \,, \qquad (23)$$

where $a = f(v)$, unknown for now.

Let v_{max} be represented by the same fixed value for both K, and k by a suitable reparameterization in each reference frame. Let attribute information transfer with attribute velocity v_{max} over a positive attribute distance y^1,

$$y^1 = v_{\text{max}} \times t' , \tag{24}$$

and

$$y^1 = av_{\text{max}} \left(t - \frac{v}{v_{\text{max}}^2 - v^2} x^{1'} \right) ; \tag{25}$$

then, with reference to the frame K, an ensemble expressed in the system k has attribute velocity $v_{\text{max}} - v$, or

$$\frac{x^{1'}}{v_{\text{max}} - v} = t . \tag{26}$$

So

$$y^1 = a \frac{v_{\text{max}}^2}{v_{\text{max}}^2 - v^2} x^{1'} , \tag{27}$$

and

$$y^2 = v_{\text{max}} t' = av_{\text{max}} \left(t - \frac{v}{v_{\text{max}}^2 - v^2} x^{1'} \right) , \tag{28}$$

where

$$t = \frac{x_2}{[v_{\text{max}}^2 - v^2]^{1/2}}, \qquad x^{1'} = 0 ; \tag{29}$$

thus,

$$y^2 = a \left(\frac{v_{\text{max}}}{[v_{\text{max}}^2 - v^2]^{1/2}} \right) x^2 . \tag{30}$$

By substitution for $x^{1'}$, we obtain

$$t' = f(v) \, \gamma \left(t - \frac{vx^1}{v_{\text{max}}^2} \right) , \tag{31}$$

and

$$y^1 = f(v) \, \gamma (x^1 - vt) , \tag{32}$$

$$y^1 = f(v) \, x^2 , \tag{33}$$

where

$$\gamma = \frac{1}{\frac{1-v^2}{v_{max}^2}^{1/2}} \ .$$ (34)

To find $f(v)$, introduce K' with coordinates x^1, $x^{2'}$ and t' in parallel translation relative to x, such that the origin of k moves with attribute velocity $-v$. Assume the origins coincident. Applying the transformations we obtain

$$t' = f(-v) \ \gamma(-v) \left(t' + \frac{vy^1}{v_{max}^2}\right) = f(v) \ f(-v)t \ ,$$ (35)

$$x^{1'} = f(-v)\gamma(-v)(y^1 + vt') = f(v)f(-v)x^1 \ ,$$ (36)

$$x^{2'} = f(-v)y^2 = f(v)f(-v)x^2 \ .$$ (37)

Since the transforms from K' to K are independent of t, it follows that K and K' are relatively at rest. Therefore,

$$f(v) \ f(-v) = 1 \ .$$ (38)

Now, let there be an attribute distance of value ℓ, given independent of x^1 and $x^{1'}$; call this x^2 and $x^{2'}$, in k and K, respectively; then ℓ in k, with reference to K, is just

$$x^2 = \frac{1}{f(v)} \ .$$ (39)

Since, from symmetry, attribute distance can depend only on v, and not on direction or the sense of attribute speed, it follows that the interchange of v and $-v$ does not change ℓ. Hence,

$$\frac{1}{f(v)} = \frac{1}{f(-v)} \quad \text{or} \quad f(v) = f(-v) \ .$$ (40)

Thus, from Eqs. (37) and (39), it follows that $f(v) = 1$. Therefore, we have

$$t' = \gamma \left(t - \frac{vx^1}{v_{max}^2}\right) \ ,$$

$$y^1 = \gamma(x^1 - vt) ,$$

$$y^2 = x^2 ,$$

and

$$y^3 = x^3 ,$$

where $\gamma = 1/[1 - v^2/v_{\max}^2]^{1/2}$; these being Eqs. (13), (14), (15) and (16), respectively.[*]
QED

Theorem 24: If $u = 0$, then $I' = I$ and $D' = D$; that is, an ensemble with zero attribute velocity induces the identity transformation.

Argument:

Trivially, from the definition of attribute distance, an ensemble with zero attribute velocity, with respect to some reference ensemble, is indistinguishable from the reference ensemble.
QED

Theorem 25: If $u = v$ and attribute speed < 1, then $I' = D'$; i.e., if ensemble A (which we may interpret as an observer) has the same attribute velocity as ensemble B, their relative attribute velocity is 0.

Argument:

Trivially, from the definitions of attribute distance and velocity.
QED

Theorem 26: If $I = D$,

$$v' = \frac{(I' - D')}{(I' + D')} = -u ; \tag{41}$$

i.e., with respect to a reference ensemble A with nonzero attribute velocity, an ensemble B with zero attribute velocity is an ensemble with the same attribute speed, but with opposite sign (direction).

Theorem 27: If the reference attribute speed is less than 1, a reference ensemble A with attribute $-u$ induces the inverse of the transformation induced by changing to a reference ensemble B with attribute velocity u.
corollary 27A:

Reversing ensemble A attribute velocity sign (direction) inverts the transformation induced on the attribute velocity of ensemble B.

---

[*] Note that, although our derivation is finite and discrete, we have deliberately followed the derivation of the Lorentz transformation developed by Einstein. We wish to emphasize that, contrary to common belief, the derivation of these transformations are not dependent upon the continuum. Where Einstein used derivatives, we use finite derivates, Eqs. (20) and (21). Where he allowed for a continuum of coordinates and velocities, we are restricted to the rational fractions which suffice per Pauli and Brodsky.

Theorem 28:   The relative attribute velocity in the frame of ensemble A is bounded from below by the speed of ensemble B.

Theorem 29:   The limiting attribute velocities for an ensemble are invariant under the transformation induced by nonzero attribute velocity; i.e., $[I,0]' = [I'',0]$ for some number I''.

Argument:

If the sign of the relative attribute velocity is positive, this follows from lower bound. If negative, the inverse transformation corresponds to positive relative attribute velocity, so that D must remain invariant.

QED

30:   The increment quotient q is a function only of u.

Argument:

$[I,0]' = [I'',0]$, where I'' depends only on I and u. However, by Theorem 14, $[I',D] = [I,0]' + [0,D]'$, hence $I' = I''$; thus $q = I'/I$ depends only on I and u. However, q cannot depend on I, since otherwise $[2I,2D]'$ would have a different attribute velocity than $[I,D]$.

QED

Theorem 31:   The inverse transformation induced by an ensemble with attribute velocity $-u$ has an increment quotient of $1/q$.

Argument:

The inverse transformation is I/I'.

QED

Theorem 32:   The decrement quotient is the inverse of the increment quotient:

$$\frac{D'}{D} = \frac{1}{q}.$$ (42)

Argument:

First, reverse I and D to get $-v$, then take inverse transformation associated with $-u$, which multiplies the increment (which is now D) by $1/q$ to get $-v'$, then reversing I and D again to get v'; Thus, D' results from multiplying D by $1/q$, and it follows that $D'/D = 1/q$.

QED

Theorem 33:

$$q = \frac{(1-u)}{\gamma},$$

where

$$\gamma^2 = 1 - (u^2).$$ (43)

Argument:

By the definition of the decrement quotient [Eq. (42)], $D' = D/q$, and from the increment quotient [Eq. (7)], $I' = qI$, so that from the definition of $v = (I' - D')/(I' + D')$ [Eq. (41)], we can write $v == (qI - D/q)/(qI + D/q)$. Since q is a function only of u, we can choose any values of D and I that lead to an equation in q and u, and its solution will define the general functional dependency. Assume $I = D$ so $v = 0$ and $v' = -u$; then, from Eq. (41),

$$-u = \frac{(qI - I/q)}{(qI + I/q)} = \frac{(q^2) - 1}{(q^2) + 1} .$$

Solving for q results in the relationship to be proved.

QED

Theorem 34: Relative to the zero velocity frame v = 0, the **size change** δm of an ensemble with attribute velocity v' is

$$\delta m = \frac{S}{\gamma} . \tag{44}$$

Argument:

Multiplying in the first part of (43) by $(1 + u)$ gives $1/q = (1 + u)/\gamma$ and $D = Ix$ for an ensemble with zero attribute velocity, this follows immediately.

QED

Theorem 35: Attribute velocities combine according to

$$v' = \frac{v - u}{1 - vu} . \tag{45}$$

Argument:

By definition,

$$v = \frac{r - 1}{r + 1}, \qquad r' = \frac{I'}{D'} = \frac{r(1 - u)}{(1 + u)} ,$$

and

$$v' = \frac{r' - 1}{r' + 1} .$$

Then, by substitution and recollection of terms, we have

$$v' = \frac{v - u}{1 - vu} . \tag{46}$$

QED

Theorem 36: For an ordering operator O of cardinality N and for each run of cardinality k, the **minimal attribute distance increment** i is

$$i(O) = \frac{1}{k!} \,. \tag{47}$$

Argument:

Consider a sequence of productions from an unspecified ordering operator of cardinality N to be used as a coordinate basis. We can compute the minimal attribute distance increment which can be generated in a given run of cardinality k of the operator, straightforwardly: it is the ratio of the number of (order) distinguishable states C (i.e., combinations—by excluding order, we take only those states that are distinguishable under a particular ordering operator) to the number of states P (i.e., permutations—by including order, we include all states, even those which are not distinguishable under a particular ordering operator).

$$C(k; N) = \frac{N!}{k!(N-k)!} \,, \tag{48}$$

$$P(k; N) = \frac{N!}{(N-k)!} \,, \tag{49}$$

$$i(O) = \frac{C(k; N)}{P(k; N)} = \left[\frac{N!}{k!(N-k)!} \right] \times \left[\frac{(N-k)!}{N!} \right] = \frac{1}{k!} \,.$$

QED

In general, C gives all the possible attribute states that could produce a sequence of state ensembles of the proper cardinality k, while P gives the number of ensembles of cardinality k possible in the same total space of cardinality N. This is, of course, subject to the constraint $k < N$.

Theorem 37: The total attribute distance $d(k; I; N)$ for an ensemble of cardinality k implied by I increments of i in a total space of cardinality N is

$$d(k; I; N) = \frac{I^k}{k!} \,. \tag{50}$$

Argument:

Suppose that we want to generate I increments in the attribute distance; then we want to turn the crank of the ordering operator which produces each attribute state I times. In the absence of further knowledge about the specifics of the ordering operator generation, we cannot enforce sequence so that the increments are disjoint; this is equivalent to sampling k objects from a population of I objects, with repetition allowed. Call this $R(k; I)$; then, in general, for an ordering operator to generate an attribute distance d equal to I increments from a run of cardinality k on a space of cardinality N, we have:

$$d(k; I; N) = R(k; I) \times \frac{C(k; N)}{P(k; N)} = \frac{I^k}{k!} ,$$

where

$$R(k; I) = I^k .$$

QED

Theorem 38:   The sum of all values of Eq. (50) from $k = 0$ to $k = K$ approaches e^I (for any expression of I) as K becomes large. We call this **the combinatoric definition of** $e(K)$.

$$e^I \approx \sum_{k=0}^{K} \frac{I^k}{k!} = e^I(K) . \tag{51}$$

Argument:

From the identity of definition of terms of the power series for e^I and the combinatoric definition of $I^k/k!$, the result follows for all discrete, finite values of k, N and I.

QED

Theorem 39:   The attribute distance, given a distance function g transformed by reparameterization from a distance function f, is just:

$$d[k; I; g(M)] = \sum_{k=0}^{K} d[k; I; f(N)] \times D(f; k; N) , \tag{52}$$

where $D(f; k; N)$ are the kth derivates of f.

Argument:

Consider a reparameterization of $d(k; I; N)$ from a distance function f on a d–space of cardinality N to a distance function g on a d–space of cardinality M, where the attribute is first order for both f and g. This is given by multiplying the attribute distance increment for $D(k; I; N)$ by a conversion factor (rational fraction), D. Since the attribute distance increment is inversely proportional to N, we have:

$$d[k; I; g(M)] = d[k; I; f(N)] \times D \approx d[k; I; f(N)] \times (N/M) . \tag{53}$$

Now, examine a general distance function $f(I; N)$ defined on a d–space S. By Principles I (finiteness), II (discreteness) and III (finite computability), $f(I; N)$ may be expressed as some ordering operator O, which generates **attribute states of an attribute of some order.**[*] Call this order K. To express the generation of O in terms of the underlying discretum of cardinality N, we must take into account the possible contributions from all orders k from 0 to K. In general, D is not constant, but is dependent on f, N and k. Thus, for a general distance function $f(I; N)$, we have:

$$d[k; I; g(M)] = \sum_{k=0}^{K} d[k; I; f(N)] \times D(f; k; N) \ .$$

Note that the $D(f; k; N)$ may be solved for by the method of difference quotients[26]. These are the kth derivates of f. The series is always finite (and, hence, there is no question of "divergence" for a given evaluation of the series) since N is fixed. For sufficiently large N, the series Eq. (52) approaches the Taylor series with arbitrary precision.

QED

The Lagrange form of the remainder is of particular interest here, since it gives a measure of the deviation from the discrete form by the analytic form of the truncated Taylor series Eq. (52).

$$R_n(x) = \frac{f^{n+1}(\epsilon)}{(n+1)!} \ (x - a)^{n+1} \ , \tag{54}$$

where

$$x < \epsilon < a \ .$$

For sufficiently complex attributes and large N, this approaches the usual form of the exponential operator, as normally used to describe transport along a parameter. The sum may be understood as the contributions to distinguishability by successively more complex aspects of the attribute, weighted by the probability that a particular sequence that can generate the required distance is the correct one.

Theorem 40:   The **incremental transport** x_0 along a basis x^i at x parameterized on t is just

$$d[k; I; f(x + x_0)] = \sum_{k=0}^{K} \left(\frac{(x_0)^k}{k!} \right) \times D(f; k; t) \ , \tag{55}$$

where $D(f; k; t)$ are the kth derivate operators on x with respect to t.

---

[*] Recall that an attribute of an attribute is called an attribute of second order, an attribute of an attribute of an attribute is called an attribute of third order, etc.

Argument:

We wish to compute the incremental transport δx along a given coordinate basis x in terms of the above formulation. This is equivalent to a reparameterization from f to g, in which f and g are related as follows:

$$g(x) = f(x + x_0) \,, \tag{56}$$

with x_0 being the minimum attribute distance increment.

Since we do not know the particular ordering operator, but only the ultimate cardinality of the ensemble and the cardinality of the space, we must use the general form of reparameterization, Eq. (52). The result follows from substitution of Eq. (56) in Eq. (52).

QED

If the ordering operator produces a sequence which is of first order (linear in the ordering parameter), then the rate of change of attribute distance with respect to the ordering parameter is constant. This is, of course, just the first discrete derivative (derivate). If the ordering operator produces a sequence which is of second order, then the rate of change of attribute distance with respect to the ordering parameter is a first order function of the ordering parameter, i.e., the second derivate. Similar arguments hold for ordering operators of higher order.[*] In order to compute the transport along x^i from x to $x + x_0$, we must take into account the contributions of each order up to the order of the operator.

Theorem 41: Given reference frames F and F', coordinate transformations between unsynchronized events satisfy Eqs. (13), (14), (15), (16) of Theorem 23, statistically.

Argument:

Consider two reference frames, F and F', given by two sets of independent generations S_1, S_2 and S_3, and S_1', S_2' and S_3'. Again, we initially synchronize each set of three and let them go independently (Theorem 13). We count the occurrence of an attribute state which may be used as a metric mark in one of the generations as a 1, and any other attribute state as a 0, for purposes of analyzing the statistics.

Now, however, we have two ordering operators which we label O and O', global to F and F', respectively. In the absence of further information regarding the ordering operator, we will assume a normal distribution (Principle IV) of distinguishable states

---

[*] This analysis is consistent with the requirement that the kth derivate may be obtained from confluent divided differences of k arguments. The k arguments are order independent and, hence, are "sampled from a population of cardinality I with repetition allowed," as previously noted.

about a metric mark in either F or F' (i.e., generated by independent O and O' as per Theorem 13).[†]

Consider a discrete mapping from F to F'

From the combinatoric definition of the base of the natural logarithm and the definition of the normal distribution, a sample size of two standard deviations around an attribute state, taken as the mean or center of the distribution, will consist of all the distinguishable states around the mean[28], and, therefore, a metric mark, with a probability equal to the ratio of distinguishable states to all states, summed over all possible attribute states that might be selected in F' as a metric mark. However, this is just $1/e(N)$; thus, for a well-defined "metric mark" in F, a arbitrary transformation to F' results in a measure in F' which deviates from a metric mark by $\pm\sigma$. For a normal distribution, 2σ is just the transport for a minimum attribute distance increment. Computing the population variance σ^2 is then, for population of size N

$$\sigma^2 = \sum \frac{(U - \mu)^2}{N} , \tag{57}$$

where μ is the average of U. Suppose U is just the attribute distance in F; then the "mean attribute distance" μ is just the attribute velocity multiplied by the number of generations over which the attribute velocity has evolved. This is equivalent to giving the number of increments minus the number of decrements in terms of a global ordering operator spanning both F and F'. In other words, in the frame of the minimum of the maximum attribute velocities, the maximum number of generations for the ordering operator producing the attribute will be N, the cardinality of the universe. The normalized variable x^* corresponding to x with mean 0 and variance 1 is just

$$x^* = \frac{(x - \mu)}{\sigma} . \tag{58}$$

Here, $x - \mu$ is just the difference in the global frame between the increment and the decrement. Sigma is then the probability of obtaining an attribute increment corresponding to the ordering operator, which produces metric marks in F' relative to the ordering operator, which produces metric marks in F. Thus, Eq. (58) is equivalent to going to dimensionless (i.e., frame independent) quantities[29].

Note that for a discrete function on finite domain, this x^* is always bounded and finite; i.e., sigma is never 0 whenever $x - \mu$ is not 0. In addition, since the fluctuations in x are bounded and finite, it makes no sense to speak of specifying x beyond that discrete step length which results in the smallest fluctuation.

---

† In the absence of large N, we could as easily use the binomial distribution justified by the combinatorics to reflect finite N, and use the appropriate Yates adjustment in which $y_0 - \frac{1}{2}$ is substituted for y_0 in the computation of the probability $Pr(y > y_0)$, so that the unit normal variable probability $Pr(z > z_0)$ is just $z_0 = (y_0 - \frac{1}{2} - Np)/\sqrt{Npq}$, where p is the probability of a 1 and q is the probability of a 0. However, we assume here that the normal approximation is adequate in the light of the usual criteria that $N > 5$ and the absolute value of $[(1/\sqrt{N})(\sqrt{q/p} - \sqrt{p/q}]$ is less than 0.3[27].

Now, let μ be a d-velocity v multiplied by the number of generations t over which it is measured, and let x^1 be the attribute distance in F, and y the attribute distance in F'; then, from Eq. (58),

$$y^1 = (x^1 - vt)\,\gamma\;,\tag{59}$$

so that

$$\sigma = 1/\gamma \qquad \text{and} \qquad \mu = vt\;;\tag{60}$$

then y^1 is interpretable as the normalized variable associated with x^1. Clearly, as long as β is defined as v/v_{max} as in the derivation of Theorem 23, we have recovered the coordinate transformation in the absence of synchronization. Therefore, the co-ordinate transformations, Eqs. (13), (14), (15) and (16), are applicable at all rational scales, for all frames and for all attributes.

QED

It is important to understand that the mean attribute distance increment computed by going to dimensionless coordinates and transformed from a metric mark in F arbitrarily to F' (i.e., in the absence of synchronization between F and F'), is, thus, identical to the minimal attribute distance increment, transformed under synchronized frames for metric marks. This result may also be taken as proof by construction that the combination of the minimum attribute distance increment and the coordinate transformation of Theorem 23 has bounded (i.e., over the range of meaningful rational fractions which may be defined by reparameterization on the d–space) scale invariant significance.*

Theorem 42: Let $P = Prob\,(I \to I{+}1)$ and $Q = Prob\,(D \to D{+}1)$ for $N \to N{+}1$. The **uncertainty** associated with a coordinate transformation satisfying Theorem 41 between meter marks is given by:

$$1 - (P - Q)^2(\Delta x)^2 = 4PQ(\Delta x)^2 > 1\;.\tag{61}$$

Argument:

Now, since the variance is given by

$$1 - (P - Q)^2 = 4PQ\;,\tag{62}$$

and with

$$P = \tfrac{1}{2}\,(1 + \beta)\;,\tag{63}$$

$$Q = \tfrac{1}{2}\,(1 - \beta)\;,\tag{64}$$

---

* This analysis shows why the random walk derivation of the Lorentz transformation, as presented by Stein, works[30].

the probabilities of I and D, respectively, for N generations, we obtain

$$\sigma = (NPQ)^{1/2} = \left[\frac{N}{4(1-\beta^2)}\right]^{1/2} , \tag{65}$$

so that

$$\sigma L = L\left[\frac{N}{4(1-\beta^2)}\right]^{1/2} = \left(\frac{L\gamma}{2}\right) N^{1/2} , \tag{66}$$

where L represents the discrete increment for the variable and

$$(P-Q)L = \beta L . \tag{67}$$

Thus, we arrive at an interpretation of the coordinate transform between reference frames and between metric marks. Note that, because N is finite, the variance is finite, i.e., bounded. This provides normalization of the transform, as well as a "maximal velocity." We have simply applied a consistency requirement to all allowed (i.e., rational) velocity frame transformations, namely bounded scale invariance.

Furthermore, because σ is bounded from below by one generation, it follows that the minimum deviation is always 1 between metric marks. Fluctuations between metric marks are thus bounded above and below. Letting Δx represent the discrete increment in x, the bound from below gives the uncertainty in the region directly from the variance:[†]

$$1 - (P-Q)^2(\Delta x)^2 = 4PQ(\Delta x)^2 > 1 .$$

QED

4.1 MULTIPLY CONNECTED ATTRIBUTE SPACES

We now show how a d–space can be multiply connected, and derive some consequences of this multiple-connection. Unlike other notions of nonlocality, a multiply-connected d–space has a sequence of maximal attribute velocities.

Theorem 43: In a multiple attribute d–space, the sequence of maximal attribute velocities V_i has at least one value which is a least upper bound V_{max} and at least one value which is a greatest lower bound V_{min}.

[†] As we will see in the physical interpretation, this fact implies that we do not require the concept of the wave function. Our "collapse" is nothing more than the attainment of more information about the specific ordering operator involved in the "evolution" of the discrete system. The uncertainty is nothing more than a quantification of the amount of detail expressible, given the selected basis having rational fraction values; i.e., as "meter marks."

Argument:

Trivially, from the definition of maximum attribute velocity, indistinguishable, attribute state and Principle I.

QED

Theorem 44: A multiple attribute d-space has relationships between attribute distance functions satisfying Eqs. (13), (14), (15) and (16), which display nonlocal correlations (i.e., require more generations than allowed by the ordering operator for v_{max}) and indeterminate relation (i.e., cannot be expressed as a function of N and the attribute states alone) to at least one of the attributes.

Argument:

Consider a discrete d-space U of cardinality N, with attributes E and P such that the number of attribute states of E is much greater than the number of attribute states of P. Further, consider d-subspaces L, R and S of U.

For a particular attribute A, we will represent the attribute distance from one d-subspace X to another d-subspace Y by $d(A : XY)$. By $V(A)$, we will mean the maximum of an attribute velocity $v(A)$ in the attribute A. By $C(X : EP)$, we will mean the minimum computational power necessary to represent the relationship between the attributes E and P in the d-subspace (or d-space) X.

Let the combined cardinalities of L, R and S be represented by M, and suppose that the number of attribute states of E in U is greater than $M+\log_2 M$ (Theorem 13); then there exist sequences of attribute states of E algorithmically producible within U, which cannot be differentiated from randomly distributed sequences of attribute states from within L, R or S, or any combination of L, R and S. Now consider the further relationship between E and P within U. Suppose that E is related to P via a function F which, by virtue of the fact that the number of attribute states of E is much greater than P, is a many-to-one d-map. It follows that the relationship F cannot be known within L, R or S, even when well-defined on U. Clearly, such a system is capable of exhibiting local "random" behavior.

Furthermore, it is clear that there must exist correlations (or anticorrelations) of P in $L + S$ and P in $R + S$, since this relationship is completely determined by F and incompletely expressible to either $L + S$ or $R + S$.

L, R and/or S are not large enough to discern the algorithmic relationship between P and E. By hypothesis, the maximum attribute velocity of P, $V(P)$, is greater than the maximum attribute velocity of E, $V(E)$. It follows that the correlation of P between L and R in U is limited by the velocity $V(P)$ rather than $V(E)$, and is thus nonlocal. Within the context of describing the system via the attribute E with maximum attribute velocity $V(E)$, these correlations appear instantaneous, based upon measurements of $d(E : SR)$ and $d(E : SL)$.

QED

Theorem 45: The attribute having the infimum of d-set of maximal attribute velocities for a maximal attribute velocity also has the smallest of the corresponding minimal attribute distance increments.

Theorem 46: The attribute having the infimum of d-set of maximal attribute velocities for a maximal attribute velocity corresponds to the attribute having the largest number of possible attribute states.

Theorem 47: The maximal range of attribute velocities over which relationships may be specified between arbitrarily selected attributes defined on some d–space is bounded from below by 0 and from above by the infimum of the d–set of maximal attribute velocities $V_{\min}$.

Argument:

Trivially, from the fact, if the zero attribute velocities are identified equal, then d–maps between attribute velocities can only be 1–1 over the interval $[0, V_{\min}]$.

QED

4.2 A COMBINATORIC CONSTRUCTION OF COMMUTATION RELATIONS

The commutation relations as normally understood in quantum mechanics actually involve two quite distinct principles. The first is the principle that noncoordinate bases do not commute. Given a coordinate system x^i, one can adopt the derivate operator d/dx^i as a basis for the vector field. However, any linearly independent set of vector fields can serve as a basis, and one can easily show that not all of them are derivable from coordinate systems. This is because the operators d/dx^i and d/dx^j commute for all i, j, while two arbitrary vector fields do not commute.

The Exponentiation of the Derivate Operator d/dp

Theorem 48: The transport $p_0 + \epsilon$ along $x^i(p)$ may be given as

$$e(N)^{\epsilon d/dp} x^i \big|_{p_0} . \tag{68}$$

Argument:

Let $D = d/dp$ evaluated at some point p_0 on a particular coordinate parameterization. Suppose the coordinate values $x^i(p)$ of points along the integral "curves" of a "vector field" d/dp are discrete functions of p; then the coordinates of two points with parameters p_0 and $p_0 + \epsilon$ are related by Eq. (52):

$$x^i(p_0 + \epsilon) = x^i(p_0) + \epsilon \left(\frac{dx^i}{dp} \right)\bigg|_{p_0} + \left(\frac{1}{K!} \right) \epsilon^K \left(\frac{d^K x^i}{dp^K} \right)\bigg|_{p_0}$$

$$= e(N)^{\epsilon \left(\frac{d}{dp} \right)} x^i \bigg|_{p_0} \tag{51}$$

where $e(N)$ is just the power series expansion of e truncated at the Nth term by the definition of $e(N)$.

QED

Discrete Geometric Interpretation of Generalized Commutation

We will use the shorthand notation for Eq. (52) developed in the previous theorem in the derivation of the discrete commutation relations which follows[*]

Theorem 49: The order dependence $x(B) - x(A)$ of the derivate operators d/dp, d/dq is given by

$$x(B) - X(A) = \left[\frac{d}{dp}, \frac{d}{dq}\right] + \bigcirc\left(\epsilon^3 \frac{d^2}{dp^2} \frac{d^2}{dq^2}\right) .$$ (69)

Argument:

Notice that by definition of a coordinate basis (orthonormality), x^1 is constant along the lines of x^2, which are the integral curves of the derivate operator d/dx^2. That is why the derivate operators d/dx^1 and d/dx^2 commute: each is a derivate along a line on which the other is fixed.

Consider a basis d/dp combinatorially produced by Bernoulli trials vis-a-vis an ordering operator. Consider a second basis d/dq similarly, but independently produced. Now consider a transformation from one basis to the other; i.e., we seek a transformation which takes us a distance ϵ from a point P to a point R in x^i, using λ for transport; see Figure 14.

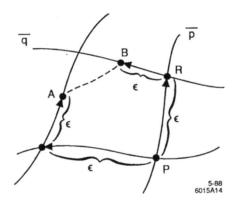

Figure 14   *Transport from a point P to a point R, using d.*

The two arbitrary vector fields V and W are defined by $V = d/dp$ and $W = d/dq$. Even the fact that the parameterizations look like that of a coordinate system is an

---

[*] Adapted from B. Schutz.[31]

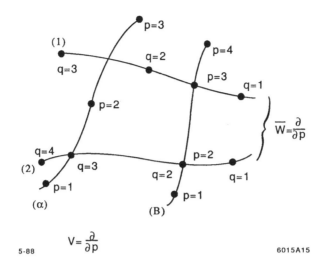

Figure 15   *Relation between parameterization and transport (see text).*

artifact of 2–space; in 3–space it may happen that curve 2 intersects curves a and b, but that curve 1 only intersects curve a; see Figure 15.

We obtain a picture of the vector $[V, W]$ in the following manner. Consider a starting point P, moving $dp = e$ along the V curve through P, and then moving $dq = e$ along the W curve. One winds up at A. Starting again at P and going first along the W curve, and then along the V curve, takes one to B. The vector stretching from A to B is $\epsilon^2[V, W]$, to lowest order in ϵ; see Figure 14.

The transport along x from P to R in discrete step lengths is just:

$$x(R) = e(N)^{[\epsilon d/dp]}x \quad \text{at} \quad P . \tag{70}$$

Now assume that we have similar relationships for d/dq. For a point A in x, ϵ distance away from P along d/dp and ϵ distance further along d/dq, the transformation is just the product of the two operators (i.e., transform along d/dp, then along d/dq).

$$x(A) = e(N)^{[\epsilon d/dp]} \times e(N)^{[\epsilon d/dq]} x \quad \text{at} \quad P . \tag{71}$$

Similarly, we may travel from P to a point B, which is located by just changing the order of the transforms. We then obtain

$$x(B) = e(N)^{[\epsilon d/dq]} \times e(N)^{[\epsilon d/dp]} x \quad \text{at} \quad P . \tag{72}$$

Now find the distance from B to A:

$$x(B) - x(A) = [e(N)^{[\epsilon d/dp]} \times e(N)^{[\epsilon d/dq]}$$
$$- e(N)^{[\epsilon d/dq]} \times e(N)^{[\epsilon d/dp]} \, x \quad \text{at} \quad P \, . \tag{73}$$

Now we undo our shorthand notation for Eq. (52), in order to multiply out the terms explicitly, and explicitly ignore higher-ordered terms which result. Expanding, we have the right-hand side of Eq. (73) as:

$$\left[1 + \frac{\epsilon d}{dp} + \tfrac{1}{2} \frac{\epsilon^2 d^2}{dp^2} + \bigcirc(\epsilon^3) \quad , \quad 1 + \frac{\epsilon d}{dq} + \tfrac{1}{2} \frac{\epsilon^2 d^2}{dq^2} + \bigcirc(\epsilon^3) \right] \, . \tag{74}$$

This is just

$$= \epsilon^2 \left[\frac{d}{dq}, \frac{d}{dq} \right] + O(\epsilon^3) \, . \tag{75}$$

Thus, for two discrete operators ("vector fields") d/dp, d/dq which are not part of the coordinate d–basis x, the commutator is just the open part of an incomplete parallelogram, whose other sides are equal parameter increments along the integral curves of the vector fields. Note that the parallelogram is complete if and only if d/dp, d/dq are one to one with the coordinate d–basis; see Figure 16.

QED·

It is important to understand how the operators which generate discrete distance functions might not be a part of the coordinate d–basis. Earlier, we noted that two ensembles A and B with increment and decrement I, D and I', D', respectively, were said to be independent if and only if all the defining states for A and B were distinguishable.

Theorem 50: For any two bases P and Q, the commutator of P and Q vanishes if and only if P and Q are independent; i.e., if and only if P and Q are coordinate bases.

Suppose that not all the defining states for A and B are distinguishable; then for some generation of the ordering operator, a redundant attribute state (instance) is generated. As a result, the additive law for attribute distance must fail; i.e., the sum of the total sizes for A and B does not equal $S' + S$. The sign of the deviation depends upon whether the deviation from $S' + S$ is accounted for by a deviation from $D + D'$ or by a deviation from $I + I'$ in the summation. Although the deviation can be treated as an attribute distance in its own right (indeed the inverse function of the additive law encourages us to do this), the ordering operator required to generate this deviation is clearly not independent of the generation of the two ensembles (consisting of a mixture of distinguishable and indistinguishable states), and is absolutely independent of the representation of both ensembles as being strictly independent (i.e., incorporating only distinguishable states); thus, it may be counted as a basis which behaves locally as an independent dimension.

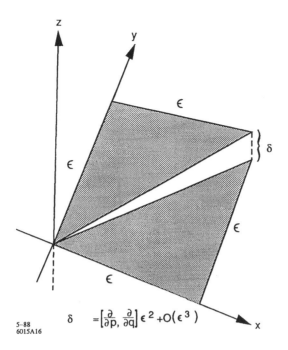

$$\delta = \left[\frac{\partial}{\partial p}, \frac{\partial}{\partial q}\right]\epsilon^2 + O(\epsilon^3)$$

5–88
6015A16

Figure 16   *Incomplete closure for parameters that are not part of a coordinate d–basis.*

If any distinguishable states are shared between the two coordinate parameters (i.e., one parameter is a function of the other), the product of the transports becomes order dependent: the computation of attribute distance for the first basis transport consumes the state and, thus, alters the ratio of distinguishable to total states for the second basis transport. Since the derivates for the basis are not in general the same, this results in a nonvanishing commutator. On the other hand, if the bases are independent, the commutator will clearly vanish.

Theorem 51:   The commutator is bounded above and below.

Argument:

In a finite system, the commutator can clearly be no larger than the absolute maximum attribute distance representable in the dependent basis, where we assume that a dependent basis provides less information than the independent basis. Hence, the commutator is bounded. If the dependent basis has cyclicity ξ with respect to the independent basis, mapping each successive ξ distinguishable attributes of the independent basis to the same ξ attributes of the dependent basis, then the commutator is bounded by ξL (and in fact is equal to ξL), where L is the "conversion length" between bases. Based upon arguments previously given regarding dimensionality, it is clear that fluctuations of the commutator less than ξL are not consistently representable within the n–space (i.e., they occur between meter marks).

QED

Theorem 52:   If $P = P(Q)$ is a first order derivate, then Eq. (75) is exact without higher-ordered terms.

Argument:

Since higher-ordered terms in Eq. (75) depend on higher-order derivates not vanishing, the theorem follows immediately.

QED

Theorem 53:   For bases P and Q, if P is cyclic in Q (an angle variable), then

$$[P,Q] = \pm \frac{i \, \text{Constant}}{2\pi(N)} \,, \tag{76}$$

where $\pi(N)$ is just the discrete computation of π by the combinatoric method in a d-space of cardinality N, as given above.

Argument:

If the indistinguishable attribute states involved combine to behave as distinguishable attributes in the proper manner, this independent dimension will behave mathematically just as though it were imaginary. Suppose, as in Theorem 52, that one of the two bases P is a function of the other:

$$P = P(Q) \,. \tag{77}$$

Furthermore, suppose that $P(Q)$ describes either a closed "orbit" or a periodic function of Q. If one of the bases is cyclic, its "conjugate" basis is constant. The corresponding orbit in the QP discrete 2–space is then just a "horizontal straight line." Following Goldstein[32], the "motion" may then be considered as the limiting case of a rotation type of periodicity, in which Q may be assigned an arbitrarily long period (subject to N, of course). This is just a change of coordinates from the real coordinate P to an imaginary coordinate J in a complex discrete 2–space, following the usual practice of using complex plane to represent such a change of coordinates; see Figure 17.

Since the coordinate in a rotation periodicity is invariably an angle, such a cyclic Q always has a natural period of $2\pi(N)$. Accordingly, the length of the path in QP discrete 2–space evaluated from 0 to $2\pi(N)$ is just $2\pi(N)$ and QP becomes:

$$J = 2\pi(N) \times i \times p \,, \tag{78}$$

for all cyclic variables. Note that we evaluate π for cardinality N here. That is, we construct the combinatoric valuation of π on the global d–space of cardinality N, and not on the local d–subspace; then we map minimum Q to 0 and maximum Q to $2\pi(N)$. The value of Q measured as an angle is then discretized in increments of $2\pi/N$ from 0 to 2π by the mapping.

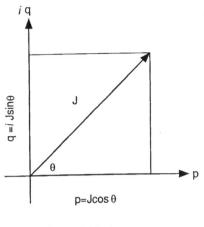

Action variable J
5-88   Complex angle variable θ   6015A17

Figure 17   *Relation between p, q and angle-action variables.*

Given Eqs. (74) and (76), we may now express the commutation relation between J and Q:

$$[J, Q] = [2\pi(N)iP, Q] = \text{Constant} ,$$

or

$$[P, Q] = +i(\text{Constant})/2\pi(N) .$$

QED

From our earlier result, however, the general commutation to first order is just

$$[P, Q] = \frac{L}{\epsilon^2} . \tag{79}$$

If P and Q are linearly related, then the higher-ordered derivates vanish and Eq. (79) is exact. If we then take ϵ to be the minimum nonvanishing discrete value, with suitable reparameterization, we have

$$[P, Q] = L , \tag{80}$$

for the least increment in the complex angle variable.

5. A DISCRETE CONSTRUCTIVE MODELING METHODOLOGY

5.1 DEFINITIONS

Having developed the elements of a discrete, finite and computational formalism via the ordering operator calculus, we proceed to a mathematical foundation for a discrete and constructive modeling methodology. Such a methodology will allow us to use the ordering operator calculus to model various phenomena which do not have the intrinsic properties required by continuum mathematics.

We motivate the modeling methodology through a variation of a dictum issued by Bastin and Kilmister[33] in 1973 concerning the separability of syntax and semantics in a mathematical system, which we refer to here as the Separability Lemma.

> *Separability Lemma:*
> *A system has a mathematical structure (syntax) which can be expounded separately from the interpretation of it (semantics), provided that it is understood that the mathematics describes a process which can be represented as a computer program.*

Clearly, the ordering operator calculus meets the criteria of the Separability Lemma as demanded by Principles I–IV. We are now ready to define a **modeling methodology** which consists of three broadly-defined structures: an epistemological framework, a representational framework and a procedural framework.

An **epistemological framework** or **E–frame** is a d–set of loosely-defined agreements made explicit by those engaging in the process of modeling (i.e., by injecting information into the model formulation).

1) *Agreement Upon Intent*

> The intent of the modeling effort must be agreed upon. The practice being modeled must be identified. It is also desirable to establish agreement regarding the conditions under which the effort will have been determined to fail, means of validation, the degree of accuracy required of the model (a stop rule) and rules for evaluation.

2) *Agreement On Observations*

> The ensemble of objects O, which constitutes the observations of and about the practice must be agreed upon.

3) *Agreement of Cooperative Communications*

> • commonly defined terms as fundamental

>> Fundamental terms, as used in describing the practice, must be understood. They CANNOT be defined.

> • fundamental versus derived terms

>> An operational distinction between fundamental and derived terms must be practiced.

- <u>agreement of pertinence</u>

  Engaging in attempts to communicate about the practice being modeled must be founded on an agreement to assume and attempt pertinence.

4) *Agreement of Explicit Assumptions*

   There must be an agreement to make assumptions explicit, rather than allowing them to be implicit.

5) *The Razor*

   - <u>agreement of minimal generality</u>

     The "scope" of the modeling effort at any point in the evolution of the model should be constrained to manageable proportions.

   - <u>agreement of elegance</u>

     The model should display a consistent and transparent structure, which minimizes the statement (size) of the model, while maximizing its explanatory (and in the event of a theory, its predictive) power.

   - <u>agreement of parsimony</u>

     The model should contain as little as possible that is either (a) sufficient but not necessary, or (b) necessary but not sufficient in modeling the intended practice.

A **representational framework** or **R–frame** is an abstract formalism FS, consisting of a set of symbols F and a set of rules of manipulation I. It is an uninterpreted typography.

A **procedural framework** or **P–frame** is an algorithm which serves to establish rules of correspondence C between the observations O (as agreed upon in the E–frame) and the symbols of the R–frame F, and which then, through recursion, serves to modify the rules of correspondence and the E–frame and R–frame, until a sufficient level of agreement concerning accuracy is achieved or the model fails. Kuhn[34] would call such a failure a "crisis," which in the fullness of time will lead to a "paradigm shift."

Thus, we see a relationship between two d–sets being established (the O and F), with two d–sets of rules (I and C) for modification and/or information extraction.

We now cast this in terms of the ordering operator calculus and, specifically, of the finite differential geometry which we have constructed within it.

An **ob set** O is an ensemble of observations. The obs are differentiated (altered from a d–sort to a d–set) by one or more ordering operators, which serve to establish the lattice structure of the obs.

An **ob subset** is a d–set of obs, defined by at least one ordering operator. They may be multiordered and multiply-connected.

244

A **d–sort of formal symbols** F is an ensemble of labels which may be ordered (converted to a d–set) by a d–set of rules of manipulation I. The resulting d–set FS of formal symbols F with rules of manipulation I is called a **formalism** or **representational framework**, and may be either closed or open under the rules. Generally, this serves to form a abstract combinatorial system.

A **P–frame rule of correspondence** is a binary d–map between an element of F and an element of O. In practice the d–set of all rules of correspondence established up to some step in the modeling procedure are expressed as a *dictionary:* given an element of O one may look up a procedure for finding the corresponding element of F.

A **procedure** P is a bounded, recursive algorithm which (a) provides a recursive and exhaustive enumeration of the elements of O and the elements of F, such that there exists a smooth d–map between O and F in the sense given above, constructed from the d–set of rules of correspondence and which (b) provides a recursive reparameterization of the d–map, such that there exists a 1–1 d–map between a d–subset of O and a d–subset of F.

Ideally, the cardinality of these d–subsets increments with each recursion of the P–frame procedure, up to the cardinality of O itself.

5.2 OBSERVATION SPACE

We begin with a number of observations which may be clustered[*] (grouped into prearranged classes) into d–sets O_i. These observation d–sets are said to cover the observation d–space O in the sense that $\bigcup_i O_i = O$. Because our O must have boundaries—for any hypothetical O_u (O is a d–subset of O_u)—and is discrete, O is non–Hausdorf.

Clearly, for any finite O, there are a finite number of possible disjoint partitions of O; namely, $\sum_k k!/n!(k-n)!$, where n is the cardinality of O, and k ranges from 0 to n. However, the partitions need not be disjoint—we allow dependent observations, and any ob to be in more than one partition. Thus, the number of partitions may be as large as we wish, being determined by the bound we place on the combinatorics of repetitive sampling with replacement.

It is often convenient, in the absence of any constraints, to take a discrete version of R^N as the image space, thus allowing an analytic interpolation for functions defined on the space. We map each partition O_i of O to some subspace S_i of R^N by some d–map R_i. If each such subspace S_i of R^N is arbitrarily "labeled" with some formal symbol F_i, then the partitions O_i of O may be taken as "objects" in O and referred to by the F_i. The R_i then form rules of correspondence.

We define relationships between the O_i objects in terms of the coordinate transformations between the S_i.

---

[*] We will discuss methods of clustering compatible with the modeling methodology and the ordering operator calculus in a later paper. Note that if a distance function or a norm is definable on the O, the method of minimal distances may be used as a clustering algorithm for the partitioning. Methods based upon a general attribute distance function are closely connected to a general theory of computational measurement, in development.

Note that our definitions tell us immediately that there is no *a priori* **parameter-**ization on S which gives a preferred reference frame. In fact, there is no structure at all on S without a parameterization. There exists no metric, only local topology induced on O by F via R. The global topology is given by the cardinalities of O and F and by the partitioning of O, as well as by coordinate transformations between partitions induced by requiring that the formal rules of manipulation I_i map isomorphically to O, via the rules of correspondence R_i, giving connectivity to the topology. The image in O under R of I may leave invariant certain attributes of O, the study of which provide an understanding of the structure of the formal model of O.

5.3 The Modeling Methodology Algorithm, Models and Theories

We now give a specific P–frame algorithm, which meets the criteria established in the preceding section, and which establishes and guides the evolution of the model.

1. Choose the ob set O with n elements. This is a recursively enumerable d–set with cardinality n.

2. Partition the ob set O.

    a. Define the n obs (labels) by partitioning the d–set O into disjoint d–subsets.

    b. Choose a set of symbols O_i for these partitions, labeling them.

3. Select or develop an abstract formalism FS meeting the criteria of an R–frame.

4. Choose a set of rules of correspondence R between the symbols O_i of O and the formal symbols F.

5. Map to some space such as R^N. (We can always choose our discrete version of R^N locally for the d–map, although we must then define the obs on open d–sets.)

6. Determine relationship between obs vis-a-vis the formalism. In particular, determine the image of the I in O under R.

7. Establish a set of coordinate transformations and determine the induced structural invariances, in order to identify the interpreted global properties of the model.

8. We say that this procedure establishes a **model**, if the cardinality of the O_i is the same as that of F and if R is an isomorphism between O and FS. If the isomorphism fails, we call the result a **theory**, in that it has predictive power. In empirical practice, we will rarely obtain a formal model.

9. If a model is not established because the isomorphism fails, then recursive application of the P–frame procedure is required to evolve the model. While no deterministic algorithm may be given which prescribes how the model should be altered, given a certain failure of the isomorphism, the P–frame procedure P allows one to develop heuristic knowledge about the modeling practice and how best to proceed in modifying the model. This heuristic knowledge may be

made explicit within the E–frame from the outset and, indeed, becomes a part of the E–frame via P–frame recursion.[†]

Keep in mind that through P–frame recursion, one has many options: we may alter the partitions of O, the range of the maps R, the coordinate parameterizations on O, the d–maps R, the rules I, and so on. Each recursion of the procedure P modifies one and only one such aspect of the model; in so doing, the entire model must be reexamined for consistency and completeness of the representation, as each change alters the definition of one or more ordering operators. These modifications are necessarily inductive, and therefore have unpredictable consequences.

The revolutionary step is taken based on an inductive decision that a Kuhnian crisis[35] has developed. This is largely based upon *subjective* criteria concerning the viability of the model and, in some sense, an *intuitive* measure of the relative benefits of proceeding, starting over or opting for a radical revision. It is important to note that such criteria can be agreed upon as part of the E–frame; namely, agreeing in advance how much and what kind of deviation from the required isomorphism will be tolerated and how the validity of the modeling effort will be judged.

We halt the classical infinite regression of analysis of terms in modeling by recognizing the effect of the epistemological framework. We deny the validity and the value of any attempt to analyze "theory-laden"[36] language as used in the E–frame. Such an analysis lies outside the purported task of generating a specific model, and would require us to generate a model containing the specific model, as an instance. In particular, analysis of fundamental terms involves treating these terms as the ob set for a modeling effort. In keeping with the agreed upon intent of the modeling practice and our methodology, we can not engage in such analysis. The practice would necessarily involve nonconstructive methods: the analyst would have to work from the specific model by generalization, having failed to construct the general model first. The transition from the specific to the general is not only inductive in nature, but not recursively definable, and constitutes a revolutionary redefinition of the modeling effort as specified in the agreement of intent.

Note the implication here that it is possible to work from the general to the specific. It is possible to constructively "model the model" or even the modeling process. Indeed, part of the power of our modeling methodology lies in the constructive and recursive nature of the process.

In practice, we always bootstrap into the modeling process with a set of loose agreements and definitions (we don't really know what we are talking about), but the ordering operator calculus gives us a consistent mechanics of typography and the procedural framework gives us a recursive method of evolving toward an acceptable model and definitions. Once the process has begun, each pass through the P–frame may generate a modified, but nonetheless well-founded and well-defined, E–frame and R–frame. Constructively, we may keep records of our efforts and review these at will. On starting the effort, we have no record of earlier effort and no way of (re)constructing one; we may make no constructive claims regarding either the earlier

---

[†] We call the process of exercising P *recursion*, rather than *iteration*, because it operates on itself, as well as the model. In some sense, P, together with the modeling participant operating on the model, constitute a self-organizing system.

effort or the results of that earlier effort. In some sense, we, thus, have a "fixed past and uncertain future," but with a fixed starting point.

5.4 HIERARCHICAL MODELS

We will frequently have cause to deal with hierarchical structures. For this reason, we give a P–frame algorithm for constructing **hierarchical models** as a constructive definition.

1. Start with a model.

2. Specify a many-to-one d–map from the formalism F to ob labels O_i.

3. Redefine the partitioning via the process of refinement, mapping from the image in O to the representation d–set F with new mutually disjoint partitions, using inverse d–map of the R_i. This insures consistency for next step.

4. Remap the formalism from new partitions induced in F under the inverse of the R_i to the image space O, using old mapping R.

5. Keep in mind the constraints of a many-to-one d–map. This d–map provides inclusion relations on the d–set F; thus partitions contain partitions or parts thereof, forming a lattice of partitions.

Theorem 54:   For each model with multiple partitions mapped to a representational framework without disjoint refinement, there exists a hierarchical model with an equivalent local topology.

6. AN INTERPRETATION: LABORATORY PHYSICS

6.1 ESTABLISHING THE E–FRAME

We start on the route to physical interpretation by adopting the constructive modeling methodology developed in the previous chapter. We must, therefore, state explicitly the E–frame, the R- frame and the P–frame. Within the E–frame, we adopt as our agreed upon intent the modeling of the current practice of physics. We take as fundamental the commonly defined terms of laboratory physics, treating terms denoting nonobservables as derived or theoretical terms. Our understanding of the current practice of laboratory physics is guided by the "counter paradigm"[37].

Any **elementary laboratory event,** *under circumstances which it is the task of the experimental physicist to investigate, can lead to the firing of a counter.*

In this context, by "can lead to the firing of a counter," we implicitly allow for any measurement apparatus which involves discrete and finite measures, i.e., counting. Inasmuch as all laboratory measurements are normally viewed as bound by limitations of precision and resources—which bounds for us are evidence of the intrinsic finite and discrete character of the practice—few, if any, laboratory measurements are excluded by the counter paradigm; one must make the connection to counting explicit. We take laboratory events as a sufficient set of observations to be modeled, without requiring the standard theoretical interpretation. We take as understood that an experimental (laboratory) measurement may encompass many acts of observation and, thus, that our obs may be complex (e.g., multiply-connected). In other words, we are not committed to accept the how and why of the observations, only the observations themselves, operationally understood.[*] If the internal structure of an act of measurement is to be examined, then there must exist a finite procedure for carrying out the measurement (i.e., the measurement must be operational), so that the internal structure is transparent. Otherwise, we are required by Principle I to plead ignorance of the apparent internal structure.

We have now satisfied the requirements of establishing an E–frame, inasmuch as the requirements have to do with making explicit various aspects of the modeling effort. As to whether or not we are faithful to the other strictures of the E–frame, we shall leave it to the reader to decide, this being the very nature of consensual validation of the value of our effort.

6.2 ESTABLISHING THE R–FRAME

As our R–frame formalism, we adopt the ordering operator calculus. Inasmuch as quantum events, as understood within the current practice of physics, are unique, discrete, irreversible, nonlocal and yet indivisible, the principles upon which development of the ordering operator calculus was based make this an appropriate formalism.

6.3 ESTABLISHING THE P–FRAME

As our P–frame procedure, we select the algorithm given in the preceding chapter. We note in advance that some detailed aspects of the model are evolving. In particular, we are in the process of refining the specification of the d–space generator required by our formalism. This will have consequences regarding the detailed specification of any distance function on any attribute we identify. In addition, any global invariants are likely to be affected. Thus, the detailed identification of physically conserved quantities within the theory is tentative, though their existence is not.

As noted in Section 5.1, the rules of correspondence may now be elucidated in the form of a dictionary. If we establish rules of correspondence between obs from the E–frame and symbols in the R–frame, any relationship between the symbols in the R–frame must reflect relationships within the context of the E–frame, whether known at this time or not. We, therefore, adopt rules of correspondence which are more useful than current practice in relating observations to the R–frame, and then see how the practice of *discrete physics* will differ from the current wisdom. In other words, we hope to see how the E–frame (and perhaps the R–frame) should be modified. Bridgman tried long ago to get rid of the representational framework by "operational" rules of procedure that reflected directly back into the E–frame. We expect that it would be conceded by most physicists that this heroic effort failed in its initial intent, and even Bridgman was led to modify it by including "mathematical operations" within the allowed procedures. One related effort was to reduce everything in physics to "pointer readings." Our methodology is even stricter in that sense, since we require every E–frame procedure and every R–frame construct to be reducible, at least in principle, to *counting* and finitely computable algorithms. We hope to have accounted for the philosophical and technical problems which led to the failure of Bridgman's operationalism.

---

[*] Note the distinction between E-terms and R-terms. Von Neumanns "observation" is, at best, only an R-term. Criticism of von Neumann's representation of quantum mechanics can start there, because his R-term is not necessarily consistent with Schrodinger continuity.

Spatial Distance

For us, an attribute distance is the only thing in the R–frame that can correspond to a datum (E–frame) achieved by an experimental measurement within the practice of physics (E–frame). From the R–frame, however, we see that attribute distance has no computational meaning or significance outside the context of a particular reference frame, or without some ordering parameter (R–frame symbols). We do not make an absolute rule of correspondence between attribute distance and spatial distance; spatial distance will be a particular attribute distance. For us, however, any quantifiable experimental measurement must correspond to some attribute distance.

Cosmological and Proper Time

As noted above, we take the notion of sequence and counting in the laboratory as fundamental, so that the very character of observation in time (E–frame) is bound to the R–frame notions of counting, synchronization and both local and global ordering. We establish a rule of correspondence between laboratory proper time (E–frame) and the ordering parameter t_i (R–frame), associated with the generation of any particular reference frame F_i, via an ordering operator O_i. Similarly, we must establish a rule of correspondence between cosmological time and the global ordering T, associated with the generation of all reference frames within the model. That the global ordering may be specified in terms of the R–frame synchronization of attributes identically to the E–frame synchronization of events, establishes a requirement that events be specifiable as some particular kind of attribute. A significant portion of this section will be devoted to establishing the required nature of event attributes.

Three Dimensional Physical Space

As seen in Theorems 43 and 44, for any attribute space, no matter how simple or complex, there is some attribute which has the greatest number of attribute states of all the attributes which may be defined on the d–space. From Theorem 46, it is also clear that the corresponding attribute velocity for this attribute will be the infimum of the d–set of maximal attribute velocities. Finally, from Theorem 44 and by definition, this maximum attribute velocity will be the first bound encountered in any function involving more than one attribute. For these reasons, we identify this unique attribute velocity with the (E–frame) speed of light c, and the corresponding attribute states with the points or "4–positions" of physical space. Note that these points are events in the sense of the geometric view of general relativity.

As demonstrated in Theorem 13, for any attribute distance function, there are at most three independent runs of the ordering operator which generates these attribute states, if the global character of the d–space so generated is that it not have a preferred coordinate. Thus, the d–dimensionality of the attribute space is three, and we establish a rule of correspondence with the three–dimensionality of laboratory space.

The Global Structure of d-Space Generator

The next rule of correspondence must specify an ordering operator U, which generates the coordinate d-bases and a reference frame (R–frame) suitable for identification with the spatio-temporal reference frame (E–frame). This ordering operator U must provide the appropriate global invariances, if the identification is to be successful.

The relevant E–frame global invariances include the fundamental constants, the scale constants and the quantum numbers. For these invariances to be generated via a discrete algorithm suggests a hierarchical structure with a stop rule. For further justification of these requirements, see Bastin, 1966[38] and Bastin, 1956[39]. We may interpret the generators of each level of such a hierarchy to be coupled ordering operators; then the coupling scale may be calculated by definition, together with probabilities of coupling between the levels, which must be the coupling constants of laboratory physics.

We allow multiple, independent, but synchronized, runs of the U in order to generate a discrete space, without a preferred axis, and preserving translational invariance (i.e., having a homogenous distance function). By Theorem 13, the dimensionality of this d–space will then be three; that is, we need only three independent runs of U or any other generator of the d–space, as additional runs will not produce additional global structure. The unobservable universal (cosmological) and locally consequential (proper) time will then* be given by the universal ordering parameter associated with U.

As noted previously, an ordering operator U may be understood as generating bit strings, instead of labels which we take as abstract representations of physical attribute states. Here, we invoke the principles requiring that any specified attributes of a finite and discrete ensemble can be mapped onto an ordered sequence of 1's and 0's, by asking whether they are present or absent in a reference ensemble. Such an ordered sequence is called a **bit string**, and may combine with other sequences of the same bit length by an operation such as XOR ("exclusive or"), symmetric difference, addition (mod 2), $+_2$ .... When Noyes treats the symbols "0," "1" as bits and/or as integers, the more general *discrimination* operation "$\oplus$" defined by

$$S^a \oplus S^b \equiv (ab)_n \equiv \left[\ldots, \left(b^a_i - b^b_i \right)^2, \ldots \right]_n$$

$$= (\ldots, b^a_i +_2 b^b_i, \ldots)_n; \ b^\ell_i \in 0, 1; \ i \in 1, 2, \ldots, n; \ \ell \in a, b, \ldots,$$

is used. Note that discrimination meets succinctly the requirements for combining serializable ordering operators, if the bit strings are linearly independent; i.e., there is no information loss regarding the distance function on discrimination, if the resultant bit string is given a dual Hamming measure—one counts the 0's instead of the 1's—and discrimination is then a length preserving operation. For us, this is a required property of U.

U is further required, by the definition of ordering operator, to consist of an incompletely specified (though, in principle, specifiable) part, and a completely specified part. The incompletely specified part must not have an effect on the global structure, nor on the combinatoric complexity of its generation. As long as the structure generated, and the order in which it is generated, is compatible with the knowable Universe (E–frame), the unspecified part can be any algorithm, whatsoever.

---

* This result was anticipated conceptually by E. W. Bastin[40].

Non–Local, Discrete Events

What is now required is an R–frame definition of event. For us, this definition must follow the geometrodynamic point of view, in that the existence of an event depends upon an operation defined on strings (or similar representation of attribute states) and a distance function defined on these strings, which satisfies the so-called "triangle inequality." That is, the distance function must be a norm (see Section 2.4). Note that the definition of a norm requires a minimum of three independent strings. We establish a rule of correspondence which identifies the satisfaction of these conditions on the points of our physical space representation with the unique, nonlocal, yet indivisible and irreversible, events of quantum mechanics, since they meet the minimal conditions for nonlocalized operations on localized d–points which have a norm.

Defining $k_i^x(n) = \Sigma_{i=1}^n b_i^x$, $x \in a, b, c$, with n being the number of generations of the ordering operator, from the definition of a norm (Section 2.4), it is easy to see, for any three strings $(a)_n\,(b)_n\,(c)_n$ which satisfy the constraint $(abc)_n = (0)_n$ where $b_i^0 = 0; \in 1, \ldots, n$, that $|k^a - k^b| \le k^c \le k^a + k^b$ (cyclic on a, b, c) for any event. Thus, k, the number of "1's" in a string, can serve as a discrete distance function; in fact, this is just the Hamming distance. Note that the our definition of events necessarily will make them nonlocal. That is, a minimum of three independent and distance function ordered bit strings is required, although some attribute distance exists between them.

In order to locate the required reference frame "origin" (which in the R–frame corresponds to a reference ensemble) of our metric symmetrically in the finite and discrete interval allowed, we define an attribute distance $q_a \equiv f(k, n, \lambda_a)$—a linear function of k, n, λ_a, where λ_a has the dimensions of attribute distance and is identified via a rule of correspondence with a physical length. At each generation of the ordering operator, q_a changes by $\pm\lambda_a$, which we associate via a rule of correspondence with the minimum attribute distance increment of the R–frame, with the sign $+$ or $-$ being determined by whether a "1" or a "0" is concatenated with the extant string; i.e., whether the distance is increasing or decreasing, with respect to the reference ensemble. Note that if perfect synchronization is possible, λ_a is just $1/n$. This factor serves to normalize the distance on the [-1, 1] interval.

If we define the local event time (proper time) as a linear function of the ordering parameter $t = n\Delta t$, we see that we can define a velocity $v_a \equiv f(k, n, \lambda_a)V_x = \beta_a V_x$ where $V_x = \lambda_a/\Delta t$ is a *maximal velocity* of magnitude identified with the speed of light c, achieved when all the steps have the same sign (i.e., are in the same direction) and $f(k, n, \lambda_a)$ is a linear transformation of the Hamming distance k. We also have an *event horizon* that grows with the number of steps the generating operator has taken.

Lorentz Invariance

It is clear that q satisfies the definition of an attribute distance and satisfies β, as required in Theorems 23 and 41. We formally establish a rule of correspondence between that β and the usual β of special relativity. The specific dependence of λ on the generation of attribute states in the sequence given by the ordering operator is unknown, and, for our purposes, not required, as long as sufficient variety is produced. From Chapter 3 (and independent of the particular generator of the d–space), we have immediately a $3+1^\dagger$ discrete—and locally flat—space with distance function, which is invariant with respect to the coordinate transformations of Theorems 23 and 41 and with the previously stated rule of correspondence that the maximal attribute velocity for this "position" attribute corresponds to the velocity of light c; i.e., to the minimum of the maximal attribute velocities. We now identify the coordinate transformations of Theorems 23 and 41, *when applied to the position attribute*, as the Lorentz Transformations.

That the definition of velocity is indeed a first derivate of the position q is obvious. If q is linear in t, then we have $(q/n) \times (\lambda/\Delta t)$, where $\lambda/\Delta t$ is just the "slope." If q is not linear in t, then λ is a function of t, so that we obtain $(q/n) \times (\Delta\lambda(t)/\Delta t)$, which (evaluated at some q and t) give the "instantaneous" velocities. Furthermore, not only these velocities, but any attribute velocities, thus satisfy Theorem 35, which is now identified as the relativistic composition law for velocities.

Persistence Effects and de Broglie Wavelengths

By evolution of a system, we mean that some attribute states are invariant under some transformations on the system, and nothing more. When such attribute states are jointly identified and are invariant together, we say that they constitute an "object" which persists or is stable. We now note that if we consider a system that evolves with constant velocity—i.e., by a linear d–map, $\beta_0 \equiv f(k_0, n_0, \lambda_0)$—strings which grow subject to this constraint—i.e., $n = n_T n_0$, $k = n_T k_0$, $1 \le n_T \le n/n_0$— will have a periodicity $T \equiv n_T \Delta t = n_T \lambda/V_x$, specifying the events in which this condition can be met. Hence, in more complicated situations, where there can be more than one "path" connecting strings with the same velocity to a single event, this event can occur only when the paths differ by an integral number of attribute distance increments. We, therefore, establish a rule of correspondence between λ and the "de Broglie wavelengths." Thus, our construction already contains the seeds of "interference" and an explanation of the "double slit experiment."

The Relativistic Doppler Shift

From Theorem 33, and independent of the particular d–space generator, we obtain the relativistic doppler shift, as required from the laboratory evidence.

Supraluminal Correlations

Because the derivations in the development of the ordering operator calculus do not depend upon any particular interpretation, particularly those which could be read as referring to "physical distance," it is clear that the principles and axioms suffice to imply relativistic and quantum effects which could be identified with physical characteristics other than distance.

---

† Our use of 3+1, here, is meant only to emphasize the evolution of the ordering operator which locally distinguishes the ordering parameter, and not to deny the validity of the 4–space geometric view which is globally valid after the generation has taken place.

On the face of it, this is a surprising conclusion. However, for us, it demands that we treat the universe as a multiply-connected attribute space. If it is not the case that nonspatial attribute distances behave as does the spatial attribute distance, then either conventional or discrete theory must supply some reason for this difference. To our knowledge, making such a distinction has yet to be motivated in current analyses. Clearly, not all the attributes which may be generated in a discrete space will satisfy the precise definition given for q. Therefore, regardless of the generator of the d–space, we must conclude that the d–space is multiply-connected, with the consequences derived in Theorems 43–46. We show in this section that the theory encourages us to accept as "obviously possible" the disturbing facts demonstrated by the laboratory experiments of Clauser, Frye, Aspect and others[41]. Indeed, the theory predicts that such results could be obtained for quantum attributes other than spin and polarization. These results are predicted in the following way.

Theorem 43 describes the essential character of Aspects EPR experiments, where E is electromagnetic and P is polarization, S represents the source, L the left detector and R the right detector systems. The time-of-flight experiment does not alter the model, since this only serves to verify the "instantaneous" character of the anticorrelations. The results of such experiments are readily understood in this context.

Note that supraluminal communication is not allowed, since the connection between E and P is not 1:1 and is, in fact, locally "random." Furthermore, the theory is not a hidden variable theory, nor is it a nonlocal theory in the usual sense in which these are understood. We do not provide hidden variable extensions to quantum mechanics or to special relativity in order to understand the correlations: we provide a theory which reduces to quantum mechanics or special relativity under certain restricted interpretations (e.g., the existence of the continuum). We do not postulate an absolute nonlocal quantum multiple-connectedness, as is implied, for example, by Bohm's implicate order. Neither is the multiple-connectedness like that proposed by the branching universe of Wheeler and DeWitt. Rather, we postulate a topology which admits multiple, usually independent, distance functions and metrics.

For Aspect's experiments in particular, the global relation between polarization angle and electromagnetic propagation must be identified as some cosine-squared function. This function must be independent of the electromagnetic attribute distance identified as q, but dependent upon the polarization attribute distance—i.e., the difference between the polarization angles—by hypothesis. Since the least increment for polarization angle is defined by the event horizon N (i.e., from a computation of $\pi(N)$ via the method given in Chapter 1), we may expect that the number of spatial attribute states is approximately the square of the number of polarization attribute states. This suggests that the correlations seen by Aspect will fall off as the time for propagation of changes in the optical switches approaches the square root of the propagation delay for light.

We are led by the formalism to predict that there is a correlated rate of change of the optical switch, which destroys the correlation between the arms; namely, $V(P)$. That is, when the time T between switching in one arm versus switching in the other arm is short compared to $d(P : LR)/V(P)$, the correlation should be destroyed by our analysis. An examination of the correlation with T would show stronger correlation as T approaches $d(E : LR)/V(E)$ from below. One might reasonably expect the distribution to be exponential. Unfortunately, T is likely to be extremely short for any practical distance $d(E : LR)$.

The global topology of the discrete finite attribute space is multiply-connected. There is a unique attribute which serves to define a global metric; in our case, conventional 3–space as provided by the electromagnetic attribute. Globally, our d–space is necessarily limited to 3–space. However, locally a nonisotropic n–space may be defined. That is, if we no longer require translational invariance, there is no preferential coordinate, or if synchronization is not required locally in transforming between reference frames, one may define more than three independent, short runs of the parameterized bases which will behave (locally) as coordinates.

This topology, together with the fact that events as defined have intrinsic quantum interference properties, leads one to suspect that superluminal correlations should display quantum interference; namely, the "measurement" in the right and left detectors constitute events in both E and P attribute space. Suppose that the events are arranged in such a way that they are separated in E–space, but not in P–space. Furthermore, suppose that in P–space the events have wavelengths such that interference can occur. This interference should then modulate the correlation in E–space. Such a "correlation interference pattern" would be striking evidence of the proposed topology, since this cannot occur in distant (in E) events in the conventional theories.

Computer Models

We may model our system with the required topology on a computer.[*] In particular, the violation of Bell's Inequalities and related effects may be demonstrated in the computer model, since our formalism is strictly computable. Care must be taken in establishing the functional connection between E and P in the computer model, however. The connection must be sufficiently complex computationally to lead to the appearance of local (i.e., restricted memory) "random" behavior. This is just the problem of precision in computer modeling, used in reverse to establish certain statistical properties of the model. Indeed, it would appear that the model may be set up to demonstrate physical supraluminal correlations between physically separated computer systems in a distributed processing, shared memory environment.[†]

---

[*] As has been partially done for a particular 3–space generator.[42].

[†] Related Work: The relationship between this model and cryptographic techniques is interesting as well. A recent paper by Goldreich[43] considers a constructive approach to random bit strings based on computational complexity which is similar, though more specific and restrictive than that introduced in the present paper. In particular, the authors introduce programs that run in polynomial time and which lead to identical results when fed with either a set S of strings or elements randomly selected from the set of all strings.

Such poly-random collections can be shown to enable many parties to share efficiently a random function f in a distributed environment, by which we mean that if f is evaluated at different times by different parties on the same argument x, the same value $f(x)$ will be obtained. Such sharing can be achieved by selecting k-bits to specify a function in a poly-random collection. These k-bits are then communicated to and stored by each party. No further messages need be exchanged between parties to share f. It is a trivial matter to make the sharing either correlated or anticorrelated if $f(x)$ is two-valued. The physical communication of the k-bits may be dispensed with in a multiply-connected attribute space, as the k-bits may be "local" through some particular attribute. Thus, the k-bits are always available in "local" shared memory.

Mass and the Law of Relativistic Mass Change

We can associate a parameter m with the total size S [Eq. (44)] of the ensemble, and establish a rule of correspondence which identifies m (R–frame) with *mass equivalent* or *energy* (E–frame). Note that we differentiate between the mass and the energy. For a bit string in an evolving system to have an invariant mass at constant attribute velocity, the mass may be defined as the energy divided by some normalization factor, which depends on the cardinality of the attribute states which might be generated, and on the cardinality of the attribute Universe (R–frame). In this way, adding a distinguishable state (a '1') to the Universe and to the bit string do not alter the "mass" parameter in a measurable way, and results in a statistically invariant mass. For consistency with our finite principle, we must require $0 < k < 1$; thus, no massive event can lie on the event horizon.[‡] Independent of the particular generator of the d–space, Theorem 34 is interpreted as showing that the definition of this parameter follows the law of relativistic mass change.

Momentum Conserving Events

We require the existence of a norm for an attribute which can be identified with momentum, and in this way obtain momentum conservation. Once we have shown that the attributes of position and momentum can be identified (or equivalently, position, velocity and an invariant mass), and a norm in each of these spaces defined for a configuration which we identify as a quantum mechanical event, the generator of the d–space can be any algorithm whatsoever.

Defining $p_a \equiv m_a v_a = m_a \beta_a V_x = \beta_a m_a \lambda_a / \Delta t$ and establishing the rule of correspondence which identifies this as momentum, we see that $|p_a - p_b| \leq p_c \leq p_a + p_b$, provided only (as is required for consistency) $m_a \lambda_a / \Delta t$ is any finite constant independent of a. Thus, there is a norm in momentum attribute space. As Noyes would put it, the "triangle" thus closes in "momentum space," as well as "configuration space." Our d–events can now be interpreted as 3–momentum conserving, 3–particle scattering events in the zero momentum frame, with the "center-of-mass" of laboratory physics at rest.

These results are, of course, familiar in terms of so-called public key encryption systems. Here, a public key is distributed for encryption of messages to the key distributor. Although the encryption key is public, the cryptographic function does not allow decryption without access to the private key. And the number of possible private keys is too large to be determined by trial and error.

Actually, the entire scheme of shared random number generators has been put into effect. One can purchase a plastic card which contains a microprocessor. This processor produces an apparently random sequence of bit strings. When interrogated by a system which shares the random function, a match is produced and thus the card serves as a "key." Each card contains a k-bit code for the particular function and this serves to identify the particular user. Clearly, two cards with the same k-bit code would be perfectly correlated regardless of separation and yet would produce apparently random output.

‡ From the definition of maximal attribute velocity, we should be led to the mass conversion law.

256

Zitterbewegung

We have already seen that any system with "constant velocity" (i.e., at those generations of the ordering operator when events can occur) evolves by discrete increments $\pm\lambda$ in q *between* d–events. These steps occur in the *void* where space and time are undefined. Since $\lambda/\Delta t = V_x$, each step occurs forward or backward with the limiting velocity. Thus, we deduce a discrete *Zitterbewegung* from our theory. If we think of this as a "trajectory" in the traditional pq phase space, each time step induces a step $\pm\lambda$ in q correlated with a step $\pm mV_x$ in p. Even in the case of a particle "at rest," this must be followed by two steps of the opposite sign to return the system to "rest;" see Figure 18.

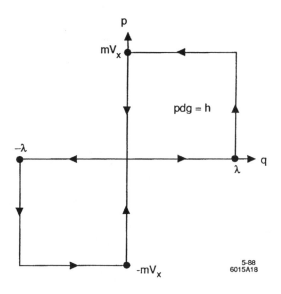

Figure 18   Zitterbewegung in phase space for a particle "at rest."

Thus, there is, minimally, a four-fold symmetry to the "trajectory" in phase space, corresponding to the generation periodicity we discovered above.

Commutation Relations, Uncertainty, Planck's Constant

From the E–frame definitions of the obs corresponding to p and q, and consistent with the present example, we see that p and q are not independent. It follows from Theorem 50 that p and q do not commute, and from Theorem 42 that there is an uncertainty associated with the product of the variances in p and q. We establish a rule of correspondence between the constant in Eq. (76) and Plancks constant. By definition, the least step in p is just mc, since this step occurs at the maximal attribute velocity. Once again, these results are independent of the particular d–space generator chosen.

Since the least change in the product of the variances is h by the rule of correspondence, it follows that the least step in q is appropriately identified as just $L = h/mc$. To go on to the commutation relations, we take the usual step in the geometrical description of periodic functions, of taking the $q_J J$ plane to be the complex plane $(q, 2\pi i p)$; then the steps around the cycle in the order $qpqp$ are proportional to $\pm 2\pi(1, i, -1, -i)$, where $\pm$ depends on whether the first step is in the positive or negative direction or, equivalently, whether the circulation is counterclockwise or clockwise.

We have now shown that $qp - pq = \pm i\hbar$ for free particles; this result holds for any theory which uses a discrete free particle basis.

The Angular Momentum Commutation Relations

Going to three dimensions, the commutation relations for angular momentum (as usually defined) follow immediately. Following T. F. Jordan[44], we may now derive the angular momentum commutation relations. Suppose we have P and Q in a discrete 3-space (i, j, k), related by a basis vector L, which we will call the angular momentum:

$$L = Q \times P ,$$

which is shorthand for three equations

$$L^i = Q^j P^k - P^j Q^k ,$$

with i, j, k taking all values from 1 to 3, and not equal to each other.

From the previous derivation of the P, Q commutation relations, we have

$$Q^j P^j - P^j Q^j = \frac{ih}{2\pi(N)} ,$$

$$L^i \times L^j = \frac{ihL^k}{2\pi(N)} .$$

For example,

$$L^1 L^2 - L^2 L^1 = (Q^2 P^3 - Q^3 P^2)(Q^3 P^1 - P^1 Q^3) - (Q^3 P^1 - Q^1 P^2)(Q^2 P^3 - Q^3 P^2) ,$$

$$= Q^2 P^3 Q^3 P^1 + Q^3 P^2 Q^1 P^3 - Q^3 P^1 Q^2 P^3 - Q^1 P^3 Q^3 P^2 ,$$

$$= Q^1 P^2 (Q^3 P^3 - P^3 Q^3) + Q^2 P^1 (P^3 Q^3 - P^3 Q^3) ,$$

$$= (Q^1 P^2 - Q^2 P^1) \frac{ih}{2\pi(N)} = \frac{ihL^3}{2\pi(N)} .$$

Similar results follow for each of the relationships involving other coordinates $(Q^1, Q^2, Q^3, P^1, P^2, P^3, L^1, L^2, L^3)$.

We have now shown that

$$L^i \times L^j = \frac{ihL^k}{2\pi(N)} \; ,$$

for free particles; this results holds for any theory which uses a discrete free particle basis.

Complete Identification of Laboratory Units

Now that we have shown, once given a specific generator of the 3+1–space, how to compute two ($\hbar$ and c) of the three dimensional constants needed to connect a fundamental theory to experiment in the 3–space in which physics operates, and which we have proved must be the asymptotic space of our theory, all that remains is to determine a unit of mass. Theorem 34 allows us to specify that the mass of an object is just the size S, although it does not tell us what object determines the fundamental unit. This can only be done once a specific generator of the d–space has been selected.

Scattering Range Computation

Once a unit of mass has been identified, we can show how to compute the classico-quantum scattering range from attribute distance. Note that for $h/(2m_p c)$, from the existing rules of correspondence for c, m_p and h, one obtains the following. Define an attribute such that the minimum attribute distance increment is I, with the following definitions holding: $h = I^2$ (minimum possible "area" in "phase space," $m_p = S = I + D$, and $c = v_{max} = I - D/I + D$. Thus, $h/(2m_p c) = I^2/2(I - D)$ when $v = c = v_{max}$; i.e. when $D = 0$. Therefore, we have $I^2/2(I) = I/2$, where I is just the minimum attribute distance increment for the attribute corresponding to the ensemble A with invariant size (mass) $I+D$. Clearly, since the 3–space is homogenous, we may interpret I as a diameter. Suppose a second ensemble B "approaches" with the first. Take two cases for the (generalized) attribute distance between them; $r > I$ and $r \leq I$. If $r > I$, then ensemble A may "travel" a distance I without any states in the generalized attribute distance being shared with the attribute states of B. If, however, $r \leq I$, then there exists the possibility of shared generalized attribute states between A and B, and thus nonindependence, exactly as described in the explanation of commutivity.

Note further, that if ensembles A and B do not have the same size and the same attribute distance definition for velocity computation, then the minimum interaction distance is not just the minimum of the "minimum attribute distance increments" for A and B, as compared via the generalized attribute distance. This is because the operation of addition is no longer well-defined: ensembles A and B are no longer independent, and this alters the generalized attribute distance definition.

Wheeler–Feynman and Massless Particles

Along other lines, we also have indications of how to compute transition probabilities from the ratios of the number of ensembles in given states, as determined by the combinatorial hierarchy. Let there be two attributes, such that the enumeration of states generated by the corresponding ordering operators are just the inverse enumeration of each other; that is, the last state generated by one is indistinguishable from the first state generated by the other, the next to the last state indistinguishable from the second state, etc. Further, let the representation of the states be duals ('0' in one represents the same thing as '1' in the other, and vice versa). From our rules of correspondence, these then correspond to particle and antiparticle.

This geometry suggests that zero mass particles are anomalous: no photon can be observed without both emission and absorption, and the path length in the photon frame is zero.[*] In the rest frame of the photon, any point on the photon trajectory can be treated as an electron/positron pair without violating relativity or the conservation laws. It would appear that photon emission/absorption is then modeled in our formalism as an electron emitted by the "emitter" and a positron (i.e., electron traveling backward in time) emitted by the "absorber," so that the photon can be treated as a virtual particle. From the reference frame of the photon, this exchange, and the evolution of the corresponding state vector, takes place atemporally. It is outside of time, happening everywhere along the photon path "at once." There is a difference in the energy of the two ends of the trajectory which is given by the torsion of the space—this being related to the constant identified above as Planck's constant, and to the minimal attribute distance increment exactly as in the (five–dimensional) Kaluza–Klein model. Thus, there is an apparent "transfer of energy" in the electron/positron pair exchange. This structure can *not* be detected locally. A similar argument holds for massless particles, in general.

6.4 RELATED RESULTS: THE COMBINATORIAL HIERARCHY

AND PROGRAM UNIVERSE

Bastin, Kilmister, Amson, Noyes and Parker–Rhodes have shown that there exists a unique finite hierarchy, combinatorially generated, which constructs at least some of the properties we require. This structure is referred to in the literature as the **combinatorial hierarchy**. Without developing the details here, we point out the essential features which make this structure interesting. First, the cardinalities of the primary objects (discriminately closed subsets) at each level of the structure are identifiable with the number of (E–frame) objects which may participate in the fundamental forces: to first order (which in our terms assumes first degree coupling only), they are the scale constants of laboratory physics (which we would identify computationally with the coupling scale of the relevant ordering operators). Second,

---

[*] This is just the Wheeler–Feynman rule, as was pointed out to us by H. P. Noyes. Indeed, the work of Cramer's transactional interpretation is in full agreement, and is an extension of the Wheeler–Feynman interpretation. That such an interpretation results in a time-symmetric, self-renormalizing QED with no singularities or second-quantization problem is indeed encouraging[45].

Parker–Rhodes has shown that the construction leads to an amazingly accurate *computation* of the ratio of the mass of the proton to the mass of the electron (consistent with the present work). Third, Noyes et al. have developed a particular algorithm, known as **Program Universe**, for generating the combinatorial hierarchy, and have shown that the quantum numbers may be specified in such a manner as to make appropriate identification with the first generation of leptons and quarks. We refer to this algorithm as *PU*.

This last algorithm is of particular interest for our purposes, since it has all the characteristics of an ordering operator, including the fact that it is too complex to be deterministicly knowable from partial generation. The algorithm has two degrees of freedom, that is, two points at which an appeal to an arbitrariness generator is necessary. These two steps in the algorithm do not affect the global structure of the combinatorial hierarchy thus ultimately produced. Rather, the specifics of these steps will determine the dynamic evolution of the structure and the statistics during this evolution. Once the structure has been completely generated, the statistics are no longer affected.

For these reasons, we point out that PU[46] is, as an algorithmic definition, exemplary of the type of ordering operator which will generate the three-dimensional d–space, as described in Theorem 13. We caution the reader, however, to keep in mind that PU, the specific distance functions which are defined on it and the related derivations *are simply an example* of how we may proceed in detail. We identify U with PU, subject to falsification and subsequent modification. We are not dependent upon these details for the results presented here, which deal primarily with a physical interpretation of the ordering operator calculus. Nonetheless, we believe that either the details are valid, or that these aspects of the model can evolve smoothly (via the P–frame) to become valid.

For example, PU generates a universe of such strings which grows, sequentially, in either number (SU) or length (NU). The main program starts with PICK, an arbitrariness generator that picks two arbitrary strings from memory and discriminates them. This is one of the degrees of freedom mentioned. If this produces a novel string, an operation called ADJOIN results, which adjoins the string to the universe (SU:=SU+1). If the string produced by PICK is already in the universe, an arbitrariness generator called TICK is triggered which increases each string independently, by concatenating it with one arbitrary bit (NU:=NU+1). After either ADJOIN or TICK, the algorithm then recourses to PICK. The arbitrariness which occurs in selected strings from memory (in PICK) or in selecting bits to concatenate (in TICK), serves to guarantee that the algorithm represented by Program Universe is incompletely specified (though in principle specifiable) and, hence, we may treat the output as a Bernoulli trial (as required by Theorem 13), and PU as an ordering operator. If these are fully specified in an algorithmic sense, PU becomes deterministic, and the full evolution of the cosmology becomes known. However, much of the phenomena of laboratory physics arises specifically because we do not have the information. Indeed, we claim that the finite system represented by laboratory physics lacks the space complexity required to fully represent such an algorithm. Thus, some free parameters in the algorithm may not be determined from the recorded output of PU to date. At best then, PU represents a class of algorithms, each of which is sufficient, but not necessary, to account for the phenomena of laboratory physics. We propose that the class encompasses the necessary conditions.

That PU meets the conditions outlined in previous chapters for an ordering operator which is a metric generator, is easy to see. When the operation TICK of PU occurs, there will be three strings connected with the generation process which satisfy the conditions

$$S^a \oplus S^b \oplus S^c = (0, 0, ..., 0)_{NU} .$$

When NU is large, these conditions will be satisfied by many combinations. We can now identify the free function $f(k, n, \lambda)$, presented in the discussion entitled Nonlocal, Discrete Events in defining the attribute distance q_a, subject to a possible scaling factor. For PU, $f(k, n, \lambda) \equiv [2k^a(n) - n]\lambda_a$, and the conditions required in the preceding paragraphs are satisfied automatically. Therefore, PU is consistent with—and can legitimately appeal to—the results presented in this paper, without further derivation.

In an earlier work by Noyes et al.[47], a propagator for relativistic quantum scattering theory was derived. Now that we have shown how to explicitly construct the commutation relations, the interpretation or use of complex notation and how to construct the exponentiation operator, we claim that this work is well founded in all its detail.

Noyes has subsequently shown how to provide the interaction terms of the theory, by identifying our 3–momentum conserving events as "Yukawa vertices." Additionally, a tentative identification has been given of the first three levels of the hierarchy with (1) chiral electron-type neutrinos, (2) electrons, positrons and photons and (3) up and down quarks in a color octet, and with level four to provide weak-electromagnetic unification, with weak coupling to the first three levels.

That the overall mass scheme should come out right, is clearly suggested by the success of the Parker–Rhodes calculation:[48] $m_p/m_e = 137\pi/[(3/14)[1 + 2/7 + (2/7)^2](4/5)] = 1836.151497 \ldots$, which was later reformulated by Noyes to be consistent with the present theory. As Noyes has pointed out[49], the cosmology of Program Universe appears to have a charged lepton and a baryon number consistent with current observation, and, hence, with a locally flat space. These results can be understood as following immediately from establishing rules of correspondence between laboratory practice in high energy physics and performing the appropriate computations. Indeed, this author believes that there are few degrees of freedom available in establishing that interpretation, and perhaps none whatsoever. For example, if PU is selected, we must compute the largest to the smallest mass ratio; but this has already been done for us by the combinatorial hierarchy result $2^{127} + 136 \simeq 1.7 \times 10^{38} \simeq \hbar c/Gm_p^2 = (M_{Planck}/m_p)^2$, which tells us that we can either identify the unit of mass in the theory as the proton mass—in which case we can calculate, to about 1% in this first approximation, Newton's gravitational constant—or, if we take the Planck mass as fundamental, calculate the proton mass.

CONCLUSIONS

The ordering operator calculus has provided a formalism compatible with, and having explanatory and predictive power regarding, the current practice of physics. Indeed, a discrete and unified model of quantum mechanics and special relativity has been made possible.

Much work remains to be done. Not only is considerable effort required in establishing and validating the rules of correspondence, but extensions of the ordering operator calculus to other domains of mathematical investigation are desirable—we have mentioned some of these efforts along the way—and, of course, we would clearly like to incorporate a discrete version of general relativity in our theory. We have laid the foundation for doing so with the definitions of manifolds, neighborhoods, one-forms and other relevant mathematical objects. The reader should note that ours is always a "locally Lorentz invariant" theory and that local frames are, by construction, "inertial," meaning that the geometry is locally flat and exhibits no accelerations. Indeed, accelerations can only arise between the kinds of events we have constructed nonlocally, via the global topology (the connection), even though any dynamics are completely determined from the local geometry. Also in keeping with the geometric picture, our coordinate space has been constructed (from the beginning) from attribute "events," which locate an event by "what happens there," the ordering operator calculus being context sensitive. We already have some indication that "local (gravitational) distortion" of our distance function by a mass can be shown, and work we have recently encountered in the domain of cellular automata is relevant to, our corresponding notion of a field.

A number of experimental predictions have been made. According to P. Suppes[50], there are many generalized inequalities concerning joint probabilities, among which Bell's Inequalities are but a specific example. We have suggested a means of using these inequalities to test whether the nonlocality which violation of the inequalities demonstrates is absolute (along the lines of Bohm's Implicate Order), or, in fact, due to a multiply-connected topology.

We also suggest several other tests of the topology. Our theory predicts that the correlation in Aspect's time-of-flight experiments must be sensitive to the time between changes in the randomly shifted Brewster mirrors, and that the correlation will disappear for data taken arbitrarily close in time to one or the other shift. We should also be able to calculate the shape of an expected distribution curve for the fall-off in correlation, and might be able measure the slope experimentally. These experiments will be quite difficult because of the accuracy in measurement required.

Finally, we have suggested that this phenomena is NOT necessarily microscopic, or limited to spin and polarization quantum variables. The theory is sufficiently general that macroscopic violations of Bell's Inequalities should be constructable. Certainly, the effect can be modeled on computers and, indeed, is used today in publicly key encrypted security (access) cards.

As pointed out in the introduction, the ordering operator calculus is intended as a formalism for modeling diverse phenomena, and not just physical phenomena. Work along these lines is proceeding, and as yet unpublished applications to computational linguistics and computer science have been quite successful.

ACKNOWLEDGMENTS

The authors would like to acknowledge the contributions of the following, without whom this work could not have progressed: Tom Etter, C. W. Kilmister, Ted Bastin, John Amson, M. J. Manthey, C. Gefwert and, finally, our deceased colleague, F. Parker–Rhodes.

REFERENCES

1. Godel, K., *On Formally Undecidable Propositions of Principia Mathematica and Related Systems I*, in Von Hiejenoort, J. ed., *From Frege to Godel: A Sourcebook in Mathematical Logic, 1879–1931*, Harvard University Press, Cambridge, MA (1967), pp. 592–616.

2. Dirac, P. A. M., *Principles of Quantum Mechanics*, Clarendon Press (1981), p. 48.

3. Bishop, E. and Bridges, D., *Constructive Analysis*, Springer–Verlag (1985), p. 2.

4. Bishop, E., op. cit., pp. 4–13.

5. Bishop, E., op. cit.

6. Bishop, E., op. cit.

7. Bishop, E., op. cit.

8. Beeson, M., *Foundations of Constructive Mathematics*, New York, Springer–Verlag (1985).

9. Parker–Rhodes, A. F., *Theory of Indistinguishables*, D. Reidel, Dordrecht, Holland (1981).

10. Kilmister, C. W., *The Mathematics Needed for Self-Organization*, in *Disequilibrium and Self-Organization*, C.W.K., ed., (Reidel, 1986), pp. 11–17.

11. Noyes H. P., C. Gefwert and M. J. Manthey, in *Symposium on the Foundation of Modern Physics*, June 16–20, 1985, Joensuu, Finland; Lahti, P. and Mittelstaedt, P., eds., pp. 511–524 (World Scientific, Singapore); see also *Toward a Constructive Physics*, SLAC–PUB–3116 (1983) and references therein.

12. Parker–Rhodes, A. F., op. cit., p. 66.

13. Blizzard, W. D., *Theory of Multisets with Indiscernables*, Oxford University (1987).

14. Birkhoff, G., Lattice Theory, Amer. Math. Soc., Providence, Rhode Island (1967).

15. Conway, J. H., *On Numbers and Games*, Academic Press, London (1976).

16. McGoveran, D., *The Importance of Being Discrete: Four Alegories* (1984) unpublished.

17. Kolmogorov, A., *Three Approaches to the Concept of "The Amount of Information."*, Prob. Inf. Trnsm. 1, 1 (1965).

18. Chaiten, G. J., *On the Length of Programs for Computing Finite Binary Sequences*, J. ACM 13, 4 (1966), pp. 547–570.

19. Milne–Thomson, L. M., *The Calculus of Finite Differences*, Chelsea Publishing Co., New York, 2nd ed. (1981).

20. McGoveran, D., *Getting Into Paradox*, 1981, in Proceedings, ANPA West (1985).

21. Huffman, D., *A Method for the Construction of Minimum Redundancy Codes*, Proc. IRE, vol. 40 (1952) pp. 1098–1101.

22. Wootters, Willian K., *The Acquisition of Information from Quantum Measurements*, Thesis, University of Texas at Austin (1980).

23. Feller, W., *Probability Theory and Its Applications*, Vol. 1, Wiley (1950) p. 247.

24. Stapp, H., to Noyes, H. P., private communication (1987).

25. Milne–Thomson, L. M., op. cit.

26. Milne–Thomson, L. M., op. cit., p. 23.

27. Box, G., Hunter, W. and Hunter, J., *Statistics for Experimenters*, Wiley Interscience, New York (1978) pp. 43–50 and 123–131.

28. Box, G., Hunter, W. and Hunter, J., op. cit., pp. 43–50.

29. Feller, W., op. cit., p. 179.

30. Stein, I., seminar at Stanford (1978) and papers at ANPA 2 (1980) and ANPA 3 (1981), King's College, Cambridge.

31. Schutz, B., *Geometrical Methods of Mathematical Physics*, Cambridge University Press (1980) pp. 43–47.

32. Goldstein, H., *Classical Mechanics,* Addison–Wesley, 2nd ed., (1980); see the discussion on cyclic coordinates and angle variables in physical systems, pp. 445–471.

33. Bastin, Ted and Kilmister, C. W., *Fields and Particles in a Discrete Physical Model*, Work paper (1973).

34. Kuhn, T. S., *The Structure of Scientific Revolutions*, University of Chicago Press (1962).

35. Kuhn, T., op. cit.

36. Suppe, F,. ed., *The Structure of Scientific Theories*, University of Illinois Press (1979) pp. 191–192.

37. Noyes, H. P., Gefwert, C. and Manthey, M. J., *A Research Program with No Measurement Problem*, in *New Techniques and Ideas in Quantum Measurement Theory,* Greenberger, D. M., ed., Annals of the New York Academy of Sciences, Vol. 480 (1986) pp. 553f.

38. Bastin, E. W., *On the Origin of the Scale-Constants of Physics*, Studia Philosophica Gandensia, v. 4 (1966) pp. 77.

39. Bastin, E. W., *A General Property of Hierarchies*, Proc. Royal Camb. Soc. (1956).

40. Bastin, E. W., op. cit.

41. Clauser, J. F. and Shimony, A., Rep. Prog. Phys., v. 41 (1970) pp. 1881f.

42. Manthey, M. J., McGoveran, D. and Noyes, H. P, in preparation.

43. Goldreich et al., *How to Construct Random Functions*, Journal of the Association for Computing Machinery, **33**, No. 4 (1986) pp. 792–807.

44. Jordan, T. F. *Quantum Mechanics In Simple Matrix Form*, John Wiley and Sons, New York (1986) p. 161.

45. Cramer, J. G, *The Transactional Interpretation of Quantum Mechanics*, Rev. Mod. Phys. Vol. 58, No. 3 (1986).

46. Noyes, H. P., Gefwert, C. and Manthey, M. J., in *Symposium on the Foundation of Modern Physics*, June 16–20, 1985, Joensuu, Finland, Lahti, P and Mittelstaedt, P., eds., World Scientific, Singapore, pp. 511–524; see also *Toward a Constructive Physics*, SLAC–PUB–3116 (1983) and references therein.

47. Noyes, H. P., Gefwert, C. and Manthey, M. J., *A Research Program with No 'Measurement Problem'*, Annals of the N. Y. Acad. Sci. **480** (1986) pp. 553–563.

48. Parker–Rhodes, A, F,m op. cit., pp. 184–185

49. Noyes, H. P., McGoveran, D. O. et al., *A Discrete Relativistic Quantum Physics*, SLAC–PUB–4327 (1987).

50. Suppes, P. and Zanotti, M., *New Bell-type Inequalities for $N > 4$ Necessary for the Existence of a Hidden Variable*, submitted to Phys. Rev. Letters (1987).

Physics Essays   volume 2, number 1, 1989

An Essay on Discrete Foundations for Physics[1]

H. Pierre Noyes and David O. McGoveran

Abstract

We base our theory of physics and cosmology on the five principles of finiteness, discreteness, finite computability, absolute nonuniqueness, and strict construction. Our modeling methodology starts from the current practice of physics, constructs a self-consistent representation based on the ordering operator calculus *and provides rules of correspondence that allow us to test the theory by experiment. We use* PROGRAM UNIVERSE *to construct a growing collection of bit strings whose initial portions (labels) provide the quantum numbers that are conserved in the* events *defined by the construction. The labels are followed by* content strings, *which are used to construct event-based finite aned discrete coordinates. On general grounds such a theory has a limiting velocity, and positions and velocities dₒ not commute. We therefore reconcile quantum mechanics with relativity at an appropriately fundamental stage in the construction. We show that 1) events in different coordinate systems are connected by the appropriate finite and discrete version of the Lorentz transformation, 2) three-momentum is conserved in events, and 3) this conservation law is the same as the requirement that different paths can "interfere" only when they differ by an integral number of de Broglie wavelengths.*

The labels are organized into the four levels of the combinatorial hierarchy *characterized by the cumulative cardinals 3, 10, 137, $2^{127} + 136 \simeq 1.7 \times 10^{38}$. We justify the identification of the last two cardinals as a first approximation to $\hbar c/e^2$ and $\hbar c/Gm_p^2 = (M_{Planck}/m_p)^2$ respectively. We show that the quantum numbers associated with the first three levels can be rigorously identified with the quantum numbers of the first generation of the standard model of quarks and leptons, with color confinement and a first approximation to weak-electromagnetic unification. Our cosmology provides an event horizon, a zero-velocity frame for the background radiation, a fireball time of about 3.5×10^6 years, about the right amount of visible matter, and 12.7 times as much dark matter. A preliminary calculation of the fine structure spectrum of hydrogen gives the Sommerfeld formula and a correction to our first approximation for the fine structure constant, which leads to $1/\alpha = 137.035\,967\,4\ldots$. We can now justify the earlier results $m_p/m_e = 1836.151\,497\ldots$ and $m_\pi/m_e \lesssim\, = 274$. Our estimate of the weak angle is $\sin^2\theta_{Weak} = 1/4$ and of the fermi constant $G_F \times m_p^2 = 1/\sqrt{2}(256)^2$. Our finite particle number relativistic scattering theory should allow us to systematically extend these results. Eteris paribus, caveat lector.*

Key words: finite, discrete foundations: relativity, elementary quantum particles, cosmology, dark matter

1. INTRODUCTION

Physics is an *experimental* science that relies on *counting*. For instance, Galileo counted the number (equal "by construction" and presumably by experiential comparison) of intervals passed by a ball rolling down a smooth groove in an inclined plane while water flowed into a receptacle during the same interval. He then *counted* the number (equal "by construction" and presumably by experiential comparison) of weights that would balance the water content of the receptacle. We could say that from these experiments he proved the invariance of the local acceleration due to

gravity. We start our discussion by insisting that finite and discrete counting is the proper starting point for any fundamental theory of physics.

Physicists have long known that counting is not enough to achieve consensus. Sometimes the counts differ under the "same" circumstances; the scatter in the results is not always easy to understand after the fact, let alone to allow for before. So a "theory of errors" has grown up, which is partly pragmatic, and more recently relies on "statistical theory." As a first-rate experimental physicist has remarked, "you can't measure errors." Current practice in high-energy physics tries to estimate errors by simulating the experimental setup on a computer and making a finite number of pseudorandom runs to compare with the "real time" data. In this specific practice, the estimate of errors is also based on finite counting.

Until recently, the legacy inherited by physicists from continuum mathematics, which some of their most illustrious predecessors had helped to create, dominated thinking about measurement and errors. In particular, continuum models for probability – which can never be tested in a finite amount of time – dominated the theory of errors just as Euclidean geometry and its multidimensional extensions dominated the model space into which physical theories were thrown. Bridgman made a heroic effort to get out of this trap (but never went as far as to abandon the continuum). Eddington attacked the problem from a point of view historically connected to the approach adopted here but was never able to carry any substantial body of physicists along with him. Much that is relevant to our work was going on in minority views about the foundations of mathematics at the same time. We leave the investigation of that background to others.

Computer scientists do not have the luxury of relying on "existence proofs" which they cannot demonstrate on a computer within budget and within a deadline. They have evolved a new science, which differs in significant ways from conventional continuum mathematics, in order to meet their specific needs. It is from this background that the most productive work in the theory presented here has arisen. We leave that aspect of the scientific revolution we hope to help initiate to other papers. This paper is addressed to physicists.

In the next section we review those aspects of the historical practice of physics that we find most relevant to our enterprise. Cosmology relies on particle physics for most of its quantitative observational data. So Sec. 3 sketches those aspects of elementary particle physics we feel need to be modeled accurately if our alternative theory is to be taken seriously by particle physicists and cosmologists. The basic methodology for our alternative approach is presented in Sec. 4. What in an older terminology might be called the formal structure, and in ours is called the representational framework, follows in Sec. 5. Here, we find that many of the ad hoc attempts to fit relativity and quantum mechanics into the historically established framework – attempts which some distinguished physicists still find fall far short of their conceptual requirements – can be replaced by a finite and discrete alternative. Section 6 compares our results with experience. Section 7 steps back and looks at what we have and have not accomplished as part of a research program that has been going on for some of us for over three decades. It is here that we try to justify, or at least explain, the claims made in the abstract.

2. THE HISTORICAL PRACTICE OF PHYSICS
2.1 Scale Invariant Physics
Physics was a minor branch of philosophy until the seventeenth century. Galileo started "physics" in the contemporary sense. He emphasized both mathematical deduction and precise experiments. Some later commentators have criticized his a priori approach to physics without appreciating his superb grasp of the experimental method which he created, including reports of his experiments that still allow replication of his accuracy using his methods. He firmly based physics on the measurement of length and time; from our current perspective he established the uniform acceleration of bodies falling freely near the surface of the Earth.

A century later, Newton entitled what became the paradigm for "classical" physics The Mathematical Principles of Natural Philosophy, recognizing the roots that physics has in both disciplines. He also was a superb experimentalist.[2] To a greater extent than Galileo, Newton had to create "new mathematics" in order to express his insight into the peculiar connection between experience, formalism, and methodology that still remains the core of physics. To length and time, he added the concept of mass in both its inertial and its gravitational aspect and tied physics firmly to astronomy through universal gravitation. For philosophical reasons he introduced the concepts of absolute space and time and thought of actual measurements as some practical approximation to these concepts.

It is often thought that Einstein's special relativity rejects the concept of absolute space-time until it is smuggled back in through the need for boundary conditions in setting up a general relativistic cosmology. In fact, the concept of the homogeneity and isotropy of space used by Einstein to analyze the meaning of distant simultaneity in the presence of a limiting signal velocity is very close to Newton's absolute space and time. What Einstein shows is that it is possible to use local, consequential time to replace Newton's formulation of the concept. This was pointed out to the author (Noyes – HPN) by McGoveran[1] in the context of our fully finite and discrete approach to the foundations of physics, and our derivation of the Lorentz transformations using "information-transfer" velocities that are rational fractions of the limiting velocity. This same analysis shows that in a discrete physics, the universe has to be multiply connected. The spacelike separated "supraluminal" correlations predicted by quantum mechanics – and recently demonstrated experimentally to the satisfaction of many physicists – can be anticipated for spins and for any set of countable degrees of freedom more impoverished that those needed to specify a "material object."

2.2 Breaking Scale Invariance
Nineteenth-century physicists saw the triumph of the electromagnetic field theory. Classical physics was still firmly based on historical units of mass, length, and time. Quantized atomic masses had been discovered by chemists early in the century and quantized charges related to them by Faraday, but physicists only began to take them seriously after the discovery of the electron and "canal rays." Prior to the discovery of Planck's constant and the recognition that mass and charge were separately quantized, together with the understanding that the propagation velocity in free space required by Maxwell's equations was a universal limiting velocity, classical physics provided no way to question scale invariance.

Quantum theory and relativity were born at the beginning of this century. Quantum mechanics did not take on its current form until nearly three decades of work had passed. Although one route to quantum mechanics (that followed by de Broglie and Schrödinger) started from the continuum relativistic wave theory, the currently accepted form breaks the continuity by an interpretive postulate due to von Neumann sometimes called "the collapse of the wave function."

Criticism of this postulate as conceptually inconsistent with the time reversal invariant continuum dynamics of wave mechanics has continued ever since. This criticism was somewhat muted for awhile by the near consensus of physicists that Bohr had "won" the Einstein-Bohr debate and the continuing dramatic technical successes of quantum mechanics. Scale invariance is gone because of the quantized units of mass, action, and electric charge. These specify in absolute (i.e., countable) terms what is meant by "small." Explicitly $r_{Bohr} = \hbar^2/m_e e^2$ (with m_e the electron mass)

An Essay on Discrete Foundations for Physics

specifies the atomic scale, $\lambda_{Compton} = (e^2/\hbar c)r_{Bohr} = \hbar/m_e c$ specifies the quantum electrodynamic scale, and the "classical electron radius" $e^2/m_e c^2 = (e^2/\hbar c) \lambda_{Compton} \simeq 2\hbar/m_e c \simeq 14\hbar/m_p c$ specifies the nuclear scale; here, m_p is the proton mass, and $m_\pi \lesssim 2 \times 137 m_e$ is the pion mass. The elementary particle scale $\hbar/m_p c$ is related to the gravitational scale by $\lambda_G = (G\hbar^3/c)^{1/2} = \hbar/M_{Planck}c = (Gm_p^2/\hbar c)^{1/2}(\hbar/m_p c)$.

The expanding universe and event horizon specify what is meant by "large." Here, the critical numbers any *fundamental* theory must explain are: "mass" of the universe as about $3 \times 10^{76} m_p$ – or at least ten times that number if one includes current estimates for dark matter; "size" of the universe or *event horizon* – naively the maximum radius that any signal can attain (or arrive from) transmitted at the limiting signal velocity c during the age of the universe; "age" of the universe as about 15 billion (15×10^9) years. Backward extrapolation using contemporary laws of physics to the energy and matter density when the radiation breaks away from the matter (size of the fireball) is consistent with the observed 2.7°K cosmic background radiation. The cosmological parameters are numerically related to the elementary particle scale by the fact that the visible mass in the currently observable universe is approximately given by $M_{vis.U} \simeq (\hbar c/Gm_p^2)^2 m_p$, and that linearly extrapolating backward from the fireball to the start of the big bang gives a time $T_{fireball} \simeq (\hbar c/Gm_p^2) \times (\hbar/m_p c^2) = 3.5$ million years. Any theory that can calculate all these numbers has a claim to being a fundamental theory.

For a while, it appeared that reconciliation between quantum mechanics and special relativity would resist solution; the uncertainty principle and second quantization of classical fields gave an infinite energy to each point in space-time. During World War II, Tomonaga, and afterwards Schwinger and Feynman, developed formal methods to manipulate away these infinites and obtain finite predictions in fantastically precise agreement with experiment. Recently, the non-Abelian gauge theories have made everything calculated in the "standard model" finite. Weinberg asserted at the Schrödinger Centennial in London that there is a practical consensus – but no proof – that second quantized field theory is the *only* way to reconcile quantum mechanics with special relativity. He also pointed out that the finite energy due to vacuum fluctuations is then 10^{120} too large compared to the cosmological requirements; the universe should wrap itself up and shut itself down almost as soon as it starts expanding.[2] Anyone who is willing to swallow this camel will still have to strain at the gnats of inflationary scenarios and the difficulties associated with including strong gravitational fields in any quantum theory. Continued attention to foundations seems fully justified.

2.3 Events and the Void: An Alternative?

The concept on which most of elementary particle physics rests has moved a long way from the mass points of post-Newtonian dynamics. For us, a paraphrase of the concept used by Eddington is more useful: a particle is "a conceptual carrier of conserved three-momentum and quantum numbers between events."[3]

This definition applies in the practice of elementary particle physics 1) in the high-energy particle physics laboratory and in the theoretical formulations of either 2) second quantized field theory or 3) analytic S-matrix theory. In 1), the experimental application, "events" refer to the detection of any finite number of incoming and outgoing "particles" localized in macroscopic space-time volumes called "counters" or some conceptual equivalent. In 2), "events" start out as loci in the classical Minkowski four-space continuum at which the "interaction Lagrangian"

acting on a state vector creates and destroys particle states in Foch space. Since this prescription, naively interpreted, assigns an infinite energy and momentum to each space-time point, considerable formal manipulation and reinterpretation is needed before these "events" can be connected to laboratory practice. In 3), "events" refer to momentum-energy space "vertices" which conserve four-momentum in "Feynman diagrams." These diagrams were originally introduced in context 2) as an aid to the systematic calculation of renormalized perturbation theory. S-matrix theory makes a strong case for viewing continuous space-time as a mathematical artifact produced by Fourier transformation. Like any scattering theory, or any application of second quantized field theory to discrete and finite particle scattering experiments, S-matrix theory includes rules for connecting amplitudes calculated from these diagrams directly to laboratory practice 1).

An alternative approach to the problem, which is beginning to be called *discrete* and/or *combinatorial* physics, is focused on constructed, discrete processes.[4] A quick characterization of the theory could be: chance, events, and the void suffice. Only discrete, finitely computable, combinatorial connectivities are allowed. But the multiple connectivity and the indistinguishables that our approach requires introduce subtle differences from conventional mathematics and physics at an early stage.

The connectivity can be provided by a growing universe of bit strings. The events generated by PROGRAM UNIVERSE[5] connecting bit strings use part of the string, called the *label*, to define conserved quantum numbers. The bits not used as the label can be called the *content* of the string. Looking back to our first pass at what we mean by a particle, one "carrier" connecting events is the evolving labeled string. Yet, once the universe is mature enough to allow a meaningful discrimination between label and content, there are many strings with the same label. The arbitrary evolution connects shorter to longer strings, or for strings of the same length connects two three-events to form a four-event. Thanks to the "counter paradigm," this discrete model also accounts for the conservation of three-momentum and quantum numbers consistent with laboratory practice 1) and serves the same purposes as the theoretical constructs in second quantized relativistic field theory 2) or analytic S-matrix theory 3).

The next section reviews the language – supposedly adequate to describe the relevant pheonomena – that elementary particle physicists employ and expect others to employ when entering on their turf.

3. CONTEMPORARY PARTICLE PHYSICS

3.1 Yukawa Vertices

With the exception of gluons, the standard model of quarks and leptons starts from conventional interaction Lagrangians of the form $g\bar\psi\psi\phi$, into which various finite spin, isospin, etc., operators may be inserted. Here, g is the coupling constant that measures the strength of the interaction relative to the mass terms in the free particle part of the Lagrangian, ψ ($\bar\psi$) is a fermion (antifermion) second quantized field, and ϕ a boson or quantum field. All three fields can be expanded in terms of creation and destruction operators acting on particle or Foch space states, which in the momentum space representation, contain separate four-momentum vector variables for each fermion, antifermion, or quantum.

Fortunately for us, in one of the first successful efforts to tame the infinities in this theory, Feynman introduced a diagrammatic representation for the terms generated by such interaction Lagrangians in a perturbation theory (powers of g), expansion of the terms which need to be calculated and summed in order to obtain a finite approximation for the predictions of the theory. These Feynman diagrams have taken on a life of their own; they bring out the symmetries and conservation laws of the theory in a graphic way. This can be a trap, particularly if they are reified as representing actual happenings in space-time. If used with care they can short-circuit a lot of tedious calculation (or suggest viable additional approximations) and provide a powerful aid to the imagination.

In the usual theory, Minkowski continuum space-time is assumed, and any interaction Lagrangian is constructed to be a Lorentz scalar. Consequently, the quantum theory conserves four-momentum at each three-vertex. Here, one must use care because of the uncertainty principle. If four-momentum is precisely specified, the uncertainty principle prevents any specification of position; the vertex can be anywhere in space-time. This is the most obvious way in which the extreme nonlocality of quantum mechanics shows up in quantum field theory. If we use a momentum space basis, we can still have precise conservation laws at the vertices for which the masses have an unambiguous interpretation. In practical applications of the theory momentum cannot be precisely known; quasi-localization is allowed as long as the restrictions imposed by the uncertainty principle are respected. In a careful treatment, this is called "constructing the wave packet"; actually specifying this construction requires some care as can be seen, for instance, by consulting Goldberger and Watson's *Collision Theory*. In practice, one usually works entirely in momentum space, knowing that the orthogonality and completeness of the basis states will allow the construction of appropriate wave packets in any currently encountered experimental situation. We have made a start on the corresponding construction in our theory.[4]

Although four-momentum conservation is insured in the conventional treatment, this is not the end of the problem. For a particle state with energy ϵ and three-momentum $\mathbf{p}$ the formalism insures that $\epsilon^2 - \mathbf{p} \cdot \mathbf{p} = M^2$; here, M is any invariant with the dimensions of mass and need not correspond to the rest mass of the particle m. In the usual perturbation theory this is simply accepted. The dynamical calculations are made "off mass shell," and the specialization to physical values appropriate to the actual laboratory situations envisaged is reserved to the end of the calculation. S-matrix theory sticks closer to experiment in that all amplitudes refer to physical (realizable) processes with all particles "on mass shell." The dynamics is then supposed to be supplied by imposing the requirement of flux conservation (unitarity) – a nonlinear constraint – and by relating particle and antiparticle processes through "crossing." The analytic continuation of the amplitudes for distinct physical processes which gives dynamical content to the equations then makes S-matrix theory into a self-consistent or "bootstrap" formalism. There is no known way to guarantee a solution of this bootstrap problem short of including an infinite number of degrees of freedom – if then; of course, it is also well known that there is no known way to prove that quantum field theory possesses any rigorous solutions of physical interest. One must have recourse to finite approximations that may or may not prove adequate to particular situations.

The finite particle number scattering theory[6]-[9] keeps all particles on mass shell and hence has three-momentum conservation at three-vertices. This theory insures unitarity for finite particle number systems by the form of the integral equations; these also provide the dynamics. The uncertainty principle is respected because of the "off-energy-shell" propagator, as it is in nonrelativistic scattering theory; the approximation is the truncation in the number of particulate degrees of freedom.

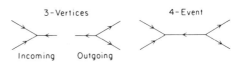

Figure 1. The connection between three-vertices and four-events.

If we put the Feynman diagrams of the second quantized perturbation theory on mass shell, we can talk about three-vertices and four-events using a common language for all three theories. The rules are easy to state, particularly if we do so in the (cosmological) "zero momentum frame." We are justified in using any description derived from this cosmological frame within the mathematical models, because we have restricted ourselves to free particle, mass shell kinematics. We can use a corresponding statement in the laboratory, because this frame is empirically specified as the frame at rest with respect to the 2.7°K background radiation. Then the Poincaré invariance of the theories allows us to go from this description to any other convenient Galilean frame.

As we show in Sec. 5.4, the three-momenta at a three-vertex add to zero. Diagrammatically we have three "vectors" that are "incoming" or "outgoing." By putting one of each together we obtain the generic 2-2 channel four-event, as indicated in Fig. 1. Clearly, for four-events the total momentum of the two outgoing lines has to equal the total momentum of the two incoming lines, but the plane of the outgoing three-event can be any plane obtained by rotating the outgoing vectors in the planar figure about the axis defined by the single line connecting them. By associating quantum numbers with each line, we can extend this description of three-momentum conservation in Yukawa vertices and the four-events constructed from them to the conservation of quantum numbers that "flow" along the lines.

The idea of associating physical particles with the lines as carriers of both momentum and quantum numbers that comes from this pictorial representation is almost irresistible. The reader is warned once again to resist this temptation. The diagram is in $3 + 1$ momentum-energy space and *not* in space-time. In fact, if we insist on interpreting it as a space-time diagram representing the motion of particles, the quantum theory will blow up. It will force us to assign an infinite energy and momentum to each point of that space-time, and simplicity of interpretation becomes elusive.

Once we have this picture in hand, "crossing" is easy to define: If reversing a line and at the same time changing all its quantum numbers to their negatives does not alter the conservation laws, the new diagram also represents a possible physical process. The "particle" whose quantum numbers are the negative of another is called its "antiparticle." So crossing can also be stated as the requirement that the reversal of a reference direction and the simultaneous change from particle to antiparticle represents another possible physical process. The manner in which a single diagram in which momenta and quantum numbers add to zero at a general three-vertex generates emission, absorption, decay, and annihilation vertices by successive applications of this rule is illustrated in Fig. 2. The manner in which a single diagram in which momenta and quantum numbers add to zero in a general four-event generates six physically observable processes by this rule is illustrated in Fig. 3.

An Essay on Discrete Foundations for Physics

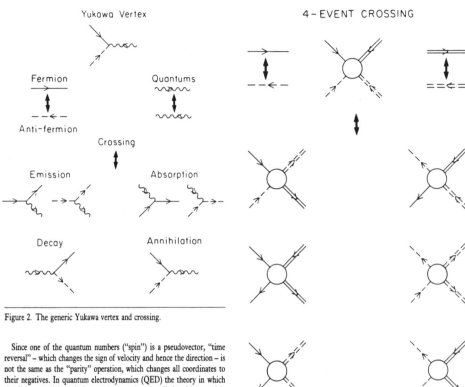

Figure 2. The generic Yukawa vertex and crossing.

Since one of the quantum numbers ("spin") is a pseudovector, "time reversal" – which changes the sign of velocity and hence the direction – is not the same as the "parity" operation, which changes all coordinates to their negatives. In quantum electrodynamics (QED) the theory in which the diagrams originated, the quantum number that distinguishes particle from antiparticle is electric charge; these rules are a consequence of the "CPT invariance" of the theory. They generalize to other types of "charge," e.g. "color charge" in quantum chromodynamics (QCD). Spin is of great interest since it has a space-time significance as well as sharing the discrete, quantized character of other quantum numbers.

Before going on to the other quantum numbers, we note that the form of the Yukawa vertex couples the particle and antiparticle field in such a way that in the time-ordered interpretation of the diagrams the number of fermions minus the number of antifermions is conserved; this is called the conservation of fermion number. Clearly, the diagrams respect this conservation law; as far as we know, f-number conservation is followed in nature.

3.2 The Standard Model

The fermions encountered in nature fall into two classes: leptons and baryons. As far as we know, lepton number and baryon number are separately conserved. The lifetime for the decay of the proton into leptons and other particles has been shown to be greater than 10^{35} years; the experimental upper limit for the value depends on which decay mode was searched for. This fact has already ruled out many proposed schemes for grand unification.

Figure 3. Four-leg crossing.

The existence of the enormous underground detectors constructed to test the hypothesis of proton decay had an unexpected payoff when two of them detected, simultaneously, neutrino bursts from a supernova explosion 50 000 pc (1 pc = 3.3 light-years) away. Individual neutrinos within the burst were cleanly resolved, but the time spread of the burst itself was so short that information about upper limits for the masses of the neutrinos could be obtained only by sophisticated statistical analysis. Although the time for the actual production of the neutrinos is supposed to be very short, the spread induced by the subsequent diffusion of the neutrinos out through the bulk of the star makes the calculation sensitive to the model used for calculating the explosion. Empirically, we can take the three types of neutrinos to be massless with an upper limit of 30 eV/c^2.

The quanta that couple via elementary Yukawa vertices in the standard model all have spin one. The earliest coupling explored in quantum field

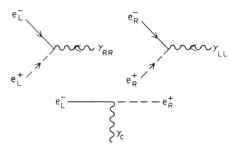

Figure 4. Quantum electrodynamics.

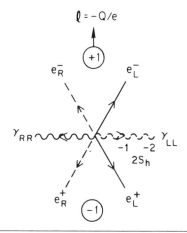

Figure 5. Quantum electrodynamic conservation laws as planar vectors.

theory was the electromagnetic coupling between electrons (e^-), positrons (e^+), and the massless electromagnetic quanta; the theory, which can be extended to other charged fermions, is called quantum electrodynamics (QED). The masslessness of the electromagnetic quanta is imposed within the second quantized relativistic field theory by requiring the theory to be gauge invariant. A lower limit to the mass of either fermions or quanta with specified quantum numbers defines a well-understood experimental problem; if all such lower limits had to be finite, this would kill gauge invariance. The requirement of gauge invariance is not compelling for us prior to some rough consensus as to what additional, independent tests (at an accuracy specified in advance) are relevant. We know of no proposed experimental program that could test gauge invariance within realistic error bounds; the concept of gauge invariance does not meet Popper's requirement. The upper limits on the mass of electromagnetic quanta are very good; empirically, we can assume photons to be massless.

The skepticism just implied makes our explanatory problem difficult. The current fashion in high-energy elementary particle physics starts from non-Abelian gauge theories. Their broken "symmetries" generate "mass" from a "spontaneous breakdown of the vacuum." With care, this mechanism is claimed to be a guaranteed way to remove the infinities from a tightly constrained version of second quantized field theory. Without those constraints, which start from the necessity to get rid of the classical infinity of the e^2/r potential (infrared divergence) and the infinity of energy-momentum at each space-time point forced on us classically by point particles and retained in the quantized field theory in spite of the uncertainty principle (ultraviolet divergence), these theories are *prima facie* nonsensical. Self-consistency *within* the mathematical theory is disputed by some who take the rigor of continuum mathematics seriously.

Following a conventional route in a four-dimensional formalism one runs into trouble because a massless photon with momentum has only two chiral states (γ_{LL} and γ_{RR}) while the formalism requires four components for a four-vector. For a massive spin one particle (i.e., something that can carry three-momentum between two events in any coordinate system and whose mass defines a rest system) there is no problem. The three states that quantum mechanics requires for spin one can be resolved along, against, or perpendicular to the direction of motion, while the fourth component of the four-vector is related to these three components "on shell" by the invariant mass. When the invariant mass is zero, we are left

with only two chiral three-momentum carrying states. For fermions this is no problem, once parity conservation is abandoned. But for spin one massless bosons, the third and fourth component of the four-vector have to combine to yield an undirected $1/r$ coulomb potential in a gauge-invariant and manifestly covariant manner. In a classical theory with extended sources this was no problem, because the transformation between the four-vector notation and the coulomb gauge was always well defined, although coordinate-system dependent. But in second quantized field theory achieving consistency between the classical substrate and the Feynman rules requires all kinds of technical artifices (indefinite metrics and the like). In a finite particle number theory one can avoid some of these technical difficulties by always using transverse photons and the coulomb interaction in a well-defined coordinate system, provided the (no longer manifest) covariance can be maintained. Of course, this removes some of the (we believe superficial) formal simplicity of the manifestly covariant four-vector formalism. Since the theory we have developed commits us to three-momentum conservation as fundamental, this is a natural route for us to take.

Once this is understood, the particular crossing symmetric Yukawa vertices $e_L^-(Q = -e, s_h\hbar = -\frac{1}{2}\hbar)$, $e_R^-(Q = -e, s_h\hbar = +\frac{1}{2}\hbar)$ specifying massive leptonic QED for a single flavor (in this case e) coupled to $\gamma_{LL}, \gamma_{RR}, \gamma_c$ are given in Fig. 4. We note that for electromagnetic coupling, charge and lepton number go together; the conservation law for one implies the conservation law for the other. We represent the combined conservation laws of $2s_h \in 0, \pm 1, \pm 2$, and $l = -Q/e \in 0, \pm 1$, by the vector states in a plane in Fig. 5. A Yukawa (QED) vertex requires three quantum number vectors consisting of a fermion, an antifermion, and a quantum that add to zero, plus the temporally ordered processes derived from the fundamental diagram by crossing. The field theory notation for this QED coupling is[10] $-i Q \bar{e} \gamma_\lambda e A_\lambda$, with $Q^2/\hbar c = e^2/\hbar c \simeq 1/137$.

In contrast to the parity-conserving electromagnetic vertices, the weak interactions violate parity conservation maximally. An easy way to

272

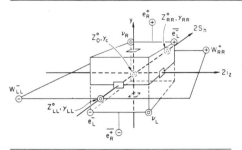

Figure 6. Weak-electromagnetic unification in terms of weak hypercharge, weak isospin, and helicity.

Table I: Quantum Numbers for Weak-electromagnetic Unification

Particle		Q	Y	$2i_z$	l	$2h$	m in GeV/c^2
Fermion	ν_L	0	−1	+1	−1	−1	0
	$\bar{\nu}_L$	0	+1	−1	+1	+1	0
	e_L^-	−1	−1	−1	−1	−1	0.511×10^{-3}
	$\bar{e}_L^+$	+1	+1	+1	+1	+1	"
	e_R^-	−1	−2	0	−1	−1	"
	$\bar{e}_R^+$	+1	+2	0	+1	+1	"
Quantum	W_{LL}^-	−1	0	−2	0	−2	$37.3/\sin\theta_W$
	$\bar{W}_{LL}^-$	+1	0	+2	0	+2	"
	Z_{LL}^0, γ_{LL}	0	0	0	0	−2	$37.3/\sin\theta_W \cos\theta_W, 0$
	$\bar{Z}_{LL}^0, \bar{\gamma}_{LL}$	0	0	0	0	+2	"
	Z_0^0, γ_c	0	0	0	0	0	"

represent this is to use a massless neutrino (ν_L), conventionally called left-handed. Consider an arrow in front of you with the head on the right. If you slip your right hand under the arrow to pick it up, your thumb will point in the same direction as the head; if you pick it up by slipping your left hand under the arrow, your thumb will point in the opposite direction to the head. The latter case is called left-handed. By the Feynman rule the antineutrino $\bar{\nu}_L$ is then right-handed. The charged quantum that couples to the electron and neutrino is called W (the weak vector boson) and is also chiral, since in the zero momentum frame $e_L^- + \bar{\nu}_L \rightarrow W_{LL}^-$; in field theory notation the coupling is

$$-i(G_F M_W^2/\sqrt{2})^{1/2}\bar{\nu}\gamma_\lambda(1 - \gamma_5)eW_\lambda.$$

The Weinberg-Salam-Glashow "weak-electromagnetic unification" requires in addition to this electrically charged weak boson, which was a convenient way to parametrize the parity-non-conserving theory of β decay, the neutral weak boson Z_0 responsible for "neutral weak currents." The reasons had to do initially with the removal of infinities from the theory, and go through a complicated sequence of arguments that predict, in addition, one or more scalar "Higgs bosons," for which there is at present no laboratory evidence. Since our theory is born finite and cannot produce the infinities of second quantized field theory, we have no need for these hypothetical particles in the first place. If they should be discovered (thanks to current efforts at many laboratories that are now consuming a large fraction of their experimental and computational resources), we will be faced with some difficult conceptual problems in our discrete theory. Fortunately, for the moment we can ignore them, which makes our presentation of the conservation laws in the leptonic sector considerably simpler.

The coupling of the Z^0 to neutrinos is chiral and is given by

$$(-i/\sqrt{2})(G_F M_Z^2/\sqrt{2})^{1/2}\bar{\nu}\gamma_\lambda(1 - \gamma_5)eZ_\lambda.$$

The coupling to electrons is more complicated, because it brings in the "weak angle" θ_W that distinguishes the coupling to left- and right-handed electrons in the following way:

$$(-i/\sqrt{2})(G_F M_Z^2/\sqrt{2})^{1/2}\bar{e}\gamma_\lambda[R_e(1 + \gamma_5) + L_e(1 - \gamma_5)]eZ_\lambda.$$

Here, $R_e = 2\sin^2\theta_W$, $L_e = 2\sin^2\theta_W - 1$. If $\sin^2\theta_W = 1/4$, which is not too bad an approximation to the experimental value, Z couples to electrons

like a heavy gamma ray, except that it is a pseudovector rather than a vector. The mixing angle is not independent of the masses of the weak bosons, because

$$M_W \sin\theta_W = [\pi e^2/\hbar c G_F\sqrt{2}]^{1/2} = 37.3 \ GeV/c^2 = M_Z \sin\theta_W \cos\theta_W.$$

Since there were estimates of the weak mixing angle available before the discovery of the weak bosons, their masses could be estimated to be around 84 and 94 GeV/c^2 respectively, which aided greatly in their experimental isolation. Since the W's are charged, they couple to photons and also directly to the Z. These couplings are given in Ref. 10, p. 116. Eventually, the more complicated four-vertices given in the same reference should provide a critical test of the standard model and, conceivably, might even distinguish between our theory and the standard model even in the absence of experimental evidence for the Higgses. We ignore this complexity in what follows.

The conservation law situation is now considerably more complicated than it was for electromagnetic quanta. Charge, lepton number, and helicity are still conserved, but the pattern is not easy to follow if written in those terms. Following a strategy that was first introduced into nuclear physics to describe the approximate symmetry between neutron and proton as an "isospin doublet," we form a "weak isospin doublet" from the left-handed electron ($i_z = -\frac{1}{2}$) and left-handed neutrino ($i_z = +\frac{1}{2}$), and assuming lepton number conservation, can talk about either charge conservation or "z component of isospin conservation" by introducing an appropriate version of the Gell Mann-Nishijima formula, namely $Q = 1/2 + i_z$, for the left-handed doublet. To include the right-handed electron, which does not couple to neutrinos, we make it an isospin singlet. To couple it to γ rays, we assign it a "weak hypercharge" $Y = -2$ and modify the Gell Mann-Nishijima formula to read $Q = Y/2 + i_z$. Our quantum numbers are now conveniently described in the three-space picture given in Fig. 6. The numerical specifications are given in Table I.

Although the type of spacial representation of the quantum numbers presented in Fig. 6 suggests that there might be rotational invariance in this space, actually only the values on the axes have precise meaning in terms of conservation laws. Total isospin is only approximately conserved; it is a broken symmetry. Perhaps this should not be a surprise in a relativistic theory; if we take the four independent generators of the

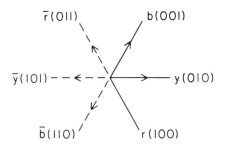

Figure 7. Colors and anticolors as discrete vectors.

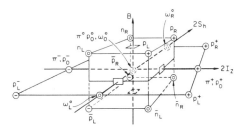

Figure 8. Spin, isospin, and baryon number conservation for color singlet neutrons and protons $p = u(ud)$, $n = d(ud)$.

Poincaré group to be mass, parallel, and perpendicular components of three-momentum and helicity (i.e., the component of angular momentum along the parallel direction), the total angular momentum cannot be simultaneously diagonalized. People often forget that total spin is not a well-defined concept in a relativistic theory.

Now that we have looked at the weak-electromagnetic unification of electrons, whose mass is 0.511 MeV/c^2, and their associated massless neutrinos, the full weak-electromagnetic unification scheme is easy to state. In addition to the electrons, we have two systems of leptons with much larger masses: the muon with mass 105.66 MeV/c^2 and the tau lepton with mass 1784 MeV/c^2. Associated with each are left-handed $(\nu_\mu)_L$ and $(\nu_\tau)_L$ neutrinos whose interactions can be experimentally distinguished from those of the electron neutrinos $(\nu_e)_L$ and from each other. They may well be massless, but the upper limits on their masses were much higher than for the electron-type neutrinos prior to the supernova measurement. As already noted, all three upper limits are now comparable. The coupling scheme is the same as that we have already discussed above within each generation (e, μ, τ = first, second, third). The coupling between generations, specified by the Kobiyashi-Maskawa mixing angles, is weak.

To complete the scheme for the weak interactions we must bring in the quarks. There are two "flavors" (up and down) for the first (electron) generation, and two (charmed and strange) for the second (muon) generation; there are supposed to be two more in the third (tau) generation to complete the picture. The existence of the beautiful (or bottom) quark is well established, but searches for the true (or top) quark are still under way. It is the only particle missing from the scheme, other than the Higgses, if you stick to three generations. The quarks are fermions and have electric charge $Q_{u,c,t} = +\frac{2}{3}$, $Q_{d,s,b} = -\frac{1}{3}$ and baryon number $\frac{1}{3}$. Each forms a weak isodoublet and an isosinglet in the now familiar pattern. This completes the weak interaction picture at the level we will discuss here.

The quarks differ markedly from the leptons in several respects. To begin with, they carry a conserved "color charge" with three colors, three anticolors, and an eightfold symmetry we will describe in more detail in Sec. 5.5. They couple strongly at low energy to eight spin one colored gluons. Color conservation is given a vector representation in Fig. 7.

Remarkably, both quarks and gluons are "confined": they show up like

internal particulate degrees of freedom in high-energy experiments (parton model) but never have been liberated to be studied as free particles. Hence the definition of their masses is indirect; recent calculations would seem to indicate that the mass of an up or down quark is about $\frac{1}{3}$ the mass of a proton at low energy, but falls off like $1/p^2$ as the momentum with which they interact increases.[11] One up quark combined with an up-down pair in a spin singlet state to form an overall color singlet state form a proton with charge 1, while a down quark combined with the pair in the same way forms a neutron with charge 0. Consequently, the β decay properties of the neutron can be related to the weak isodoublet description given above.

As far as quantum number conservation goes, we can talk about baryon number (B) spin and (strong) isospin with charge conservation given by $Q = B/2 + I_z$ in the same way we talked about weak hypercharge and weak isospin conservation above. Quark-anti-quark pairs describe the mesons (pions, etc.) which older theories used to explain nuclear forces, but the details of how the quark-nuclear physics interface actually works quantitatively is a very controversial field of research. The easiest way to picture all this is to write the "color" vertices separately as vectors in a plane and assume that they add to form a color singlet (which can be a neutral colored or anticolored triplet, or any one of the color-anti-color pairs). Then we can return to the familiar picture of neutron, proton, their antiparticles, and associated mesons in the (s_h, I_z, B) space pictured in Fig. 8. Note the symmetry of the diagram for these parity-conserving strong interactions in contrast to the asymmetric diagram, which pictures the parity nonconserving weak-electromagnetic unification.

For any theory to get the quantitative details right is obviously a major research program. A useful reference that gives some idea of the magnitude of the task is the proceedings of the 1986 SLAC Summer Institute.[12] Clearly, we must stop at some point short of that effort here.

4. AN ALTERNATIVE STARTING POINT?

The previous section has only skimmed the surface of the phenomena that elementary particle physicists expect to be discussed, quantitatively, in their own terms before they will take a rival approach seriously. Since cosmology, condensed matter physics, etc., rest on the same foundations and must confront much richer experiential detail, a serious alternative appears to be difficult to construct. Nevertheless, a start has been made.

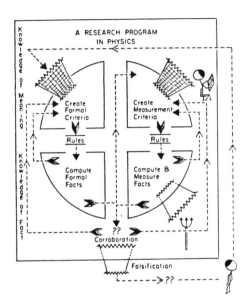

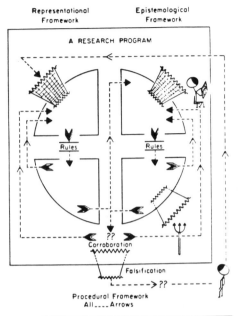

Figure 9(a). The participator model for a research program in physics.

Figure 9(b). Comparison with McGoveran's modeling methodology.

4.1 Modeling Methodology for Physics

The practice of physics cannot get off the ground without essential agreement among the practitioners as to what they are about, how to go about it, and what constitutes progress in their common effort. Often this is clear enough to the inside group, but in times of change the boundary shifts to include others. Then more formal – and more discursive – attention to these essential aspects of practice can be helpful. Keep in mind that the basic presenting problem we are tackling is to find a common origin for the structure of both quantum mechanics and relativity.

We adopt McGoveran's modeling methodology.[1] This has three critical elements:

(1) *Epistemological framework* (E frame). This is a set of loosely defined agreements made explicit by those injecting information into the model formulation; Gefwert[13] would call this a practical understanding of physics.

(2) *Representational framework* (R frame). This is an abstract formalism consisting of a set of symbols and a set of rules for manipulation; to formulate such a frame is, for Gefwert, to practice syntax.

(3) *Procedural framework* (P frame). This is an algorithm that serves to establish rules of correspondence between the observations agreed on in the E frame and the symbols of the R frame. Gefwert would describe this activity as the practice of semantics. Through recursion the P frame serves to modify the rules of correspondence,

the E frame, and the R frame until a sufficient level of agreement concerning accuracy is achieved – or the model fails. Kuhn[14] would call such a failure a crisis, which in the fullness of time could lead to a paradigm shift.

Note that we halt the infinite regress of the analysis of terminology in constructive modeling by recognizing the epistemology. We deny the validity and the value of any attempt to analyze theory-laden language. Such an analysis lies outside our task when we engage in generating a specific model. Attempting to make such an analysis would require us to generate a model that would contain the specific model as an instance. We *cannot* do so within our methodology. Analysis of that sort would involve nonconstructive methods: The analyst *must* work from a specific model by generalization – having failed to construct the general model first.

In an earlier paper[15] we illustrated Gefwert's analysis of the role of the participator in a research program as is shown in Fig. 9. The comparison with McGoveran's modeling methodology (Fig. 9b) is supposed to bring out the fact that the possible legal walks of the diagram are the same, but that the research program is contained *within* the methodology and that the methodology contains routes (arrows) that are *outside* the program. Thus the entry of the participator from a direction outside the box, and of the empirical confrontation (represented by Poseidon's trident Ψ) from a different direction remain the same; so does the fact that corroboration leaves the participator inside, while falsification takes him outside, in yet another direction. The methodology implies iteration in the EPR or ERP sequence or any interleaving of such sequences. The practitioner (and hopefully the reader of our papers) should keep on asking after each iteration how far our E frame has gone toward expressing the aspects of contemporary physics which he can accept as a starting point.

The modeling methodology presupposes that the community adopting it commits itself, individually and collectively, to:

(1) Agreement of cooperative communications:
    (a) commonly defined terms as fundamental;
    (b) fundamental vs derived terms;
    (c) agreement of pertinence.
(2) Agreement of intent.
(3) Agreement on observations.
(4) Agreement of explicit assumptions.
(5) The razor:
    (a) agreement of minimal generality;
    (b) agreement of elegance;
    (c) agreement of parsimony.

Our agreed-upon intent is to model the practice of physics. We take as fundamental the commonly defined terms of laboratory physics, treating terms denoting nonobservables as derived or theoretical terms. We recognize that it is very unlikely that agreement on the distinction between observable and theoretical terms can be reached before several passes through the whole scheme have been made. We take laboratory events as a sufficient set of observations to be modeled without requiring the standard theoretical interpretation. We take as understood that an experimental (laboratory) measurement may encompass many acts of observation. In other words, we are not committed to accept the how and why of the observations, only the observations themselves, operationally understood.

4.2 Five Principles

In the last section we spelled out our modeling methodology with more attention to underlying ideas than physicists usually employ. We believe that this methodology is close to that customarily employed by the best physicists. Where we part company with standard practice in contemporary theoretical physics – and much of the mathematics physicists employ – is that we reject, from the start, the concept of the continuum. Physics has always rested on counting when it came to experiment; we are being conservative in taking discrete, numerical practice as our starting point.

The R frame theory is constructed with the intent to meet the following five principles:

Principle 1: The theory possesses the property of strict finiteness.

Principle 2: The theory possesses the property of discreteness.

Principle 3: The theory possesses the property of finite computability.

Principle 4: The theory possesses the property of absolute nonuniqueness.

Principle 5: The formalism used in the theory is strictly constructive.

McGoveran[1] has chosen these five principles and the order in which they are presented with particular care; their current form came into

existence after an all-day discussion of the theory with Kilmister, in which he remarked: "The reader should be warned that this is a damned subtle theory." Since HPN has never before known Kilmister to use such strong language, this warning should be taken to heart. We will not attempt here to give the precision to these principles which mathematicians, philosophers, and computer scientists require; consult McGoveran's discussion if you desire that.

A few casual remarks for physicists are in order. Finiteness comes before discreteness. This requires us to specify in advance how far we intend to count; there is always some finite ordinal N_{Max}. If we exceed this initial bound, *all* arguments must be reexamined. Finite computability requires all algorithms to terminate within this N_{Max} and require no more memory for the storage of their coding and results than can be bounded by some cardinal $\log_2 N + N$. Absolute nonuniqueness requires us to assign equal prior probabilities to cases in the absence of further information; it also introduces indistinguishables whose cardinal number can exceed their ordinal number. Strict constructivism puts us firmly on one side of many debates about the foundations of mathematics. All these requirements make sense to practicing computer scientists and should also appeal to high-energy experimental physicists who get frustrated by the vagueness of the predictions their theoretical colleagues often make.

McGoveran goes on to use these principles in the construction of an ordering operator calculus and a finite and discrete geometry based on derivates (i.e., finite differences) rather than the derivatives of continuum theories. Since one of us (DMcG) has broken new ground for the construction, and his methods are unfamiliar, we do not attempt here to match his precision of thought. This paper is aimed at being introductory; unfortunately, it cannot, under the circumstances, be obvious.

5. EVENTS, CONSERVATION LAWS, AND "(ANTI-) PARTICLES"

Our next task is to actually construct a self-consistent representational framework that embodies our principles. As Gefwert would put it, we will now practice syntax. Our intent is to reconcile quantum mechanics with relativity in a consistent way. We should exercise care *not* to introduce theory-laden language into the representational frame. The self-consistency must not rely on intuitive ideas drawn from physics. I (HPN) fear that I have not succeeded in avoiding this trap altogether; forty years of practicing theoretical physics in a conventional way has left me with some bad habits. The formalism presented here has been scrutinized by Kilmister and Bastin, who are more sensitive than HPN to this trap; they support at least the essential aspects of the result.

5.1 The Combinatorial Hierarchy

Historically, the line of research that has led to the results presented here began with Eddington and Bastin's thinking about Eddington's fundamental theory. Bastin realized that when we go to the very large (distant galaxies, early times, etc.) or the very small (quantum events, elementary particles, etc.), the information available to us becomes extremely impoverished compared to the phenomena modeled by classical physics. He concluded that this fact should be reflected in the theory in such a way that this restriction is respected.

The route into the theory initially followed by Bastin and Kilmister concentrated on the problem of modeling discrete events.[16],[17] Ordered strings of zeros and ones gave a powerful starting point for analyzing this problem. Attention eventually centered on the question of whether bit

An Essay on Discrete Foundations for Physics

Table II: The Combinatorial Hierarchy

l	$B(l+1)$ $= H(l)$	$H(l)$ $= 2^{B(l)} - 1$	$M(l+1)$ $= [M(l)]^2$	$C(l)$ $= \sum_{j=1}^{l} H(j)$
Hierarchy level				
(0)	$\cdots$	2	(2)	$\cdots$
1	2	3	4	3
2	3	7	16	10
3	7	127	256	137
4	127	$2^{127} - 1$	$(256)^2$	$2^{127} - 1 + 137$

Level 5 cannot be constructed because $M(4) < H(4)$

strings were the same or different. Define a bit string by

$$(a)_n \equiv (\ldots, b_i^a, \ldots)_n; \quad b \in 0, 1; \quad i \in 1, 2, \ldots, n.$$

An economical way to compare an ordered sequence of two distinct symbols with other sequences of the same bit length is to use the operator XOR ("exclusive or," symmetric difference, addition (mod 2) $= +_2$, OREX, ...). Since we sum (or count) the 1's in the string to specify a measure we can treat the symbols "0" and "1" as integers, and only in some contexts can we think of them as bits; hence our bit strings can be more complicated conceptually than those encountered in standard computer practice. We therefore use the more general discrimination operation "$\oplus$," and a shorthand notation for it. Define the symbol $(ab)_n$ and the discrimination operation $\oplus$ by

$$(ab)_n \equiv S^a \oplus S^b = (\ldots, (b_i^a - b_i^b)^2, \ldots)_n = (\ldots, b_i^a +_2 b_i^b, \ldots)_n.$$

The name comes from the fact that the same strings combined by discrimination yield the null string, but when they differ and $n \geqq 2$ they yield a third distinct string that differs from either; thus the operation discriminates between two strings in the sense that it tells us whether they are the same or different.

We define the null string $(0)_n$ by $b_i^0 = 0$, $i \in 1, 2, \ldots, n$ and the antinull string $(1)_n$ by $b_i^1 = 1$, $i \in 1, 2, \ldots, n$. Since the operation $\oplus$ is only defined for strings of the same length we can usually omit the subscript n without ambiguity. The definition of discrimination implies that

$$(aa) = (0); \quad (ab) = (ba); \quad ((ab)c) = (a(bc)) \equiv (abc),$$

and so on.

The importance of closure under this operation was recognized by John Amson. It rests on the obvious fact that $(a(ab)) = (b)$ and so on. We say that any finite and denumerable collection of strings, where all strings in the collection have a distinct tag $i, j, k \ldots$, are discriminately independent (d.i.) iff

$$(i) \neq (0) : (ij) \neq (0), (ijk) \neq (0), \ldots (ijk \ldots) \neq (0).$$

We define a discriminately closed subset of non-null strings $\{(a), (b), \ldots\}$ as the set with a single non-null string as member or by the requirement that any two different strings in the subset give another member of the subset on discrimination. Then two discriminately independent strings generate three discriminately closed subsets (DCs S's), namely

$$\{(a)\}, \{(b)\}, \{(a), (b), (ab)\}.$$

Three discriminately independent strings give seven discriminately closed subsets, namely

$$\{(a)\}, \{(b)\}, \{(c)\}$$

$$\{(a), (b), (ab)\}, \{(b), (c), (bc)\}, \{(c), (a), (ca)\}$$

$$\{(a), (b), (c), (ab), (bc), (ca), (abc)\}.$$

In fact, x discriminately independent strings generate $2^x - 1$ discriminately closed subsets, because this is simply the number of ways one can take x distinct things one, two, three, $\ldots$, x at a time.

The discovery of the combinatorial hierarchy[18] was made by Parker-Rhodes in 1961. The history is fast receding.[19] Parker-Rhodes did indeed generate the sequence 3, 10, 137, $2^{127} + 136 \simeq 1.7 \times 10^{38}$ in suspiciously accurate agreement with the scale constants of physics. This was a genuine discovery; the termination is at least as significant. The sequence is simply $(2 \Rightarrow 2^2 - 1 = 3)$, $(3 \Rightarrow 2^3 - 1 = 7)$ $[3 + 7 = 10]$, $(7 \Rightarrow 2^7 - 1 = 127)$ $[10 + 127 = 137]$, $(127 \Rightarrow 2^{127} - 1 \simeq 1.7 \times 10^{38})$. The real problem is to find some stop rule that terminates the construction.

The original stop rule was due to Parker-Rhodes. He saw that if the discriminately closed subsets at one level, treated as sets of vectors, could be mapped by nonsingular (so as not to map onto zero) square matrices having uniquely those vectors as eigenvectors, and if these mapping matrices were themselves linearly independent, they could be rearranged as vectors and used as a basis for the next level. In this way, the first sequence is mapped by the second sequence $(2 \Rightarrow 2^2 = 4)$, $(4 \Rightarrow 4^2 = 16)$, $(16 \Rightarrow 16^2 = 256)$, $(256 \Rightarrow 256^2)$. The process terminates because there are only $256^2 = 65,536 = 6.5536 \times 10^4$ d.i. matrices available to map the fourth level, which are many too few to map the $2^{127} - 1 = 1.7016 \ldots \times 10^{38}$ DCsS's of that level. By now, there are many ways to achieve and look at this construction and its termination.[20]-[23] The (unique) combinatorial hierarchy is exhibited in Table II.

5.2 The Label-content Schema

For some time, the only operation used in the theory was discrimination. Kilmister eventually realized that one should also think about where the strings came from in the first place. He met this problem by introducing a second operation which he called generation. As he and HPN realized, this operation eventually generates a universe that goes beyond the bounds of the combinatorial hierarchy. Once this happens, we can separate the strings into some finite initial segment that represents an element of the hierarchy, which we call the label, and the portion of the string beyond the label which we now call the content. It is clear that from then on the content ensemble for each label grows in both number and length as the generation operation continues. Since it takes $2 + 3 + 7 + 127 = 139$ linearly independent basis strings to construct the four levels of the combinatorial hierarchy, the labels will be of at least this length; if we use the mapping matrix construction, they will be of length 256. Call this fixed length L, the length of any content string n, and the total length at any TICK (see next section) in the evolution of the universe $N_U = L + n$. Then the strings will have the structure $S^a = (L_a)_L \| (A_x^a)_n$, where a designates some string of the $2^{127} + 136$ which provide a representation of the hierarchy, and x designates one of the 2^n possible strings of length n; the symbol "$\|$" denotes string concatenation.

PROGRAM UNIVERSE 1

PROGRAM UNIVERSE 2

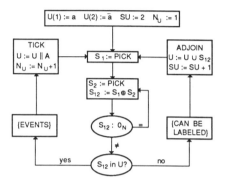

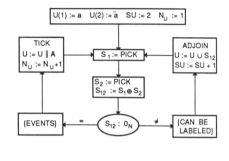

Figure 10. PROGRAM UNIVERSE 1 and 2 compared.

5.3 PROGRAM UNIVERSE

In order to generate a universe of strings that grows, sequentially, in either number (SU) or length (N_U) Manthey and HPN created PROGRAM UNIVERSE. Recently, Manthey realized that the criterion we used to increase the string length (TICK) was unjustifiably selective. The previously published version of the program,[5] called PROGRAM UNIVERSE 1, is compared with Manthey's new proposal, called PROGRAM UNIVERSE 2, in Fig. 10. A potentially significant effect of the change is to allow the bit string universe to contain, ephemerally in many cases, distinct strings that are indistinguishable under discrimination. The difference between PU1 and PU2 does not affect anything in this paper but might eventually provide alternative cosmological models that make observationally different predictions.

The program is initiated by the arbitrary choice of two distinct bits, which become the first two strings in the universe. Whether insisting that one be "0" and the other "1," as is done in the flow chart, rather than allowing both to be arbitrary will eventually produce a significantly different cosmology (or choice among cosmologies) at our epoch is an open question.

Entering the main routine at PICK, we choose two strings (i) and (j) and discriminate them: $(ij) \equiv (i) \oplus (j)$. Whenever the two strings picked are identical, $(ij) = (0)_{N_U}$ and we go to TICK. TICK concatenates a single bit, arbitrarily chosen for each string, to the growing end, notes the increase in string length, and the program returns to PICK. The alternative route, which occurs when discrimination generates a non-null string, simply ADJOIN's the newly created string to the universe, and the program returns to PICK.

In the older version we proved that TICK had to be "caused" (in the computer simulation) either by the occurrence of the three-event configuration $S^a \oplus S^b \oplus S^c = 0_{N_U}$ or by the configuration $S^a \oplus S^b \oplus S^c \oplus S^d = 0_{N_U}$, which we called a four-event. But this implied a uniqueness that has no known demonstrable counterpart in events as modeled by contemporary physics; there can be many simultaneous events. At the ninth Alternative Natural Philosophy Association Meeting (ANPA 9) the author HPN extended the definition of event to include all cases in which, at a given string length (or TICK), three or four strings combine under discrimination to produce the null string. This definition of event is retained here, but in PROGRAM UNIVERSE 2 is no longer the cause of TICK. Instead we TICK whenever two strings interact without producing any novelty. This is as close as we need to get to defining what would be called a point in a continuum theory. We will see in Sec. 6.2 that this construction of a point is consistent with our development of Einstein synchronization and, hence, to the extent possible in our discrete theory, consistent with the conventional use of the term event in relativity theory.

The method Manthey and HPN used to construct the hierarchy is much simpler than the original matrix construction given by Parker-Rhodes; in fact, some might call it simple-minded. The objection we now find cogent is that the method is nonconstructive and hence violates our fundamental principles; new efforts to meet this objection are under way. Manthey and Noyes claimed that all we had to do was to demonstrate explicitly (i.e., by providing the coding) that any run of PROGRAM UNIVERSE contained (if we entered the program at appropriate points during the sequence) all we needed to extract some representation of the hierarchy and the label content scheme from the computer memory *without* affecting the running of the program. (Subsequently, DMcG has pointed out that this way of meeting the problem is not strictly constructive and should be replaced by a generation scheme that develops the hierarchy constructively.) The obvious intervention point exists where a new string is generated, i.e., at ADJOIN. The subtlety here is that if we assign the tag i to the string $U[i]$ as a pointer to the spot in memory where that string is stored, this pointer can be left unaltered from then on. It is of course simply the integer value of $SU + 1$ at the time in the simulation (sequential step in the execution of that run of the program) when that memory slot was first needed. Of course, we must take care in setting up the memory that *all* memory slots are of length $N_{max} > N_U$, i.e., can accommodate the longest string we can encounter during the (necessarily finite) time our budget will allow us to run the program. Then, each time the program TICK's, the bits that were present at that point in the sequential execution of the program when the slot [i] was first assigned will remain unaltered; only the growing head of the string will change. Thus if the strings $i, j, k \ldots$ tagged by these slots are discriminately independent at the time when the latest one is assigned, they will remain discriminately independent from then on.

Once this is understood the coding Manthey and HPN gave for our labeling routine should be easy to follow. We take the first two discriminately independent strings and call these the basis vectors for level 1. The next vector that is discriminately independent of these two starts the basis array for level 2, which closes when we have three basis vectors discriminately independent of each other and of the basis for level 1, and so on until we have found exactly $2 + 3 + 7 + 127$ discriminately independent strings. The string length when this happens is then the label length L; it remains fixed from then on. During this part of the construction we may have encountered strings that were *not* discriminately independent of the others, which up to now we could safely ignore. Now, we make one *mammoth* search through the memory and assign each of these strings to one of the four levels of the hierarchy; it is easy to see that this assignment (if made sequentially passing through level 1 to level 4) has to be unique.

From now on, when the program generates a new string, we look at the first L bits and see if they correspond to any label already in memory. If so, we assign the content string to the content ensemble carrying that label. If the new string also has a new label, we simply find (by upward sequential search as before) what level of the hierarchy it belongs to and start a new labeled content ensemble. Because of discriminate closure, the program must eventually generate $2^{127} + 136$ distinct labels, which can be organized into the four levels of the hierarchy. Once this happens, the label set cannot change, and the parameters i for these labels will retain an *invariant* significance no matter how long the program continues to TICK. It is this invariance that will later provide us with the formal justification for assigning an invariant mass parameter to each string. We emphasize once more that *what* specific representation of the hierarchy we generate in this way is irrelevant; any run of PROGRAM UNIVERSE will be good enough for us.

What was *not* realized when this program was created was that this simple algorithm provides us with the minimal elements needed to construct a finite particle number scattering theory. The increase in the number of strings in the universe by the creation of novel strings from discrimination is our replacement for the particle creation of quantum field theory. It is not the same, because it is both finite and irreversible; it

also changes the state space. Note that the string length N_U is simply the number of TICK's that have occurred since the start-up of the universe; this order parameter is irreversible and monotonically increasing like the cosmological time of conventional theories. Our events are unique, indivisible, and global, in the computer sense; consequently, events cannot be localized and will be supraluminally correlated.

5.4 "Vector" Conservation Laws

So far, we have a gross structure based on bit strings and two operations that generate them via a specific program: 1) ADJOIN, which adjoins a non-null string produced by discrimination to the extant bit string universe, and 2) TICK, which increases the string length by concatenating a single bit, arbitrarily chosen for each string at the growing head of each string. We have two kinds of connectivity that result from this construction. One is the label-content schema. Once the label basis has closed under discrimination to form $2 + 3 + 7 + 127$ linearly independent strings, PROGRAM UNIVERSE will necessarily generate some representation of the combinatorial hierarchy at that label length; this will close with $3 + 7 + 127 + 2^{127} - 1$ labels of that length. Once the label basis (and label string length) is fixed, PROGRAM UNIVERSE assigns each novel content string to a specific label when it is created by discrimination and augments each content string by an arbitrary bit at each TICK. The second is the connectivity between strings of the same length (i.e., "between ticks"), which we have characterized as three-vertices $(abc)_{L+n} = (0)_{L+n}$ and four-events $(abcd)_{L+n} = (0)_{L+n}$.

To come closer to what we need for physics in the sense of relating the (R frame) model to measurement (counting) in the laboratory, we need to introduce a quantitative measure and a norm for such measures. Once we have done this, we can introduce a third operation connecting bit strings (inner product) that allows us to derive conservation laws. Define a measure $\|x\|$ on (x) by

$$\|x\| \equiv \sum_{i=1}^{n} b_i^x, x \in a, b, c \ldots.$$

This is the usual Hamming measure. This is McGovern's normalized attribute distance relative to the reference string (0) $(b_i^0 = 0$ for all i; $\|0\| = 0)$, and $(n - \|x\|)/n$ is the distance relative to the antinull string (1) $(b_i^1 = 1$ for all i; $\|1\| = n)$.

Consider a three-vertex defined by $(abc) = (0)$, or equivalently by $\|abc\| = 0$.

Theorem 1. The measure $\|x\|$ is a norm, i.e.,

$$(abc) = (0) \Rightarrow |\, \|a\| - \|b\| \,| \leq \|c\| \leq |\, \|a\| + \|b\| \,|, \text{ cyclic on } a, b, c.$$

Argument: From the definition of discrimination, if we consider the three bits at any ordered position i in the three strings of a three vertex, we can only have either one 0 and two 1's in the three strings, or three zeros. If the single zero is $b_i^a = 0$, call the number of times this occurs n_{bc} (cyclic on a, b, c), and the number of times we have three 0's n_0. Clearly, $n_{bc} + n_{ca} + n_{ab} + n_0 = n$ and $\|a\| = n_{bc} + n_{ca}$, cyclic on a, b, c, from which the desired inequalities follow.

Note that this theorem depends on a computer memory. It is static in that it depends only on a particular type of configuration that is "wired in" by the program. It is dynamic, in the sense that the three strings are brought together as a consequence of past sequences that are arbitrary from the point of view of the local vertex. It is global in that any single three-vertex (or four-event) *could* lead to a TICK, which affects the whole bit string universe.

If we now define the inner product $\langle (x) \cdot (y) \rangle$ between two strings (a), (b) connected by a three vertex $(abc) = (0)$ with the equality

$$2 \langle (a) \cdot (b) \rangle \equiv ||a||^2 + ||b||^2 - ||c||^2,$$

it follows immediately that

Corollary 1.1

$$||ab||^2 = \langle (a) \cdot (ab) \rangle + \langle (b) \cdot (ab) \rangle = \langle (ab) \cdot (ab) \rangle$$
$$||a||^2 = \langle (ab) \cdot (a) \rangle + \langle (b) \cdot (a) \rangle = \langle (a) \cdot (a) \rangle$$
$$||b||^2 = \langle (ab) \cdot (b) \rangle + \langle (a) \cdot (b) \rangle = \langle (b) \cdot (b) \rangle$$

If we define a four-vertex by $(abcd) = (0)$, or equivalently by $||abcd|| = 0$, with an obvious extension of the notation it also follows that

Theorem 2

$$(abcd) = (0) \Rightarrow ||a|| = ||bcd||, \text{ cyclic on } abcd$$
$$||ab|| = ||cd||; \, ||ac|| = ||db||; \, ||ad|| = ||bc||.$$

Argument: $(abcd) = (0) \Rightarrow (abc) = (d)$, etc., and $\Rightarrow (ab) = (cd)$ etc., from which the result follows.

Corollary 2.1 For any pair taken from the ensemble $abcd$ the appropriate version of Corollary 1.1 follows.

Corollary 2.2

$$\langle (a) \cdot (cd) \rangle + \langle (b) \cdot (cd) \rangle = ||ab||^2 = ||cd||^2$$
$$= \langle (c) \cdot (ab) \rangle + \langle (d) \cdot (ab) \rangle$$

and so on for any of the three pairs. It follows that we can put two three-events together to make a four-event in the six different ways required by 2-2 crossing, as discussed in our presentation of the practice of particle physics.

As Kilmister has pointed out to us, this is not sufficient for us to go from these results and our earlier definition of the inner product to the conclusion that a four-vertex defines the vector conservation law

$$a + b + c + d = 0$$

in all cases. Fortunately, all we need for the physics we develop below is the 2-2 crossing in observable events, which *does* follow from what we have developed above when clothed with the appropriate rules of correspondence; that is, we can justify what in a vector theory would be written as the three interpretations

$$a + b = c + d; \, a + c = b + d; \, a + d = b + c.$$

Since a four-vertex $(abcd) = (0)$ can be decomposed in seven different ways, namely

$$(ab) = (cd); \quad (ac) = (bd); \quad (ad) = (bc)$$
$$(a) = (bcd); \quad (b) = (cda); \quad (c) = (dab); \quad (d) = (abc)$$

we can – under appropriate circumstances – still make seven different temporally ordered interpretations of the single four-vertex given above: three (2,2) channels, four (3,1) channels, and the unobservable (4,0) channel. Note that all eight relationships are generated by one four-vertex.

5.5 The Standard Model for Quarks and Leptons Using Combinatorial Hierarchy Labels

Our next step is to recall that we can always separate a string into two strings $(a)_{L+n} = (L_a)_L || (A_a)_n$ where "||" denotes string concatenation. We call the first piece the label and the second the content. There is a simple correlation between the two pieces. If we take some content string A_a and call its velocity $\beta_a = 2||A_a||/n - 1$, the string $(a1)$ has the opposite velocity. Further, if we use the string (a) as the reference string for a conservation law defined by the inner product relations given above, the reversal of the velocity achieved by discrimination with the antinull string can be correlated with the definition of label quantum numbers and conservation laws in such a way that physically observable crossing symmetry is respected. Then the theory is invariant under the arbitrary choice of reference direction.

It can be seen that the string for which both label and address are the antinull string plays a special role in the theory, since it specifies the relationship between particle and antiparticle, and interacts with everything whether it is massive or massless. Since it is unique among the $2^{127} + 136$ labels, it is readily identified as the Newtonian gravitational interaction. It is the only level four label we will refer to explicitly, for reasons discussed below.

Physical interpretation of the labels naturally starts with the simplest structures, which are the weak and electromagnetic interactions. We can get quite a long way just by looking at the leading terms in a perturbation theory in powers of $e^2/\hbar c \simeq 1/137$ for quantum electrodynamics and of $G_F \simeq 10^{-5}/m_p^2$ for the low-energy weak interactions such as β decay. As Lee and Yang saw, if the neutrino is massless and chiral, the Fermi β-decay theory will violate parity conservation maximally; this is still the simplest accurate description of low-energy weak interactions.

Since level 1 has only two basic entities, we identify these with the neutrino ν and the anti-neutrino $\bar{\nu}$. One might think that their closure would be the zero helicity component of the spin one neutral weak boson Z^0, but if we take the neutrinos to be massless, and hence their content strings to be null or antinull, they cannot form a three-vertex with a massive particle. Actually, the Z_0 and W must couple to all of the first three levels and hence must be assigned to level 4, which we are not attempting to model in detail in this article. Further, although massive, they are all unstable – as are all massive level 4 – and hence require us to go beyond the simple modeling of Yukawa vertices for stable, elementary particles developed in this article. If we follow the usual convention of defining the chirality of the neutrino as left-handed, once we have added content strings and defined directions, we still need a convention as to whether the label is to be concatenated with the string $(1)_n$ with velocity $+c$, or the string $(0)_n$ with velocity $-c$. We can take the bit string state $(\nu_L)_{L+n} = (\nu_\lambda)_L || (1)_n$ and the right-handed (i.e., anti-) neutrino $(\nu_R)_{L+n} = (\bar{\nu}_\rho)_L || (0)_n$. Then if we use a representation in which $(\nu_\rho)_L = (1\nu_\lambda)_L$, the Feynman rules will be obeyed. The vertex can be interpreted as the gravitational interaction of a neutrino or an anti-neutrino. Note that for massless particles $(\beta = \pm c)$, we cannot specify a direction until we connect them to slower particles whose directions can be assigned. Thus we are forced to adopt a Wheeler-Feynman type of theory in which all massless "radiation" emitted by charged particles must be absorbed.

Interpretation of level 2 as modeling the vertices of quantum electrodynamics for electrons, positrons, and photons follows the fol-

An Essay on Discrete Foundations for Physics

lowing scheme. We take as the linearly independent basis strings (e_λ^+), (e_λ^-), $(\Gamma_{\lambda\lambda})$ and define the non-null string which guarantees their independence as $(\Gamma_c) = (e_\lambda^+ e_\lambda^- \Gamma_{\lambda\lambda})$. The remaining three label strings which close level 2 are then defined by

$$(e_p^+) = (\Gamma_c e_\lambda^-); \quad (e_p^-) = (\Gamma_c e_\lambda^+); \quad (\Gamma_{pp}) = (\Gamma_c \Gamma_{\lambda\lambda}).$$

We take the same convention for positive direction and chirality as we did for level 1, using the negative, left-handed electron as our reference string and the velocity $\beta_{e_L^-} = 2k_{e_L^-}/n - 1$ as positive when this number is positive. The physical states, where we omit the subscripts on β, are then given by

$$(\gamma_c)_{L+n} = (\Gamma_c)_L \|(1)_n; \quad (e_L^-) = (e_\lambda^-)\|(-\beta)_n; \quad (e_L^+) = (e_\lambda^+)\|(-\beta)_n$$
$$(e_R^+) = (e_p^+)\|(\beta)_n = (\gamma_c e_L^-); (e_R^-) = (e_p^-)\|(\beta)_n = (\gamma_c e_L^+)$$
$$(\gamma_{RR}) = (\Gamma_{pp})\|(1)_n; \quad (\gamma_{LL}) = (\Gamma_{\lambda\lambda})\|(0)_n = (\gamma_c \gamma_{RR}),$$

and the Feynman rules are obeyed for all three-vertices.

The four-vertex $(e\bar{e}\,\gamma\gamma_c) = (0)$ cannot be readily discussed until we have the configuration space theory nailed down. It is related to our finite treatment of bremsstrahlung in a coulomb field. The vertex $(\gamma_{LL}\gamma_{RR}\gamma_c) = (0)$ would seem to imply an interaction between photons and the coulomb field – a vertex that vanishes in the conventional theory because of the masslessness of the photon and gauge invariance.

A related problem arises with the vertices implied by our connection between particles and antiparticles, namely

$$(\nu\bar{\nu}1) = (0); \quad (e\bar{e}1) = (0); \quad (\gamma\bar{\gamma}1) = (0).$$

A little thought shows that such vertices will occur for *any* particle-anti-particle pair. Hence the antinull label string "interacts" with everything and must be assigned to level 4. This unique label string, which occurs with probability $1/(2^{127} + 136)$, is identified with Newtonian gravitation. It leads to the bending of light in a gravitational field. Of course, to get the experimentally observed result, we will have to identify the spin two gravitons as well and show that they double this deflection.

We conclude this section by identifying the level 3 structure with the quarks and gluons of quantum chromodynamics. This discussion follows along the lines already laid down in discussing the first two levels. We take as our basis label strings a quark part (u^+), (u^-), (d^+), or (d^-) concatenated with a color part (r), (y), (b) which gives us the seven independent strings needed to form level 3. The color strings are linearly independent, so we can define (analogous to what we did at level 2)

$$(ryb) = (w); \quad (\bar{r}) = (rw); \quad (\bar{y}) = (yw); \quad (\bar{b}) = (bw),$$

from which it follows that

$$(r y \bar{b}) = (0); \quad (r\bar{y}b) = (0); \quad (\bar{r}yb) = (0); \quad (\bar{r}\bar{y}\bar{b}) = (0).$$

Similarly, the linear independence of the quark parts allows us to define

$$(u^+ u^- d^+ d^-) = (Q); \quad (\bar{q}) = (qQ), q \in u^+., u^-, d^+, d^-.$$

Then a colored quark label $(q_c^\pm) = (q^\pm)\|(c)$ and a colored gluon label $(g_c) = (Q)\|(c), c \in r, y, b$, allow us to recognize the label part of the Yukawa vertex for QCD as $(q_c \bar{q}_c g_{c_3}) = (0)$. The essential point here is that, as proved above, $(c_1 \bar{c}_2 c_3) = (0)$ for any three distinct colors. We can then attach content labels and helicity in the same way as we did in QED, and once again the Feynman rules apply. Anyone familiar with lowest-order QCD can now immediately derive from our formalism the

"valence quark" structure of the proton and neutron in terms of three quarks, and the structure of the π, ρ, and ω in terms of quark-anti-quark pairs. In contrast to the level 2 situation, the three gluon vertex does not vanish and implies a four-gluon vertex. So we find that we have constructed *all* the lowest-order vertices of QCD with the correct conservation laws.

The problem of color confinement is solved, in principle, by McGoveran's theorem[24],[25]; i.e., the conclusion that in any finite and discrete theory there can be no more than three "homogeneous and isotropic dimensions" that remain indistinguishable as the (finite and discrete) cardinals and ordinals keep on increasing. (We discuss this theorem with more care in Sec. 6.1.) Because our labels are tied to contents, and hence via the counter paradigm to macroscopic directions, we can only have three quantum number "dimensions" asymptotically. These are saturated by the three absolutely (as far as we know currently) conserved quantum numbers: lepton number, baryon number, and charge (or z component of isospin), leaving no room for free quarks or gluons conserving asymptotic color charge. They can occur at short distance as degrees of freedom in the scattering theory – as we showed above – but eventually they have to "compactify" and become distinguishable from free particle quantum numbers. We can conclude this immediately without any detailed dynamical argument.

6. COMPARISON WITH EXPERIMENT

We use the traditional phrase for the title of this section. In McGoveran's terminology, we provide here the rules of correspondence, or a procedural framework by means of which we can connect our formal representational framework (Gefwert's syntax developed in the last section to the informal epistemological framework – the practice of experimental and theoretical physics in the laboratory – which it is our intent to model. In so doing we provide meaning, or as Gefwert would put it, practice semantics.

6.1 The "Counter Paradigm"

Bastin has insisted for decades that the primal contact between a (computable) formalism and the empirical "world" can only be made once. This was a basic reason why he and Kilmister[16],[17] fastened on steps of a scattering process as a likely point at which to investigate the connection between finite mathematics and physical theory. HPN started thinking of the elementary scattering process as fundamental thanks to his early involvement in Chew's S-matrix theory; for him this gave specific content to Bridgman's operationalism and Heisenberg's very early ideas. At ANPA 2 and 3 some of us saw that Stein's random walk derivation of the Lorentz transformation and the uncertainty principle[26] must somehow connect to scattering processes; others recognized the seminal nature of his work because of his ontological viewpoint.

The specific genesis of the counter paradigm occurred after HPN's presentation[27] at the conference honoring de Broglie's ninetieth birthday. Fortunately, HPN had an opportunity to start working on the final version of that paper[28] in consultation with Bastin before it was published. HPN realized that if he thought of Stein's random walk as a model for two sequential events in two spatially separated laboratory counters with the discrete step length being the de Broglie relativistic phase wavelength that, by representing Stein's random walks as bit strings with the bit one taken as a step toward the final counter and the bit zero a step away from it, he had the right point of contact between the bit strings used in the combinatorial hierarchy and the start of a scattering theory.

So far, we have only discussed three- and four-vertices for a fixed value of n. But each time PROGRAM UNIVERSE TICK's, each content string in each labeled ensemble acquires an arbitrary bit at the growing end. In the absence of further information, each content string therefore represents a sequence of Bernoulli trials with 0 and 1 representing the two possibilities. This has an extremely important consequence, which we call McGoveran's theorem.[24],[25] As has been noted by Feller[29], if we have D independent sequences of Bernoulli trials, the probability that after n trials we will have accumulated the same number (k) of 1's is

$$p_D(n) = \left(\frac{1}{2^{nD}}\right) \sum_{k=0}^{n} \left(\frac{n}{k}\right)^D.$$

He then shows that the probability that this situation will repeat N times is strictly bounded by

$$P_D(N) = \sum_{n=1}^{N} p_D(n) < \left[\frac{2}{\pi D}\right]^{1/2} N^{-1/2 \, (D-1)}.$$

Consequently, for $D = 2, 3$, where $p_D(n) < n^{-1/2}, n^{-1}$, such repetitions can keep on occurring with finite probability, but for four or more independent sequences, this probability is strictly bounded by zero in the sense of the law of large numbers.

McGoveran uses finite attributes, which can always be mapped onto ordered strings of 0's and 1's, as the starting point for his ordering operator calculus. As is discussed in more detail in Ref. 1, these can be used to construct a finite and discrete metric space. In order to introduce the concept of dimensionality into this space, he notes that we need some metric criterion that does not in any way distinguish one dimension from another. (In a continuum theory, we would call this the property of "homogeneity and isotropy"; we need it in our theory for the same reason Einstein did in his development of special relativity.) McGoveran discovered that by interpreting the coincidences $n = 1, 2, \ldots, N$ in Feller's construction as "metric marks" the metric space so constructed has precisely the discrete property corresponding to homogeneity and isotropy as just defined. Consequently, Feller's result shows that in *any* finite and discrete theory, the number of independent homogeneous and isotropic dimensions is bounded by three. If we start from a larger number of independent dimensions using *any* discrete and finite generating process for the attribute ensembles, we find that the metric will, for large numbers, continue to apply to only three of them, and that what may have looked like another dimension is not; the probability of generating the next metric mark in any of the others (let alone all of them) is strictly bounded by $1/N_{MAX}$.

Of course, the argument depends on the theory containing a universal ordering operator which is isomorphic to the ordinal integers. Further, since we know empirically that elementary particles are chiral, we will need three rather than two "spacial" dimensions. Thus *any* discrete and finite theory such as ours when applied to physics must be globally described by three dimensions and a monotonically increasing order parameter. Consequently, we are justified in constructing a rule of correspondence for our theory which connects the large number properties of our R frame to laboratory (E frame) $3 + 1$ space-time.

We begin with the paradigmatic case of a single particle entering a space-time volume (detector) $\Delta V \Delta T$, causing a count, and a time T later entering a second detector with similar resolution a macroscopic distance L from the first and causing a second count. We then say that the (average)

velocity of the particle between the two detectors is $V = L/T$; empirically this number is always less than or indistinguishable from the limiting velocity c.

This language is well understood by the particle physics experimentalist but raises a number of problems for others. To begin with, he uses "cause" in a philosophically vague but methodologically precise sense, which includes a host of practical experience about "background," "spurious counts," "real counts," "goofs," "GOK's" (i.e., "God only knows"), etc.

The actual practice of experimental particle physics implies the concept of indistinguishability in a critical way; the experimentalist uses, often without conscious analysis, finite collections whose cardinal number may exceed their ordinal number; this fact is diagnostic for sorts that are not reducible to sets.[23] To put it more formally in terms of background and counts, in the absence of a constructive definition of the two subsets – which is often unavailable in practice, and in our theory we would claim can be unavailable in principle – the two collections are sorts rather than sets.

The rule of correspondence in the counter paradigm case (two sequential counts spatially separated) applies to a labeled string with label L_a which at the TICK with the content string length n_0 was part of a three- or four-vertex and again part of a vertex at content string length $n_0 + n_a$, *and which is appropriately assigned to theoretically relevant data rather than to background*. We ask how many 1's were added to the content string; we call these k_a. We identify the (average) laboratory velocity of the particle ($V = L/T$) with the R frame quantity by the equation $V = [(2k_a/n_a) - 1]c$. The sign of this velocity defines the positive or negative sense of the direction between the counters in the laboratory (or vice versa: a choice must be made *once*). Since the evolution of the bit string universe will provide many candidates for the strings which meet these criteria within the time and space resolution of the counters, we will have to provide more and more precise definitions of these criteria as the analysis develops.

6.2 Event-based Coordinates and the Lorentz Transformations

As is discussed with much more care in Ref. 1, any theory satisfying our principles can be mapped onto ensembles of bit strings simply because, with respect to *any* attribute, we can say whether a collection has that attribute or does not. To introduce a metric, we need a distance function relative to some reference ensemble. Because of our finite and discrete principles, any allowed program can only take a finite number of steps to bring any ensemble into local isomorphism with the reference ensemble *in respect to that attribute*. Note that there can be many attributes, many distance functions, and that the space can be multiply connected. Note that this definition also provides a (dichotomous, e.g. $\pm$) *sense* to the computation steps: They must increase the attribute distance or decrease it. Calling the number of increments I and the number of decrements D, using a well-defined computational procedure, the attribute distance is, clearly, $d_A = I - D$, and the total number of steps $N = I + D$. Then we can also define the attribute velocity with which the two ensembles are separating or coming together $v_A = (I - D)/(I + D)$. Thus there always is a limiting velocity for each attribute, which is attained when all steps are taken in the same direction.

If we wish to model the events of which contemporary physics takes cognizance, we know that all physical attributes are directly or indirectly coupled to electromagnetism. Therefore, the limiting velocity of physics,

An Essay on Discrete Foundations for Physics

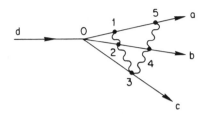

Figure 11. A four-event followed by five events involving limiting velocity signals which can be used to establish the Lorentz transformations for event 3.

c, will be the *smallest* of these limiting attribute velocities, simply because it refers to the attribute with the maximum cardinality. Any ensemble of attributes specified by a more limited description involves a supraluminal velocity without allowing supraluminal communication of information. Hence we can expect to find correlation between and synchronization of events in spacelike separated regions; from our discrete point of view the existence of the effects demonstrated in Aspect's and other EPR-Bohm experiments is anticipated and in no way paradoxical. We guarantee Einstein locality for causal events; that is, for those initiated by the transfer of physical information.[30]

In order to go from this general proof of the limiting velocity to the laboratory practice of relativistic particle quantum mechanics, we need a more specific formalism than the general derivation given in Ref. 1. We start from the three- and four-vertices already mentioned and consider how they can be used to model the laboratory situation given in Fig. 11. The initial four-vertex $(abcd)_{L+n_0} = 0$ is followed sequentially by five vertices involving "soft" photons, as is explained below. In the laboratory neither vertices, nor elementary events, nor soft photons can be observed; limiting cases in which the disturbance caused by the firing of counters connected with these five events is negligibly small are easy to envisage. We use a specific example of labels that can, if we wish, be given a specific interpretation in which particles a, b, c have spin $\frac{1}{2}$ and the photons have left or right spin one helicity.

We assume that it takes n_i TICK's of PROGRAM UNIVERSE beyond $L + n_0$ to generate the strings involved in the ith event. Since all strings will have the portion through content string length n_0 unaltered, we need use only these relative values: $n_i = N_U(i) - L - n_0$ and the corresponding terminal pieces of the strings for our contents. For event 1, we take the three strings to be

$$(a) = (1000)\|(A_1^a)_{n_1}; \quad (a') = (0100)\|(A_1^a)_{n_1}; \quad (\overline{\gamma}) = (1100)\|(0)_{n_1}.$$

Hence $(aa'\overline{\gamma}) = (0)$ defines a three-vertex in which the velocity of a does not change; we could call it a "soft photon" vertex. By crossing (cf. Secs. 3.1 and 5.5) this also can be interpreted as a vertex in which a flips its spin and emits a photon with the appropriate helicity, i.e., $(\gamma) = (0011)\|(1)_{n_1}$. The laboratory direction between events 1 and 2 then defines the reference direction for all subsequent discussion. The remaining vertices can be consistently represented by using

$$(b) = (1000)\|(A_2^b)_{n_2}; \quad (\gamma) = (0011)\|(1)_{n_2}; \quad (b') = (0111)\|(A_2^b)_{n_2}$$
$$(\gamma') = (1100)\|(1)_{n_2}$$
$$(c) = (1000)\|(A_3^c)_{n_3}; \quad (\gamma') = (1100)\|(1)_{n_3}; \quad (c') = (0111)\|(A_3^c)_{n_3}$$
$$(\overline{\gamma}) = (0011)\|(0)_{n_3}$$
$$(b') = (0111)\|(A_4^{b'})_{n_4}; \quad (\overline{\gamma}) = (0011)\|(0)_{n_4}; \quad (b'') = (1000)\|(A_4^{b'})_{n_4}$$
$$(\gamma) = (1100)\|(0)_{n_4}$$
$$(a') = (0100)\|(A_5^{a'})_{n_5}; \quad (a'') = (1000)\|(A_5^{a'})_{n_5}; \quad (\overline{\gamma}) = (1100)\|(0)_{n_5}.$$

We now trust that our rule of correspondence between three- and four-vertices and a standard laboratory situation used in the derivation of the Lorentz transformations is clear.

For simplicity, we consider here that particle a is, on the average, at rest between events 0, 1 and between events 1, 5:

$$k_0^a = n_0/2; \quad k_1^a = n_1/2; \quad k_5^a = n_5/2.$$

We also assume, again on the average, that b and c have constant velocity over the approprite intervals:

$$\beta_b = 2k_0^b/n_0 - 1 = 2k_2^b/n_2 - 1 = 2k_4^b/n_4 - 1$$
$$\beta_c = \beta = 2k_0^c/n_0 - 1 = 2k_3^c/n_3 - 1.$$

Our next simplification is to assume that all the events lie on a single line, reducing this to a $1 + 1$ dimensional problem. None of these simplifications are needed, as can be seen from the general discussion in Ref. 1.

In conventional terms, we are asking the question of how the coordinates of an event at $x = \beta ct$ in one coordinate system (the one in which particle a is at rest) transform to the coordinate system in which particle b is at rest. We are forced by our principles to assume, as in conventional treatments, that the velocity of light is the same in all coordinate systems and that the time at which event 3 occurs is the average between when the light signal that defines event 3 was emitted by a and returns to it. Introducing a parameter with the dimensions of length, whose value we will discuss later, these statements follow immediately from the definitions of attribute distance and velocity, since

$$x/\lambda = 2k - n; \quad ct/\lambda = n; \quad \beta = (2k/n) - 1$$

for any particle, and $k = 0$ or n specifies a connection with the limiting velocity for any set of strings. This is even clearer when we introduce light-cone coordinates:

$$d_+ = n + (2k - n) = 2k; \quad d_- = n - (2k - n) = 2(n - k).$$

The relationship between the two descriptions is illustrated in Fig. 12.

One way to derive the Lorentz transformations is to require that the interval s between events 0 and 3 be invariant, where

$$s^2/\lambda^2 = (c^2 t^2 - x^2)/\lambda^2 = n^2 - (2k - n)^2 = 4k(n - k).$$

In light-cone coordinates this relationship becomes

$$d_+ d_- = 4k(n - k) = s^2/\lambda^2,$$

which makes one way of insuring the invariance requirement particularly simple, namely

$$k' = \rho k, \quad n' - k' = \rho^{-1}(n - k) \Rightarrow 4k'(n' - k') = 4k(n - k).$$

Note that if we are to compare the integer bit string coordinates, this restricts k' to be a rational multiple of k. One of the great successes of our theory is precisely this restriction that keeps events an integral number of de Broglie wavelengths apart. A fundamental explanation of why our theory can contain interference phenomena starts here.

If we now note that

$$d_\pm = (1 \pm \beta)n,$$

the invariance requirement gives us that

$$(k'/k)[(n - k)/(n' - k')] = \rho^2 = [(1 + \beta')/(1 + \beta)][(1 - \beta)/(1 + \beta')].$$

Hence

$$\beta_\rho = (\beta' - \beta)/(1 - \beta\beta') \Leftrightarrow \rho^2 = [1 + \beta_\rho]/[1 - \beta_\rho].$$

From the fact that when transforming from a system at rest $d_+/d_- = 1$, we see that the relative velocity between the two systems is simply β_ρ. We have derived the velocity composition law for rational fraction velocities in any system. Etter arrived at this composition law for attribute velocities on general grounds, as is discussed in Ref. 1. With

$$\gamma = (1/2)[\rho + \rho^{-1}],$$

we have that

$$x' = \gamma(x + \beta_\rho ct); \quad t' = \gamma(ct + \beta_\rho x) \qquad \text{Q.E.D.}$$

6.3 Quantum Mechanics

PROGRAM UNIVERSE provides an invariant significance for the label strings, once they close (in some length with at least 139 bits) to form some basis for some realization of the combinatorial hierarchy. For each of the $2^{127} + 136$ labels L_l we can assign a dimensional parameter λ_0^l which is the step length when the particle is at rest; i.e. when, on the average $2k_l = n_l$. Since PROGRAM UNIVERSE increases the string length one arbitrary bit at a time, this requirement can at best be satisfied only at every other step. We have seen that when all steps are in the same direction (i.e., when the content string is either the null string or the antinull string), this corresponds to a light signal. In any string evolution all steps are executed at the limiting velocity c – a finite and discrete zitterbewegung. The invariance of λ_0^l allows us to associate with each label an invariant parameter with the dimensions of mass m_0^l, and relate the two by $\lambda_0^l = h/m_0^l c$, where h is a universal constant with the dimensions of action. We will now show that h can, indeed, be identified with Planck's constant.

The extension of our Lorentz transformations to momentum space is now immediate. We simply define $E = \gamma m_0 c^2$, $p = \gamma\beta m_0 c$. For $p_\pm = E/c \pm p$ we have that $p_+ p_- = m_0^2 c^2$, $p_+/p_- = k/(n - k)$ and $\frac{1}{2}(p_+ x_- + p_- x_+) = Et - px$. The justification of calling this momentum is more than definitional; we showed above that three- and four-vertices support vector conservation laws and crossing symmetry. We have three-momentum conservation in any allowed event-based reference frame. Clearly, $m_0 c \lambda_0 = h = E\lambda/c$ in any allowed coordinate system, and we have recovered the initial identification of the step length in the random walk as $\lambda = hc/E$, the de Broglie phase wavelength with which our initial statement of the counter paradigm began. We can now *derive* the quantum mechanical commutation relations from our model.

We note that if we consider a system that evolves with constant velocity $\beta_0 \equiv 2k_0/n_0 - 1$, strings that grow subject to this constraint,

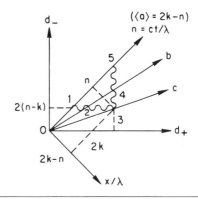

Figure 12. The connection between space-time and light-cone coordinates in terms of bit string distances and velocities for the physical situation envisaged in Fig. 11.

i.e., $n = n_T n_0$, $k = n_T k_0$, $1 \leq n_T \leq n/n_0$ will have a periodicity $T \equiv n_T \Delta t = n_T \lambda/c$ specifying the events in which this condition can be met. Hence, in more complicated situations where there can be more than one "path" connecting strings with the same velocity to a single event, this event can occur only when the paths differ by an integral number of d-wavelengths λ. Thus our construction already contains the seeds of interference and a conceptual explanation of the double-slit experiment.

We have already seen that any system with constant velocity – at those TICK's when events can occur – evolves by discrete steps $\pm\lambda_a$ in $x = q_a$ *between* TICK's. McGoveran's ordering operator calculus[1] which specifies the connectivity between events allows these discrete happenings to occur in a *void* where space and time are meaningless. Since $\lambda/\Delta t = c$, each step occurs forward or backward with the limiting velocity. Thus we deduce a discrete zitterbewegung from our theory. If we think of this as a trajectory in the pq phase space, each time step induces a step $\pm\lambda$ in q correlated with a step $\pm mc$ in p. Even in the case of a particle at rest, this must be followed by two steps of the opposite sign to return the system to rest. Thus there is, minimally, a fourfold symmetry to the trajectory in phase space corresponding to the generation periodicity we discovered above.

If we now recall from classical mechanics[31] that for any momentum that is a constant of the motion we can transform to angle and action variables with $\oint p_j dq_j = J$, where J has the dimensions of action, $p_J = J/2\pi$ and q_J is cyclic, we have an immediate interpretation. In the classical case, the period goes to infinity for a free particle; for us, we have already seen that we have a *finite* period $T = \lambda/c$. Therefore, we can immediately identify $m_a c \lambda_a = J = n_T h$; we have constructed Bohr-Sommerfeld quantization within our theory.

To go on to the commutation relations, we can replace the geometrical description of periodic trajectories in phase space by using complex coordinates $z = (q, ip)$ [or by $(q_j, in_T h/2\pi)$], where q_j is restricted to $2n + 1$ values with $-n_T \leq n \leq +n_T$. Then the steps around the cycle in the order $qpqp$ are proportional to $\pm 2\pi(1, i, -1, -i)$ where $\pm$ depends on whether the first step is in the positive or negative direction or equivalently whether the circulation is counterclockwise or clockwise. We have now shown why $qp - pq = \pm i\hbar$ for free particles in our theory; this result holds for any theory satisfying our principles which uses a discrete free particle basis.

In order to go to a detailed, three-dimensional description, we must supply three discriminately independent reference strings, define inner products with respect to them (cf. Sec. 5.4), and go to a coordinate description. There will then be three independent periodicities (velocities and momenta) that will commute with each other but not with their conjugate position variable. The commutation relations for angular momentum follow immediately. Since this has already been shown in general terms in Ref. 1, we will leave the details to future publications. An alternative is to develop the radial coordinate (n, l, m) description using bound states as the basis.

Now that we have two ($\hbar$ and c) of the three-dimensional constants needed to connect a fundamental theory to experiment in the three-space in which physics operates, and which we have proved must be the asymptotic space of our theory, all that remains is to determine the unit of mass. But this has already been done for us by the combinatorial hierarchy result $2^{127} + 136 \simeq 1.7 \times 10^{38} \simeq \hbar c/Gm_p^2 = (M_{\text{Planck}}/m_p)^2$ which tells us that we can either identify the unit of mass in the theory as the proton mass, in which case we can calculate (to about 1% in this first approximation) Newton's gravitational constant, or, if we take the Planck mass as fundamental, calculate the proton mass. From now on, we have to compute everything else. If we fail to agree with experiment to the appropriate accuracy (one of the rules of correspondence), we must either revise or abandon the theory.

6.4 A Discrete Model for the Bohr Atom

We have seen that any bit string has the de Broglie periodicity h/mc^2 for each digital time step $\Delta n = 1$, and that when it evolves with constant velocity also has the longer digital period n_0 connected to the velocity by $\beta = 2k_0/n_0 - 1$ at each finite position $N_{ph}n_0\beta = N_{ph}(2k_0 - n_0)$ where an event can (but need not) occur after the initial vertex at $N_{ph} = 0$. We define $\Delta k_0 = k_0 - n_0/2$ and hence $\beta = \Delta k_0/n_0$. Only one integer can be added to the string at each step. This must happen Δk_0 times before the periodic pattern can be completed. Therefore, the number of step lengths in the periodic pattern – the coherence length – is $n_0 = 1/\beta$. Since, as we saw above, the step length is $\lambda = hc/E$, we find that the coherence length required for periodic phenomena at constant velocity is $\lambda_c = hc/\beta E = h/p$.

By adding a constraint representing a second periodicity we can now model the periodicity representing a "closed orbit around some fixed center." Clearly, this periodicity must use the coherence length derived above if we are to have a stable, repeating pattern that starts from some origin and closes after N_B coherence lengths. This model, which only describes the average motion, will persist from the time when we start the model off to the time when some vertex – for example, the absorption of a "hard" photon – ends the finite sequence of periods. Of course, this can only occur at one of the positions allowed for events. In the average sense we can image this "trajectory" as a regular polygon with N_B sides of length λ_c. With the usual geometrical image in mind, we call the distance traversed in this period $2\pi R = N_B\lambda_c$ and hence $mvR = N_B\hbar$. Afficionados of the early history of quantum mechanics will recognize that

we have constructed a digital version of de Broglie's analysis of the geometry of the Bohr atom and produced a reason for angular momentum quantization. For the meaning of π in a discrete and finite theory, refer to the discussion in Ref. 1.

Although this part of the derivation of the Bohr atom should be reasonably familiar, our introduction of the electromagnetic interaction will be radically different from the conventional approach. We have seen above that the coulomb interaction is represented by only one out of 137 labels in the combinatorial hierarchy construction, and that strings evolve by the arbitrary selection of strings from memory to calculate the vertices; thanks to the counter paradigm, these vertices have now become events. In the case at hand, 136 of these choices can only provide a "background" which will cause fluctuations of the position of our particle; on the average these must cancel out. Only once in 137 times will the step correspond to the vertex that serves to keep the particle in its orbit. We can think of this as happening at the vertices of the polygon; i.e., N_B times in one full period. So, compared to the basic electromagnetic orbit, $\beta = 1/137N_B$. Making the hierarchy identification $137 = \hbar c/e^2$, our quantization condition derived above then gives us the standard result $R = N_B^2\hbar^2/me^2$, and an explanation of the old puzzle of why the Bohr radius is 137 times the Compton wavelength!

To calculate the binding energy, consider the energy change between this average motion and the particle at rest caused, for example by the emission or absorption of a photon. We must use the average velocity, because in the absence of other information, we cannot know "where" in the orbit the interaction occurs. Our theory can readily accommodate emission and absorption of photons, conserving both momentum and energy, as we have seen in our derivation of the Lorentz transformations, and can include the usual recoil correction if we so desire. Thus we argue that the binding energy ϵ_{N_B} is related to the velocity $\beta_{N_B} = 1/137N_B$ by $(\epsilon_{N_B} + m_0c^2)^2 = m_0^2c^4/(1 - \beta_{N_B}^2)$ from which all the usual results for the Bohr atom follow to order β^2.

6.5 Scattering Theory

To construct a scattering theory, we need to provide the connectivity between events. To obtain a statistical connection between events, we start from our counter paradigm and note that because of the macroscopic size of laboratory counters, there will always be some uncertainty $\Delta\beta$ in measured velocities, reflected in our integers k_a by $\Delta k = \frac{1}{2}N\Delta\beta > 0$. A measurement that gives a value of β outside this interval will have to be interpreted as a result of some scattering that occurred among the TICK's that separate the event (firing of the exit counter in the counter telescope that measures the initial value of $\beta = \beta_0$ to accuracy $\Delta\beta$) which defines the problem and the event which terminates the free particle propagation; we must exclude such observable scatterings from consideration.

What we are interested in is the probability distribution of finding two values k, k' within this allowed interval and how this correlated probability changes as we tick away. If $k = k'$ it is clear that, when we start, both lie in the interval of integral length $2\Delta k$ about the central value $k_0 = (N/2)(1 + \beta_0)$. When $k \neq k'$ the interval in which both can lie will be smaller, and given by

$$[(k + \Delta k) - (k' - \Delta k)] = 2\Delta k - (k' - k),$$

when $k' > k$ or by $2\Delta k + (k' - k)$ in the other case. Consequently, the correlated probability of encountering both k and k' in the "window" defined by the velocity resolution, normalized to unity when they are the same, is $f(k, k') = [2\Delta k \mp (k' - k)]/[2\Delta k \pm (k' - k)]$, where the positive sign corresponds to $k' > k$. The correlated probability of finding two values k_T, k'_T after T ticks in an event with the same labels and same normalization is $[f(k_T, k'_T)]/[f(k, k')]$. This is one if $k' = k$ and $k'_T = k_T$. However, when $k' \neq k$, a little algebra allows us to write this ratio as

$$\frac{1 \pm \dfrac{2(\Delta k - \Delta k_T)}{(k' - k)} + \dfrac{4\Delta k\Delta k_T}{(k' - k)^2}}{1 \mp \dfrac{2(\Delta k - \Delta k_T)}{(k' - k)} + \dfrac{4\Delta k\Delta k_T}{(k' - k)^2}}.$$

If the second measurement has the same velocity resolution $\Delta\beta$ as the first, since $T > 0$ we have that $\Delta k_T < \Delta k$. Thus, if we start with some specified spread of events corresponding to laboratory boundary conditions and tick away, the fraction of connected events we need consider diminishes. If we now ask for the correlated probability of finding the value β' starting from the value β for the sharp resolution approximation (i.e., ignoring terms smaller than $1/T$ or proportional to $1/T$ and smaller) this is 1 if $\beta = \beta'$ and bounded by $\pm 1/T$ otherwise. That is, we have shown that in our theory a free particle propagates with constant velocity with overwhelming probability – our version of Newton's first law and Descartes' principle of inertia.

Were it not for the $\pm$, the propagator in a continuum theory would simply be a δ function. In our theory we have already established relativistic point particle scattering kinematics for discrete and finite vertices connecting finite strings. We also showed that the order in which we specify position and velocity introduces a sign that depends on which velocity is greater, which in turn depends on the choice of positive direction in our laboratory coordinate system, and hence in terms of the general description on whether the state is incoming or outgoing. In order to preserve this critical distinction in our propagator, and keep away from the undefined (and undefinable for us) expression const/0, we write the propagator as

$$P(\beta, \beta') = [-i\eta\lambda/(\beta' - \beta \mp i\eta/T)],$$

where η is a positive constant less than T. The normalization of the propagator depends on the normalization of states and is best explored in a more technical context, such as the relativistic Faddeev equations for a finite particle number scattering theory in the momentum space continuum approximation being developed elsewhere.[6]-[9]

7. CONCLUSIONS AND A LOOK FORWARD

The research program discussed here started, as far as some current participators go, in the 1950s – and earlier if you look back to Eddington. By now, there is a solid body of results, both conceptual and numerical. One aspect that conventional physicists find puzzling is that we can reach some fundamental results very easily – results that for them require enormously complicated calculations and a generous (though often unrecognized) input of empirical data. For instance, to prove the $3 + 1$ asymptotic structure of space-time starting from conventional string theory requires the "compactification" of an initially 26-dimensional structure whose uniqueness can, mildly speaking, be questioned. For us, this $3 + 1$ structure for events follows directly from McGoveran's theorem, once our basic principles and rules of correspondence are understood. For those familiar with Kuhn's model for scientific revolutions, this should come as no surprise. Any new fundamental theory

finds some problems easier to solve, and for other problems loses (sometimes for a long while) some of the explanatory power of the theory it is attempting to replace.

At a somewhat less fundamental level than the global irreversibility of time and the three-dimensionality of space, all conventional theories take the existence of a limiting velocity and the quantization of action as a just-so story. We show why any theory satisfying our principles has to have both a limiting velocity and noncommutativity. We show that our positions and velocities for our events must be connected by a discrete form of the Lorentz transformations. We derive three-momentum conservation, quantum number conservation, and on-shell four-momentum conservation at our elementary vertices. We also show that when one compares position and velocity in the connected circumstances implicit in the physics of conjugate variables, the resultant noncommutative structure can be mapped onto that employed in quantum mechanics.

Moving on up to more concrete aspects of conventional theories, given c and $\hbar$ – and the scale-invariant laboratory methods of relating them to arbitrary standards of mass, length, and time – conventional physicists need some mass or coupling constant that has to be taken from experiment. Once again, the existence of this unique constant – let alone a means of computing it within the theory – is not an obvious structural requirement of conventional practice. In contrast, we obtain a first-order estimate $\hbar c/e^2 \simeq 137$ and $\hbar c/Gm_p^2 \simeq (M_{\text{Planck}}/m_p)^2 \simeq 2^{127} + 136$. As has been emphasized above, any fundamental theory of MLT physics must compute everything else as physically dimensionless ratios once these constants are fixed.

It is sometimes suggested that ours is a "Pythagorean" or a priori theory. This criticism implies a lack of understanding of our modeling methodology. We start from the current practice of physics, both theoretical and experimental, and try to construct 1) a self-consistent formal structure guided by that prior knowledge, and 2) rules of correspondence that bring us back to laboratory practice, including empirical tests. In this sense, we are trodding a well-worn path followed by many physicists engaged in constructing fundamental theories.

Another related criticism assumes that the high degree of structural information we must ascribe to counting finite integers is a very loose mesh. Changes of interpretation seemed possible before the program produced a coherent lump of concepts and structure and numerical correlations. Bastin was often able to be sure that some of HPN's early attempts at interpretation had to be wrong; unfortunately, these objections had to be made at a level of generality that prevented the specific technical line of argument from being developed. We now have a 35-year track record of meeting honest criticism and modifying our ideas to meet the challenges posed. Some challenges come from the explosion of precise information provided by contemporary high-energy particle physics and observational cosmology. Others come from questions of self-consistency and coherence that can only be met by a paradigm shift. Perhaps the best way to meet these challenges is to summarize the positive predictions that stem from our program – predictions whose failure would require us to modify or abandon the theory. We summarize these predictions in Table 3.

The conventional physicist accepts all the structural results we have listed; in his practice he uses numbers that satisfy (to an accuracy discussed below) the numerical consequences of the algebraic relations given. At this point we would like to ask this conventional reader why he

An Essay on Discrete Foundations for Physics

Table III: Predictions Made by Discrete and Combinatorial Physics

Structural Predictions

- 3 + 1 asymptotic space
- limiting velocity
- discrete events
- supraluminal synchronization and correlation *without* supraluminal signaling
- discrete Lorentz transformations (for event-based coordinates)
- noncommutativity between position and velocity (for event-based coordinates)
- transport (exponentiation) operator
- recognizable conservation laws for three- and four-events
- quantum numbers of the standard model for quarks and leptons
- event horizon
- zero-velocity frame for the cosmic background radiation
- color confinement – quark and gluon masses not directly observable

Algebraic Cosmological Predictions	*Algebraic Elementary Particle Predictions*
$\left(\dfrac{M_{\text{Planck}}}{m_p}\right)^2 \simeq 2^{127} + 136 \simeq \dfrac{\hbar c}{G m_p^2}$	$\dfrac{1}{\alpha} = \dfrac{137}{\left[1 - \dfrac{1}{127 \times 30}\right]}$
$M_{\text{Vis.U}} \simeq (2^{127} + 136)^2\, m_p$	$\dfrac{m_p}{m_e} = \dfrac{137\pi}{\dfrac{3}{14}\left(1 + \dfrac{2}{7} + \dfrac{4}{49}\right)\dfrac{4}{5}}$
Fireball time $\simeq (2^{127} + 136)\, \dfrac{\hbar}{m_p c^2}$	$m_\pi \lesssim 2 \times 137\, m_e$
$M_{\text{Dark}} \simeq 12.7\, M_{\text{Vis.U}}$	$\sin^2 \theta_{\text{Weak}} = \dfrac{1}{4}$
	$G_F \times m_p^2 = \dfrac{1}{\sqrt{2}\,(256)^2}$
	$m_{u,d}(\rho_0) = \dfrac{1}{3} m_p$

accepts the structural results we have predicted from our principles. The unconventional reader may accept some but not all of our structural results; we ask him *how* he makes that selection. We ask either type of reader what would cause them to *reject* any of these results which they now accept. We also ask them to *explain* why they accept or reject any of our results.

Many people are uncomfortable with a theory that rests on what appears to be so little empirical foundation. Of course, there are tried-and-true routes out of the problems our theory poses: naked empiricism, just-so stories, laws of thought, uncontrolled skepticism, solopsism, logic, quantum logic, infinity, etc. We believe we are close to the current practice of physics when we reject such escape hatches as likely to dump us in a still more unfortunate situation. We part company with most contemporary practice only by insisting that it is important to ask these fundamental questions. We are comfortable with the ways, sketched in this

paper, we arrive at our conclusions. We are prepared to scrub the theory if there is clear evidence that any piece of this structure fails and will look to such failures for clues to where to look in starting a new approach.

Physicists tend to be impatient with philosophical challenges. We turn next to the cosmological predictions. Ours have both a structural and a quantitative aspect. Conventional cosmology breaks into two parts: the evolution of the universe after the radiation breaks away from the matter, which we call fireball time, and the model-sensitive earlier history. Since the combinatorial hierarchy result set the gravitational and electromagnetic scales back in 1961, and we have subsequently given detailed proof that we can calculate atomic and nuclear problems in close enough agreement with experiment for most cosmological purposes, conventional extrapolation of the 2.7° K background radiation back to that time works as well for us as for anyone else – given the 50% empirical uncertainties in the critical parameters. There is an event horizon beyond which even radio galaxies disappear, and behind that the fireball; this backward extrapolation is reasonably consistent with contemporary physics as it works here and now.

All of this works for us, because our estimate of the visible matter within the event horizon is an order of magnitude smaller than the amount of matter needed to close the universe in conventional (general) relativistic cosmologies. Since we have established the conservation laws of the standard model, and our labels are created either by discrimination or TICK in order to form the labels in the first place, we can estimate the number of vertices in which two different labels participate for the first time as $2^{127} + 136$. Once the labels are formed, the construction retains each of them independently as labels for content ensembles. Hence there are something like $(2^{127} + 136)^2$ quantum number conserving labels generated before the space-time content has much meaning. In the absence of further information, the average mass must be our unit mass m_p, from which the estimate follows. This prediction is in agreement with observation since the observed visible mass within the event horizon is about what we estimate. A more precise estimate will require a more detailed statistical calculation of the probability of formation of lepton and baryon labels. With such an open universe, Newtonian gravitation is good enough for postfireball time cosmological calculations.

This same estimate gives us our next prediction. Manthey noted that the fact that it takes $2^{127} + 136$ TICK's to form the labels defines a time, and HPN identified it as fireball time. The problem here is an old one. As we go back earlier, we have to rely more and more on what we mean by the laws of physics or whatever phrase describes the methodology used for extrapolation. Once one tries to extrapolate backward from fireball time using a linear time scale, one rapidly approaches extreme conditions that currently occur only in the interiors of stars, in the cosmic radiation when it interacts with matter, in the neighborhood of massive black holes or in high-energy physics laboratories. When one tries to get back inside of the first three minutes the empirical evidence vanishes and only disciplined conjecture provides a guide. We simply assert that time loses any useful model-independent meaning somewhere between fireball time and the first three minutes. In our model, if we use the appropriate unit of time $(\hbar/m_p c^2)$, our backward extrapolation gives us roughly three and one-half million years back to the first discrimination. Other models give roughly similar results back to the first three minutes. Before that I see no way for a physicist to make testable statements as to whether the universe always existed or came into being at a finite time. As we have already commented on above, the conventional wisdom is in much worse shape here than we are. Most of their model universes are unsubstantiated.

The prejudice of most cosmologists is that the universe should be closed or just closed, for reasons that escape me. (I find an open universe much more satisfactory, particularly after reading Dyson's scientific eschatological analysis.[32]) The deficit from the conventional perspective is now to be made up by dark matter. Here, they have a good observational case in that ten times as much of the mass of galaxies, as measured by Newtonian gravitation and the Doppler shift, is dark rather than electromagnetically visible. How much more there is depends, once again, on details of the cosmological model rather than on observation.

Here our theory makes a new prediction. Visible matter can only be understood by us in terms of the 137 labels for the first three levels of the hierarchy. But there are $3 + 7 = 10$ labels that cannot be interpreted prior to the formation of the background of the 127 labels that make up level 3. Whatever they are, they must be electrically neutral and will occur, statistically, 12.7 times more frequently than the level 3 labels. They could form electromagnetically inert structures at any scale compatible with our finite scheme (quantum geons?). So our estimate of the amount of dark matter left over from the big bang to the visible matter is 12.7; a better estimate will depend on what version of the early stages of PROGRAM UNIVERSE we use. Quantitatively, the prediction for the gravitational constant (using m_p, c, and $\hbar$ to connect our units of mass length and time to experiment) fails by a little less than 1%. We anticipate a correction of order α, and hope to be able to compute it once we have sorted out the experimental effects usually ascribed to general relativity.

Most of this cosmology is predicated on the assumption that we have got the atomic and nuclear physics right. If one believes the six results given in Table 3, which we compare with experiment below, as elementary particle predictions and accepts our finite particle number scattering theory as both unitary and crossing symmetric, we can do at least as well as most practitioners in reproducing one or another currently accepted phenomenology for atomic, nuclear, and high-energy particle physics. This will be obvious to readers with an S-matrix background; we will never be able to convince some physicists who are not used to that type of practice. So we concentrate here on where these six numbers come from, what estimate of theoretical uncertainty we ascribe to them, and how they compare with experimental values.

The calculation of the fine structure constant is due to McGoveran.[33] It is preliminary and was discussed at ANPA 10. The calculation came out of an examination of the Sommerfeld formula for the fine structure spectrum.[34] HPN argued that since we now have a fully relativistic theory, including angular momentum conservation and noncommutativity, a nonrelativistic combinatorial model for the Bohr atom (Sec. 6.4), and Bohr-Sommerfeld quantization (Sec. 6.3), we should be able to get this relativistic correction by including two different periods in the calculation. This is indeed the case, but then HPN realized that our approximation of 137 for $1/\alpha$ is no longer good enough. He feared we would have to do all of QED to order α^2 in order to sort this out, but McGoveran realized that the existence of two frequencies in the problem gave us a combinatorial argument that leads to the result quoted above. Numerically this formula predicts $1/\alpha = 137.035\ 967\ 4\ldots$ as compared to the two values quoted in the particle properties data booklet[35]: (old) 137.036 04(11) and (new) 137.035 963(15). As far as we can see, any correction to our prediction should be of order G_F/M_W^2 or that number times $(\sin^2\theta_{Weak} - \frac{1}{4})$; if this

estimate of the uncertainty is correct, we do not find the close agreement with experiment surprising.

The m_p/m_e formula is due to Parker-Rhodes.[23] Since our theory differs from his, in the past we could only provide heuristic justification for the calculation. Now that we have a fully developed relativistic quantum mechanics, with three-momentum conservation, these past arguments become rigorous when we view the calculation as a calculation of the mass in the electron propagator – for us, a finite self-energy. One puzzle was the extreme accuracy of the result, using 137 rather than the empirical value for $1/\alpha$. However, now that we have found that the empirical value comes about in systems that lack spherical symmetry, or in combinatorial terms have two independent frequencies, and recognize that in the m_p/m_e calculation there is no way to define a second frequency, we have a rigorous justification for the formula as it stands. Numerically, we predict $m_p/m_e = 1836.151\ 497\ldots$ as compared with[35]: (old) 1836.151 52(70) and (new) 1836.152 701(100). We see that the proposed revision in the fundamental constants has moved the empirical value outside of our prediction by a presumably significant amount. For the m_p/m_e calculation the correction due to nonelectromagnetic interactions could be large enough to affect our results.

The calculation of the neutral pion mass was made long ago.[36] The model is due to our interpretation of Dyson's argument[37] that the maximum number of charged particle pairs which can be counted within their own Compton wavelength is 137. Taking these to be electron-positron pairs, we get the result. The argument in the past rested on the use of the coulomb potential. Now that we have a combinatorial calculation of the Bohr atom, we no longer need this extraneous element. If one looks at the content strings minimally needed to describe the possible states of the bound system, the saturation at 137 pairs emerges. As we can see from the Bohr atom calculation (e.g. by considering one electron or positron interacting with the average charge of the rest of the system), the first approximation for the binding energy is nonrelativistic. Consequently, the estimate for the system mass, interpreted as the neutral pion mass, is just the sum of the masses, or $274 m_e$, in agreement with experiment to better than ten electron masses. It will be interesting to calculate the α relativistic corrections (including the virtual electron-positron annihilation) and the neutral pion lifetime. Adding an electron-anti-neutrino pair to get the π^-, or a positron-neutrino pair to get the π^+, will be a good problem for sorting out our understanding of weak-electromagnetic unification.

The weak-electromagnetic unification needs more work, as has already been indicated. The first order prediction of $\sin^2\theta_{Weak} = 0.25$ as compared to the experimental result[38] of 0.229 ± 0.004 is firm, and reasonably satisfactory at the current stage of development. The identification of the weak coupling constant (without the factor of $\sqrt{2}$) was suggested by Bastin long ago[18]; our formula predicts a result which is about 7% too large. Since this is roughly the amount by which we fail to get the weak angle, the two discrepancies might find a common explanation.

The quantum number structure of the quarks has been discussed in Sec. 5.5 and does lead to the usual three "valence" quark structure of the baryons, which gives us the usual nonrelativistic quark model as a starting point. As already noted, McGoveran's theorem does not allow more than three asymptotic degrees of freedom, so we do predict color confinement. This means that we cannot use our standard free-particle states to describe quarks or gluons and define their mass. We suspect that we can eventually obtain

An Essay on Discrete Foundations for Physics

running masses analogous to those Namyslowski[11] gets out of the conventional theories but have only just started thinking about the problem. Another challenge will be to relate the pion model discussed above to a quark-anti-quark pair.

By now, we hope the reader will grant that we have made a case for discrete physics as a fundamental theory. We have been led to many conceptual and numerical results that can only be obtained with difficulty, or not at all, by more conventional approaches. We believe the program will prove to be useful even if it ultimately fails. So far we have run into no insuperable barriers – frankly, somewhat to HPN's surprise. We have nailed down the quantum numbers in agreement with the standard model and have computed reasonable values for the basic masses and coupling constants. Thanks to the high degree of overdetermination of elementary particle physics due to crossing and unitarity – Chew's bootstrap – we can expect to do about as well as conventional strong interaction theories. This means that when a difficulty *does* arise, it will suggest an area of phenomena that will deserve detailed experimental and theoretical examination. Again, we share this strategy with more conventional approaches.

Acknowledgments

We are most grateful to Ted Bastin and Clive Kilmister for extended discussions while this work was in progress and a careful perusal of what we had hoped would be the final manuscript. Several critical changes resulted from the discussion following their work. More generally, this paper results from a continuing attempt to reconstruct quantum mechanics on a discrete basis initiated several years ago to which Amson, Bastin, Etter, Gefwert, Karmanov, Kilmister, Manthey, DMcG, HPN, Parker-Rhodes (deceased), and Stein have all contributed crucial elements. Still more generally, the discussions at the ten annual international meetings of the Alternative Natural Philosophy Association in Cambridge, England and the four annual meetings of ANPA West at Stanford have played a very constructive role in sharpening up the arguments. Some differences of opinion remain within the group, so positions taken here – and more particularly any errors that remain – are the sole responsibility of HPN and DMcG.

This work was supported by the Department of Energy, contract DE-AC03-76SF00515.

Received on 5 September 1988.

Résumé

Nous basons notre théorie physique et cosmologique sur cinq principes : la finitude, la séparation, la capacité de mesure, la non-unicité absolue, et l'élaboration rigoureuse. Notre méthode de modélisation s'inspire des pratiques actuelles propres à la physique, elle élabore une représentation cohérente basée sur le calcul des opérateurs ordonnés et procure des règles de correspondance qui nous permettent de vérifier la théorie par l'expérience. Nous employons le PROGRAM UNIVERSE pour construire une collection croissante d'objets dont des portions initiales (étiquettes) produisent les nombres quantiques qui sont conservés dans des événements définis par la construction. Les étiquettes sont accompagnées des éléments de contenu qui servent à construire des coordonnées basées sur des événements finis et discrets. En général, une telle théorie a une vitesse limite, et positions et vitesses ne sont pas commutatifs. Nous pouvons donc concilier mécanique quantique et relativité à une étape fondamentale opportune de la construction. Nous montrons que des événements survenant dans différents systèmes de coordonnées peuvent être reliés entre eux par une forme finie et discrète de la transformation de Lorentz; nous montrons aussi que le moment trois est conservé dans des événements, et que cette loi de conservation est la même que celle exigeant que différentes voies ne peuvent "interférer" que quand la différence entre elles est constituée par un nombre entier de longueurs d'onde de de Broglie.

Les étiquettes sont organisées selon les quatre niveaux de la hiérarchie combinatoire caractérisées par les nombres cardinaux cumulatifs 3, 10, 137, $2^{127} + 136 \simeq 1,7 \times 10^{38}$. Nous justifions l'identification des deux derniers cardinaux: ils sont une première approximation de $\hbar c / e^2$ et de $\hbar c / G m_p^2 = (M_{Planck}/m_p)^2$, respectivement. Nous montrons que les nombres quantiques associés aux trois premiers niveaux peuvent être rigoureusement identifiés avec les nombres quantiques de la première génération du modèle standard des quarks et des leptons, avec confinement des couleurs et une première approximation du champ unifié électromagnétique faible. Notre cosmologie comprend un horizon événementiel, un système de référence de vitesse nulle du front d'onde continu, une période de répartition de l'énergie dans les particules d'à peu près $3,5 \times 10^6$ ans, à peu près la quantité correcte de matière visible, et 12,7 fois autant de "matière opaque." Un calcul préliminaire de la structure fine du spectre de l'hydrogène donne la formule de Sommerfeld et une correction de notre première approximation de la constante de structure fine conduit à $1/\alpha = 137,035 967 4 \ldots$. Nous pouvons ainsi justifier les résultats antérieurs, à savoir $m_p/m_e =

1836,151 497 ... et $m_\pi/m_e \lesssim = 274$. *Nous estimons que l'angle faible est* $sin^2\sigma_{Faible} = \frac{1}{4}$ *et que la constante de Fermi est* $G_F \times m_p^2 = 1/\sqrt{2}(256)^2$. *Notre théorie de la diffusion relativiste d'un nombre fini de particules devrait nous permettre d'étendre systématiquement ces résultats.* Eteris paribus, caveat lector.

Endnotes

[1] A preliminary version of some of this material was presented at the Ninth Annual International Meeting of the Alternative Natural Philosophy Association, Department of the History and Philosophy of Science, Cambridge University, 23-28 September 1987, in "Discrete and Combinatorial Physics," edited by H.P. Noyes, published by ANPA WEST 25 Buena Vista, Mill Valley, Calif. 94941; hereinafter referred to as Proc. ANPA 9 under the title "Discrete Physics: Practice, Representation and Rules of Correspondence," p. 105.

[2] Consider, for instance, his demonstration that gravitational and inertial mass are proportional using pendulum bobs of the same weight and exterior size and shape but composed of different materials. Eötvös had to rely on two centuries of technological development to construct a better technique; some physicists are still struggling to go beyond Eötvös (cf. Phys. Today, (7) **1988**).

References

1. D.O. McGoveran, "Foundations for a Discrete Physics," in Proc. ANPA 9, edited by H.P. Noyes, p. 37.
2. A. Linde, in Phys. Today, (9), **1987**, 61, estimates this same number as 10^{125}; his attempt to "save the phenomena" is more speculative than ours.
3. Kilmister supplies the following reference for Eddington's version: "in *Fundamental Theory*, p. 30: 'The term "particle" survives in modern physics, but very little of its classical meaning remains. A particle can best be defined as *the conceptual carrier of a set of variates*. [Footnote: It is also conceived as *the occupant of a state* defined by the same set of variates.]' That is, of course, CUP 1946. But already in *Proc. Cambridge Philos. Soc.* **40**, 37 (1944) he says: 'A "conceptual carrier of variates" is generally called a particle.' "
4. "Discrete and Combinatorial Physics," Proc. ANPA 9.
5. M.J. Manthey, "PROGRAM UNIVERSE," in Proc. ANPA 7, 1986, published by F. Abdullah, City University, Northampton EC1V OHB, England; see also SLAC. Pub. 4008 and references and appendices therein.
6. H.P. Noyes and J.V. Lindesay, Aust. J. Phys. **36**, 601 (1983).
7. J.V. Lindesay, A.J. Markevich, H.P. Noyes, and G. Pastrana, Phys. Rev. D **33**, 2339 (1986).
8. H.P. Noyes, "A Minimal Relativistic Model for the Three Nucleon System," in *The Three Body Forces in the Three Nucleon System*, edited by B.L. Berman and B.F. Gibson (Springer-Verlag, Berlin, 1986), p. 495.
9. J.V. Lindesay and H.P. Noyes, Ph.D. thesis, Stanford University, SLAC Report No. 243, 1981.
10. C. Quigg, *Gauge Theories of the Strong, Weak and Electromagnetic Interactions* (Benjamin, Menlo Park, 1983), p. 113.
11. J.M. Namyslowski, Phys. Lett. B **192**, 170 (1987).
12. *Probing the Standard Model*, edited by Eileen C. Brennan SLAC Report No. 312.
13. C. Gefwert, "Prephysics," in Proc. ANPA 9, p. 1.
14. T.S. Kuhn, *The Structure of Scientific Revolutions* (University of Chicago Press, 1962).
15. H.P. Noyes, C. Gefwert, and M.J. Manthey, *Ann. N.Y. Acad. Sci.* **480**, 553 (1986).
16. E.W. Bastin and C.W. Kilmister, Proc. R. Soc. London, A **212**, 559 (1952).
17. *Idem*, Proc. Cambridge Philos. Soc. **50**, 254, 278 (1954); *Ibid.* **51**, 454 (1955); *Ibid.* **53**, 462 (1957); *Ibid.* **55**, 66 (1959).
18. T. Bastin, Studia Philos. Gandensia **4**, 77 (1966).
19. For one attempt to reconstruct it and some of the controversy it evoked, see the article by HPN in Proc. ANPA 9, p. 117.
20. T. Bastin, H.P. Noyes, J. Amson, and C.W. Kilmister, Int. J. Theor. Phys. **18**, 455 (1979).
21. C.W. Kilmister, "The Mathematics needed for self-organization," in *Disequilibrium and Self Organization* edited by C.W. Kilmister (Reidel, Dordrecht, 1986), p. 11; see also "Brouwerian Foundations for the Hierarchy" (based on a paper presented at ANPA 2, Cambridge, 1980), Appendix II.1; and "On Generation and Discrimination" (based on a paper presented at ANPA 4, Cambridge, 1982), Appendix II.2 in H.P. Noyes, C. Gefwert, and M.J. Manthey, SLAC. Pub. 3116, rev. September, 1983.
22. *Idem*, "A Final Foundational Discussion?" in Proc. ANPA 7 and as an appendix in H.P. Noyes, SLAC. Pub. 4008, June, 1986.
23. A.F. Parker-Rhodes, *The Theory of Indistinguishables* (Synthese Library, Reidel, Dordrecht, 1981), p. 150.
24. Reference 1 gives McGoveran's current development of this point.
25. H.P. Noyes, D. McGoveran, T. Etter, M. Manthey, and C. Gefwert, in "A Paradigm For Discrete Physics?" Abstracts, 8th Int. Congress of Logic, Methodology and Philosophy of Science, compiled by V.L. Rabinovich, Inst. Phil. Acad. Sci. USSR, Moscow, 1987, Vol. 2, Sec. 8, p. 98.
26. I. Stein, seminar at Stanford, 1978, and presentations at ANPA 2 and 3, Cambridge, 1980 and 1981; the latest version of this work appeared in Phys. Essays **1**, 155 (1988).
27. H.P. Noyes, SLAC. Pub. 2906, 1982.
28. *Idem*, "A Finite Particle Number Approach to Physics," in *The Wave-Particle Dualism*, edited by S. Diner, D. Fargue, G. Lochak, and F. Selleri (Reidel, Dordrecht, 1984).
29. W. Feller, *An Introduction to Probability Theory and its Applications*, 3rd ed. (Wiley, N.Y.), Vol. 1, p. 316. For more detail see the first edition, p. 247.
30. H.P. Noyes, remark in Proc. ISQM 2, edited by M. Namiki, Y.

Ohnuki, Y. Murayama, and S. Nomura, Phys. Soc. Japan, 1986, p.78; A more detailed explanation is given in Ref. 1 and also in D.O. McGoveran and H.P. Noyes, SLAC. Pub. 4729, submitted to Bell's Theorem Workshop, George Mason University, October, 1988.

31. H. Goldstein, *Classical Mechanics*, 2nd ed. (Addison-Wesley, 1980), p. 445.

32. F.J. Dyson, Rev. Mod. Phys. **51**, 447 (1979). This article was written assuming the proton is stable, which is the best bet so far. Subsequently, Dyson, in a seminar at Stanford, demonstrated that even if the proton decays with a lifetime of 10^{35} years – the current lower limit – by that time the universe (containing only electrons, positrons, and electromagnetic quanta) is so dilute that the electrons and the positrons continue to persist. Consequently, large-scale electromagnetic structures, cultural centers, communication, etc., can go on evolving. He shows in the paper quoted that the problem will not be isolation from communication but selection among the riches available.

33. D.O. McGoveran and H.P. Noyes, SLAC. Pub. 4730, to be submitted to Phys. Rev. Lett.

34. A. Sommerfeld, Ann. Phys. (Leipzig) **17**, 1 (1916).

35. M. Aguilar-Benitez, *et al.*, Phys. Lett. B **170** (1986).

36. H.P. Noyes, SLAC. Pub. 1405, 1975 (unpublished) p. 42.

37. F.J. Dyson, Phys. Rev. **51**, 631 (1952).

38. G. Barbiellini, SLAC Report No. 312, p. 418 (Ref. 12 above).

H. Pierre Noyes
Stanford Linear Accelerator Center
Stanford University
Stanford, California 94309 U.S.A.
and
David O. McGoveran
Alternative Technologies
150 Felker St., Suite E
Santa Cruz, California 95060 U.S.A.

Physics Essays   volume 4, number 1, 1991

On the Fine-Structure Spectrum of Hydrogen

David O. McGoveran and H. Pierre Noyes

Abstract

Using our discrete relativistic combinatorial bit-string theory of physics in the context of the hydrogen spectrum, we calculate *our first two approximations for the fine-structure constant as* $\alpha(1) = 1/137$ *and* $\alpha(2) = [1 - 1/(30 \times 127)]/137 = 1/137.035\,967\,4\,\ldots$; *we can then* derive *the Sommerfeld formula.*

Key words: discrete physics, combinatorial hierarchy, fine-structure constant, Sommerfeld formula

We present here a novel calculation based on an unconventional but actively developing physical theory[1]; a reasonably complete overview of this theory has been published in this journal.[2] The theory asserts that to order α the fine structure constant used to describe the energy spectrum of the hydrogen atom should have the value 1/137. We go on to predict on the same basis that the second-order value in the same context is $[1 - 1/(30 \times 127)]/137 = 1/137.0359\,674\,\ldots$, close to the currently accepted value given[3] as "1/137.0359 895(61) [at $Q^2 = m_e^2$; at Q^2 of order m_W^2 the value is approximately 1/128]." Both the derivation and the calculation will require corrections of order α^3 and $(m_e/m_p)^2$ when extended beyond the context of the hydrogen atom. Since most current theories do not contemplate the possibility of calculating α, although Weinberg[4] has indicated that this should be possible in principle, we must justify our method before presenting the calculation.

Conventional theories take the structure of relativistic quantum mechanics as given. The two empirical constants c and $\hbar$ are connected to the arbitrary historical standards of mass, length, and time by various, hopefully self-consistent, means. A third fundamental constant, such as the square of the electronic charge or the electron, proton, Planck, and so on, mass has to be taken from experiment before theoretical "predictions" can be attempted. Often the resulting comparisons with experiment can remain very rough, until supplemented by a generous amount of additional empirical input and theoretical structure. For instance, the high-dimension Kaluza-Klein theories coupled to a large number of Yang-Mills fields, when compactified, in effect

take the Planck mass $(\hbar c/G)^{1/2}$ as the third-dimensional parameter. In this context Weinberg[4] calculates the coupling constants of the fields, which are supposed to include the equivalent of α, G_F, g_s, and so on. Numerical results are quantitatively inadequate for comparison with experiment. Further modifications of this type of theory, needed to close the gap, often have only a tenuous connection to algorithmic precision or actual laboratory practice. Starting from one of the four "empirical" numbers mentioned above (i.e., e^2, m_e, m_p, M_{Planck}), there is no consensus on how to calculate the other three – a clear requirement for any fundamental physical theory that allows only empirical standards for mass, length, and time, or some equivalent like c, $\hbar$, and m_p to dictate the common units for the intercomparison of experiments between laboratories. We have recently provided a systematic discussion of how our theory can start from c, $\hbar$, and the Planck mass.[5]

Our theory differs in that we claim to be able to *calculate* a first approximation to the ratio of the Planck mass to the proton mass, the ratio of the proton mass to the electron mass, and the ratio of the square of the elementary electromagnetic charge to the product of the unit of action and the limiting velocity. Therefore, we can connect our theory to experiment by taking any one of the four accepted values from experiment and calculating a first approximation for the other three. From then on our iterative improvement of the theory is, in principle, much the same as for any other fundamental theory, such as the currently popular "string theories."

Although our methodology looks almost conventional when we describe it above, our practice is significantly different in several ways. In contrast to most "elementary particle" theories, we do *not* take relativistic quantum mechanics for granted. Our "mathematics" relies on the *ordering operator calculus*.[6] We accept the principles of finiteness, discreteness, finite computability, absolute nonuniqueness, and require the formalism to be strictly constructive.[7] The fact that we are able to use our fundamental principles to *construct* (rather than postulate) the limiting velocity and discrete events, and then to *derive* the Lorentz transformations and the noncommutativity of position and velocity gives our theory more explanatory power than the conventional approach. We start from the current practice of physics, construct an uninterpreted (but motivated) model that stands on its own feet as a piece of mathematics, and then construct *rules of correspondence*[7] which allows us to compare this structure and calculations made from it with current theoretical practice and experimental results. Anomalies, ambiguities, and discrepancies then call for iteration of the procedure starting anywhere in the chain and moving in either direction, keeping past experience in mind.

The theory that we are iteratively developing started in the 1950s, motivated in part by a search for a hierarchical structure that would give clues as to how the scale constants of physics and cosmology might be constructed. This research effort led to the discovery of the *combinatorial hierarchy*[8],[9] by Parker-Rhodes in 1961. The hierarchy is constructed from two recursively generated sequences: $n_{i+1} = 2^{n_i} - 1$ and $m_{i+1} = m_i^2$ starting from $n_0 = 2 = m_0$, which terminate at $i = 4$ because the mapping (see below) connecting the second sequence to the first cannot be constructed beyond that term. This discovery supported no obvious "rule of correspondence" connecting the cumulative number of elements in play at the third $(3 + 7 + 10 = 137)$ and fourth $(137 + 2^{127} - 1 \simeq 1.7 \times 10^{38})$ levels as good approximations to the known scale constants $137 \simeq \hbar c/e^2$ and $1.7 \times 10^{38} \simeq \hbar c/Gm_p^2 = (M_{\text{Planck}}/m_p)^2$.

The model is conveniently represented by ordered strings of the symbols 0 and 1 (bit-strings):

$$\mathbf{a}(S) = (\ldots, b_s^a, \ldots\ldots)_S; \quad b_s^a \in 0, 1;$$

$$s \in 1, 2, \ldots S; \quad 0, 1, \ldots, S \in \text{ordinal integers} \tag{1}$$

which can combine by *discrimination* (XOR) symbolized by "$\oplus$":

$$\mathbf{a} \oplus \mathbf{b} = (\ldots, b_i^{a \oplus b}, \ldots)_S = \mathbf{b} \oplus \mathbf{a}; \quad b_i^{a \oplus b} = (b_i^a - b_i^b)^2 \tag{2}$$

or *concatenation* symbolized by "$\|$":

$$\mathbf{a}(S_a) \| \mathbf{b}(S_b) = (\ldots b_i \ldots)_{S_a} \| (\ldots b_j^b \ldots)_{S_b}$$

$$= (\ldots\ldots, b_k^{a\|b}, \ldots\ldots)_{S_a + S_b}$$

$$b_k^{a\|b} = b_i^a, \ i \in 1, 2, \ldots, S_a; \ b_k^{a\|b} = b_j^b,$$

$$j \in 1, 2, \ldots, S_b, \ k = S_a + j. \tag{3}$$

We take as our model for generating these strings the class of algorithms called "program universe".[10],[11] These pick two arbitrary strings from a universe containing strings of length S, discriminate them, and if the result is not the null string ($b_s^0 = 0$ for all s) adjoin it to the universe; otherwise

they concatenate an arbitrary bit, separately chosen for each string, to the growing end of each string. If we think of this bit-string universe as a block of strings of length S and height H, the second operation (called TICK) amounts to adjoining an arbitrary column (Bernoulli sequence) and hence $S \rightarrow S + 1$. The first operation (called PICK) generates a string from the extant content and adds it as a new horizontal row ($H \rightarrow H + 1$).

Finite sets of non-null bit-strings which *close* under discrimination are called "discriminately closed subsets" (DCS's). For example, two *discriminately independent* bits-strings (i.e., $a \oplus b \neq 0$) generate three DCS's: $\{a\}, \{b\}, \{a, b, a \oplus b\}$. The three-member set closes under discrimination because any two members discriminate to the third. Similarly, three discriminately independent bit-strings generate seven DCS's:

$$\{\mathbf{a}\}, \{\mathbf{b}\}, \{\mathbf{c}\}$$

$$\{\mathbf{a}, \mathbf{b}, \mathbf{a} \oplus \mathbf{b}\}; \ \{\mathbf{b}, \mathbf{c}, \mathbf{b} \oplus \mathbf{c}\}; \ \{\mathbf{c}, \mathbf{a}, \mathbf{c} \oplus \mathbf{a}\} \tag{4}$$

$$\{\mathbf{a}, \mathbf{b}, \mathbf{c}, \mathbf{a} \oplus \mathbf{b}, \mathbf{b} \oplus \mathbf{c}, \mathbf{c} \oplus \mathbf{a}, \mathbf{a} \oplus \mathbf{b} \oplus \mathbf{c}\}.$$

Clearly, given j non-null discriminately independent strings one can form $2^j - 1$ DCS's. If one starts with two discriminately independent bit-strings of length 2 [(01), (10) or (01), (11) or (11), (10)] and forms the three DCS's, these can be mapped by three nonsingular 2×2 matrices which have them as their only eigenvectors and which are discriminately independent to provide three basis elements for a new level. This mapping can be repeated using 4×4 matrices with $7 = 2^3 - 1 < 16$ nonsingular and discriminately independent exemplars, and once again using 16×16 matrices because $127 = 2^7 - 1 < 256$; however, the mapping cannot be carried further because 256×256 matrices have only 256^2 discriminately independent exemplars and $256^2 \ll 2^{127} - 1$. This is still the simplest way to explain how the combinatorial hierarchy can be generated and why it terminates.

Although our program universe algorithm need not explictly contain the matrix mapping proposed by Parker-Rhodes, the fact that the strings grow by concatenation of bits at only one end and the property of discriminate closure explained above ensure that we will automatically generate many different bit-string representations of the combinatorial hierarchy in the early parts of the strings. We use these early parts of the string as *labels* for the rest of the string. We employ these labels to construct quantum numbers and the rest of the string to construct our discrete version of space-time, as we explain in more detail in Ref. 2 and later work.[12]

Events are defined by the constraint that either three of four strings combine to the null string. If we take as our measure the number of 1's in a string of length S (the Hamming measure), this together with our definition of event ensures that these measures for the three or four strings satisfy a triangle inequality and can be used to define a metric. For two independently generated measures in a locally flat *discrete* space (d-space), these combine in quadrature to a third measure $c^2 = a^2 + b^2$, but the value of c as a "square root" may not exist. However, we can always define symmetric factors $c^2 = (c'+f)(c'-f) = (c')^2 - f^2$ where f is a rational fraction less than c' which has to be consistently assigned in context.

Once we have constructed the label-content concatenation, we can interpret the situations where PICK leads to a non-null string (i.e., $\mathbf{c} = \mathbf{a} \oplus \mathbf{b}$, or equivalently $\mathbf{a} \oplus \mathbf{b} \oplus \mathbf{c} = 0$) as the production (e.g., by pair annihilation or bremsstrahlung) or absorption of a single label which either initiates or terminates a propagation of the label that continues for (or ends after) some

David O. McGoveran and H. Pierre Noyes

finite number of TICK's. This is a discrete model for a Feynman vertex. The completed process combining two such vertices models a four-leg diagram $\mathbf{a} \oplus \mathbf{b} \oplus \mathbf{c} \oplus \mathbf{d} = \mathbf{0}$ which we call a 4-event. The choice of this criterion is not arbitrary. McGoveran (Ref. 6, Theorem 13) has shown that any discrete space of D "homogeneous and isotropic" dimensions synchronized by a universal ordering operator can have no more than *three* indefinitely continuable dimensions; three separate out and the others "compactify" after a surprisingly small number of constructive operations. This theorem is also discussed in Ref. 2.

A tentative rule of correspondence between the last two cardinals of the combinatorial hierarchy and a known result in relativistic quantum field theory was suggested by Noyes[13] (HPN) in 1973. HPN argued that Dyson's calculation[14] of the maximum number of terms in the renormalized QED perturbation theory series in $\alpha = e^2/\hbar c \approx 1/137$ which are meaningful (137, because the series with $\alpha \to -\alpha$ diverges beyond that point) shows that the maximum number of charged particle pairs which can be *counted* within their own Compton wavelength is $137 \simeq (2mc^2)^{-1} [e^2/(\hbar/2mc)]^{-1} = \hbar c/e^2$. The same argument applied to gravitation shows that the maximum number of gravitating baryons of protonic mass which can be counted within their own Compton wavelength is $\hbar c/Gm_p^2 = 1.7 \times 10^{38} \simeq 2^{127}$. Thus the two largest combinatorial hierarchy integers can be interpreted as counting numbers of particles in an appropriate physical context; *why* this should be so remained a mystery until a full physical context had been worked out.

Once the bit-string representation of quantum events was connected to discrete quantum number conservation laws, relativistic three-momentum conservation and relativistic Bohr-Sommerfeld quantization, it became possible[2] to construct a rule of correspondence connecting the first three levels of the hierarchy to the first generation of the standard model of quarks and leptons. The current rule is that the first level has two chiral neutrinos and an associated quantum (3 labels), that the second level has electrons, positrons, gamma rays, and the coulomb interaction (7 labels), and that the third level has up and down quarks with associated gluons in a color-confined octet ($16 \times 8 - 1 = 127$ labels). (Color confinement is proved in our context by the extension of McGoveran's theorem to label-content "space".) In the absence of further information, the coulomb interaction will occur with a probability of 1/137 in all events that must contain the first three labels. Corrections will occur when we must also consider less probable complexities brought in, for example, by the 256^2 possibilities that occur in the mapping of level 3 when four fermions engage directly in an event. These we associate with weak-electromagnetic unification and calculate a first approximation for the Fermi "coupling constant" to be $\sqrt{2}G_F m_p^2 = 1/256^2$; our first approximation for the weak angle is $\sin^2 \theta_{\text{Weak}} = 1/4$. The overall context in which the calculations we now describe is summarized in the Table of Results.

In our bit-string model, as we have already explained, part of the string (the label) represents the quantum numbers generated by the combinatorial hierarchy as discussed above and the remainder of the string (of *content* length n) can represent a biased random walk between events in which the 1's represent k steps in one direction and the 0's represent $n - k$ steps in the other. Generalizing from Stein[15] we use a rule of correspondence that requires each step in any content string or strings allowed in context to be executed at the limiting velocity c and have length h/mc; hence the velocity between events is $\beta c = (2k/n - 1)c$. If we wish to model "constant velocity," this restricts content strings of length nN to have kN

Table of Results, June, 1990

General structural results

- 3 + 1 asymptotic space-time
- combinatorial free particle Dirac wave functions
- supraluminal synchronization and correlation *without* supraluminal signalling
- discrete Lorentz transformations for event-based coordinates
- relativistic Bohr-Sommerfeld quantization
- noncommutativity between position and velocity
- conservation laws for Yukawa vertices and 4-events
- crossing symmetry, CPT, spin, and statistics

Gravitation and Cosmology

- the equivalence principle
- electromagnetic and gravitational unification
- the three traditional tests of general relativity
- event horizon
- zero-velocity frame for the cosmic background radiation
- mass of the visible universe: $(2^{127})^2 m_p = 4.84 \times 10^{52}$ g
- fireball time: $(2^{127})\hbar/m_p c^2 = 3.5$ million years
- critical density: of $\Omega_{\text{vis}} = \rho/\rho_c = 0.011\ 75$ [$0.005 \le \Omega_{\text{vis}} \le 0.02$]
- dark matter = 12.7 times visible matter [10?]
- baryons per photon = $1/256^4 = 2.328 \ldots \times 10^{-10}$ [2×10^{-10}?]

Unified theory of elementary particles

- quantum numbers of the standard model for quarks and leptons with confined quarks and exactly three weakly coupled generations
- gravitation: $\hbar c/Gm_p^2 = [2^{127} + 136] \times [1 - 1/(3 \cdot 7 \cdot 10)] = 1.701\ 47 \ldots [1 - 1/(3 \cdot 7 \cdot 10)] \times 10^{38} = 1.6934 \ldots \times 10^{38}$ [$1.6937(10) \times 10^{38}$]
- weak-electromagnetic unification:
  $G_F m_p^2/\hbar c = (1 - 1/(3 \cdot 7))/256^2\sqrt{2} = 1.02\ 758 \ldots \times 10^{-5}$ [$1.02\ 684(2) \times 10^{-5}$];
  $\sin^2 \theta_{\text{Weak}} = 0.25(1 - 1/(3 \cdot 7))^2 = 0.226\ 7 \ldots$ [$0.229(4)$]
  $M_W^2 = \pi\alpha/\sqrt{2}G_F \sin^2 \theta_W = (37.3\ Gev/c^2 \sin \theta_W)^2$;
  $M_Z \cos \theta_W = M_W$
- the hydrogen atom: $(E/\mu c^2)^2 [1 + 137N_B)^2] = 1$
- The Sommerfeld formula: $(E/\mu c^2)^2 \{1 + a^2/[n + \sqrt{(j^2 - a^2)^2}]\} = 1$
- the fine-structure constant: $\frac{1}{\alpha} = \frac{137}{1 - 1/(30 \times 127)} = 137.0359\ 674 \ldots$ [$137.0359\ 895(61)$]
- $m_p/m_e = \frac{137\pi}{(3/14) \cdot (1 + 2/7 + 4/49) \cdot (4/5)} = 1836.15\ 1497 \ldots$ [$1836.15\ 2701(37)$]
- $m_{\pi^{\pm}}/m_e = 275[1 - 2/(2 \cdot 3 \cdot 7 \cdot 7)] = 273.1292 \ldots$ [$273.12\ 63(76)$]
- $m_{\pi^0}/m_e = 274[1 - 3/(2 \cdot 3 \cdot 7 \cdot 2)] = 264.2\ 1428 \ldots$ [$264.1\ 160(76)$]
- $(G_{\pi N}^2 m_{\pi^0})^2 = (2m_p)^2 - m_{\pi^0}^2 = (13.868\ 11 m_{\pi^0})^2$
  [( ) = empirical value (error) or (range)]

1's, defining the de Broglie wavelength periodicity N as the "positions" where events *could* (but need not) occur. Because the step-wise change in "position" h/mc implies a change in momentum mc, both of which can be reversed at the next step enclosing an area h in phase space, or more generally enclosing an area nNh when we return to a cyclic starting point after nN steps, we have derived relativistic Bohr-Sommerfeld quantization

294

from our model, including the zitterbewegung associated with the string mass specified by the system label.

Most of this background is not directly invoked in the algebraic steps needed to obtain our result. But we have found that, without such an explanation, most physicists cannot see why these algebraic steps lead to a *physical* and not just a mathematical result. It may be easier to follow our reasoning if one goes back to the stage in quantum mechanics when Bohr computed the relativistic formula for the energy levels of the hydrogen atom.[16] Sommerfeld was able to extend this result[17] to compute, in agreement with experiment, the known fine-structure splitting. Bohr had made use of the correspondence principle to tie his model to "classical orbits" at *large* space-time separations between electron and proton, but was well aware of the fact that these classical ideas did not apply to the low-lying states. He was also well aware of the fact that his "circular orbits" in such cases did *not* imply "flat atoms" but in fact described spherically symmetric systems in 3-space — a realization that is all too often lost in elementary discussions of the "Bohr atom." Because he could not rely on his classical space-time intuition, his calculation concentrated on the quantum rules that connect an abstract *energy level model* of the atom to the *observed* transition frequencies as interpreted from the wavelengths of the line spectrum. It was this concentration on observed frequencies rather than spacial models that, in the hands of Heisenberg, led to matrix mechanics. Our calculation is made in the same spirit, but employs our labeled bit-string construction rather than the correspondence principle to insert the quantum rules into the calculation.

We consider a system composed of two masses, m_p and m_e — which we claim to have computed from first principles[5] in terms of $\hbar$, c and G — and identified by their labels using our quantum number mapping onto the combinatorial hierarchy.[2] In this framework their mass ratio [to order α^3 and $(m_e/m_p)^2$] has also been computed using only $\hbar$, c, and 137. However, to put us in a situation more analogous to that of Bohr, we can take m_p and m_e from experiment, and treat 1/137 as a counting number representing the coulomb interaction; we recognize that corrections of the order of the square of this number *may* become important when we have to include degrees of freedom involving electron-positron pairs. We attribute the binding of m_e to m_p in the hydrogen atom to coulomb events, i.e., only to those which involve a specific one of the 137 labels at level 3 and hence occur with probability 1/137; the changes due to other events average out (are *indistinguishable* in the absence of additional information). We can have any periodicity of the form 137j, where j is any positive integer. So long as this is the only periodicity, we can write this restriction as 137j *steps* = 1 *coulomb event*. Since the internal frequency 1/137j is generated independently from the *zitterbewegung* frequency which specifies the mass scale, the normalization condition combining the two must be in quadrature. We meet the bound state requirement that the energy E be less than the system rest energy $m_{ep}c^2$ (where $\mu = m_{ep} = m_e m_p/(m_e + m_p)$ is used to take account of 3-momentum conservation) by requiring that $(E/\mu c^2)^2[1 + (1/137N_B)^2] = 1$. If we take $e^2/\hbar c = 1/137$, this is just the relativistic Bohr formula[16] with N_B the principle quantum number.

Since most of our readers have never encountered this formula, and might have trouble chasing through the units in which it is expressed starting from the 1915 paper, we derive it in a modern way[5] by treating the bound state as a pole in the relativistic wave function — or S matrix. The basic S matrix point of view associates a bound or resonant state of any two-particle system with a pole at invariant 4-momentum squared s_0 in the two-particle momentum-space wave function $\phi(s, s_0)$ whose residue

defines the "coupling constant" f^2. In the narrow-width approximation this translates to

$$\phi(s, s_0) = f^2 \mu/(s - s_0). \tag{5}$$

Assuming the state contains only two particles of mass m_1, m_2 yields the normalization condition

$$\int_{(m_1+m_2)^2}^{\infty} ds \, |\phi(s, s_0)|^2 = 1, \tag{6}$$

which forces us, for dimensional reasons, to include some mass μ in the definition of the residue if we wish (in analogy with $e^2/\hbar c$) to keep the coupling constant f^2 dimensionless. By performing the integration we obtain a simple connection between masses and coupling constants

$$(f^2\mu)^2 = (m_1 + m_2)^2 - s_0. \tag{7}$$

Note that the *magnitude* of f^2 is not seriously restricted by this algebraic connection until we have inserted more information. We assert that this starting point is *nonperturbative* and rests only on *unitarity* and relativistic quantum mechanics in a finite particle number space. If we take $f^2 = e^2/\hbar c = \alpha \simeq 1/137$, take the free system mass equal to the reduced mass "$m_1 + m_2$"$\to m_{ep} = m_e m_p/(m_e + m_p)$ [which implies that $s_0 = (m_{ep} - \epsilon_{Bohr})^2$] and use this also as the reference mass μ, we obtain once again the relativistic Bohr formula.

The Sommerfeld model for the hydrogen atom (and, for superficially different but profoundly similar reasons,[18] the Dirac model as well) requires two *independent* periodicities. If we take our reference period j to be an integer and the second period s to differ from an integer by some rational fraction Δ, there will be two minimum values $s_0^{\pm} = 1 \pm \Delta$, and other values of s will differ from one or the other of these values by integers: $s_n = n + s_0$. This means that we can relate ("synchronize") the fundamental period j to this second period in two different ways, namely to

$$137j \frac{\text{steps}}{\text{(coulomb event)}} + 137s_0 \frac{\text{steps}}{\text{(coulomb event)}} = 1 + e = b_+, \tag{8}$$

or to

$$137j \frac{\text{steps}}{\text{(coulomb event)}} - 137s_0 \frac{\text{steps}}{\text{(coulomb event)}} = 1 - e = b_-, \tag{9}$$

where e is an event probability. Hence we can form

$$a^2 = j^2 - s_0^2 = (b_+/137)(b_-/137) = (1 - e^2)/137^2. \tag{10}$$

Note that if we want a finite numerical value for a, we cannot simply take a square root, but must determine from context which of the symmetric factors [i.e., $(1-e)$ or $(1+e)$] we should take (cf. the discussion about factoring a quadratic above). With this understood, we write $s_n = n + (j^2 - a^2)^{1/2}$.

We must now compute the probability e that j and s are mapped to the same label, using a single basis representation constructed within the combinatorial hierarchy. We can consider the quantity a as an event probability corresponding to an event **A** generated by a global ordering operator which ultimately generates the entire structure under consideration. Each of the two events j and s can be thought of as derived by sampling from the same population. That population consists of 127 strings defined at level 3 of the hierarchy. In order that j and s be independent, at least the last

118

David O. McGoveran and H. Pierre Noyes

of the 127 strings generated in the construction of s (thus completing level three for s) must not coincide with any string generated in the construction of j. There are 127 ways in which this can happen.

There is an additional constraint. Prior to the completion of level 3 for s, we have available the $m_2 = 16$ possible strings constructed as a level 2 representation basis to map (i.e., represent) level 3. One of these is the null string and cannot be used, so there are 15 possibilities from which the actual construction of the label for s that completes level 3 are drawn. The level can be completed just before or just after some j cycle is completed. So, employing the usual frequency theory of probability, the expectation e that j and s as constructed will be indistinguishable is $e = 1/(30 \times 127)$.

In accordance with the symmetric factors $(1 - e)$ or $(1 + e)$, the value e can either subtract from or add to the probability of a coulomb event. These two cases correspond to two different combinatorial paths by which the independently generated sequences of events may close (the "relative phase" may be either positive or negative). However, we require only the probability that all s_0 events are generated within one period of j, which is $1 - e$. Hence the difference between j^2 and s^2 is to be computed as the "square" of this "root," $j^2 - s_0^2 = (1 - e)^2$. Thus for a system dynamically bound by the coulomb interaction with two internal periodicities, as in the Sommerfeld or Dirac models for the hydrogen atom, we conclude that the value of the fine-structure constant to be used should be

$$\frac{1}{a} = \frac{137}{1 - 1/(30 \times 127)} = 137.0359\ 674\ \ldots$$

in comparison to the accepted empirical value of[3]

$$1/\alpha \simeq 137.0359\ 895(61).$$

Now that we have the relationship between s, j, and a, we consider a quantity H' interpreted as the energy attribute expressed in dynamical variables at the $137j$ value of the system containing two periods. We represent H' in units of the invariant system energy μc^2. The independent additional energy due to the shift of s_n relative to j for a period can then be given as a fraction of this energy by $(a/s_n)H'$, and can be added or subtracted, giving us the two factors $[1 - (a/s_n)H']$ and $[1 + (a/s_n)H']$. These are to be multiplied just as we multiplied the factors of a above, giving the (elliptic) equation $(H')^2/(\mu^2 c^4) + (a^2/s_n^2)(H')^2/(\mu^2 c^4) = 1$. Thanks to the previously derived expression of $s = n + s_0$ this can be rearranged to give us the Sommerfeld formula[17]

$$H'/\mu c^2 = \left\{ 1 + \frac{a^2}{[n + (j^2 - a^2)^{1/2}]^2} \right\}^{-1/2}.$$

Several corrections to our calculated value for α can be anticipated, depending on context. As noted above, we have a first approximation for the Fermi constant and for the weak angle, each of which disagrees with currently accepted values by a few per cent. Just how weak-electromagnetic unification takes shape in our developing theory could lead to problems at this level of accuracy. More immediately, the Sommerfeld formula uses the "system mass" which, naively, is $\mu = m_e m_p/(m_e + m_p)$, and hence suggests that there may be corrections of order $(m_e/m_p)^2$. At the level of accuracy of our calculation, these can be ignored in hydrogen, but will be significant in muonium. In QED these corrections depend on the sum of the squares of the two finite masses, which is consistent with our rule that requires two independent quantities to add in quadrature. If we cannot derive the appropriate correction, our theory could be in serious trouble. Another point that bears watching is whether there is spin dependence at this level of accuracy. Clearly, the calculation presented here cannot apply to positronium or to the "Lamb shift," but these effects go beyond order α^2 in conventional QED, and presumably for us as well. Our theory is frangible if we cannot meet these challenges, just as QED was in the late 1940s. We ask the reader to consider whether the conceptual advantages and quantitative results our approach has already demonstrated provide sufficient incentive for engaging in the hard work needed to extend our theory to higher approximation and to the systematic calculation of other physically dimensionless parameters.

Acknowledgment

This work was supported in part by the Department of Energy, contract DE-AC03-76SF00515.

Received on 14 March 1990.

Résumé

Par notre théorie combinatoriale discrète relativiste à bit-string dans le contexte du spectre de l'hydrogène, nous calculons nos deux premières approximations de la constante de structure fine comme $\alpha(1) = 1/137$ et $\alpha(2) = [1 - 1/(30 \times 127)]/137 = 1/137.035\ 967\ 4\ \ldots$; nous pouvons ensuite dériver la formule de Sommerfeld.

296

References

1. *Discrete and Combinatorial Physics: Proceedings of the 9th Annual International Meeting of the Alternative Natural Philosophy Association*, edited by H.P. Noyes (ANPA WEST, 25 Buena Vista Way, Mill Valley, CA, 94941, 1988).
2. H. Pierre Noyes and David O. McGoveran, Phys. Essays **2**, 76 (1989).
3. M. Aguilar-Benitz *et al.*, *Particle Properties Data Booklet* (North Holland, Amsterdam, April 1988), pp. 2, 4.
4. S. Weinberg, Phys. Lett. B **125**, 265 (1983); P. Candelas and S. Weinberg, Nucl. Phys. B **237**, 393 (1984).
5. H.P. Noyes, in *Physical Interpretations of Relativity Theory II*, edited by M.C. Duffy (Conf. at Imperial College, London, 1990), p. 196 and SLAC-PUB-5218, July, 1990.
6. D.O. McGoveran, in *Discrete and Combinatorial Physics*, p. 37; see, also, SLAC-PUB-4526, June, 1989.
7. *Ibid.*, p. 39.

8. T. Bastin, Studia Philosophica Gandensia **4**, 77 (1966).
9. T. Bastin, H.P. Noyes, J. Amson, and C.W. Kilmister, Int. J. Theor. Phys. **18**, 455 (1979).
10. M.J. Manthey, in *Proc. of ANPA 7 (1986)* (published by Dr. F. Abdullah, Room E517, The City University, Northampton Square, London EC1V 0HB); see, also, SLAC-PUB-4008.
11. Noyes and McGoveran, Phys. Essays **2**, 76 (1989), p. 87.
12. H.P. Noyes, in *Proc. of ANPA 11 (1989)* (published by Dr. F. Abdullah, Room E517, The City University, Northampton Square, London EC1V 0HB); see, also, SLAC-PUB-5085, Jan. 1990.
13. *Idem*, SLAC-PUB-1405, rev. November 1975, unpublished.
14. F.J. Dyson, Phys. Rev. **51**, 631 (1952).
15. I. Stein, Phys. Essays **1**, 155 (1988).
16. N. Bohr, Philos. Mag., 332, Feb. 1915.
17. A. Sommerfeld, Ann. Phys. IV **17**, 1 (1916).
18. L.C. Biedenharn, Found. Phys. **13**, 13 (1983).

David O. McGoveran
Alternative Technologies
15905 Bear Creek Road
Boulder Creek, California 95006 U.S.A.

H. Pierre Noyes
Stanford Linear Accelerator Center
Stanford University
Stanford, California 94309 U.S.A.

Comment on *"Our Joint Work"*
David McGoveran (2000)

When my friend, mentor, and colleague asked if he could include our joint papers in this volume of collected works, I must admit I had some misgivings. These papers, both those in which I was the primary author and those in which Pierre was the primary author, have been the source of no end of misunderstanding. In trying to explain their content, I have been confronted by physicists of great skill who expressed their discomfort by resorting to petty objection or by deflecting the point at hand by introducing questions on extraneous topics. The task of providing convincing argument in favor of a new approach to physics (let alone simply a new theory about a particular phenomenon) is a demanding one, requiring a full time commitment, and a great deal of time devoted to explanatory writing. Above all, it requires answering a never ending stream of objections. Anything less is a *de facto* admission of being wrong. Being unable to take on the task previously, I admit to hesitation in opening up this Pandora' Box of skepticism.

Foundations of a Discrete Physics was primarily an attempt to capture and express the results of a particular thread of over twenty years of my personal research efforts. With the exception of Foundations of a Discrete Physics, the papers contained in this volume which I co-authored with Pierre Noyes contain material that was derived from other of my papers and talks from a period of approximately fifteen years (1974-1990). For the most part, those papers have never been published except in ANPA and ANPA West Proceedings and I am still not in a position to make those papers otherwise available. To make matters worse, my contribution to most of the papers contained herein was one of providing a mathematical approach (or perhaps an occasional result) rather than being representative of a personal commitment to the particular philosophy: I see bit-string physics almost as a shadow (or more accurately, a projection) of a richer and more complex form of discrete physics. Just as the geometry of a two-dimensional projection tells us something of three-dimensional solid, so bit-string models tell us something of the discretum. These facts will undoubtedly make it difficult (if not impossible) for the typical reader to perceive any continuity in my own work simply from reading the papers contained in this volume. Indeed, I suspect the claims, constructions, and results which appear herein will convince no one of the research's intrinsic value.

Although I do not insist on bit string physics as the best description of the physical world, for the record, I do concur with (and helped elucidate or develop) several concepts introduced in these papers:

- What we call the physical world is a discretum and is best described as such.
- Particles are virtual carriers of events.
- Measurement is ultimately reducible to counting.
- The combinatorial hierarchy is an important mathematical structure to which there are corresponding physical structures.
- Dimensionless invariants (a.k.a. structure constants) must be reproduced by any combinatorial theory of physics.
- A process view in which there is no inherently differentiated observer will be a crucial aspect of any "correct" physical theory (although a partitioning into observer/observed should reproduce certain phenomena).

Foundations of a Discrete Physics is a dense work, consisting mostly of the work I did on developing a purely discrete mathematics and which I call the *ordering operator calculus*. The development is tedious, and is intended to make contact with the more traditional mathematics of differential geometry in a kind of mathematical correspondence principle. The results are then applied to physics. The last sections of this work are further focused on showing that the work of Noyes and others on a bit-string model of physics can be understood in terms of ordering operators. Foundations was not expected to be tutorial or even narrative in character. For this reason, the reader needs considerable help in understanding the point of view and motivation taken not only in Foundations, but also in the papers which followed. To this end I would like to add a few prefacing comments.

Foundations is essentially a work of abstract mathematics, albeit highly motivated by the desire to have a rich mathematical tool that would be purely discrete and yet could accommodate notions of process. It remains my strong belief that much of what works well in applied mathematics should be recoverable from a discrete approach. Furthermore, (I was (and remain) unhappy with traditional discrete mathematical approaches: I find the philosophical motivations and much of the reasoning to be contaminated with continuum ideas. To say that the same is true of modern physics would be an understatement: a truly discrete theory of physics has never obtained any reasonable acceptance let alone popularity among physicists. Even the inventors and leaders in quantum physics and its successors were unable to accept a purely discrete approach.

Physicists rely heavily on measurement. In order to be considered for acceptance, a theory must ultimately produce quantitative predictions. Sadly, most physicists are so deeply steeped in traditional theories and have so little knowledge of discrete mathematics that only theories with continuous or semi-continuous metrics would ever be considered. In contradiction to this requirement, (the only metrics that are possible with a purely discrete or combinatoric theory will involve counting. If there is sufficient structure in the theory, then certain characteristic numbers will appear. A standard way of applying an abstract combinatoric theory is to map these structures and their characteristic numbers to empirical facts. Traditional physicists have a nasty name for this otherwise respectable approach: numerology.

But how is this approach different from any type of mapping that is done between mathematics and experimental evidence? If the truth be spoken, it isn't. One of the most fundamental techniques of the laboratory physicist is curve fitting. A series of measurements are obtained, a relationship is hypothesized, and a best fit mathematical description is obtained by a process of successive approximation. If dimensionless units are used, certain numeric ratios appear which partially describe the relationship. These invariants are given great importance and few physicists would question their validity. Indeed, much of what we now recognize as laws of physics must reproduce these pure numbers: just about everything else is illusion induced by a particular frame of reference or set of units. Whether these numbers are the invariants of relativity, eigenvalues of quantum mechanics, or the group numbers of quantum field theory, measurement is determined by our ability to map pure numbers to physical concepts.

As a particular combinatorial structure, the so-called *combinatorial hierarchy* has proven to be a great interest in applying a discrete theory to physics. The reasons for that interest are too numerous to discuss here, but suffice it to say that its combinatorial invariants happen to be suspiciously close to certain physical invariants. Perhaps this is coincidence; perhaps not. But far less than "coincidence" such as this has been seized upon by others over the centuries as possible evidence of discovery of an important theory, and both physicists and mathematicians have spent decades in pursuit of an elusive discovery with less reasonable motivation.

When examined from the point of view of the ordering operator calculus, a certain structure is common in physics. That structure is most easily understood as a joint probability. Suppose that we have a discrete set of size $2N$ which is equi-partitioned. The probability of selecting a particular member of one of the partitions is just $1/N$. The probability of selecting that member and a particular member from the other partition is $(1/N)(1/N)$. Just as interesting is the inverse of this joint probability, namely the number of combinations that can be formed from the two partitions, in this case $N \times N$.

Now suppose that the set is not equi-partitioned, but is "biased" by an amount n. Then we have $(N+n)(N-n)$ possible combinations. This general scenario of a partitioned set is quite common in nature. But what is interesting is the way in which things work out under a fairly simple assumption that the partition is not fixed, but perturbed. In particular, suppose that the set can be partitioned in either of two ways (but only these two ways), namely at $(+n)$ or at $(-n)$ and that it is found partitioned in these two ways with equal probability. Then, the size of each subset is just N on the average and the average number of combinations is $N \times N$. If we now normalize the number of combinations by the average number we obtain a very important and very simple combinatorial (or probabilistic) equation:

$$Q = (N+n)(N-n)(1/N \times N) = (N \times N - n \times n)(1/N \times N) = (1 - n \times n/N \times N) \quad (1)$$

This equation is referred to by Noyes as the "handy-dandy formula" and plays a central role in the computation of the fine structure constant (see Chapter 6, "On the Fine...", p. 118, discussion leading to Eq. 8, 9, and 10), as well as a number of other computed physical invariants (some of which are included in the appended table). For the sake of historical accuracy, the fine structure constant computation was derived on purely theoretical grounds and without access to the then known empirical value. The other computed invariants (see appended table) were likewise computed on theoretical grounds up to a single rational coefficient which itself was severely constrained to a small set.[†]

This equation is also important in a different way: Whenever a physicist assumes that an "average" condition applies to a physical situation, a mathematical condition of symmetry is imposed which may not in fact exist. This assumption of symmetry leads one to assume that the Q above (actually a joint probability) is a probability squared ($P \times P$), with the result that the square root of Q is taken to find P. Explicitly, this assumption replaces Eq.(1) by $P^2 + n^2/N^2 = 1$. Unfortunately, this permits non-physical algebraic solutions of the left hand side of the equation using the factors $P + in/N$ and $P - in/N$, and therefore introducing the miraculous concept of complex or (if the difference of complex conjugates is taken) imaginary probability amplitudes. [‡]

[†]HPN: The invariants computed by David which appear in this table were published in *Proc. ANPA 11*, together with the calculations and arguments leading to them. These published calculations are reprinted below as: *Other Second Order Corrections* and *Some Further Speculations*. Although the numerical value of the end result for the electron-proton mass ratio calculation which this line of reasoning lead to is the same as Parker-Rhodes' value, David's new way of arriving at it — as published in *Proc. ANPA 12* — is also profound. This new caclulation is reproduced here under the heading *The Electron-Proton Mass Ratio*.

[‡]HPN: This situation occurs in the first version of David McGoveran's calculation, "The Fine Structure of Hydrogen", *Proc. ANPA 10*, pp 28-52, and is presented again in our joint paper (Ch. 6). At ANPA 10 (and subsequently) various members of ANPA have had difficulty following David's argument. To meet this problem, we append here, with his permission, his paper "Adding in Quadrature: Why?", which occurs as Section 2 (pp 83-88) in his paper "Advances in the Foundation" published in *Proc. ANPA 11*. The phrase "...must be added in quadrature..." occurs in our joint paper presented at the 1988 Physical Interpretations of Relativity Conference in London (H.P.Noyes and D.O.McGoveran, "Observable gravitational and electromagnetic orbits and trajectories in discrete physics", SLAC-PUB-4690, November 28, 1988, pp 13-14) in the paragraph: "To

Lets look at another way in which this equation becomes important. It was noted above that in a purely discrete theory, the only metrics are derived by counting. When the elements of a set are partially ordered, there is a convergence between our notions of counting cases (where the elements of the set are counted in order) and of counting units of "distance". This convergence of probability and measurement is an extremely important result of the ordering operator calculus: it allows us to move back and forth between a probabilistic or combinatorial interpretation and an abstract spatial interpretation of any result. It also provides an important correspondence principle between a discrete model of physics and a continuum model. Measurement in laboratory physics is ultimately dependent on counting, and on finite counts at that. Unfortunately, the physicist's accepted notions of the continuum are abstract and have little bearing on laboratory practice. Although a full account of the measurements that a physicist makes in terms of counting is quite involved (the establishment of standard units, of performing comparisons, of calibration, and so on must all be explained in detail although they are normally taken for granted), it is nonetheless possible to understand all the empirical measurements upon which laboratory physics rests in a purely discrete manner.

Without arguing the point further, lets take a look at another place in which equation (1) appears in physics and one with which every physicist is familiar: the Lorentz transformation. Most physicists are also (or were at one time) familiar with the derivation of the Lorentz transformation via the basic laboratory set up of the

---

extend this analysis to a coulomb bound state with a system [reduced] mass $\mu = \frac{mM}{m+M}$ we take from our interpretation of the combinatorial hierarchy the fact that only 1 in 137 of the events will be a coulomb event, the others averaging out in the first stage of the analysis; in other words 137 N_B steps = 1 coulomb event. This means that we now have two frequencies (in units of $\mu c^2/h$), the zitterbewegung frequency corresponding to the rest mass, which we take to be unity, and the coulomb frequency $1/137N_B$. Since these two motions are incoherent, the frequencies must be added in quadrature [emphasis supplied] subject to the constraint on the energy E defining a bound state that in the rest system $E/\mu c^2 < 1$ Hence $(E/\mu c^2)^2[1 + (1/137N_B)^2] = 1$. In the language of the ordering operator calculus, this is simply the normalization of the metric corresponding to the energy attribute under the appropriate constraint. If we take $e^2/\hbar c = 1/137$, this is just the relativistic Bohr formula (N.Bohr, *Phil. Mag*, Feb. 1915)."

Michelson-Morley experiment. What few physicists realize is that any discrete theory of velocity in which the discretum is finite must necessarily obey such a transformation (a fact initially brought to my attention by Tom Etter). Here the reader is asked to refer to my derivation of the Lorentz transformation (Ch 4. "Foundations..." pp. 48-54.) Despite the obvious asymmetry of the derivation of $(c - v)(c + v)$, it is not uncommon to find expositions in which symmetry is forced with further assumption of an imaginary signature (e.g., by using the Minkowski metric) when the transformation is factored into $(1 - iv/c)$ and $(1 + iv/c)$ and symmetric roots are found, any of which may be treated as physical solutions, depending on the particular interpretation of the theory. This forced symmetry conceals important structure that is represented by the real, asymmetric factors of the transform. An obvious conclusion of treating the Lorentz as a discretum transformation is that it mirrors the partitioning of a set (i.e., the spacetime discretum itself), in this case the possible partitions being $(+v)$ and $(-v)$ and equally probable.

With a "bit" (pun intended) of luck , these introductory notes will provide some insight into a reasonable justification for what might otherwise be considered the outrageous claims that have been put forth in this collection of papers. Justifying the work more carefully would require perhaps a lifetime of work. Unfortunately, I have neither that lifetime nor the resources to further elucidate these matters. For this I apologize to those who find the approach intriguing, let alone those who find it of value. I hope that these papers will encourage you to think outside the constraints of traditional continuum approaches, and take you on a journey of a thousand discoveries, one "bit" at a time.

David McGoveran

Boulder Creek, California

July 10, 2000

Predictions as of 1997 — see *A Short Introduction to Bit-String Physics*

$$G_N^{-1}\frac{\hbar c}{m_p^2} = [2^{127} + 136] \times [1 - \frac{1}{3 \cdot 7 \cdot 10}] = 1.693\ 31\ldots \times 10^{38}$$

$$experiment = 1.693\ \mathbf{58}(21) \times 10^{38}$$

$$\alpha^{-1}(m_e) = 137 \times [1 - \frac{1}{30 \times 127}]^{-1} = 137.0359\ \mathbf{674}....$$

$$experiment = 137.0359\ 895(61)$$

$$G_F m_p^2/\hbar c = [256^2\sqrt{2}]^{-1} \times [1 - \frac{1}{3 \cdot 7}] = 1.02\ \mathbf{758}\ldots \times 10^{-5}$$

$$experiment = 1.02\ 682(2) \times 10^{-5}$$

$$sin^2\theta_{Weak} = 0.25[1 - \frac{1}{3 \cdot 7}]^2 = 0.2267\ldots$$

$$experiment = 0.2259(46)]$$

$$\frac{m_p}{m_e} = \frac{137\pi}{< x(1-x) >< \frac{1}{y} >} = \frac{137\pi}{(\frac{3}{14})[1 + \frac{2}{7} + \frac{4}{49}](\frac{4}{5})} = 1836.15\ \mathbf{1497}\ldots$$

$$experiment = 1836.15\ 2701(37)$$

$$m_\pi^\pm/m_e = 275[1 - \frac{2}{2 \cdot 3 \cdot 7 \cdot 7}] = 273.12\ \mathbf{92}\ldots$$

$$experiment = 273.12\ 67(4)$$

$$m_{\pi^0}/m_e = 274[1 - \frac{3}{2 \cdot 3 \cdot 7 \cdot 2}] = 264.2\ \mathbf{143}\ldots$$

$$experiment = 264.1\ 373(6)]$$

$$m_\mu/m_e = 3 \cdot 7 \cdot 10[1 - \frac{3}{3 \cdot 7 \cdot 10}] = 207$$

$$experiment = 206.768\ 26(13)$$

$$G_{\pi N\bar{N}}^2 = [(\frac{2M_N}{m_\pi})^2 - 1]^{\frac{1}{2}} = [195]^{\frac{1}{2}} = 13.96....$$

$$experiment = 13.3(3),\ or\ greater\ than\ 13.9$$

[From Proc. ANPA 11 pp 89-93]
David McGoveran*

3. Other Second Order Corrections

The currently accepted empirical value of the fine structure constant is $1/(137.035963(15))$. According to the combinatorial hierarchy and Program Universe, the calculated value to first order would be $1/137$. Note that this is the value obtained if (A') is $\frac{m}{2}$—i.e. if $n = 0$ and events A and B are in fact identical. When the structure of A and B, and the possibility that they are not independent $(n = 1)$, are taken into account, this yields a second order correction. The value obtained is given by

$$\left(\frac{1}{137}\right)\left(1 - \frac{1}{2*15*127}\right) = \frac{1}{(137.0359674)}$$

in close agreement with the empirical value.

By following an argument directly analogous to that presented in computing the fine structure constant, one can obtain corrections to the weak or Weinberg angle and to the Fermi coupling. This should not be surprising since there is a relationship between the "weak" structure constant and the fine structure, as introduced by Glashow:

$$g_W = \frac{e}{\sin \theta_W}$$

Thus, given that the definition of the fine structure constant is $e^2/\hbar c$ and weak structure constant analogously by $g_W^2/\hbar c$, we have the ratio of the fine structure constant to the weak structure constant:

$$\frac{a}{a_W} = \sin^2 \theta_W$$

The weak coupling g_W and the Fermi coupling G_F are related by

$$G_F = 2^{1/2} g_W^2 r_W^2$$

where r_W is the range of the weak force.

Just as the Coulomb event depends on 127 cases of level 3 of the combinatorial hierarchy and the 16 - 1 possible labels to represent them, the weak event depends on the 7 cases of level 2 and the 4 - 1 possible labels to represent them. This gives the number of cases as 3 * 7 analogous to 15 * 127. However, the measurement of the weak event is not in the context of a bound system like that of the hydrogen atom, but rather is understood in the context of a decay process. There is no reference frame in which to distinguish event A from event B. As a result, the cases for each of two events are not orderable: the factor of two that occurs in $\frac{m}{2}$ for the Coulomb event does not occur for the weak event. Thus $\frac{m}{2}$ is 3 * 7 and the correction term (A') is $(1 - \frac{1}{3*7})$.

Fermi Coupling Constant

The empirical value of the Fermi coupling (as given in terms of the proton mass and factoring out the square root of two and the proton mass squared that would otherwise appear) appears as

$$G_F * (2)^{1/2} = 1.02684(2) * 10^{-5} .$$

To first order, this value is calculated from the combinatorics of Program Universe as

$$\frac{1}{(256 * 256)} = 1.07896 * 10^{-5}$$

The second order correction gives

$$(1.07896 * 10^{-5}) \left(1 - \frac{1}{3 * 7}\right) = 1.0275808 * 10^{-5}$$

again in close agreement with the empirical value.

The Weak Angle

The currently accepted weak or Weinberg angle squared empirical value is

$$\sin^2 \theta = 0.229(4)$$

Again, according to Program Universe the ratio of weak to Coulomb events is 2:1 so that the calculated first order value of the weak angle is 1/2 and the square of the weak angle is then

0.25.

The second order correction applied to the weak angle (not to the square) and then squaring gives

$$\left(0.5 * \left[1 - \frac{1}{3 * 7}\right]\right)^2 = 0.2267573$$

again in good agreement with the empirical value. .

Note that we have no "running constants" in our theory nor do we have perturbative approximations. Ours are true corrections due to well–defined, finite system effects. We do anticipate a proper correction factor correlated with the energy of the system.

4. Some Further Speculations

Having had reasonably good success with computing these physical values, I am inclined to make a few conjectures under the assumption that similar corrections would work for other physical values. Three have been presented by Program Universe to–date: the charged pion/electron mass ratio, the neutral pion/electron mass ratio, and the gravitational structure constant.

Charged Pion/Electron Mass Ratio

The empirical value of the charged pion/electron mass ratio is

273.13

The pion is represented in Program Universe as 137 electron positron pairs plus either an electron—antineutrino or a positron–neutrino pair, suggesting a mass ratio of

275.

First note that this is a bound system of two sets of 137, i.e. electron type events and positron type events. Thus ordering is important and the factor of 2 discounted in the correction of the weak structure can not be discounted here. Suppose that the charged pion electron/positron pairs are weakly coupled ($3 * 7$ cases) via the 7 labels required to represent level 2. Furthermore, suppose that there are two cases (via exchange of weak labels) which can not be distinguished as belonged to either the electron or the positron set (i.e. $n = 2$) for each pair.

The second order correction is then

$$275 * \left[1 - \frac{2}{2 * 3 * 7 * 7}\right] = 273.12925$$

in excellent agreement with the empirical value.

Neutral Pion/Electron Mass Ratio

The empirical value of the neutral pion/electron mass ratio is

264.10.

Program Universe suggests that the pion consists of 137 electron/positron pairs, so that the first order computed value is

274.

Suppose the neutral pion is a more complicated system. If there are three indistinguishable cases instead of two, and only manifesting half the time, and that it does not couple back to the 7 labels in level 2 as the charged pion does, the second order correction is[*]

$$274 * \left[1 - \frac{3}{2 * 2 * 3 * 7}\right] = 264.21428$$

Gravitational Structure Constant

The empirical value of the gravitational analogue to the fine structure constant is

$$\hbar c / Gm^2 = 1.6937(10) * 10^{38}$$

where m is the proton mass.

The first order value computed by Program Universe is

$$(2^{127} - 1 + 137) = 1.70147 * 10^{38} .$$

Suppose that there is a coupling between a combinatorial hierarchy level 2 weak event with $3 * 7$ cases and a compound level 1 - level 2 event with $3 + 7 = 10$ cases. These then couple to give $3 * 7 * 10$ possible cases. If one of these cases is indistinguishable ($n = 1$) and order is unimportant, one obtains for a second order correction

$$(1.70147 * 10^{38}) \left[1 - \frac{1}{3 * 7 * 10}\right] = 1.6933675 * 10^{38}$$

again in good agreement with the empirical value.

---

[*] Compare this with Noyes recent argument in "Bit String Scattering Theory" – SLAC PUB 5085, January 16, 1990 – from estimation of the pion-nucleon coupling constant.

310

[From   Proc. ANPA 12 pp 15-16]
David McGoveran*

Part II
The Electron-Proton Mass Ratio

I. Introduction

The computation of the electron-proton mass ratio produced by
Frederick Parker-Rhodes was based on the combinatorial hierarchy
numbers, in particular 137. However, the model he used was
strictly a continuum probability model. He evaluated the
distribution of a charge in order to come up with certain
weighting factors. In this part of the paper we show how the
same calculation can be given a purely combinatorial basis and
briefly discuss the model of the proton and electron that this
basis implies.

NOTE: Due to typographical problems, I will use p for the
constant "pi" and a for alpha.

II. The Electron Mass

Let $a_1 = 137*2p$. We interpret 137 as an "angular" frequency (in
the probability sense), suggesting that the usual fine structure
already has a $2p$ embedded in it.   The value a_1 is the
"linearized" counterpart to the first order approximation of the
fine structure constant.

Model electron "mass" as:

1)   primarily due to Coulomb events (a_1).

2)   due to a level 2 self-interaction: 6 of every 7 events
      are indistinguishable between the electron and itself)

3)   the effects due to virtual electron "generation" are
      with respect to that without self-interaction and so we
      normalize with respect to that effect.

The electron picture suggested by this model is that of a "bag"
of 137 different randomly occuring events, but for which 6 out of
7 of these event are equivalent under some set of operations
(namely those defined at Level 2 by the combinatorial hierarchy).
The boundary of this "bag" defines the electron Compton
wavelength in terms of the possible space-time parameters than t
can be consistently given to these events while maintaining a
causal structure.

Although this is a Level 2 process, we write it out in Level
1 "units" -- i.e. since there are three Level 1 events for every
seven Level 2 events, multiply by 3/7 -- this will allow us to
combine the computations for electron and proton:

$$m_e = (3/7)*(a_1)*[1-(6/3*7)^N] / [1-(6/3*7)] \qquad (i)$$

where N is the degree of self interaction (this is an important interpretive point -- see Appendix A below).

III. The Proton Mass

Now treat m_p, the proton observed "mass", as:

1) Consisting of two parts:

- a "fundamental" "mass" m_p(fundamental)

- a portion due to Level 1/Level2 coupling:

m_p(coupling) = 3/7 m_p

2) There is no self-interaction term (the proton is always distinguishable from itself)

3) As with the electron, normalize with respect to the level 2 term as given for the electron.

Unlike the model of the electron, this proton model treats the proton as more fundamental. It is entirely due to coupling between Level 1 and Level 2.

Write this as

m_p(fundamental) = [m_p - m_p (coupling)] / m_p (virtual)

or

$$m_p\text{(fundamental)} = m_p * [1-(3/7)] / [1-(6/3*7)] \qquad (ii)$$

Combining (i) and (ii) we obtain:

$$m_p/m_e = ([1-(6/3*7)] / [1-(3/7)]) /$$

$$\{(3/7)(1/2p)(1/137)[1-(6/3*7)^N]/[1-(6/3*7)]\} \qquad (iii)$$

which, for N=3, is algebraically identical to:

$$m_p/m_e = 137p / (3/14)[1+(2/7)+(4/49)]*(4/5) \qquad (iv)$$

$$= 1836.151497$$

as derived by Parker-Rhodes using integrals. Note that the value of N just corresponds to the number of expansion terms in the series solution to one of the Parker-Rhodes integrals.

312

[From Proc. ANPA 11 pp 83–88]
David McGoveran*

2. Adding In Quadrature: Why?

Over the last year, the one aspect of my computation of the second order correction to the fine structure constant which was most difficult to explain was what Noyes has referred to as adding in quadrature. As I have pointed out, adding in quadrature is actually a consequence of having what I think of as simultaneous and independent "paths" in the system. In Feynman's terminology these are alternative "paths". My own, largely non–verbal, conception of the systems involved has finally found expression in terms of a certain complex system and the frequency with which certain alternatives will be manifested.

The terminology I shall use is that of mathematical probability (see Uspensky). This is important to remember since the meanings of the terms in the physical sciences can be quite different. So as not to burden the reader with having to look up the definitions of terms in the references, I will define a few key ones here.

By an event I mean a well–defined, abstract arrangement or class of arrangements. The event (arrangement) may manifest in either space or time or both. An arrangement is identified by its properties or an equivalence class of properties. In fact, a manifestation of an event may be purely abstract. It may be defined only in some mathematical context and so be not physical at all. I will use the term *statistical* event when I mean this kind of event, since this differs from physical events.

By an occurrence of a statistical event I mean a particular manifestation of that statistical event. Each particular way (one detailed arrangement among the arrangements of an equivalence class of arrangements) in which the statistical event can be manifested is called a case. For example, the results of throwing a die is an occurrence and each of the possible results (a face with 1, 2, 3, 4, 5, or 6 points) is a case. The number of ways in which a case can occur is called a case count—these are the number of favorable cases for a particular outcome.

The confluence of one or more statistical events will be said to form a *system.* Exactly how these multiple statistical events are arranged in the system also defines a statistical event, albeit at a higher level of complexity. I will refer to the partitions of the system, each consisting of a component statistical event as the sub–systems.

The system under investigation consists of two or more statistical events. For convenience, I will speak of only two. These two events can occur in various ways, called the cases for each event. A particular set of possible cases provides a *representation* of the sub–systems; any given representation may or may not be orthogonal and complete in the sense that it allows us to distinguish (a) each of the cases and (b) the sub–systems.

We can think then of a particular manifestation of a statistical event as sampling from a population consisting of p copies of all the possible cases. The populations (in particular, the individual cases) for two or more statistical events are not necessarily distinct. The information available about the system is **(1)** the number of (not necessarily distinct) members of each population, **(2)** the relative frequency with which each case occurs, and **(3)** the representation system. — why q^2?

For example, we might know that there are 2^q cases and q^2 possible labels to represent those cases for each sub–system. The problem is then as follows: Is there a way of representing the system in terms of the 2^q cases with the q^2 labels (*i.e.* a map) of each sub–system such that (1) the maximum amount of information about an event in the total system is obtained and (2) any known constraints regarding relative frequencies, cases, or populations are respected?

The answer is sometimes yes, given information about the structure of the system. Following a constructive point–of–view, we are required to construct the statistics about the system from the cases for each of the statistical events.

Consider two statistical events A and B. Let these two statistical events jointly have a total of m cases (which we write as $(a) + (b)$) subject to two conditions: (1) The statistical events A and B are *statistically equivalent* in the sense that there is a mapping for which pairwise mappings of cases have the same case count. (2) They are also *independent*, however, in the sense that the (a) and (b) cases are distinguishable, *i.e.* assignable to A and B respectively, except for n cases. These

n cases must be assigned to either A or B arbitrarily.* For example if $n = 1$, we do not know whether A consists of either $\left(\frac{m}{2} - 1\right)$ or $\left(\frac{m}{2} + 1\right)$ distinguishable cases. Thus, whenever one of the n indistinguishable cases occur, we must recognize that a statistical event of both types (i.e. A and B) has manifested. So in this sense, n is a measure of the degree to which A and B are not independent. Note that the number of cases *must* be non–integral for odd numbers of cases. However, this is not a problem if, as here, we have assumed that A and B are statistically equivalent in terms of the number of case counts (condition (1) in the preceeding paragraph). If A and B are not statistically equivalent, then we do not simply divide m by 2 in the formula.

The necessity of assigning the n cases to A or B is a problem if we assume the independence of A and B. When we count cases, there is an implied assumption that we know what we are counting cases of ... that the cases are classifiable into those which should be counted for A and those which should be counted for B. We have no place for indistinguishable cases, and so must find a way to assign the n cases to either A or B. Since we know m and we know that A and B are identical except for the n cases, we know that if we assign n to either A or B, then $\left(\frac{m}{2} - n\right) + \left(\frac{m}{2} + n\right) = m$, and then the value of (a) is $\left(\frac{m}{2} - n\right)$ and that of (b) is $\left(\frac{m}{2} + n\right)$ or vice–versa.

As an example, consider a special die. The faces on this die are either green or blue with the exception of n faces. These n faces are turquoise. An omniscient observer can distinguish green and blue from turquoise. Our real observer is not even aware that the color turquoise exists and so always sees either green or blue, even when the face is turquoise. In this example, I assume that there is no particular bias toward either green or blue and so the real observer says a turquoise face is either green or blue with equal probability. This assumption of equal a

---

* Note that n is integer unless A and B are periodically repeating statistical events. Then it is possible that x cases out of every y repetitions of A and/or B are indistinguishable. Then we may take $\frac{x}{y}$ as the expectation value of n. If A and B are not strictly repeating, the relationship must be analyzed more carefully and the expectation value of n must be computed accordingly.

priori probabilities is the only reasonable one when there is no information to the contrary.

Suppose we have one such die so that there are 6 faces total with one turquoise face on the die, three of green (or blue) and two blue (or green). A green face is a favorable case for the statistical event A and a blue face is a favorable case for the statistical event B. Our real observer assumes that the die contain no turquoise faces and in fact that green and blue are equally distributed. Whenever our real observer throws a die, if the turquoise face comes up, it can be identified as either green or blue. If it is identified as blue, this simply re–establishes the statistical equivalence of A and B with $(a) = (b) = \frac{m}{2}$. A and B are treated then as completely independent. Note that this is equivalent to there being no turquoise faces at all. Otherwise one obtains $(\frac{m}{2} + 1)$ for (a) and $(\frac{m}{2} - 1)$ for (b).

We can made the situation more interesting, if less physical, by using two dice, one of which has a turquoise face and both of which are otherwise either totally green or totally blue. In this system, the turquoise face couples A and B. This couples the two dice when the judgment call is "bad"—e.g. identifying as green a turquoise face on an otherwise blue face die. When the judgment call is "good", each of the two die always contributes to either statistical event A or statistical event B (but never to both) and so they are independent in the context of A and B.

If there is no way of telling whether or not the turquoise face is on the "green die" or the "blue die", this situation corresponds to the kind of system described in the computation of the fine structure constant.[*] In the physical experiment, we are always looking for data regarding a coupled system (*i.e.* an "orbit" composed of two oscillations, one corresponding to the major axis and the other to the minor axis of an ellipse. By definition, the judgment call is always "bad". When the call is "good", we obtain a system that is partitionable into two statistically equivalent

---

[*] In the ordering operator calculus, the attribute distance between the two turquoise faces can be zero so that they are truly indistinguishable. This kind of coupling is pre–supposed in the analysis of the fine structure constant.

and completely separable sub–systems. We would call the results *noise* in the experimental system and throw out the data. Such data comes from the case where the "orbit" is circular.

To continue, for event A, the probability of one of the (a) cases manifesting is then $\frac{1}{(m/2)} * (\frac{m}{2} - n)$ and similarly for event B. We write these as (A') and (B'), where (A) would be the true probability of an A case were we able to deal with the partial independence of A and B. Under the assignment of n to either A or B, the events are once again independent. The compound probability (AB) of simultaneous occurrence of cases belonging to A and B is then

$$(A')(B') = \left[\frac{1}{(m/2)}\right]^2 \left(\frac{m}{2} - n\right) \left(\frac{m}{2} + n\right)$$

Given such a compound probability, under the (real observer's) assumption of complete identity and independence of A and B, it is natural to then compute the $(A) = (B)$ as

$$(A)^2 = [(A') * (B')]$$

This, on expansion and rearrangement give the "adding in quadrature" formula"

$$\left(\frac{m}{2}\right)^2 (A)^2 + (n)^2 = \left(\frac{m}{2}\right)^2 .$$

However, the physical way in which (A') and (B') are measured in a physical system may determine whether or not both (A') and (B') contribute. For the fine structure constant measurement, it is only (A'), corresponding to $(\frac{m}{2} - n)$ that contributes to the measured compound probability:

$$(A) = [(A') * (A')]^{1/2}$$

In the case of the computing the fine structure constant, we have $m = 4*127*15$ and $n = 1$. The factor of four comes from two independent events having $127 * 15$

cases each, any pair of which may be ordered as A then B or B then A, i.e. in two more ways.[†] Each repeated occurrence of the compound event A and B is a mutually exclusive case and it takes 137 such cases to have a Coulomb event. The total probability for a set of mutually exclusive and independent cases is then just

$$137 * \alpha = \left[\frac{1}{(m/2)}\right] \quad (A')$$

But (A') is just $(\frac{m}{2} - 1) = (2 * 15 * 127 - 1)$, so that

$$\alpha = \left(\frac{1}{137}\right) \left[1 - \left(\frac{1}{2 * 15 * 127}\right)\right]$$

The two statistical factors contribute to the combined event A AND B. By multiplying the (almost) independent probabilities, one obtains a formula for "adding in quadrature" on rearrangement of the terms. The formula is, of course, different if there are more than two almost independent events.

To obtain the general case, one must consider more than two statistical events (of number k) and all the various possibilities for judging the indistinguishable cases as being favorable to one or more of these statistical events. All possible assignments of the n indistinguishables must be taken into account. Whenever n is a multiple of k, there is the possibility of an equi–probable assignment to all k statistical events. Otherwise one must obtain terms in the relative frequencies which look like binomial coefficients. I hope to have an opportunity to spell out the general case in the near future.

---

† Note that in the general case of p multiple events the factor of two is replaced by the number of permutations of p events.

Comment on

"Statistical Mechanical Origin of the Entropy

of a Rotating, Charged Black Hole"

H. Pierre Noyes [a]

Stanford Linear Accelerator Center

Stanford University, Stanford, California 94309

Submitted to Physical Review Letters

---

[a] Work supported by Department of Energy contract DE–AC03–76SF00515.

Zurek and Thorne [1] have shown that the number of bits of information lost in forming a rotating, charged black hole is equal to the area of the event horizon in Planck areas, i.e., the *Beckenstein number* [2]. Wheeler [3] has suggested that this could be a significant clue in the search through the foundations of physics for links between information theory and quantum mechanics. In this comment we show that if one accepts the conservation of baryon number, as attested by the *experimentally* unchallenged stability of the proton, one can argue that the proton is a stable, charged, rotating black hole with baryon number +1, charge $+e$, angular momentum $\frac{1}{2}\hbar$, and Beckenstein number $N = \hbar c/Gm_p c^2 \simeq 1.7 \times 10^{38}$.

Consider an assemblage of N proton-antiproton pairs with all quantum numbers zero that contains an additional proton; this system has baryon number +1, charge e, and angular momentum $\frac{1}{2}\hbar$. Suppose the average distance between each pair is $\hbar/m_p c$. Then the gravitostatic energy E is

$$
E = \frac{NGm_p^2}{\hbar/m_p c} = N\frac{Gm_p^2}{\hbar c}\left(m_p c^2\right),
$$

which is equal to the proton rest energy when $N = \hbar c/Gm_p^2$. This is analagous to the bound $N_e = 137 \simeq \hbar c/e^2$ on the number of charged particle-antiparticle pairs established by Dyson [4] when he showed that the renormalized QED perturbation series in α is not uniformly convergent. No particulate constituent of the gravitational system we envisage can escape; the escape velocity exceeds c. Yet proton-antiproton pairs can annihilate to produce Hawking radiation [5], which is not necessarily bound to the system. The predictable endpoint of this system—granted baryon number conservation, charge conservation, and quantized angular momentum conservation—is a system with mass and conserved quantum numbers indistinguishable from those of the proton. Since this system started from $N = \hbar c/m_p c^2$ indistinguishable pairs, the number of bits of information *lost* in this way can reasonably be called "the Beckenstein number of the proton."

REFERENCES

1. W. H. Zurek and K. S. Thorne, *Phys. Rev. Lett.* **54**, 2171–2175 (1985).

2. J. D. Beckenstein, *Phys. Rev.* **D 7**, 2333 (1973).

3. J. A. Wheeler, "Information, Physics, Quantum: the Search for Links," in *Proc. 3rd ISFQM,* Tokyo, 1989, pp. 334–368.

4. F. J. Dyson, *Phys. Rev.* **51**, 631 (1952).

5. S. W. Hawking, *Phys. Rev.* **D13**, 191 (1976).

Comment on This Comment
H. Pierre Noyes (2000)

This attempt to get the Dyson argument as applied to quantum gravity into the mainstream physics literature failed. The argument is considerably more general than that given in the above note, as we now show. As discussed in Ch. 1, we interpret Dyson's argument that the renormalized perturbation series for QED is not uniformly convergent beyond $\hbar c/e^2 \approx 137$ terms as due to the formation of a pion with mass $\approx 2 \times 137m_e$. Here the mass of the electron occurs because it is the lightest stable charged particle and the factor of two because the mass corresponds to concentrating 137 electron-positron pairs within a radius of $\hbar/2m_e c$. Applied to an assemblage of gravitating particles of *any* mass m concentrated within a Planck radius which add up to one Planck mass, the same argument says that the number of such masses $N(m) = \hbar c/G_N m^2 = (M_{Planck}/m)^2$. In general, such an assemblage will be ephemeral because it decays by Hawking radiation — analogous to the neutral pion decaying electromagnetically. However if the assemblage contains in addition an electron and an electron-flavored neutrino, it will be stable against electromagnetic decay because of charge conservation and can only decay through the weak interaction. Analogously the proton-antiproton assemblage considered in the note, if it contains an extra proton can decay back to a proton, but no farther — unless some super-symmetric interaction providing it with a lifetime greater than 10^{35} years exists. This suggests that the proton is a charged, rotating black hole. The same applies to the electron because there is no stable, isolated charged elementary particle of lighter mass. By the same argument, if the lightest neutrino has finite mass, it will be a neutral, rotating black hole up to possible instabilities due to the (so far unobserved) super-symmetric interactions. In this sense, we have the starting point for a unified theory of strong, electromagnetic, weak and gravitational interactions (SEWGU). As has been pointed out by E.D.Jones, the fundamental nature of the Planckton in closing off the elementary particle scheme has significant cosmological implications and can be used to show that the "cosmological constant" has the right sign and approximately the right magnitude. This is discussed further in Ch. 16.

ANTI-GRAVITY: the key to 21st Century Physics[*]

H. Pierre Noyes

Stanford Linear Accelerator Center
Stanford University, Stanford, California 94309

Invited paper presented at the 14th Annual Meeting of the
ALTERNATIVE NATURAL PHILOSOPHY ASSOCIATION
Cambridge, England; September 3-6, 1992

[*] Work supported by the Department of Energy, contract DE–AC03–76SF00515.

Abstract

The masses, coupling constants and cosmological parameters obtained using our *discrete* and *combinatorial physics* based on *discrimination* between *bit-strings* indicate that we can achieve the unification of quantum mechanics with relativity which had become the goal of twentieth century physics. To broaden our case we show that limitations on measurement of the position and velocity of an individual massive particle observed in a colliding beam scattering experiment imply real, rational commutation relations between position and velocity. Prior to this limit being pushed down to quantum effects, the lower bound is set by the available technology, but is otherwise *scale invariant*. Replacing force by force per unit mass and force per unit charge allows us to take over the Feynman-Dyson proof of the Maxwell Equations and extend it to weak gravity. The crossing symmetry of the individual scattering processes when one or more particles are replaced by anti-particles predicts both Coulomb attraction (for charged particles) *and* a Newtonian repulsion between any particle and its anti-particle. Previous quantum results remain intact, and predict the expected relativistic fine structure and spin dependencies. Experimental confirmation of this anti-gravity prediction would inaugurate the physics of the twenty-first century.

1. WE NEED A NEW STRATEGY

The ANPA program has achieved a number of quantitative successes in calculating most of the fundamental mass ratios, coupling constants, and cosmological parameters needed in elementary particle physics and physical cosmology. These are summarized in the Table which concludes this paper. One might think that this success, which conventional "theories of everything" are aiming at, but have no more than vague ideas as to how to accomplish quantitatively, would provoke some interest among physicists and cosmologists. Yet when I finally succeeded in getting a short announcement published in the magazine that goes to all US Physicists,[1] I only got *one* enquiry — a brief "What the hell is going on?"! I replied with technical details, but got no response. I have also tried to involve several elder statesman of my acquaintance, but among the theorists have garnered little interest.

The stock response is "Predict something that hasn't been observed." Of course no other theorist is doing that in our sense. They usually take a generous amount of both structure and parameters from existing experiments and attempt to compute a correction or two that might be observable. To play that game, we have to analyse nearly half a century of theoretical and experimental work by thousands of the best physicists in the world and recast it in our own terms. I am making progress along these lines, but without the help of eager and conventionally trained colleagues, cannot hope for any rapid developments. Even the top quark mass is getting pinned down; this is the last well defined parameter that could be predicted prior to experimental observation. Improvement on the values of the Kobiyashi-Maskawa mixing angles will get us little recognition. There are enough theories of neutrino masses around to insure that one of ours would be bound to have a successful conventional alternative. About all we can do is make the negative prediction that there should be no Higgs mesons with simple structure. However, we can expect many of the effects which will be used to "discover" Higgses are also contained in our theory. A clean discrimination between our theory and conventional alternatives will take even more work than getting the K-M parameters right, and will require many of the same steps.

Of course some members of the ANPA community resist the idea that we are trying to construct a new physical theory to be evaluated using the same criteria as those employed by the establishment. But, at least to the jaundiced eye of this physicist, I see no evidence that conceptual clarity and philosophical purity will do us much good. Many mathematicians and computer scientists claim just that, and I see no likelihood that we will stand out among the host of competitors. Physicists pay no attention to that vast body of literature in any case. Those of us who want to convince *physicists* that the ANPA program has led to exciting new results will have to find a new strategy.

Fortunately, experimental high energy physicists are more willing than the theorists to entertain new ideas. They are properly distrustful of theorists, and unhappy with an experimental situation that relies so heavily on intricate theoretical calculations to "measure" *anything*. Maurice Goldhaber was intrigued with the idea that anti-matter might fall up, and suggested looking into tests with muonium and anti-muonium. Direct free fall tests are impossible because of the 2.2 microsecond lifetime. But he is looking into another possibility, which — so far — he is keeping close to his chest. The experimental groups working with anti-protons and trying to produce anti-hydrogen atoms have a lively interest in the prediction that anti-protons will fall up which Starson and I made at ANPA WEST 7.[2] I have recently reviewed their experimental programs.[3]

Making a case for anti-gravity has provoked more interest in the ANPA program among physicists than anything else to date. I propose to follow it up vigorously. I believe that even if the prediction fails, we will get more constructive attention from that failure than from any improvement in our quantitative predictions, no matter how impressive. But even within ANPA, I have failed to elicit any attempts to improve on or to refute my arguments for anti-gravity. I hope that this paper will stimulate or provoke some constructive criticism.

2. BOHR-ROSENFELD REVISITED

Bohr and Rosenfeld[4,5] proved that the restrictions on measurability due to the *non-relativistic* uncertainty principle applied to the charges and currents which detect the fields can be used to *derive* the commutation relations between the electric field E and the magnetic field H which are more easily obtained by the legemanderain of "second quantization". Basically, this is possible because the theory involves only h and c, leaving it *scale invariant*. This allowed them to use as complicated an apparatus as they liked *within* a wavelength of the radiation. Their apparatus consisting of rigid rods and springs, massive charged objects and current loops. The rods and springs are used to compensate, in so far as possible, for radiation reaction, and can get pretty complicated. One post-doc who reviewed the paper at a SLAC seminar called it an exercise in nineteenth century electrical and mechanical engineering!

In the course of preparing the final version of my paper for PIRT III,[6] I came to realize that the non-commutativity of position and momentum measurements made using macroscopic counters can be cast in a scale-invariant form by making angular momentum per unit mass (area change per unit time) the basis of quantization rather than angular momentum. Then the units of quantization of length and time and the measurement of mass ratios depend only on space-time measurements. For instance they can be related to the smallest measured velocity in units of c, and a scale invariant quantity set by technological assumptions. Thus the basis for the Bohr-Rosenfeld argument can be recast *without* introducing Planck's constant provided only that the sources and sinks of the field are relativistic charged particles. This removes the restriction of their paper to non-relativistic quantum mechanics, which is obviously desirable.

Once one has scale invariant commutation relations between position, velocity and angular momentum per unit mass, the very peculiar proof of Maxwell's Equations which Feynman showed to Dyson in October, 1948 but refused to publish during his lifetime[7] becomes understandable. I had argued elsewhere that this derivation is rigorous within the framework of bit-string physics.[8,9] The new derivation presented here is scale invariant, making the proof even more general. It therefore seems worth while to present this new result before discussing the gravitational field.

3. PROOF OF THE MAXWELL EQUATIONS

The Feynman-Dyson proof of the Maxwell Equations starts with Newton's Second Law

$$m\ddot{x}_j = F_j(x, \dot{x}, t); \ j \in 1, 2, 3 \qquad\qquad D-1$$

and the commutation relations

$$[x_j, x_k] = 0 \qquad\qquad D-2$$

$$[x_j, \dot{x}_k] = i\hbar\delta_{jk} \qquad\qquad D-3$$

and proves that there exist fields $E(x, t)$ and $H(x, t)$ satisfying the Lorentz force equation

$$F_j = E_j + \epsilon_{jkl}\dot{x}_k H_l \qquad\qquad D-4$$

and the Maxwell Equations

$$div\ H = 0 \qquad\qquad D-5$$

$$\frac{\partial H}{\partial t} + curl\ E = 0 \qquad\qquad D-6$$

The proof relies on the Jacobi identities and taking a total derivative with respect to time, but involves no formal subtleties.

Because our theory is relativistic, we measure all speeds in units of c. Since, by definition, $c = 299\ 792\ 458\ m\ sec^{-1}$ we can always pick our dimensional scales in such a way that these speeds for any massive particle in any one direction are rational fractions less than unity.

Our first step is to eliminate the concept of mass from the problem in favor of mass ratios measured relative to some standard particle beam using only space, time, velocity and velocity change measurements. For this purpose we use *counter telescopes* consisting of two *counters* with thickness Δl containing recording clocks having the time between ticks Δt. We pick our units such that $\Delta l = c\Delta t = 1$, making all measurable distances and times *integers*. This commits us to insuring that we never talk about fractional space and time intervals as *measurable*. If the spacial interval between the counters is L and the time interval between two sequential counter firings is T we attribute the counter firings to the passage of a particle with velocity $V = L/T$. All data discussed here will be collected at a slow enough rates so that the interval between the passage of particles allows this measurement to be unambiguous. We also assume that, to an accuracy to be discussed, all four of the telescopes introduced below (eight counters) record the *same* speed.

Although, by hypothesis, $V = L/T$ must be a rational fraction less than unity, we will not in general be able to measure L and/or T to the nearest integer. To represent this fact we define

$$v = \frac{t_1' - t_1}{t_1' + t_1} = V = \frac{L}{T}; \quad L = N(t_1' - t_1); \quad T = N(t_1' + t_1) \tag{3.1}$$

We assume that t_1' and t_1 *are* known to the nearest integer and that N can be estimated but not directly measured. By interference techniques we do not have time to discuss in this paper, one can measure relative path lengths and determine N to the nearest integer. What remains unobservable is the time t in any interval $0 \le t = n_t \Delta t \le t_1 + t_1'$. In this finite and discrete language, we are talking about a periodic phenomenon with N periods, each of duration $T(v) = t_1 + t_1'$ whose *absolute* phase is unknown within a period. The fraction $t/T < 1$ is our conceptual equivalent of the unobservable phase of quantum mechanics in a *scale invariant* context bounded from below by measurement accuracy rather than something related to Planck's constant.

Our paradigm for position and velocity measurement is to use four counter telescopes 11',22',33', 44' all pointed at the same region. To a first approximation, the lines 11',etc. all pass through a "circle of confusion" X of radius unity. We assume that the first two counter telescopes fire in the sequence 11'22' and the second pair in the sequence 33'44', and that 1'23'4 lie on a circle of radius d centered at X. We assume that the lines 1'2 of length b and 3'4 of length b' are parallel and are bisected by a line perpendicular to them through X. That is we have two isosceles triangles with a common vertex and parallel bases. Calling their inferred heights h, h' and angles $\pi - \theta$, $\pi - \theta'$ we have, as a first approximation,

$$h^2 = d^2 - \frac{1}{4}b^2 = 2d^2 sin^2\frac{\theta}{2} = h^2[(\frac{d}{b})^2 - \frac{1}{4}]$$ (3.2)

and similarly for h', θ'. We take as our coordinate directions j parallel to the altitudes h, h' and k parallel to the bases b, b', and assign coordinates (x_j, x_k) as follows:

$$1' : (0,0); \quad X : (h, b/2); \quad 2 : (0, b)$$ (3.3)

We now relate this geometry to the common velocity v registered by all four counter telescopes and the fact that we can only measure times to an accuracy $\Delta T = t_1 + t'_1$. Then

$$1'X = X2 = d = 2v\Delta T; \quad 1'2 = b = (\frac{b}{d})2v\Delta T$$ (3.4)

and the velocity components are

$$v_j^{1'\rightarrow X} = (\frac{h}{d})v; \quad v_j^{X\rightarrow 2} = -(\frac{h}{d})v; \quad v_k^{1'2} = (\frac{b}{2d})v; \quad v_j^2 + v_k^2 = v^2$$ (3.5)

In order to make this into a scattering experiment, we assume that each time we get the sequence of firings 11'22' we also get the sequence of firings 33'44'. For sufficiently weak beams, this will be unambiguous. We attribute this confluence of events to the scattering of one particle from each beam within the region X. We can then *define* the ratio of the mass m' of the particles in the second beam to the mass m of the particles in the first beam by the equality

$$m'b' = mb \qquad (3.6)$$

It is a matter of experience that the *scale invariant* equality $m'/m = b/b'$ is independent of the commom measured velocity v for all known pairs of particles which can be compared in this way and hence defines a velocity invariant scale for all particles relative to any one type. Note that we make the comparison at the *same velocity* to avoid the complications of relativistic kinematics. This is why our theory can remove the puzzle stressed by Dyson that the Feynman derivation seems to produce peaceful coexistence between Newtonian and non-relativistic quantum mechanics and the Lorentx invariant Maxwell equations they seem toimply.

This step allows us to replace forces — which historically related masses compared inertially to masses compared using weight — by mass ratios using the relativistic equivalent of Newton's Third Law, a step we freely acknowledge was suggested to us by Mach in his *Science of Mechanics*. The advantage of using this macroscopic and operational *change* in the velocity $2v \, sin^2 \frac{\theta}{2}$ is that we can measure both the magnitude v and the scattering angle θ using macroscopic counter telescopes. Although we have described this situation as if the particles met at a point, all we can measure macroscopically are the scattering angles θ, θ' and the common rational fraction velocity to some finite accuracy. Thus, the "interaction" could well be *non-local*. As we have shown elsewhere,[10] this description is invariant under appropriate rational velocity boosts and finite angular rotations. We claim this is the appropriate starting point for a *scale invariant relativistic action at a distance* theory. Comparison of the different mass beams at the *same* speed allows us to defer discussion of relativistic kinematics to a later point in the development.

We can now start the proof, replacing Dyson's F_j by $f_j \equiv \frac{F_j}{m}$. We use a common time interval ΔT to measure all velocities $\dot{x}_i \equiv v_i$ and all velocity changes

Δv_j (rather than discussing accelerations $\ddot{x}_j$). Then Newton's Second Law, (D-1), becomes

$$f_j = \Delta v_j \qquad\qquad N-1$$

Because the eight counters occupy fixed positions, the coordinate $x_k = b/2$ of the midpoint between 1' and 2 and between 3' and 4 can be chosen independent of how we decide to interpret the measurement of position and velocity "at" X. Therefore Dyson's postulate (D-2) holds for us as well.

Our next step is to view this impulsive velocity change as a *measurement* of position x_j and velocity v_j of the particle in the first beam, to the limited accuracy allowed by the "circle of confusion" around X produced by our assumption that we cannot give meaning to distances less than Δl and times less than Δt. We have defined our units so that this is a circle of radius 1, until we fix the x_k coordinate by (D-2). Then it becomes a line segment of length one between two coordinates $x_j^1 = h_1$ and $x_j^2 = h_2$. The lines from either h_1 or h_2 to either counter will have lengths

$$d_1^2 = h_1^2 + b^2/4; \ d_2^2 = h_2^2 + b^2/4 \qquad\qquad (3.7)$$

If d_1 runs from 1' to h_1, the velocity

$$v_j^{1'\to h_1} = +(\frac{h_1}{d_1})v \qquad\qquad (3.8)$$

which differs from the v_j in 3.5 by $[(h_1/d_1) - (h/d)]v$. Since our measurement philosophy does not allow us to assign the momentum change to a point, the line d_2 must then run from h_2 to 2 and the corresponding velocity component is

$$v_j^{h_2\to 2} = -(\frac{h_2}{d_2})v \qquad\qquad (3.9)$$

which differs from the v_j in 3.5 by $[(h/d) - (h_2/d_2)]v$. We can now attribute the position measurement to position h_1 followed by a velocity measurement at position

h_2, i.e. $x_j = h_1$, $|v_j| = (h_2/d_2)v$, or in the opposite order, i.e. $|v_j| = (h_1/d_1)v$, $x_j = h_2$. Hence $[x_j, v_j] = \frac{h_1 h_2 (d_1 - d_2) v}{d_1 d_2} \neq 0$ It is of little interest what this constant is. All we need in this paper is that we can replace Dyson's (D-3) by

$$[x_j, v_j] = C \delta_{jk} \qquad\qquad N - 3$$

where C is fixed by the accuracy achieved or postulated in the technology of scattering measurements. Examining the remaining steps in the Dyson proof, we find that they only require the commutator to be a *constant* independent of x_j, v_j, and *not* on this constant being imaginary. Therefore N-3 is a satisfactory replacement for D-3 and removes Planck's constant from the problem altogether, provided only we do not encounter explicit quantum phenomena, such as the quantization of radiation independent of measurement accuracy below some threshold.

Equation D-4, interpreted as the force on a particle of charge e is a force per unit charge rather than a force per unit mass. But as used to be well known, using only macroscopic measurements of particle trajectories with static electric and magnetic fields gives us only e/m and not e or m separately. So, once again we can make a scale invariant choice of units such that F_j in (D-4) is the same as f_j in (N-1), leaving us with

$$f_j = E_j + \epsilon_{jkl} v_k H_l \qquad\qquad N - 4$$

We now have in hand all the ingredients necessary to carry through steps (D-9) to (D-21), the final step in the Dyson proof, which has now become fully algebraic. The algebra is uninformative and will not be reproduced here. The only subtlety is the interpretation of total and partial derivatives as discrete differences along the lines of our derivation of the free particle Dirac equation. We will return to this problem on another occasion. One significant fact about the algebra used in the proof is that it no where makes use of the imaginary equation $i^2 = -1$. McGoveran has remarked that the "i" in quantum mechanics a "book keeping device" of no deep significance. We have provided here a specific example of how this observation can be illustrated.

4. QUANTIZED CONIC SECTIONS

Our approach to the Bohr-Sommerfeld problem[11] starts from our basic postulate that events can, but need not, occur only when they are an integral number of deBroglie wave lengths apart. We can then approximate circular orbits by an integral number of straight line segments representing velocities $\beta_n = 1/137n$ and the closure constraint $2\pi r = j\lambda$

In that approach, we take double slit interference as primitive rather than macroscopic velocity change. If a beam of particles of some velocity v is incident on a slit with spacing w followed by a detector screen a distance D downstream and the spacing between interference fringes on the screen is s, the deBroglie wavelength λ is given in terms of laboratory length standards by the relation

$$\lambda = \frac{ws}{D} \tag{4.1}$$

Then if a beam of particles with a different mass m' but the *same* velocity v is incident on the same arrangement and produces a fringe spacing s', we can *define* the mass ratio by

$$m's' = ms \tag{4.2}$$

The similarity to Eq. 3.6 should be transparent. In discrete physics we think of 3.6 (Newton's Third Law) as *derived* from our quantum mechanical relation 4.2. The advantage of using 3.6 as basic is that we can then treat the classical theory as possessing non-commutativity, but in a scale invariant way down to the point where an *absolute* measurement of $\hbar$ or e has been made.

Once we have recognized that λ, in an appropriate context, continues to represent the minimal measurable distance between distinct events our "scale invariant" treatment of scattering limited by accuracy of measurement can be applied to any macroscopic problem which is on a scale such that Planck's constant does not need to enter the analysis. Our analysis then provides a "correspondence limit" for *relativistic* quantum mechanics in all such cases. In particular, we claim to have proved in the last chapter that Maxwell's Equations can be reinterpreted as a *necessary* consequence of any relativistic action at a distance theory which is careful to incorporate macroscopic limits on measurement, and hence *also* as the correspondence limit of our relativistic particle theory.

Since the primary focus of this paper is on gravitation in the macroscopic limit, our primary interest is in elliptical and hyperbolic orbits rather than in radiation. We will return to quantum effects in Chapter 7, but here need only the connection between Rutherford scattering analysed classically, the analagous problem for gravitating objects, and the relationship between hyperbolic and bound orbits. At the level of analysis we need to establish anti-gravity we can ignore both the gravitational "fine structure" splitting and the loss of energy due to radiation. Then the Coulomb problem differs from gravitation only in that (a) the coupling constant is much weaker and (b) there is currently no empirical evidence for anti-gravity (i.e. hyperbolic orbits corresponding to short range repulsion).

Since I spent some time at ANPA 13[12] discussing the relationship between Galileo's pendulum experiment, Newton's circular orbit paradigm, circular velocities indistinguishable from c (black holes) and our quantum theory, I will defer a detailed "scale invariant" treatment along the lines sketched above to another occasion. I simply note that if we take r_1 as the perihelion distance, $|r_1 - r_2|$ as the distance between foci and $\beta = (k_1 - k_2)/(k_1 + k_2)$ as the velocity at perihelion, we can construct a quantized theory of "conic sections" in terms of three of the four integers r_i, k_i. We then specify the fourth in terms of a macroscopic scale parameter such as the maximum path length of the periodic bound state orbits we use to establish our time resolution, or the maximum size of the scattering chamber we use to measure scattering angles. This in turn can be used to specify the accuracy in the ratio of asymptotic to perihelion velocities we can measure.

5. CROSSING SYMMETRY PREDICTS ANTIGRAVITY

In our discussion of the Maxwell Equations, we made use of a scattering chamber with two entrance and two exit counter telescopes for two particle beams of different mass. This gives us eight counter firings and four velocities determined by space-time measurements. If all four particles are charged, and we back these up by measurements of the radius of curvature in a magnetic field backed up by a third counter to insure that the same v is still valid we end up with 16 pieces of information, and can use these to form four energy-momentum 4-vectors which are conservedpairwise between the initial and the final states. These allow us to define our discrete version of the usual Mandelstam variables[13]. Then *any* 2-2 scattering process which can be reduced to a finite number of convergent Feynman diagrams can be calculated for our discrete variables, which have the limits of measurement already built in. Details will be presented elsewhere.

One important fact about 2-2 Feynman diagrams expressed in terms of relativistically invariant variables and quantum numbers is that they are *crossing symmetric*. Suppose we have a diagram that represents a process in which a particle of mass m_1, energy E_1, momentum P_1, angular momentum J_1, and some collection of discrete quantum numbers Q_1 interacts with particle 2 similarly described to produce two particles 3 and 4 similarly described. Crossing symmetry asserts that, where the free particle kinematics of the initial and final free particle states allow, the same diagram with one, two, three or four particles changed to anti-particles represents a physical process whose probability amplitude can be computed from the same diagram in that appropriate kinematic region.

In our bit-string theory, this crossing symmetry derives from the fact that, if we make the proper identification between quantum numbers and kinematic variables derived from bit-strings, interchanging 0's and 1's in a bit-string corresponds to interchanging particle and antiparticle. In particular, this is true of our representation of the standard model of quarks and leptons using strings of 16 bits, although the published demonstrations of this statement are incomplete. If we interchange the 0's and 1's in *all* the strings for a theory in which the combinatorial hierarchy construction has closed, we produce a dual theory which is formally distinct but which is indistinguishable so far as *all* physical predictions go. I have called this Amson invariance. In conventional theories this is the CPT theorem: changing all particles to anti-particles, reversing their velocities ($P_i \rightarrow -P_i$), and making a mirror reflection across three perpendicular planes ($J_i \rightarrow -J_i$) can have no observable consequences. In particular, this theorem requires particles and anti-particles to have identical *inertial* masses. But in the absence of an accepted theory of quantum gravity, gravitational mass (or better "gravitational charge") could either reverse or stay the same.

It is important to realize that crossing symmetry is more restrictive than CPT invariance. For instance, since we know that protons fall toward the earth, all it says that anti-protons fall toward an anti-earth. This is not helpful for constructing an *experiment crusis*! But crossing symmetry applied to the coulomb problem tells us that anti-particles have opposite electric charge to particles and hence that if a particle is attracted toward a center, an anti-particle will be repelled by it. This follows immediately from the conic section formalism we have developed. But for gravitation, the definition of inertial mass remains the same as for coulomb attraction, and the same crossing symmetry applies. Hence, since particles are known to attract each other gravitationally, a particle and its anti-particle should repel each other. *Our prediction of anti-gravity is that simple.* It remains to try to meet objections.

6. THE CONVENTIONAL WISDOM*

To begin with, our prediction is in flat contradiction with the equivalence principle (i.e. that there is no way to detect a difference between gravitational and inertial mass) and hence with General Relativity. For many physicists this is already sufficient reason to dismiss anti-gravity out of hand. Only particle theorists and others who believe in CPT invariance will pursue the matter further. But the usual context in which CPT invariance arises is in the second quantized relativistic field theory. In such theories the electromagnetic field has massless quanta with spin 1 while gravitation has massless quanta with spin 2. There is a general argument that, although the force between two particles which exchange spin 1 quanta is repulsive between a pair of particles or a pair of anti-particles, and attractive between a particle antiparticle pair, it is always attractive between *any* two systems which exchange spin 2 quanta.

However, if one looks at the "proof" of this theorem in more detail, one finds that it does not just depend on the spin of the quanta.[14] In the case of any pair of particles which interact by exchanging massless quanta with integral spin j (in our case $j=1$ or 2) the momentum change p (or force) must vanish like p^j as p goes to zero. This would be a disaster for the conventional theories, because the major effect observed for small p in electromagnetism is the Coulomb or electrostatic force between charges. For gravitation the only directly measured force is ordinary Newtonian gravity. The spin-2 "gravitons" which the theory predicts cannot be directly detected, and whether classical gravitational radiation has been detected or not is controversial. The way conventional theory gets around this disaster is to insist that the theory be gauge invariant as well as Lorentz invariant. The low momentum limit— if one believes the somewhat tricky mathematics — then produces the desired Coulombic and Newtonian forces out of this theorists hat. But, unlike fields which have a direct connection with the observed motions of test particles, "potentials" whether "gauge" or other, have no directly observable

---

* Quoted from Ref 3, with some modifications.

consequences. One is permitted to view them as theoretical inventions, rather than as a transcription of empirical fact into mathematics. I made the technical argument at the Münich Workshop on anti-hydrogen in April, 1992.[15]

The end conclusion is that *if* anti-protons "fall" up, one will have to abandon *both* the equivalence principle (i.e. gravitational mass is identical to inertial mass) *and* relativistic gauge invariance. Such an experimental result would kill two theories with one measurement, which is a good investment when one is looking for a crucial experiment. Fortunately experimentalists are not deterred by theoretical arguments, and are forging ahead as carefully as they can. We may have the answer in five years.

7. QUANTUM CONSIDERATIONS

Until the last chapter we could ignore "spin" because our fine structure comes from our relativistic analysis of the limitations of measurement and does not depend on the existence of an indivisible unit for orbital angular momentum. Historically it was Bohr's quantization of angular momentum via the quantization of energy that gave a first version of quantum mechanics. His relativistic treatment was an after thought, and Sommerfeld's successful extension did not require the concept of spin.

Our relativistic treatment of Kepler's Laws shows that we can define an impact parameter from the relation $2\pi r = j\lambda = jh/p$ and that if we define $j = r/\lambda$, we have that the square of the area swept out in the time it takes to move λ is $\Delta A^2 = \lambda^4(j^2 - \frac{1}{4})$. Geometrical examination of the alternatives $\ell = j \pm \frac{1}{2}$ shows that they correspond to the straight line of length λ being taken as the tangent or the chord, respectively, showing that $(\Delta A/\lambda^2)^2 = \ell(\ell+1)$ gives the quantum mechanical result (i.e. "ℓ^2" $\rightarrow \ell(\ell + 1)$ for the square of the orbital angular momentum ℓ because we are taking the geometric mean between the distinct values computed from the inscribed and circumscribed polygon. This is consistent, because the maximum linear distance between them is h/mc. If we tried to measure this difference to the accuracy of $\pm h/2mc$, we would be able to create a particle anti-particle pair, and would have to include their degrees of freedom in the analysis before we could proceed.

The same analysis shows that a transition between the two possibilities changes ℓ by $\hbar$, which is equivalent to the spin-flip transition between $\pm\frac{1}{2}\hbar$ and $\mp\frac{1}{2}\hbar$. So interchange of massless spin 1 quanta interacting with an orbiting spin $\frac{1}{2}$ particle, with probability reduced by a factor of $1/137$ compared to the Coulomb interaction, is consistent with our picture. We get the same results as QED to order $1/137^2$ without any need for gauge invariance.

The same argument shows that the Coulomb interaction, which only depends on the direction toward the attracting center and the local acceleration its field produces is spin independent and velocity independent, while the spin flip transition depends on either the traveling photon interacting with the moving charge or the magnetic field produced by the center acting on the moving charge depending on which description you wish to use. Thus we can distinguish electric from magnetic forces as static or velocity dependent as we did in Maxwell's Equations or as spin dependent or spin-independent in the quantum theory. The pictures support each other.

When we come to gravity, the positive protons, negative electrons and neutral neutrons all attract each other as well as particles of the same mass with thestandard Newtonian interaction. The velocity dependent forces only show up in the bending of starlight by the sun and the precession of the perihelion of Mercury. As we have argued elsewhere, both effects are explained by spin 2 gravitons.[16] In terms of spin, this is explicable if, as before the Newtonian term is spin independent, while the spin dependence (down by $GMm/\hbar c$) allows only the five distinct triplet-triplet transitions. In terms of velocity dependence this implies the extreme non-locality of coupling the motion of both objects by two velocity dependencies. This is, of course, another way of saying that the interaction is scalar (i.e. Newtonian) -tensor and distinct from the scalar (i.e. Coulombic) -vector electro-

340

magnetic interaction. All of this fits neatly into the crossing symmetry argument for anti-gravity and hence reinforces it.

8. PRINCIPLES AND RESULTS OF MY APPROACH

In order to summarize the position I take with respect to the establishment, I quote from a recent letter to a colleague:

Our principles are finiteness, discreteness, finite computability, absolute non-uniqueness* and our procedures must be strictly constructive. For us, the mathematics in which the Book of Nature is written is finite *and* discrete. We model nature by *context sensitive* bits of information. In this sense we are participant observers.

Physics, as a science of measurement, can expect that at least some of the structures uncovered in nature could result from the way we perform experiments. For example, Stillman Drake[17] has discovered that Galileo measured the ratio of the time it takes for a pendulum to swing to the vertical through a small arc to the time it takes a body to fall from rest through an equal distance as $948/850 = 1.108\,2....$ We now compute this ratio as $\pi/2\sqrt{2} = 1.110\,7....$ Thus Galileo *measured* this constant to about 0.3 % accuracy (Ref. 13).

We now believe that this constant will be the same "anywhere that bodies fall and pendulums oscillate" independent of the units of length and time.

In any theory satisfying our principles which counts events by a single sequence of integers, any metric when extended to large counts can have at most *three* homogeneous and isotropic dimensions in our finite and discrete sense.[18] More complex degrees of freedom, indirectly inferred to be present at "short distance" automatically "compactify". Hence we can expect to observe at most three absolutely conserved quantum numbers at macroscopic distances and times. Guided by current experience, we can take these to be lepton number, charge (or the z-component of weak isospin), and baryon number. These are reflected in the experimentally

---

* eg. In the absence of further information, all members of a (necessarily finite) collection must be given equal weight.

uncontroverted stability of the proton, electron and electron-type neutrino. This choice is empirical but not arbitrary, since structures with appropriate conservation laws isomorphic with this interpretation arise in our construction.

Take the chiral neutrino as specifying two states with lepton number ± 1 and no charge. They couple to the neutral vector boson Z_0. In the absence of additional information, these states *close*. The 4 electron states couple to two helical gamma's and the coulomb interaction. These seven states can be generated by any 3-vertex which includes two electron states and an appropriate gamma. These $3 + 7 = 10$ states when considered together then generate the $W^\pm$. This completes the leptonic sector in the first generation of the standard model of quarks and leptons. Bit-strings of length 6 provide a compact representation of these states which *closes* under *discrimination* (exclusive-or), and conserves both lepton number and the z component of weak isospin at each vertex. No unobserved states are predicted at this level of complexity, and no observed states are missing.

Two flavors of spin $\frac{1}{2}$ quarks and three colored gluons provide the seven elements of the baryonic sector which generate the inferred 127 quark-antiquark, 3 quark, 3 antiquark, 8 gluon ... states (16 fermions times a color octet minus the state with no quantum numbers) needed for the "valence level" description of the quark model. Bit-strings of length 8 provide a compact model using seven *discriminately independent* basis strings and again close producing only the appropriate states at this level of complexity. Combining them with the leptonic states allows the strings representing the vector bosons to be extended to length 14, producing all the vertices and only the vertices which occur in the standard weak-electromagnetic unification of the first generation of the standard model. Extending the whole scheme to strings of length 16 we get the three generations which are observed experimentally (and a slot with the quantum numbers of the top quark). The quarks have baryon number 1/3 and charges $\pm 1/3, \pm 2/3$ as required. The $0 \leftrightarrow 1$ bit-string symmetry makes CPT invariance automatic. As already noted, if we have only three large distance quantum numbers color (although conserved) is confined, and generation number is not conserved in weak decays.

We are now in a position to talk about the 137. *Empirically* only one of the 137 states required by the standard model of quarks and leptons corresponds to the coulomb interaction. Hence, by our principle of absolute non-uniqueness, the probability of this interaction occurring is 1/137 in the absence of further information.

Our basic quantum mechanical postulates are that (a) the square of the invariant interval between two events connected by a "particle" which carries conserved quantum numbers and conserved 3-momentum between them, is the product of two integers times $(h/mc)^2$ and that (b) space-like correlations for particle states with the same constant velocity can occur only an integer number n_λ of deBroglie wavelengths ($\lambda = h/p$) apart. These give us relativistic kinematics and the usual commutation relations for position, momentum and angular momentum.

If we model the hydrogen atom by events a distance r from a center we must have $n_\lambda \lambda = 2\pi r$. This interpretation is supported by noting that if the radius vector sweeps out equal areas in equal times, $\Delta A/\lambda^2 = (n_\lambda^2 - 1/4)(1/2\pi)^2$ and with $\ell = n_\lambda - 1/2$, the angular momentum is $\ell(\ell + 1)\hbar^2$. Since these events occur with probability $1/137n_\lambda$, we get (Ref. 11) the relativistic Bohr formula[19] for the hydrogen spectrum. When we include a second degree of freedom, and take proper account of the ambiguities in counting, we get not only the Sommerfeld formula but the formula for α to which you object. Similarly, the fact that the basic Fermi interaction involves 16 possible states of four fermions gives us $\sqrt{2}G_F = (256m_p)^{-2}$ where the square root comes from the conventional interaction Lagrangian to which experimental numbers are compared, and m_p comes from the stability of the proton.

I am willing to grant that the original Amson, Bastin, Kilmister, Parker-Rhodes sequence $3,10,137,2^{127} + 136$, STOP —discovered in 1961 after a decade of disciplined research — does *sound* like numerology. That was my own first response. I was willing to think there might be something to it *after* I had used the Dyson argument to identify the last two numbers as the maximum number of charged particle pairs or baryons one can *count* within the Compton wavelengths $h/2mc$ or $h/m_p c$ by, respectively, electrostatic or gravitostatic means. In fact one of my research objectives until the mid 1980's was to find a way to kill the theory and get on to something more promising. What convinced me that the evolving construction could be the starting point for a *new* physics and physical cosmology was McGoveran's calculation of the Sommerfeld formula and correction to α plus the fact that the same arguments applied to other coupling constants consistently improved agreement with experiment. I really don't think it fair any longer to call our theory "numerology".

When you assert that the dielectric constant of diamond can be calculated from first principles, you must assume (correct me if I am wrong) that you already *know* a number of physical constants. Of course one can relate the standards of mass, length and time as measured in the laboratory to three dimensional constants (which could be $c, \hbar$ and G) that occur, self-consistently, in several structures derived from "first principles". But to get to diamond you will also need $\alpha, m_e,$ **and** M_C in well defined relation to those units. Otherwise your calculation has no potential empirical test.

You must admit that, in your framework, these three numbers are too complicated to calculate from first principles. In fact, when Weinberg discusses *how* a finite coupling constant might emerge from currently acceptable theory, his errors are so large that he cannot even contemplate a quantitative prediction that can be confronted by experiment. In contrast my values for α, and m_e are good to six or seven significant figures, and I can argue that my "first principles" allow me to predict that the common isotopes of carbon will have masses of approximately 12 and 13 proton masses. I have systematic ways of improving these estimates, and also— thanks to my physical cosmology — of estimating the relative abundance of these two isotopes on a terrestrial-type planet with an age of 4.5×10^9 years in a solar system of the kind in which we are conducting experiments. Somewhere along this line my calculation from "first principles" would find empirical supplements useful, but I believe no where near as soon as yours.

344

I would locate the difference in point of view between us as coming from our different views of "space-time". If the "quantum vacuum" (which I would prefer to call a "quantum plenum") of renormalized second quantized relativistic field theory is the underlying concept, its properties certainly change as you "squeeze" it. The received wisdom today is that if the squeezing produces an energy density something like 10^{16} times that of the proton the "strong", "electromagnetic" and "weak" interactions come together (one basic "coupling constant" — grand unification) and that if one can extend the theory another three orders of magnitude, gravitation will find its appropriate place in the scheme. It seems to me that adopting "principles", however beautiful, that force one to go thirteen orders of magnitude beyond currently possible experimental tests to define fundamental parameters is — to say the least — a peculiar methodology for a physicist.

On the other hand, if one starts here and now with separated charges and massive particles and "empty" or "constructed" space as the first approximation, one can *measure* masses and coupling constants in a well defined way. If one can — as we claim — get good approximations for these values from "first principles" and systematically improve the predictions, I fail to see why such values cannot be considered "primordial". After the universe becomes optically thin, we predict about 2×10^{-10} baryons per photon. This both is in agreement with observation and supports our "empty space" philosophy.

I have recently succeeded[20] in deriving the solutions of the free particle Dirac equation by summing the "vacuum fluctuations" in such a way that they cancel out leaving the *physical* mass of the particle as a first approximation. The calculation is simple, and I will be happy to write it out for you if you are interested. The hydrogen atom and fine structure we already have, as noted above. "Running" coupling constants are unitarity corrections to the low energy values from which we start. We should have the Lamb shift, etc. before too long.

Since I know you are concerned about "time", I beg you to consider the proposition that, for finite beings who can count, keep records, and retrieve those records, time is simply a finite counting parameter for these recorded or remembered **events** which can be put into correspondence with the integers interpreted as irreversible counting numbers. In the absence of further information, events which are assigned to the same integer must be given equal weight. This is one way to see why "indistinguishables" must enter our theory in an essential way and lead us into new mathematical territory.

9. CONCLUSION

All we need is a major experimental success, such as anti-gravity, to put us on the map.

ON TO THE 21ST CENTURY!

REFERENCES

1. H.P.Noyes, *Physics Today*, pp 99-100, March, 1992.

2. H.P.Noyes and S.Starson, "Discrete Anti-Gravity" in *Interdisciplinary Models in Science: Proc. ANPA WEST 7*, F. Young, ed., ANPA WEST, 409 Lealand Ave., Stanford, CA 94306, 1991.

3. H.P.Noyes, "Anti-Gravity", *ANPA WEST Jour.*, **3**, No. 1, 1-4 (1992).

4. N.Bohr and L.Rosenfeld, *Det. Kgl. Danske Videnskabernes Selskab., Mat.-fys. Med.*, **XII**, 8 (1933).

5. L. Rosenfeld, in *Niels Bohr and the Development of Physics*, pp 70-95.

6. H.P.Noyes, "Electromagnetism from Counting", *Physical Interpretations of Relativity Theory. III.*, M.C. Duffy, ed, Imperial College, London, 1992 (in press) and SLAC-PUB-5858 (Dec. 1992).

7. F.J.Dyson, *Am. J. Phys.*, **58**, 209-211 (1990).

8. H.P.Noyes, "On Feynman's Proof of the Maxwell Equations", SLAC-PUB-5411, presented at the *XXX Internationale Universitätswoken für Kernphysik*, Schladming, Austria, Feb. 27- Mar.8, 1991.

9. H.P.Noyes, "Feynman's Proof of the Maxwell Equations", SLAC-PUB-5588, Nov., 1991(unpublished).

10. H.P.Noyes, "The RQM Triangle: a paradigm for Relativistic Quantum Mechanics", in *Constructing the Object; Proc. ANPA WEST 8*, F.Young, ed., ANPA WEST, 409 Lealand Ave., Palo Alto, CA 94306.

11. D.O.McGoveran and H.P.Noyes, *Physics Essays*, **4**, 115-120 (1991).

12. H.P.Noyes, "On The Measurement of π", *Proc. ANPA 13*, C.W.Kilmister, ed. and SLAC-PUB-5732, Feb. 7, 1992.

13. *Particle Properties Data Book*, June 1992, available from Berkeley and CERN, p. 162 et.seq.

14. S.Weinberg, *Phys. Rev.*, **4 B**, 76 (1965).

15. H.P.Noyes, *ANTI-HYDROGEN: The cusp between Quantum Mechanics and General Relativity*, SLAC-PUB-5856, September 1992 (unpublished).

16. H.P.Noyes, "Observable Electromagnetic and Gravitational Orbits and Trajectories in Discrete Physics", *Physical Interpretations of Relativity Theory.I.*, M.C. Duffy, ed, Imperial College, London, 1988, pp. 42-61 and SLAC-PUB-4690 (Nov. 1988).

17. Stillman Drake, *Galileo: Pioneer Scientist*, University of Toronto Press, 1990, p. 8, p. 237.

18. D.O.McGoveran and H.P.Noyes, "Foundations of a Discrete Physics", in *Discrete and Combinatorial Physics*, H.P.Noyes, ed., Theorem 13, p. 59.

19. N.Bohr, *Phil. Mag.* **332**, Feb. 1915.

20. H.P.Noyes, "Lectures on Bit-String Physics" in Philosophy 242a, Stanford University, Fall, 1991.

PREDICTIONS FROM A FUNDAMENTAL THEORY

Table: **Coupling constants, mass ratios and cosmological parameters** predicted by the finite and discrete unification of quantum mechanics and relativity. Empirical Input: $c, \hbar$ and m_p as understood in the "Review of Particle Properties", Particle Data Group, *Physics Letters*, **B 239**, 12 April 1990.

COUPLING CONSTANTS

Coupling Constant	Calculated	Observed
$G^{-1}\frac{\hbar c}{m_p^2}$	$[2^{127} + 136] \times [1 - \frac{1}{3 \cdot 7 \cdot 10}] = 1.693\ 37\ldots \times 10^{38}$	$[1.69358(21) \times 10^{38}]$
$G_F m_p^2 / \hbar c$	$[256^2 \sqrt{2}]^{-1} \times [1 - \frac{1}{3 \cdot 7}] = 1.02\ 758\ldots \times 10^{-5}$	$[1.02\ 682(2) \times 10^{-5}]$
$sin^2\theta_{Weak}$	$0.25[1 - \frac{1}{3 \cdot 7}]^2 = 0.2267\ldots$	$[0.2259(46)]$
$\alpha^{-1}(m_e)$	$137 \times [1 - \frac{1}{30 \times 127}]^{-1} = 137.0359\ 674\ldots$	$[137.0359\ 895(61)]$
$G^2_{\pi N \bar{N}}$	$[(\frac{2M_N}{m_\pi})^2 - 1]^{\frac{1}{2}} = [195]^{\frac{1}{2}} = 13.96..$	$[13, 3(3), > 13.9?]$

MASS RATIOS

Mass ratio	Calculated	Observed
m_p/m_e	$\dfrac{137\pi}{\frac{3}{14}\left(1+\frac{2}{7}+\frac{4}{49}\right)\frac{4}{5}} = 1836.15\ 1497\ldots$	$[1836.15\ 2701(37)]$
$m_\pi^{\pm}/m_e$	$275[1 - \frac{2}{2 \cdot 3 \cdot 7 \cdot 7}] = 273.12\ 92\ldots$	$[273.12\ 67(4)]$
m_{π^0}/m_e	$274[1 - \frac{3}{2 \cdot 3 \cdot 7 \cdot 2}] = 264.2\ 143\ldots$	$[264.1\ 373(6)]$
m_μ/m_e	$3 \cdot 7 \cdot 10[1 - \frac{3}{3 \cdot 7 \cdot 10}] = 207$	$[206.768\ 26(13)]$

COSMOLOGICAL PARAMETERS

Parameter	Calculated	Observed
$\Omega_{vis}/\Omega_{closure}$	0.01175	$[0.005 \lesssim \Omega \lesssim 0.02]$
Baryons/Photon	$\frac{1}{256^4} = 2.328\ldots \times 10^{-10}$	$\sim 2 \times 10^{-10}$
M_{dark}/M_{vis}	12.7	> 10

348

Comment on *"Anti-Gravity: The Key to 21st Century Physics"*
H. Pierre Noyes (2000)

Until 1988, the only connection between bit-string physics and gravitation I had thought much about was the combinatorial hierarchy calculation of Newton's constant G_N. This situation changed thanks to my learning of the conference on "Physical Interpretations of Relativity Theory" that year in London (which turned out to be the first in a series) organized by Dr. M.C.Duffy. Since I also learned that this conference encouraged the presentation of unconventional views, I thought it might prove useful to give a talk which might attract some new faces to our Alternative Natural Philosophy Association (ANPA) conference held in Cambridge at about the same time. In the paper I presented I tried to show how bit-string physics could meet the three tests of General Relativity in a new way[1], and how to test that line of reasoning by an electromagnetic experiment in a space station.

At that conference I met Scott Starson, who started from finite and discrete Planck units as the basis of a unified theory[2], and obtained intriguing results which were comparable in some sense to what the ANPA program was yielding. What really captured my imagination was the fact that his "model", in which particles are surrounded by a phase-locked gravity wave, and anti-particles by a phase locked anti-gravity wave, suggests that anti-particles will "fall" up near the surface of the earth. Tests using anti-protons then under way at CERN might have confirmed this prediction; the experimental question is still open. This encounter made me realize that bit-string physics might lead to the same conclusion. The paper reproduced above represents the tentative position I had reached on this question by January 1993, and viewed that way needs little revision.

In the section of the Table giving cosmological parameters Ω_{vis} should be replaced by $\Omega_{baryonic}$, as is discussed in Chapter 16. I discuss my current position in the Comment on Chapter 9.

References

[1] H.P.Noyes, "Observable Electromagnetic and Gravitational Orbits and Trajectories in Discrete Physics", *Physical Interpretations of Relativity Theory*, M.C.Duffy, ed., Imperial College, London, 1988, pp 42-61 and SLAC-PUB-4690 (Nov. 1988).

[2] S.Starson, "Fundamental Units, Unification, and the Age of the Universe", in *Alternatives*, K.G.Bowden, ed., 1995, pp 300-323 (*Proc. ANPA 16*).

CROSSING SYMMETRY IS INCOMPATIBLE WITH GENERAL RELATIVITY*

H. PIERRE NOYES

Stanford Linear Accelerator Center
Stanford University, Stanford, California 94309

Consider a proton moving in empty space past a positively charged earth with charge (squared) $Q_E^2 = Gm_p M_E$. In the absence of further information, we expect it to move past the earth without being deflected. The CPT theorem asserts that an anti-proton must move past a negatively charged anti-earth in precisely the same way, so gives us no new information. Conventional theories predict that an anti-proton moving past a positively charged earth at distance r would experience an acceleration toward the earth of twice the amount $Gm_p M_E/r^2 = Q_E^2/r^2$ that a neutral object of mass m_p would experience. However, if we are correct in asserting that "crossing symmetry" requires *all* "forces" (i.e. accelerations per unit inertial mass) on a particle to reverse when that particle is replaced by its anti-particle, then the fact that we can use electromagnetic forces to balance a particle in a gravitational field plus crossing symmetry *predict* that the electromagnetic fields would have to be reversed in order to balance an anti-particle in the same configuration.

Our argument from "crossing symmetry" is unconventional in that we use the observable phenomenon of "acceleration per unit mass", or in relativistic S-matrix theory the change in the space-components of 4-velocity, in our definition. The conventional second-quantized relativistic field theory starts, instead, from an interaction Lagrangian expressed in terms of a "gauge potential" which is not observable. Such theories are not problem-free in the Newtonian and Coulombic limits. To quote Weinberg[1]

"The most general covariant fields .... cannot represent real photon and graviton interactions because they give amplitudes for emission and absorption of massless particles [of spin j] which vanish as p^j for momentum $p \to 0$."

Presented at the Seventh Marcel Grossmann Meeting on General Relativity,
Stanford University, July 24–30, 1994.

---

* Work supported by Department of Energy contract DE–AC03–76SF00515.

One is permitted to question whether the "gauge invariant" prescription which Weinberg uses to meet this problem is mathematically well defined in a class of theories which (a) do not have a well defined "correspondence limit" in either non-relativistic quantum mechanics or the classical, relativistic (Maxwell and Einstein) field theories of electromagnetism and gravitation, (b) necessarily give, as Oppenheimer put it, "non-sensical [i.e. infinite] answers to sensible questions", and (c) have not reached consensus on how to formulate "quantum gravity",

In contrast, our finite and discrete reconciliation between quantum mechanics and relativity meets problem (a) by deriving the Maxwell and Einstein fields from scale-invariant measurement accuracy bounded from below in a manner reminiscent of Bohr and Rosenfeld's analysis of the measurability of electric and magnetic fields.[2] We have claimed[3] that this analysis removes the physical "paradox" in the Feynman 1948 derivation reported by Dyson[4] and extended to gravitation by Tanimura.[5] We also claim that a significant extension of the calculus of finite differences to an ordered (non-commutative) formalism provides a rigorous mathematical context for the derivation.[6] Our finite particle number S-matrix theory conserves (relativistic) 3-momentum at 3-vertices, but is off energy shell by a finite amount. Our finite and discrete kinematics satisfies discrete conservation laws for physically realizable multi-leg diagrams; it fits comfortably into the practice of high energy elementary particle physics. Because our theory is finite and discrete, and hence can identify c as the maximum velocity at which information can be transferred between distinct locations, problem (b) never arises. As to problem (c), gravitation and electromagnetism are reconciled at the bound state level by a common treatment of both which is formally equivalent to the Bohr's relativistic calculation of the energy levels of the hydrogen atom, using combinatorially calculated coupling constants; the method can be extended to yield the Sommerfeld formula for the fine structure of hydrogen[7] and its gravitational equivalent. The classic tests of GR are met, and we believe that the time tests of the pulsar data can be met as well. The quantum number structure predicts the observed particles of the standard model of quarks and leptons, and yields no unobserved particles. Given $\hbar, c$ and m_p as empirical input, we compute $G, G_F, e^2, m_p/m_e, m_\pi/m_e, m_\mu/m_e$ to an accuracy of a part in $\approx 10^4 - 10^7$, leaving enough room for improvement to make the theory interesting to pursue, or to produce a crucial conflict with experiment. Results are given in the table.

As we explain in the first paragraph, the most dramatic prediction of our theory to date, which is currently under direct experimental scrutiny at the CERN Low Energy Anti-proton Ring, and can be tested by still more sensitive techniques,[8] is that

ANTI-MATTER "FALLS" UP.

REFERENCES

1. S.Weinberg, *Phys. Rev.* **138 B**, 988 (1965).

2. H.P.Noyes, to appear in *A Gift of Prophecy; Essays in Celebration of the Life of Robert E. Marshak*, E.G.Sudarshan, ed., World Scientific (in press).

3. H.P.Noyes, "OPERATIONALISM REVISITED: Measurement Accuracy, Scale Invariance and the Combinatorial Hierarchy", *Physics Philosophy Interface* (in press), and references therein.

4. F.J.Dyson, *Amer. J. Phys.*, **58**, 209 (1990).

5. S. Tanimura, *Annals of Physics.* **220**, 229 (1992).

6. L.H.Kauffman and H.P.Noyes, "Discrete Physics and the Derivation of Electromagnetism from the Formalism of Quantum Mechanics", submitted to the *Proceedings of the Royal Society* **A**.

7. D.O.McGoveran and H.P.Noyes, *Physics Essays*, **4**, 115 (1991).

8. V.Lagomarisino, V.Lia, G.Manuzio and G.Testera, *Phys.Rev. A*, **50**, 977 (1994).

Table. **Coupling constants and mass ratios** predicted by the finite and discrete unification of quantum mechanics and relativity. Empirical Input: c, $\hbar$ and m_p as understood in the "Review of Particle Properties", Particle Data Group, *Physics Letters*, **B 239**, 12 April 1990.

COUPLING CONSTANTS

Coupling Constant	Calculated	Observed
$G^{-1}\frac{\hbar c}{m_p^2}$	$[2^{127}+136]\times[1-\frac{1}{3\cdot7\cdot10}]=1.693\,31\ldots\times10^{38}$	$[1.69358(21)\times10^{38}]$
$G_F m_p^2/\hbar c$	$[256^2\sqrt{2}]^{-1}\times[1-\frac{1}{3\cdot7}]=1.02\,758\ldots\times10^{-5}$	$[1.02\,682(2)\times10^{-5}]$
$sin^2\theta_{Weak}$	$0.25[1-\frac{1}{3\cdot7}]^2=0.2267\ldots$	$[0.2259(46)]$
$\alpha^{-1}(m_e)$	$137\times[1-\frac{1}{30\times127}]^{-1}=137.0359\,674\ldots$	$[137.0359\,895(61)]$
$G^2_{\pi N\bar{N}}$	$[(\frac{2M_N}{m_\pi})^2-1]^{\frac{1}{2}}=[195]^{\frac{1}{2}}=13.96..$	$[13,3(3),>13.9?]$

MASS RATIOS

Mass ratio	Calculated	Observed
m_p/m_e	$\dfrac{137\pi}{\frac{3}{14}\left(1+\frac{2}{7}+\frac{4}{49}\right)\frac{4}{5}}=1836.15\,1497\ldots$	$[1836.15\,2701(37)]$
$m_\pi^\pm/m_e$	$275[1-\frac{2}{2\cdot3\cdot7\cdot7}]=273.12\,92\ldots$	$[273.12\,67(4)]$
m_{π^0}/m_e	$274[1-\frac{3}{2\cdot3\cdot7\cdot2}]=264.2\,143\ldots$	$[264.1\,373(6)]$
m_μ/m_e	$3\cdot7\cdot10[1-\frac{3}{3\cdot7\cdot10}]=207$	$[206.768\,26(13)]$

COSMOLOGICAL PARAMETERS

Parameter	Calculated	Observed
N_B/N_γ	$\frac{1}{256^4}=2.328....\times10^{-10}$	$\approx2\times10^{-10}$
M_{dark}/M_{vis}	≈12.7	$M_{dark}>10M_{vis}$
$N_B-N_{\bar{B}}$	$(2^{127}+136)^2=2.89...\times10^{78}$	*compatible*
ρ/ρ_{crit}	$\approx\frac{4\times10^{79}m_p}{M_{crit}}$	$.05<\rho/\rho_{crit}<4$

Comment on "*Crossing Symmetry is incompatible with General Relativity*"
H. Pierre Noyes (2000)

One significant development since this publication is that at ANPA 20 Brian Koberlein[1] showed that it is in fact possible to introduce particulate anti-gravity into the Einstein theory in such a way that the equivalence principle and the strong gravity predictions of the theory remain untouched. This is fortunate, since it appears likely that the Tanimura derivation of the Einstein theory (Ref. 5 in Koberlein's paper) can be replaced by a rigorous DOC (discrete ordered calculus) proof along the lines of Chapter 11.

The second is that, thanks to a discovery made by James Lindesay in discussing my work with Ed Jones on the Relativistic S-wave three body problem[2], we have found it possible to construct a simple unitary and covariant two body amplitude, which under the appropriate interpretation of particle-antiparticle symmetry can start from a low lying quantum (close to zero mass) state in the particle-antiparticle channel and "cross" it to obtain an appropriate "relativistic Coulomb" amplitude in the particle-particle channel. This means that in a theory in which particles attract rather than repel (low energy gravitation), crossing will yield particle-antiparticle gravitational repulsion as argued heuristically in this paper. This also further strengthens the use of the Dyson argument to close off particulate gravitational interaction at high energy by the formation of a Planckton.

(This paper appears on pp 493-5 of Part A of the Proceedings of the Seventh Marcel Grossmann Conf.)

References

[1] B.D.Koberlein, "Crossing Symmetry and the Equivalence Principle in Einsteinian Gravity", in *Aspects II; Proc. ANPA 20*, K.G.Bowden, ed., pp 249-255.

[2] H.P.Noyes and E.D.Jones, "Solution of a Relativistic Three Body Problem". *Few Body Systems*, **27**, 123-139 (1999).

OPERATIONALISM REVISITED: Measurement Accuracy, Scale Invariance and the Combinatorial Hierarchy[*]

H. PIERRE NOYES

Stanford Linear Accelerator Center
Stanford University, Stanford, California 94309

Submitted to *Physics Philosophy Interface*

---

[*] Work supported by the Department of Energy, contract DE–AC03–76SF00515.

356

Abstract

It is claimed here that by 1936 Bridgman had developed severe criticisms of orthodox quantum mechanics from the point of view of his operational philosophy. We try to meet these criticisms by a radical analysis of the measurement of the finite and discrete length and time intervals between *particulate events*. We show that a scale invariant *counter paradigm* based on two arbitrary finite and discrete measurement intervals $\Delta\ell$, Δt offers an alternative starting point for constructing relativistic particle mechanics. Using the scale invariant definitions

$$\Delta\ell/c\Delta t \equiv 1; \ \Delta\ell^2/\kappa\Delta t = 2\pi$$

for the Einstein limiting velocity c and for "Kepler's constant" κ— proportional to the area per unit time swept out by a particle moving with constant velocity past a center — we derive the finite and discrete Lorentz invariant bracket expression $[x_i, \dot{x}_j] = \kappa\delta_{ij}$ for rectangular coordinates and velocities in three dimensions. Defining field per unit charge as the force per unit mass acting on a test particle with this constraint, we find that these fields satisfy the free field Maxwell equations provided only charge per unit mass for this test particle is a Lorentz invariant, generalizing the Feynman-Dyson-Tanimura proof. This scale invariant theory for classical relativistic fields is broken by any measurement of length with measurement accuracy $\Delta\ell < \hbar/2m_ec$ because electron-positron pairs are produced with finite probability, violating the single test particle postulate. This allows us to recover our new fundamental theory based on the *combinatorial hierarchy* and *bit-string discrimination*. Recent mass ratios, coupling constants and cosmological parameters obtained by bit-string dynamics are quoted.

...the ultimately important thing about any theory is what it actually does, not what it says it does or what its author thinks it does, for these are often very different things indeed.

— P.W.Bridgman, *The Nature of Physical Theory*

1. INTRODUCTION

This paper is dedicated to the hope that we now know enough to start re-casting modern elementary particle physics and physical cosmology in a form that would come closer to Bridgman's vision of physics. Although Bridgman's operational philosophy[1] contributed a great deal to the early discussions of the meaning of relativity and quantum theories, the "orthodox" Copenhagen interpretation of quantum mechanics as evolved by Bohr departs widely from what — in my opinion — Bridgman had in mind. He wished to base physical concepts on laboratory operations, actually carried out, or as a minimum on paradigms which are consistent as thought experiments suggested by actual laboratory practice. In later reflections he was prepared to extend his "operationalism" to sufficiently clear mathematical operations[2]; I suspect that the term "computable" would have found favor in his eyes.

In his earlier work (LMP) Bridgman criticized the concept of "light traveling". In NPT he grants that operational meaning might eventually be given to the concept of "light traveling" by measuring the scattering of light by light. Presumably he had in mind Furry's successful calculation of the finite cross section for this scattering process using the then available formulation of quantum electrodynamics (QED). But this calculation produces a finite result in QED only because of specific symmetries. Most processes predicted by the second quantized relativistic field theories are infinite, and therefore *not* related to experiment in any obvious way. It would be hard to give an operational gloss to such theories even after they are "renormalized", or restricted to the class of "non-Abelian gauge theories". I doubt that Bridgman would have found any of the theories currently used in high energy particle physics palatable. He also dismissed cosmology as operationally meaningless, a judgment that the rich body of data now available to the observational astronomer could have led him to revise.

We believe it is proving fruitful to revive Bridgman's dream using a finite and discrete reconstruction of relativistic quantum mechanics and physical cosmology. This new fundamental theory has taken shape gradually. The most easily identified starting point is the work by Bastin and Kilmister on the "Concept of Order".[3-9] The research program had its first dramatic success in the discovery of the *combinatorial hierarchy* by A.F.Parker-Rhodes in 1961.[10] An adequate, if somewhat unsystematic, survey of subsequent developments is available in the proceedings of some of the annual meetings of the Alternative Natural Philosophy Association (founded in 1979)[11] and of its western chapter.[12] We will not review this history here. Two up-to-date and systematic presentations of the theory from very different perspectives are in preparation.[13,14] The specific approach followed in this paper is adequately introduced by a short description[15] and three more technical papers.[16-18] Some additional recent work will be referenced below.

2. EXAMPLES of BRIDGMAN'S APPROACH

As an experimentalist, and a critic of the practice of theoretical physics, Bridgman had a healthy distrust of mathematics:

"[Mathematics] begins by being a most useful servant when dealing with phenomena of the ordinary scale of magnitude, but ends by dragging us by the scruff of the neck willy nilly into the inside of an electron where it forces us to repeat meaningless gibberish." (LMP, p. 149)

It is claimed in FDP[16] that McGoveran's *ordering operator calculus* provides a mathematics which could meet Bridgman's challenge. In fact, any formalism restricted to computability in McGoveran's sense might be utilized to meet that goal. For our purposes we adopt the following definition:

A physicist who claims that a problem is **computable** *must — if challenged — be* *able to produce an integer answer to that problem within a year using the computational facilities and research budget available to him.*

If we could succeed in getting this requirement adopted in the discussions of theoretical physics, it could eliminate a lot of fruitless argument. Bridgman tried to meet basically the same difficulty by pleading with physicists to stop discussing "meaningless problems". His failure leaves me little hope that the current generation of physicists — or the next — will prove to be that rational. Of course he did not intend, nor do I, to hobble *speculation* once it is properly identified as such.

Initially, Bridgman hoped that his criticism had taken root in the "new" quantum mechanics:

"This section was written early in 1926 without access to the recent literature. Our attitude toward quantum phenomena has been so much changed since then by the "new" quantum mechanics, that a number of the following statements are superseded as a statement of present opinion. However it has seemed worth while to let the section stand as written, because many of the developments actually taken in the new mechanics follow the lines that it is here urged they ought to take, and so far afford interesting confirmation of the point of view of this essay." (LMP, p. 186, footnote 1)

But he was not satisfied with the shape that quantum mechanics actually took.

"I think there is significance in the difficulty which my theoretical friends find in suggesting what sort of apparatus they would set up in the laboratory in order to answer such questions as: 'Can e, or m, or h be measured separately with unlimited precision by a single experiment, or may they be measured simultaneously in a single experiment?' Or what is the apparatus in terms of which any arbitrary 'observable' of Dirac acquires its physical meaning? I think it will be granted by most theoretical physicists that there are situations of this sort which have not yet been thought completely through. Since we are now prepared to admit that the correspondence between mathematics and experience is never a one to one correspondence, so that because a mathematical theory accomplishes successfully one-half of what we would like to have it [accomplish] there is no certainty or even a high probability that it will accomplish the other half, I think we are justified in a certain amount of disquietude in the face of any situation that has not been thought through completely.

"The mere fact that such a debate is possible as that carried out on the one hand by Einstein, Podolsky and Rosen[2], [2A.Einstein, B.Podolsky and N. Rosen,

Phys. Rev., 47,777,1935.] and on the other hand by Bohr,[3] [[3]N.Bohr, *Phys. Rev.*,48,696,1935.] increases our disquietude...." (NPT, p.118)

With regard to this specific difficulty in orthodox quantum mechanics, we can claim that our computational point of view has much to offer. In any finite and discrete theory, there is necessarily a limiting velocity for the transfer of information, and also a finite and discrete meaning for the "Lorentz transformations" (FDP, ch. 4 pp 48-54.). At the same time, this implies "supraluminal" correlations *without* supraluminal signaling.[19,20] We have also claimed that the new approach goes far toward resolving the quantum mechanical "measurement problem",[21,22] a claim which we hope is strengthened by the preliminary abstract provided as the appendix to this paper. The basic fact which make this resolution possible is that at a deep level, which is explored further in this paper, we claim to have achieved a successful reconciliation of quantum mechanics with relativity.[23]

We hope to give further evidence for the usefulness of criticizing the current practice of theoretical physics by presenting this paper. When we speak of "criticism", we follow Bridgman's usage:

"The material for the physicist as critic is the body of physical theory, just as the material of the physicist as theorist is the body of experimental knowledge." (NPT, p.2)

In contrast to the open ended and rapidly expanding task of the theorist, he saw the task of the physicist as critic as one that could be accomplished, simply because it depended on the finite mind of the critic:

"In so far as we may assume that the human mind has approximately fixed and definite properties and is not in such a rapid state of evolution that it runs away with us during the discussion, we are not here confronted with unlimited possibilities of complexity, but the field is an essentially closed one." (NPT, p.2)

Unless or until we encounter extraterrestrial species, or the claims of strong AI are convincingly demonstrated in practice, I have no quarrel with this statement. However, I find his optimistic conclusion:

"...having acquired this amount of understanding we may then pass on, leaving criticism behind us as a well rounded and more or less definite discipline." (NPT, p.3) less convincing, simply because the task has to be undertaken again in our generation.

As already noted, the task itself has become possible because of the work of a large number of people, starting with the critique of Eddington's fundamental theory in the 50's by Bastin and Kilmister. I do not discuss here — let alone evaluate the significance of — the subsequent contributions by Fredrick Parker-Rhodes, Irving Stein, Michael Manthey, and David McGoveran all of which were vital to the process which led to the current state of our new fundamental theory. Instead, I will concentrate on a recent development that — in my opinion — can revolutionize our way of thinking about the connection between relativistic quantum mechanics and classical relativistic field theory.

3. MEASUREMENT ACCURACY and SCALE INVARIANT BRACKET EXPRESSIONS

3.1 THE COUNTER PARADIGM AND THE DEFINITION OF "c"

In McGoveran's fundamental approach to finite and discrete theories based on the *ordering operator calculus* and *attribute distance* one is restricted to *derivates* (i.e. finite and discrete differences) and never encounters the limiting processes or the derivatives used in continuum mathematics. The exponentiation of the derivate operator allows him to construct a generalized commutation relation which can be given a discrete geometrical interpretation (FDP, Sec. 4.2, pp 63-69). This construction provides the bracket expressions needed to derive classical relativistic field theory from measurement accuracy, a derivation we present in the next chapter. Rather than follow McGoveran's rigorous discussion we provide a somewhat heuristic argument in the spirit of Bridgman.

We start our more limited discussion from the measurement of finite and discrete length and time intervals using methods whose accuracy is specified by some relevant technology. In the past we sometimes started our discussion of the new fundamental theory from what we called the *counter paradigm*, We had in mind the devices used in elementary physics which record whether or not some discrete "event" takes place within a finite and discrete spacial volume during a finite time interval. When tied to finite and discrete laboratory coordinates and a finite and discrete laboratory clock, we can then introduce the coupled concepts of *particle* and *event* by the informal descriptions:

A **particle** *is a conceptual carrier of conserved quantum numbers between* **events**.

An **event** *is a region which particles carring conserved quantum numbers enter and leave during a finite time interval. These quantum numbers can be iether positive or negative. The algebraic sum of the entering quantum numbers of a given type is equal to the algebraic sum of the quantum numbers of the same type which leave the region. In that sense quantum numbers are conserved across an event. However the number of particles need not be conserved across an event.*

Here the undefined term "region" can be given some content by requiring that its effective volume can be specified by three independent lengths whose magnitudes are three independent integers times a common finite and fixed "shortest length" $\Delta\ell$. The event occurs during some well specified time interval which is again an integer times a "shortest time" Δt. We postulate standard laboratory protocol for measuring these finite length and time intervals. The event itself is a NO-YES event, i.e. "NO" if it does not happen in the specified space-time volume, and "YES" if it does. The event is made manifest by the non-firing or firing of a recording counter activating some type of discrete "memory". In general it will take several such counter firings, and a fair amount of theory, to give precision to the concept of "conservation laws". We will refine earlier attempts to do so in the context of bit-string dynamics on another occasion.

What we have sometimes called the "counter paradigm" is two such YES events separated by a spacial interval L and a time interval T neither of which can be known to better accuracy that $\Delta\ell$ and Δt respectively. In high energy particle physics, this would be called a "counter telescope". It takes the physical real-ization of at least four such counter telescopes and considerable calibration and experimentation to start giving precision to the "energy-momentum" conservation laws. Other particulate conservation laws can be constructed from and tested using data so measured.[24]

Consider a counter telescope consistent with this definition, and a variety of "particle sources" with adjustable parameters more or less under our control. We pick those situations which consist of two sequential counts in the two detectors and which provide an approximately constant value for the measured ratio L/T. We find that all sources of "particles" so far explored in this way give a value which is less than or indistinguishable from a universal constant called "c", unique for any fixed units of length and time measurement. This fact has led the appropriate international committee to fix this ratio — by *definition* the velocity standard for SI units — as the integer

$$c \equiv 299\ 792\ 458\ m\ sec^{-1} \tag{3.1}$$

If there were *any* demonstrable situations in which this definition of c as a *di-mensional standard* led to contradictions between the way "c" is used in theoretical formulae and the way these formulae are used to interpret experimental results, the committee would not have introduced this *integer definition* of c. For histori-cal reasons this conventional constant is still sometimes called "the speed of light" despite the fact that is now simply a definitional dimensional standard. A better phrase from a modern point of view would be "the limiting speed for the transfer of information". Of course "supraluminal velocities" that describe correlations which *cannot* be used to transfer information — such as phase velocities — have a well defined meaning in both classical and quantum physics. In a finite and discrete theory they are in no way paradoxical, as we noted in Chapter 2.

Since we do not wish to specify our units of measurement in advance, we will build this fact into our approach by relating our minimum measurable distance and time intervals by the dimensionless constraint

$$\frac{\Delta\ell}{c\Delta t} \equiv 1 \tag{3.2}$$

3.2 FINITE AND DISCRETE LORENTZ BOOSTS

In a theory that depends on finite measurement accuracy, the measurement of the velocity of a particle — as in the counter paradigm — requires both the measurement of a finite space interval and the measurement of an ordered, finite time interval together with a discussion of the accuracy to which the ratio of these two intervals is determined by these separate measurements. We consider two YES events such as the sequential firing of the two counters in a counter telescope of length L with a time interval T, defining a velocity which we attribute to a particle. We call this a *quasi-local* measurement.

In this section we will use the two events at the ends of a counter telescope of length L with time interval T as a paradigm for the discrete measurement of the quasi-local coordinates x, t of a particle. In the next section we will "embed" this counter telescope measurement in a larger finite and discrete space-time in a way that allows us to represent finite and discrete rotations consistently with finite and discrete Lorentz boosts. In order to make that extension, it is convenient to pick some small reference region in the laboratory, which also has a standard clock associated with it and make the relation to the laboratory measurements explicit.

We specify the two spacial positions for the two events relative to this "origin" by $x_1 = n_1\Delta\ell$, $x_2 = n_2\Delta\ell$ where we assume in this example that the event at x_2 *follows* the event at x_1. Then, clearly $L = |n_2 - n_1|\Delta\ell$. Since we have already introduced the limiting velocity for information transfer, the Einstein synchronization convention allows us to specify the times at which the two events occur, according to a clock located at the "origin", as $t_1 = n_1\Delta\ell/c$ and $t_2 = n_2\Delta\ell/c$. Clearly, $T = (n_1 + n_2)\Delta\ell/c$. Note that $t_1 + t_2$ is also the time it would take a light signal emitted in coincidence with event 1 in the direction of the "origin" and reflected there to arrive at position x_2 in coincidence with the second event. We have, in effect, assumed that $t_2 > t_1$ and hence that $n_2 > n_1$. Note that if we interpret these two events as "caused by a single particle", its velocity as measured by the counter telescope is

$$v_{12} \equiv \beta_{12}c = \frac{n_2 - n_1}{n_1 + n_2}c \qquad (3.3)$$

Note further that the square of the invariant interval I between the two events isgiven by

$$I^2\Delta\ell^2 = c^2(t_1 + t_2)^2 - (x_1 - x_2)^2 = 4n_1n_2\Delta\ell^2 \qquad (3.4)$$

One point to note here is that the maximum accuracy to which we can know the two distances is fixed by requiring n_1 and n_2 to be *positive definite integers*. If there were some way we could (even by indirect or statistical means) produce a set of reliable non-integral data for any such situation, this would violate the hypothesis that $\Delta\ell$ is the *shortest* interval to which we can give well defined experimental meaning. In other words our modern operational hypothesis about length and time measurements is that *THERE IS ALWAYS A SYSTEM OF UNITS IN TERMS OF WHICH LENGTHS AND TIMES ARE INTEGERS.*

Our means of relating the counter telescope — our paradigm for an "x, t" measurement — to the laboratory coordinate system makes use of the exchange of light signals in such a way that the standard Einstein clock synchronization convention can be used in our integer environment. This allows us to separate the problem of "quasi-local" Lorentz invariance from "event horizon limited" or "coherence limited" Lorentz invariance by the following construction. First note that our velocity $v_{12} = (n_2 - n_1)c/(n_1 + n_2)$ is invariant under the transformation $n_i \to N_T k_i$ with N_T a positive definite integer *if* k_1 and k_2 are positive, definite integers. Clearly this downward scale transformation is allowed only if n_1, n_2 have N_T as a common factor. For any particular empirical situation, measurement accuracy will limit such downward scale transformations by the quasi-local velocity resolution we can achieve. We call this Δv which, as a function of $(k_1 + k_2)$ is given by $c/(k_1 + k_2)$, and the resulting limitation "scale invariance bounded from below".

For example, in high energy particle physics, we measure the momentum $p = (k_2 - k_1)\Delta mc/4k_1 k_2$ of a charged track by its radius of curvature in a magnetic field and its energy $E = (k_1 + k_2)\Delta mc^2/4k_1 k_2$ calorimetrically defining the velocity as $v = pc/E = (k_2 - k_1)/(k_1 + k_2)$, in situations where we cannot measure the distance between two counters $(n_2 - n_1)\Delta \ell$ to better than many orders of magnitude (i.e. $(n_2 - n_1) = L/\Delta \ell \approx N_L \pm \Delta N_L$, $\Delta N_L >> 1$). We have discussed on other occasions how the periodicity N_T implied by our finite and discrete measurement accuracy paradigm can lead to observable interference phenomena which *break* scale invariance. With this understood, we see that for fixed velocity resolution, (i.e., $k_1 + k_2 = K_v \in$ *positive definite integer*) we also can define a maximum coherence length $L_C = N_T^C K_v \Delta \ell$, with N_T^C the maximum number of coherent velocity periods we can show to interfere. In a "wave theory" this would be called a "coherent wave pulse".

In a finite and discrete theory, straight line motion cannot continue forever. As just discussed, we can bound it by defining the coherence length in terms of the velocity resolution K_v and the number of periods at that velocity resolution $N_{T_v}^C$ as $L_C = N_{T_v} K_v$. When we reach this limit, we can continue our coherent discussion to larger spaces by the trick of using periodic boundary conditions, or by breaking our trajectory into straight line segments and bending it around to form a closed "orbit" of length $2\pi L_C$. Taking this to be the perimeter of an orbit with discrete rotational symmetry will be discussed in the next section. The extension to elliptical orbits is straightforward (See Ref. 22).

We now return to our quasi-local position-time measurement by saying that the counter telescope measures a position $x = (k_2 - k_1)\Delta \ell$, a time $t = (k_1 + k_2)\Delta \ell/c$ and hence a velocity $\dot{x} = (k_2 - k_1)c/(k_2 + k_1)$. Clearly this places $x = 0$ half-way between the two counters in the counter telescope and $t = 0$ at a time $\frac{1}{2}(k_1 + k_2)\Delta t$ after the firing of counter 1. This language implies that in a much larger context we can measure both x_1 and x_2 to an accuracy $\Delta \ell$, or in other words measure the discrete *phase* of this combined pair of events *within* the $x_2 - x_1$ length of the telescope. But this still does *not* allow us to assign an absolute meaning to both x and $\dot{x}$. Where ever we place x within this interval, we still cannot know its value to better than $\Delta \ell$. The best we can do is to assign it to some position which is ambiguous between $x_- \in [x_1, x_1 + \Delta \ell, \ldots x_2 - \Delta \ell]$ and $x_+ = x_- + \Delta \ell$. Between these two locations the velocity is, as measured locally, $+c$. Thus, using only quasi-local information, the product "$x\dot{x}$" is ambiguous depending on whether we use x_- or x_+. Defining the difference as the bracket expression, and using c for $\dot{x}$, we have that, for local measurements,

$$[x, \dot{x}] \equiv x_+ \dot{x}(t) - x_- \dot{x}(t) = c\Delta \ell = -[\dot{x}, x] \tag{3.5}$$

It might be thought that by using a longer lever arm to improve our velocity resolution, we could improve on this limit. As we discuss in the next section, using periodic phenomena does allow us to reduce this constant to $\kappa = c\Delta \ell / 2\pi$, but no further.

It remains to show that our definition of position and velocity in the larger space allows us to define finite and discrete Lorentz transformations in the single direction "$\vec{x}$" we have so far considered. For rational fraction velocities, this amounts to showing that transformation from $v = (k_2 - k_1)c/(k_2 + k_1)$ to $v' = (k_2' - k_1')c/(k_2' + k_1')$ can be obtained from the usual velocity addition law

$$v' = \frac{v + v''}{1 + vv''/c^2} \tag{3.6}$$

with some rational fraction velocity $v'' = (k_2'' - k_1'')c/(k_1'' + k_2'')$. It is trivial to show[25] that, in the case $k_2' = k_2$,

$$k_1'' = k_1'; \ k_2'' = k_1 \tag{3.7}$$

satisfies this requirement. The problem of showing that any Lorentz transformation needed in an experimental context can always be constructed in this way *without* producing contradiction with experiment will be discussed in detail elsewhere (eg Ref. 12).

3.3 FINITE AND DISCRETE LORENTZ ROTATIONS

In order to extend our analysis from one to two spacial dimensions, it suffices initially to consider three counter firings F_1, F_2, F_3 at three fixed locations which form a triangle. [23] We assume that the three counters are at rest in the laboratory frame, and that the laboratory clock (using the usual Einstein synchronization convention) records the firings in the order 1,2,3. In order to insure that the three distances $s_{ij}\Delta\ell$ satisfy the triangle inequalities $|s_{ij} - s_{jk}| \leq s_{ki} \leq s_{ij} + s_{jk}$ it suffices to pick three positive definite times $t_i\Delta\ell/c$ with all three t_i *integers*. We then define the sums $s_{ij}\Delta\ell \equiv (t_i + t_j)\Delta\ell/c$ with the consequence that

$$|s_{ij} - s_{jk}| = |t_i - t_k| \leq s_{ki} = t_i + t_k \leq s_{ij} + s_{jk} = t_i + t_k + 2t_j \qquad (3.8)$$

We now take the counter at position 2 as the origin of coordinates, and assume that a light signal $1 \rightarrow 2 \rightarrow 3$ emitted in coincidence with F_1 arrives in coincidence with F_3. Then a counter telescope $1 - 3$ measures a velocity

$$v_{13} = \frac{(t_1 + t_3)c}{t_1 + 2t_2 + t_3} \equiv \beta c \qquad (3.9)$$

Noting that the radial distance to C1 (i.e. to counter 1, where the first event, F_1, occurs) is $s_{12}\Delta\ell$ and that the radial distance to C3 is $s_{23}\Delta\ell$, the radial velocity v_r is given by

$$v_r = \frac{(t_3 - t_1)}{t_1 + t_3 + 2t_2} \equiv \beta_r c \qquad (3.10)$$

The square of the area of the triangle is given by

$$A^2 = (t_1 + t_2 + t_3)t_1 t_2 t_3 \Delta\ell^4 = \frac{1}{16}(t_1 + 2t_2 + t_3)^4(1 - \beta^2)(\beta^2 - \beta_r^2)\Delta\ell^4 \qquad (3.11)$$

We now assume that if we make additional velocity measurements anywhere along the line defined by the counter telescope we obtain the same constant velocity *v*. Of course the shape of the triangle and the radial velocity change with time, but the area per unit time is constant, as we now prove. This is, of course just Kepler's Second Law for straight line motion past a center which exerts negligible force on the particle. Consider first the symmetric case $r_1(-t) = r_3(+t) = r(t)$ for which $v_r = 0$. The distance of closest approach to the center on this straight line, constant velocity trajectory is called the *impact parameter*. With F_2 the origin of rectangular coordinates, we take the impact parameter to have magnitude $x\Delta\ell$ in the positive x direction. Then $v = \dot{y}$ is perpendicular to it and $|t| = r(t)/c$. Since the area swept out by $r(t)$ in time $2|t|$ is $A(t) = x\Delta\ell\dot{y}|t|$, we have that

$$A(t)/|t| = \beta cx\Delta\ell = const. \tag{3.12}$$

If we now consider the general case with s_{13} in the y direction, the area of the triangle formed when the particle moves from F_1 to F_3 in time $t_{13}\Delta t = (s_{12} + s_{23})\Delta\ell/c$ with velocity $v = \dot{y} = s_{13}c/t_{13}$ divided by $t_{13}\Delta\ell$, using the half-base times altitude rule, is again $\beta cx\Delta\ell$. This proves Kepler's Second Law for straight line motion with constant velocity past a center in our discrete, relativistic model.

We have seen that we can construct our integer version of Kepler's Second Law from an arbitrary integer triangle. To demonstrate rotational invariance, all we need to do is to define transformations which keep the three sides fixed. If our triangle is to return to the same position after a finite number of planar rotations, we must use some care. For instance, we can start by fixing the impact parameter as a constant $x\Delta\ell$ and the line from the origin we have called $r(t)$ to a constant $R\Delta\ell$ in a particular, symmetric case for which the distance from C1 to C3 is $2y\Delta\ell = \dot{y}t\Delta t > 0$. Then, taking the x-axis outward from C2 along the impact parameter, in terms of dimensionless coordinates x, y, t we see that F_1 has coordinates $(x, -y; -t)$, F_2 has $(0, 0; 0)$ and F_3 has $(x, y; t)$. If we require that $R^2 = x^2 + y^2$, we must face the dilemma encountered by Pythagoras two and a half millennia ago that we cannot in general satisfy this relation in integers, and must make some decision if our theory is to remain finite and discrete. We choose to take R and y integer, and then have two choices for x, namely $x_\pm = R \pm y$ which insure that $x_+ x_- = R^2 - y^2$.[26]

We now pick our length scale in such a way that $R/y \equiv 2j$ is integer and note that this allows us to construct a regular polygon with $2(2j + 1)$ sides composed of isosceles triangles with base $2y$, slant height R and $x^2 = R^2 - y^2 = 4y^2(j^2 - \frac{1}{4})$ Clearly this construction insures rotational invariance under $2(2j + 1)$ finite and discrete rotations. The circle circumscribing this polygon has perimeter $2\pi R\Delta\ell = j(2y)$, and we find that our requirement of rotational invariance in our finite and discrete context allows us to conclude that we can always pick units such that

$$(x\dot{y})^2 = (j^2 - \frac{1}{4})\kappa^2 = l_\odot(l_\odot + 1)\kappa^2 \tag{3.13}$$

with

$$l_\odot \equiv j - \frac{1}{2}; \ \kappa \equiv \frac{\Delta\ell^2}{2\pi\Delta t} \tag{3.14}$$

and $2j$ odd.

It may seem peculiar that we have arrived at the "quantum mechanical" result $l_\odot(l_\odot + 1)$ for the quantum numbers for the square of circular orbital angular momentum correctly related to "spin 1/2" by this essentially classical construction. This has a deep significance which we will pursue on another occasion. Note that our derivation of "angular momentum per unit mass" with $\kappa \equiv \hbar/m$ requires only that space and time measurement accuracy be scale invariant but bounded from below. Physically, this is an obvious move. The single particle assumption built into our analysis breaks down if we try to measure any linear distance to an accuracy $\Delta\ell < h/2m_e c$ because at that point we produce electron-positron pairs with finite probability. That we get a "classical" version of "spin" in units of $\kappa/2$ should not be too startling if one studies our treatment of the fine structure of hydrogen (Ref. 16) and consults the discussion by L.C.Biedenharn on the "Sommerfeld Paradox" which we cite in that paper.

The extension from two to three dimensions is straightforward in terms of the finite rotations implied in our regular polygon paradigm used above. We define a "pseudovector" perpendicular to the rotational plane to represent the orbital angular momentum per unit mass with a chirality convention that (for counterclockwise rotations and a right-handed coordinate system) makes it the z axis. Then we can make the sign convention and definition

$$l_z = x\dot{y} - y\dot{x} \tag{3.15}$$

invariant for rotations in the $x - y$ plane under any choice of axes in that plane consistent with our polygonal symmetry. Since there are only $2(2j + 1)$ finite rotations defined, only this number of choices is allowed, without further injection of information into the model. As in the case of position-velocity measurement, if we include a larger system which allows us to go down to the $2y$ possible positions along the chord by measuring interference phenomena, we can define relative phases down to $\delta\phi \approx \Delta\ell/R_{max}$ but, as in conventional quantum mechanics, *absolute* phase refers to no known experimental phenomena and still eludes us.

The further articulation of the commutation relations for these finite rotations need not detain us long here. For instance, with orbital angular momenta per unit mass measured in units of κ, we can define

$$l_{\pm}^z(l_{\odot}, l_z) \equiv l_{\odot} \mp l_z \tag{3.16}$$

$$l_{+-} \equiv l_+^z(l_{\odot}, l_z - 1)l_-^z(l_{\odot}, l_z) = l_{\odot}(l_{\odot} + 1) - l_z^2 + l_z \tag{3.17}$$

$$l_{-+} \equiv l_-^z(l_{\odot}, l_z + 1)l_+^z(l_{\odot}, l_z) = l_{\odot}(l_{\odot} + 1) - l_z^2 - l_z \tag{3.18}$$

$$< l_+^z, l_-^z > \equiv \frac{1}{2}(l_{+-} + l_{-+}) = l_{\odot}(l_{\odot} + 1) - l_z^2 \tag{3.19}$$

$$[l_+^z, l_-^z] \equiv \frac{1}{2}(l_{+-} - l_{-+}) = l_z \tag{3.20}$$

372

and insure rotational invariance by the invariance of

$$< l^z_+, l^z_- > + l^2_z = l_\odot(l_\odot + 1) \equiv l^2_x + l^2_y + l^2_z \tag{3.21}$$

With

$$l_x = y\dot{z} - z\dot{y}; \quad l_y = y\dot{z} - z\dot{y} \tag{3.22}$$

the commutation relation for finite rotations is fully consistent with the commutation relation for rational fraction Lorentz boosts, (3.5) derived above. In fact, given either, the other follows (cf FDP, pp 85-86, and the reference to T.F.Jordan there cited). We conclude that the bracket expression (3.5) can be extended to the three dimensional result we need:

$$[x_i, \dot{x}_j] = \kappa\delta_{ij} = -[\dot{x}_j, x_i] \tag{3.23}$$

where $i, j, k \in 1, 2, 3$.

Since any Lorentz transformation is equivalent to a rotation and a boost, and we have now constructed integer rotations and integer boosts, we can now put the two together by the *scale invariant* definition of two constants c and κ,

$$\frac{\Delta\ell}{c\Delta t} \equiv 1; \quad \frac{c\Delta\ell}{\kappa} = 2\pi \tag{3.24}$$

3.4 CONSEQUENCES OF OUR BRACKET EXPRESSIONS

As already noted, the fundamental development in FDP derives general commutation relations from *attribute distance* and the exponentiation of the derivate operator. These bracket expressions can easily be shown to have the properties used in the next chapter to generalize the Feynman proof of the Maxwell Equations in our finite measurement accuracy context. Since we have not invoked that background, this section is devoted to showing that our more physical way of viewing the bracket expressions has the needed algebraic consequences which go beyond (3.23).

In the derivations given above we have in effect assumed that positions and velocities of a single particle in three orthogonal directions can be specified independently, subject to the constraints of invariance under finite and discrete Lorentz boosts and rotations. These constraints can be summarized by the bracket expression (3.23), where the anti-symmetry is explicitly noted. We did not explicitly record the independence of the three coordinate positions, corresponding to our free choice of counter positions in the laboratory:

$$[x_i, x_j] = 0 \tag{3.25}$$

Subject to our requirement of not going beyond the finite limits to which our measurements can refer, the fact that the x_i can be represented by integers, and the $\dot{x}_i$ by rational fractions, allows us to assume that, for λ, μ *constants* subject to the same restrictions and $A, B, C \in x_i, \dot{x}_i, i \in 1, 2, 3$, the bracket expression has the properties

$$[\lambda A + \mu B, C] = \lambda[A, C] + \mu[B, C]$$

$$[A, \lambda B + \mu C] = \lambda[A, B] + \mu[A, C] \tag{3.26}$$

$$[A, \mu] = 0$$

In the next chapter we will be concerned with functions $g(x, t)$ which are not functions of $\dot{x}$ and accelerations $\ddot{x}(x, \dot{x}, t)$ which are functions of velocity as well as position, but of no "higher derivatives". Since these are also subject to our finite integer and rational fraction restrictions, we can assume that they are polynomials whose powers have context sensitive restrictions. If n is the highest power of x which is allowed to occur, then for any component

$$[\dot{x}_i, x_j^n] = \delta_{ij}[\dot{x}_i, x_i^n] = \delta_{ij}(-\kappa x_i^{n-1} + x_i[\dot{x}_i, x_i^{n-1}]) = -n\kappa x_i^{n-1}\delta_{ij} \tag{3.27}$$

This allows us to identify the usual symbol $\partial g(x, t)/\partial x_k$ for all such functions we consider by the equality

$$[\dot{x}_k, g(x,t)] = -\kappa \partial g / \partial x_k \qquad (3.28)$$

Note also that

$$[x_i, g(x,t)] = 0 \Rightarrow g(x,t) \; independent \; of \; \dot{x} \qquad (3.29)$$

It remains to define the symbols $[\dot{x}_i, \dot{x}_j]$ and $\ddot{x}(x, \dot{x}, t)$ in our context. Since (within the restriction to polynomials mentioned above) we are now talking about *functions* of $x, \dot{x}$, and t, we can introduce the concept of a *path*

$$x(t) = (x_i(t), x_j(t), x_k(t); t) \qquad (3.30)$$

for the single particle we are considering. Then the bracket expression we derived above is equivalent to the definitions

$$x_i(t + \Delta t) \equiv x_i(t) + \dot{x}_i(t)\Delta t$$

$$[x_i, \dot{x}_j] \equiv [x_i(t + \Delta t)\dot{x}_j(t) - \dot{x}_j(t + \Delta t)x_i(t)] \qquad (3.31)$$

$$= [x_i \dot{x}_j(t) - x_j \dot{x}_i(t)] \equiv \kappa \delta_{ij}$$

Taking the obvious step of saying that if time changes by Δt, then

$$\dot{x}_j(t + \Delta t) \equiv \dot{x}_j(t) + \ddot{x}\Delta t \qquad (3.32)$$

and defining

$$[\dot{x}_i, \dot{x}_j] \equiv \dot{x}_i(t + \Delta t)\dot{x}_j(t) - \dot{x}_j(t + \Delta t)\dot{x}_i(t)$$

we have that

$$[\dot{x}_i, \dot{x}_j] + [x_i, \ddot{x}_j] = 0 \qquad (3.33)$$

We could have derived this directly when we were discussing the polygons needed for rotational invariance in the last section by noting that if x_i is the end of the chord, where the direction but not the magnitude of the velocity changes, and the change is due to the reversal of the component along that direction, while the component perpendicular to that remains unchanged, the acceleration must lie along x_j with the magnitude and sign given in the last equation above.

Now that we know what we mean by $[\dot{x}_i, \dot{x}_j]$ it is straightforward to establish the Jacobi identity

$$[A, [B, C]] + [B, [C, A]] + [C, [A, B]] = 0 \qquad (3.34)$$

for the symbols $A, B, C \in x_i, \dot{x}_i$. The same type of argument makes it easy to establish the fact that, in our context

$$g_k(x, t) = \kappa^{-1} \epsilon_{ijk} [\dot{x}_i, \dot{x}_j] \Rightarrow$$

$$\partial g_k / \partial t + [\dot{x}_j, \partial g_k / \partial x_j] = \kappa^{-1} \epsilon_{klm} [\dot{x}_l, \ddot{x}_m] \qquad (3.35)$$

a result which we will need in the next chapter.

4. SCALE INVARIANT GENERALIZATION OF THE FEYNMAN-DYSON-TANIMURA PROOF

4.1 HISTORICAL REMARKS

The development of a new fundamental theory is a tedious process, as the fact that the current effort started at least as early as 1951 clearly attests. Creating the climate of opinion which allows the professional community to make the requisite "paradigm shift" that leads to acceptance of the new theory can be even more tedious, and can be subject to various poorly understood historical delays. We discuss a possible example here. A long buried piece of work by Feynman created in 1948 (only three years earlier than Bastin and Kilmister's first publication in 1951) has recently come to light,[27] — thanks to its reconstruction by Dyson.[28] Had this been available in the 1950's, physics might have taken a different course. We recognized as soon as we saw it that Feynman's "paradoxical" proof of Maxwell's Equations from Newton's Second Law and the non-relativistic quantum mechanical commutation relations makes eminently good sense in our finite and discrete reconstruction of relativistic quantum mechanics.[29] Unfortunately, our attempt to spell this out in the regular literature failed.[30]

Fortunately, an analysis of the proof which makes some of the same points I had already noted has now been published by Tanimura.[31] We hope that this will give us another chance to attempt to get our views before a larger audience.[32] Even without our proposed generalization, Tanimura's claims are already quite startling. I quote his complete abstract:

"*R.P.Feynman showed F.J.Dyson a proof of the Lorentz force law and the homogeneous Maxwell equations, which he obtained starting from Newton's law of motion and the commutation relations between position and velocity for a single nonrelativistic particle. We formulate both a special relativistic and a general relativistic versions [sic] of Feynman's derivation. Especially in the general relativistic version we prove that the only possible fields that can consistently act on a quantum mechanical particle are scalar, gauge and gravitational fields. We also extend Feynman's scheme to the case of non-Abelian gauge theory in the special relativistic context.*"

In Tanimura's notation, the formulation of the theorem is simple:

Given

A single particle trajectory $x(t)$ in terms of three rectangular coordinates $x_i(t)$, $i \in 1, 2, 3$ subject to the constraints

$$[x_i, x_j] = 0; \quad m[x_i, \dot{x}_j] = i\hbar\delta_{ij}; \quad m\ddot{x}_k = F_k(x, \dot{x}; t) \tag{4.1}$$

then

the force components $F_k(x, \dot{x}; t)$ can be expressed in terms of two functions $E(x, t)$, $B(x, t)$ which depend only the coordinate components x_i and the time t and not on the velocities $\dot{x}_j$; these functions are related to the force by the component equations

$$F_i(x, \dot{x}; t) = E_i(x, t) + \epsilon_{ijk} < \dot{x}_j B_k(x, t) > \qquad (4.2)$$

and E, B satisfy the equations

$$div \ B = 0; \ \partial B / \partial t + rot \ E = 0 \qquad (4.3)$$

Here the Weyl ordering $<>$ is defined by

$$< ab > \equiv \frac{1}{2}[ab + ba]; \ < abc > \equiv \frac{1}{6}[abc + bca + cab + acb + cba + bac], \ etc. \quad (4.4)$$

4.2 SCALE INVARIANT POSTULATES

As was noted recently, the postulates can be made even simpler once one invokes scale invariance.[33] The Feynman postulates are independent of or linear in m. Therefore they can be replaced by the *scale invariant* postulates

$$f_k(x, \dot{x}; t) = \ddot{x}_k; \ [x_i, x_j] = 0; \ [x_i, \dot{x}_j] = \kappa \delta_{ij} \qquad (4.5)$$

where κ is any fixed constant with dimensions of area over time $[L^2/T]$ and f_k has the dimensions of acceleration $[L/T^2]$. This step is suggested by Mach's conclusion[34] that it is Newton's Third Law which allows mass *ratios* to be measured, while Newton's Second Law is simply a *definition* of force. Hence in a theory which contains only "mass points" and known mass ratios, the scale invariance of classical MLT physics reduces to a purely kinematical LT theory. Breaking scale invariance in such a theory requires not only some unique specification of a particulate mass standard, but also the requirement that this particle have some *absolute* significance.

Relativity need not change this situation. Specify c in a scale invariant way as both the maximum speed at which information can be transferred (limiting group velocity) and the minimum speed for supraluminal correlation *without* information transfer [limiting phase velocity =(coherence length)/(coherence time)]. If the unit of length is $\Delta\ell$ and the unit of time is Δt, then the equation $(\Delta\ell/c\Delta t) = 1$ has a *scale invariant* significance. Further, the interval I specified by the equation $c^2\Delta T^2 - \Delta L^2 = I^2$ can be given a *Lorentz invariant* significance. We can extend this analysis to include the scale invariant definition $\Delta E/c\Delta P = 1$ and the Lorentz invariant interval in energy-momentum space $(\Delta E^2/c^2) - \Delta P^2 = \Delta m^2$ provided we require that $\frac{\Delta P \Delta L}{\Delta m} = \frac{\Delta E \Delta T}{\Delta m}$. Then, given any arbitrary particulate mass standard, mass ratios can be measured using a Lorentz invariant *and* scale invariant LT theory. We trust that this dimensional analysis of the postulates used in the Feynman proof already removes part of the mystery about why it works.

The remaining physical point that needs to be made clear is that the "fields" referred to in classical relativistic field theory are *defined* in terms of their action on a *single* test particle. Thus, if we measure the *acceleration* of that particle in a Lorentz invariant way (force per unit rest mass) *and* the force per unit charge is also defined by acceleration *and* the charge per unit rest mass of the test particle is *also* a Lorentz invariant our electromagnetic field theory itself becomes an LT scale invariant theory. That is, once we replace the Feynman postulates by (4.5) and define $\mathcal{E}(x,t) = E/Q = F_E/m$ and $\mathcal{B}(x,t) = B/Q = F_B/m$, we need only derive the scale invariant version of equations (4.2), (4.3) obtained by the obvious notational change $F_i \to f_i$, $E_i \to \mathcal{E}_i$, $B_i \to \mathcal{B}_i$. We make a few remarks later on about the extension to gravitation, where the obvious physical postulate is that the ratio of gravitational to inertial mass of our test particle is also a Lorentz invariant.

4.3 THE PROOF

Here we essentially repeat Tanimura's version of the Feynman-Dyson proof of the Maxwell equations using our scale invariant results derived from measurement accuracy in the last chapter. There we proved that for any function g(x,t) we need consider in our finite and discrete theory (by which notation we mean that the function does not depend on $\dot{x}$ or higher derivates)

$$[x_i, x_j] = 0 \Rightarrow [x_i, g(x, t)] = 0 \tag{4.6}$$

Hence, in order to prove that a function is in this class we need only prove that it commutes with all the components x_i. For any acceleration $\ddot{x} = f(x, \dot{x}, t)$ which depends only on position, velocity and time — which Newton's second law defines as a "force per unit mass" — the result from finite measurement accuracy is that

$$[\dot{x}_i, \dot{x}_j] + [x_i, \ddot{x}_j] = 0 \Rightarrow [\dot{x}_i, \dot{x}_j] + [x_i, f_j] = 0 \tag{4.7}$$

Hence for any scale invariant "force" which has the Lorentz form

$$f_i(x, \dot{x}; t) = \mathcal{E}_i + \epsilon_{ijk} < \dot{x}_j \mathcal{B}_k > \tag{4.8}$$

the finite measurement accuracy result for the commutator has the implication

$$[x_i, \dot{x}_j] = \kappa \delta_{ij} \Rightarrow [\dot{x}_i, \dot{x}_j] = -[x_i, f_j] = \kappa \epsilon_{ijk} \mathcal{B}_k \tag{4.9}$$

This, in turn, allows us to define the scale invariant field $\mathcal{B}$ by

$$\mathcal{B}_k = \kappa^{-1} \epsilon_{ijk} [\dot{x}_i, \dot{x}_j] \tag{4.10}$$

The field $\mathcal{B}$ could still depend on velocity as well as position and time, but the fact that our bracket expression for position and velocity satisfies the Jacobi identity

$$[A, [B, C]] + [B, [C, A]] + [C, [A, B]] = 0 \tag{4.11}$$

and our central result in (4.1) allows us to show that the commutator of definition (4.10) with any coordinate vanishes, establishing the required property. Then the fact that our formalism for measurement accuracy implies that

$$[\dot{x}_k, g(x,t)] = -\kappa \partial g/\partial x_k \qquad (4.12)$$

with $\partial/\partial x_k$ interpreted as the partial derivate (finite difference) rather than the partial derivative allows us to infer that

$$[\dot{x}_k, \mathcal{B}(x,t)] = -\kappa \partial \mathcal{B}/\partial x_k \qquad (4.13)$$

But then our definition of $\mathcal{B}$ in (4.10) as proportional to the anti-symmetrized commutator together with the Jacobi identity establish the first Maxwell Equation

$$\Sigma_k \partial \mathcal{B}_k/\partial x_k \equiv div\mathcal{B} = 0 \qquad (4.14)$$

Now that we know what we mean by the magnetic field per unit charge, we can use the Lorentz "force" equation (4.8) as the definition of the electric field per unit charge. Then, taking the commutator with any component and using the definition of the Weyl ordering (4.4) together with the fact we proved above that $[x_i, \mathcal{B}_k] = 0$ we find that $[x_i, \mathcal{E}_k] = 0$ establishing that the scale invariant electric field so defined is also only a function of x and t. The final step requires us to define what we mean by "partial and total derivatives" with respect to time in the finite measurement accuracy context in concert with the space connectivity given by (4.12). The necessary result we need is noted in the last chapter and is that

$$g_k(x,t) = \kappa^{-1}\epsilon_{ijk}[\dot{x}_i, \dot{x}_j] \Rightarrow$$

$$(d/dt)g_k(x,t) = \partial g_k/\partial t + [\dot{x}_j, \partial g_k/\partial x_j] = \kappa^{-1}\epsilon_{klm}[\dot{x}_l, \ddot{x}_m] \qquad (4.15)$$

Then a sequence of algebraic steps given in Dyson's paper (Ref. 26) and summarized in Tanimura's Eq. 2.18 lead to the second Maxwell Equation

$$\partial \mathcal{B}/\partial t + rot\, \mathcal{E} = 0 \qquad (4.16)$$

We wish to emphasize that the essential formal work has already been accomplished by Tanimura. This allows us to omit algebraic details in the above rewriting of his proof, and refer the interested reader to his paper if he wishes to check them. The *physical* point that Tanimura does not mention is that the proof can be made scale invariant, and does not depend on either Planck's constant or on the bracket expression having an imaginary value. Thus it is more general that its historical "quantum mechanical" genesis might suggest.

Again, we are greatly indebted to Tanimura by analysing so carefully what properties of the bracket expression he needs. He points out that bracket expression [,] needed for the proof is not necessarily an *operator* expression. It suffices that it have the five algebraic properties

bilinearity,

$$[\lambda A + \mu B, C] = \lambda[A, C] + \mu[B, C],$$

$$[A, \lambda B + \mu C] = \lambda[A, B] + \mu[A, C];$$

anti-symmetry,

$$[A, B] = -[B, A];$$

the Jacobi identity,

$$[A, [B, C]] + [B, [C, A]] + [C, [A, B]] = 0;$$

Leibniz rule I,

$$[A, BC]] = [A, B]C + B[A, C]];$$

and Leibniz rule II,

$$\frac{d}{dt}[A, B] = [\frac{dA}{dt}, B] + [B, \frac{dA}{dt}].$$

He further notes that "*It is one of the virtues of Feynman's proof that there is no need for a priori existence of Hamiltonian, Lagrangian, canonical equation, or Heisenberg equation.*"

From the point of view of a finite and discrete theory, the most critical steps in the proof are those which depend on "derivatives". We derived all of these from measurement accuracy in the last chapter and have pointed out where they occur in our summary of Tanimura's proof.

4.4 EXTENSION TO GRAVITATION

In order to make his result manifestly covariant, Tanimura finds that he has to introduce an ordering parameter τ which is *not* the proper time. This idea has an old history going back at least to Stueckelberg, which we learned of from E.O.Alt. Alt's interest stemmed from the difficulty of formulating a relativistic quantum mechanical few body scattering theory without introducing the infinite number of degrees of freedom required in any "second quantized" formalism. We will pursue this idea elsewhere in our own context. Here we simply note that our new fundamental theory comes to us with this ordering parameter (bit-string length) built in, as will be seen on consulting the *program universe* method for generating bit-stings in the fundamental papers. Thus, as before, Tanimura's formal steps go over into our fundamental theory practically unchanged. We leave this for detailed exploration at a later stage in the development of the theory. Our further discussion of gravitation in this paper is better left until we have discussed the *combinatorial hierarchy*.

5. SUMMARY AND CONNECTION TO THE COMBINATORIAL HIERARCHY

Our new fundamental theory [13–16] models the process of "measurement" as the *counting* of finite and discrete bits of information acquired using an understood laboratory protocol. Any such theory can be modeled by the *bit strings* of contemporary computer practice, rules for combining them, and rules for relating them unambiguously (although possibly statistically) to that laboratory protocol.

Define *particles* as the *conceptual* carriers of conserved quantum numbers between events and *events* as regions across which quantum numbers are conserved. Take as the basic paradigm for two events the sequential firing of two counters separated by distance L and time interval T, where the clocks recording the firings are synchronized using the Einstein convention. Define the velocity of the "particle" connecting these two events as $v = \beta c = L/T$ where c is the limiting velocity for the transfer of *information*. If our unit of length is ΔL and our unit of time ΔT, then $c \equiv \Delta L/\Delta T$ has the same significance in any system of units; this definition is *scale invariant*. Similarly, we can define *Kepler's Constant* with the dimensions of area over time by $\kappa \equiv \Delta L^2/2\pi\Delta T = c\Delta L/2\pi$ while retaining scale invariance. In these units an event at $x = (n_1 - n_2)\Delta L$, $t = (n_1 + n_2)\Delta T$ is at a Lorentz invariant interval from the origin given by $I^2 = c^2t^2 - x^2 = 4n_1n_2\Delta L^2$. n_1, n_2 are simultaneously Lorentz invariant and scale invariant.

We assume that "fields" are to be measured by the acceleration of a "test particle" which belongs to a class of particles whose ratios of charge to mass and gravitational to inertial mass are Lorentz invariant. We relate space and time derivatives of functions of $x, \dot{x}, t$ to measurement accuracy by *deriving* the bracket expression $[x_i, \dot{x}_j] \equiv \kappa\delta_{ij}$ from an analysis of measurement accuracy and Kepler's Second Law for motion with constant velocity past a center. This also allows us to give meaning to finite and discrete accelerations through the relation $[\dot{x}_i, \dot{x}_j] + [x_i, \ddot{x}_j] = 0$ and to replace partial derivatives with finite "derivates". Then we show that Tanimura's proof [29] that the only fields which can act on such particles are structurally indistinguishable from electromagnetic and gravitational fields is a consequence of measurement accuracy alone, and does not depend on any specific assumptions about quantum mechanics.

Consider a particle bound to a center a distance r away which receives an impulsive force toward the center each time it has moved a distance λ whose square is $4n_1n_2\Delta L^2/(n_1 - n_2)^2$. If we take $2\pi r = j\lambda$, the area swept out per unit time by the radial distance to the particle is $(j^2 - \frac{1}{4})\kappa^2 = \ell(\ell+1)\kappa^2$ where we have defined $\ell = j - \frac{1}{2}$. Assuming that the probability of the impulsive force occurring after one step of length $\Delta L = h/m_e c$ is $1/137(\ell + 1)$ we obtain[16] Bohr's relativistic formula $\left(\frac{m-\epsilon_\ell}{m}\right)^2 [1 + (\frac{1}{137(\ell+1)})^2] = 1$ for the levels of the hydrogen atom[35] in the approximation $e^2/\hbar c \approx 1/137$, and hence *his* correspondence limit. Adding a second degree of freedom gives us the Sommerfeld formula and an improvement of four significant figures[16] in our value for $e^2/\hbar c$. Our scale invariant theory is the proper correspondence limit for any relativistic particle theory which breaks scale invariance by taking $m\kappa = \hbar$. [31] Note that $h/2mc$ is the longest threshold distance for the non-classical process of particle-antiparticle pair creation. For gravitational orbits of a particle of mass m about a center containing N particles of mass m, orbital velocity reaches c when $N = (\kappa c/Gm)^{\frac{1}{2}}$. Consequently the shortest distance (between *two* events!) in the theory is $\Delta L/N$. This is the "black hole radius" for mass Nm. Thanks to the fact that our Lorentz-invariant theory predicts both the (quantized) Newtonian interaction and spin 2 gravitons, it meets the three classical tests of general relativity.[36]

The numbers n_1, n_2 as integer descriptors of velocity and two more integers for angular momentum provide quantized Mandelstam variables for four-leg diagrams and discrete conservation laws. This allows a bit-string representation of particulate events. We label these "space-time descriptors" or *content* bit-strings by bit-strings of length 16. These *label* bit-strings combine by XOR (addition, mod 2) and are organized into the first three levels of the *combinatorial hierarchy:* (1) $3 = 2^2 - 1$; (2) $10 = 3 + (2^3 - 1 = 7)$; (3) $137 = 10 + (2^7 - 1 = 127) \approx \hbar c/e^2$. These 16 bits in the string specify the conserved quantum numbers of the standard model of quarks and leptons (charge, baryon number, lepton number, weak hypercharge, color) with confined color, weak-electromagnetic unification, and precisely the three observed generations.

The fourth level uses labels of length 256 and closes with $2^{127} + 136 \approx 1.7 \times 10^{38} \approx \hbar c/Gm_p^2$ possible ways they can combine pair-wise. This specifies m_p as the unit of mass. Then, since our labels conserve baryon number, lepton number and charge, any gravitating system with spin 1/2, unit charge and lepton or baryon number one collapses by emitting Hawking radiation to become a *stable* charged, rotating black hole. The number of bits of information lost in the collapse of $\hbar c/Gm^2$ is equal to the area of the event horizon in Planck areas.[37] This stabilizes the proton and electron. Then most of the electron mass is generated either electromagnetically or by the Fermi interaction. Self consistency of the electron mass calculated using either $e^2/\hbar c = 1/137$ or $G_f m_p^2/\hbar c = 1/\sqrt{2}(256)^2$ of this calculation gives weak-electromagnetic unification at the tree level. Since the labels must be generated before space-time can be constructed and takes on meaning, closure of level four plus baryon number conservation implies about $(2^{127})^2$ baryons in the universe. The resulting cosmology is good to at least first order. Current results are summarized in the following table.

Table I. **Coupling constants and mass ratios** predicted by the finite and discrete unification of quantum mechanics and relativity. Empirical Input: $c, \hbar$ and m_p as understood in the "Review of Particle Properties", Particle Data Group, *Physics Letters*, **B 239**, 12 April 1990.

COUPLING CONSTANTS

Coupling Constant	Calculated	Observed
$G^{-1}\frac{\hbar c}{m_p^2}$	$[2^{127} + 136] \times [1 - \frac{1}{3 \cdot 7 \cdot 10}] = 1.693\ 31\ldots \times 10^{38}$	$[1.69358(21) \times 10^{38}]$
$G_F m_p^2/\hbar c$	$[256^2\sqrt{2}]^{-1} \times [1 - \frac{1}{3 \cdot 7}] = 1.02\ 758\ldots \times 10^{-5}$	$[1.02\ 682(2) \times 10^{-5}]$
$sin^2\theta_{Weak}$	$0.25[1 - \frac{1}{3 \cdot 7}]^2 = 0.2267\ldots$	$[0.2259(46)]$
$\alpha^{-1}(m_e)$	$137 \times [1 - \frac{1}{30 \times 127}]^{-1} = 137.0359\ 674\ldots$	$[137.0359\ 895(61)]$
$G_{\pi N \bar{N}}^2$	$[(\frac{2M_N}{m_\pi})^2 - 1]^{\frac{1}{2}} = [195]^{\frac{1}{2}} = 13.96..$	$[13, 3(3), > 13.9?]$

MASS RATIOS

Mass ratio	Calculated	Observed
m_p/m_e	$\dfrac{137\pi}{\frac{3}{14}\left(1 + \frac{2}{7} + \frac{4}{49}\right)\frac{4}{5}} = 1836.15\ 1497\ldots$	$[1836.15\ 2701(37)]$
$m_\pi^\pm/m_e$	$275[1 - \frac{2}{2 \cdot 3 \cdot 7 \cdot 7}] = 273.12\ 92\ldots$	$[273.12\ 67(4)]$
m_{π^0}/m_e	$274[1 - \frac{3}{2 \cdot 3 \cdot 7 \cdot 2}] = 264.2\ 143\ldots$	$[264.1\ 373(6)]$
m_μ/m_e	$3 \cdot 7 \cdot 10[1 - \frac{3}{3 \cdot 7 \cdot 10}] = 207$	$[206.768\ 26(13)]$

COSMOLOGICAL PARAMETERS

Parameter	Calculated	Observed
N_B/N_γ	$\frac{1}{256^4} = 2.328\ldots \times 10^{-10}$	$\approx 2 \times 10^{-10}$
M_{dark}/M_{vis}	≈ 12.7	$M_{dark} > 10 M_{vis}$
$N_B - N_{\bar{B}}$	$(2^{127} + 136)^2 = 2.89\ldots \times 10^{78}$	*compatible*
ρ/ρ_{crit}	$\approx \frac{4 \times 10^{79} m_p}{M_{crit}}$	$.05 < \rho/\rho_{crit} < 4$

6. ACKNOWLEDGMENTS

We are indebted to V.A.Karmanov for bringing Tanimura's paper to our attention.

Appendix: COHERENCE, DETERMINISM and CHAOS

Abstract of contribution to ANPA 15, September 9-12, 1993

We assume that "fields" are to be measured by the acceleration of a "test particle" which belongs to a class of particles whose ratios of charge to mass and gravitational to inertial mass are Lorentz invariant. We relate the measurement accuracy in space, Δx, and in time, Δt, by the scale invariant definition of two constants c, and κ: $\frac{\Delta x}{c\Delta t} \equiv 1$; $\frac{\Delta x^2}{\kappa \Delta t} \equiv 2\pi$. Taking the experimental velocity resolution $\Delta v_x = \Delta x / T\Delta t = N\Delta x / NT\Delta t$ we derive the bracket expression $[x, v_x] = \kappa$ where $x = N\Delta x$. Then it is a *deductive* consequence that the only fields which can act on such particles are structurally indistinguishable from electromagnetic and gravitational fields in the sense that they satisfy the free space Maxwell Equations and Einstein geodesic equations. Such a scale invariant theory becomes the proper correspondence limit for any relativistic particle theory which breaks scale invariance by taking $m_e \kappa = \hbar$. Here we use m_e because it defines the threshold distance for position measurement, $h/2m_e c$, below which the non-classical process of electron-positron pair creation is observed, and above which that phenomenon cannot be *directly* observed. The coherence length $L = NT\Delta x$ specifies the maximum distance within which quantum mechanical interference effects can be observed. For non-overlapping "wave packets" of this length, the *deterministic* classical equations with particulate sources and sinks apply. But the characterization of a deterministic system as chaotic requires a specification of boundary conditions to a precision which violates the constraint due to measurement accuracy or electron-positron pair creation. Hence the number of degrees of freedom used in a model fixes whether the system is quantum coherent or classically decoherent but (approximately) deterministic and limits the applicability of chaos theory, removing certain paradoxes.

REFERENCES

1. P.W.Bridgman, *The Logic of Modern Physics* MacMillan, New York, 1928; hereinafter referred to as LMP.

2. P.W.Bridgman, *The Nature of Physical Theory*, Princeton University Press, 1936; hereinafter referred to as NPT.

3. E.W.Bastin and C.W.Kilmister, "The Analysis of Observations", *Proc. roy. Soc.* **A**, 559 (1952).

4. E.W.Bastin and C.W.Kilmister, "The Concept of Order I. The Space-Time Structure", *Proc. Camb. Phil.Soc.* **50**, 278 (1954).

5. E.W.Bastin and C.W.Kilmister, "Eddington's Theory in terms of The Concept of Order", *Proc. Camb. Phil.Soc.* **50**, 439 (1954).

6. E.W.Bastin and C.W.Kilmister, "The Concept of Order II. Measurement", *Proc. Camb. Phil.Soc.* **51**, 454 (1955).

7. E.W.Bastin and C.W.Kilmister, "The Concept of Order III. General Relativity as a Technique for Extrapolating over Great Distances", *Proc. Camb. Phil.Soc.* **53**, 462 (1957).

8. E.W.Bastin and C.W.Kilmister, "The Concept of Order IV. Quantum Mechanics", *Proc. Camb. Phil.Soc.* **55**, 66 (1959).

9. C.W.Kilmister, "Space, Time Discreteness", in *Modern perspectives on the Philosophy of Space and Time*, J.P. Van Bendegem, ed; *Philosophica*, **50**, 55-71 (1992,2).

10. T.Bastin, *Studia Philosophica Gandensia*, **4**, 77 (1966).

11. *Proc. ANPA*, C.W.Kilmister, or H.P.Noyes, or F. Abdullah, eds., available from F.Abdullah, City University, Northampton Square, London, EC1V0HB.

12. *Proc. ANPA WEST*, F. Young, ed., 409 Lealand Ave., Palo Alto, CA 94306.

13. T.Bastin and C.W.Kilmister, *Combinatorial Physics* (in preparation).

14. H.P.Noyes, *DISCRETE PHYSICS, A New Fundamental Theory*, J.C. van den Berg, ed. (in preparation).

15. H.P.Noyes, *Physics Today*, pp 99-100, March, 1992.

16. D.O.McGoveran and H.P.Noyes, "Foundations for a Discrete Physics", SLAC-PUB-4526 (June, 1989); hereinafter referred to as FDP.

17. H.P.Noyes and D.O.McGoveran, *Physics Essays*, **2**, 76-100 (1989).

18. D.O.McGoveran and H.P.Noyes, *Physics Essays*, **4**, 115-120 (1991).

19. FDP, pp 80-82; see also Goldreich et al., "How to Construct Random Functions", *Journal of the Association for Computing Machinery*, **33**, No. 4 (1986) pp 792-807.

20. D.O.McGoveran, H.P.Noyes, and M.J.Manthey, "On the Computer Simulation of the EPR-Bell-Bohm Experiment", presented at *BELL'S THEOREM, QUANTUM THEORY AND CONCEPTIONS OF THE UNIVERSE*, George Mason University, Fairfax, VA, 1988, and SLAC-PUB-4729 (December 1988).

21. H.P.Noyes, C.Gefwert, and M.J.Manthey, "The Measurement Problem in Program Universe", presented at the *Symposium on the Foundations of Modern Physics*, Joensuu, Finland, 1985 and SLAC-PUB-3537 rev. (July, 1985).

22. H.P.Noyes, C.Gefwert, and M.J.Manthey, "A Research Program with No 'Measurement Problem' ", *Annals of the N.Y. Acad. Sci.*, **480**, 553-563 (1985).

23. H.P.Noyes, "On the Discrete Reconciliation of Relativity and Quantum Mechanics", *Proceedings, XXVI Internationale Universitätswoken für Kernphysik*, Schladming, Austria, 1987, pp 287-290.

24. H.P.Noyes, "Quantized Conic Sections; Quantum Gravity", in *Logic, Computation and Measurement; Proc. ANPA WEST 9*, F. Young ed., ANPA WEST, 409 Lealand Ave., Palo Alto, CA 94306, 1993 and SLAC-PUB-5752 (Feb. 1993).

25. H.P.Noyes, "THE RQM TRIANGLE: A Paradigm for Relativistic Quantum Mechanics", in *Reconstructing the Object; Proc. ANPA 8*, F. Young ed., ANPA WEST, 409 Lealand Ave., Palo Alto, CA 94306, 1992 and SLAC-PUB-5752 (Feb. 1992), p.6.

26. For a more general discussion of how to "extract square roots" in the discretum cf. FDP, pp 23-24 and Ref.16, p. 116.

27. F.J.Dyson, *Physics Today*, No. 2, 32 (1989).

28. F.J.Dyson. *Am. J. Phys.* **58**, 209 (1990).

29. H.P.Noyes, "On Feynman's Proof of the Maxwell Equations", presented at the *XXX Internationale Univeristätswoken für Kernphysik*, Schladming, Austria, 1991 and SLAC-PUB-5411.

30. H.P.Noyes, "Comment on 'Feynman's Proof of the Maxwell Equations' ", SLAC-PUB-5588, Nov. 1991; rejected by *Am. J. Phys.*

31. S.Tanimura, *Annals of Physics* **220**, 229 (1992).

32. L.H.Kauffman and H.P.Noyes, "On the Proof of Classical Relativistic Field Equations using Fixed Measurement Accuracy", to be submitted to *Physical Review Letters* (in preparation).

33. H.P.Noyes, "On the Correspondence limit of Relativistic Quantum Mechanics", SLAC-PUB-6010 (May, 1993).

34. E.Mach, *Science of Mechanics*.

35. N.Bohr, *Phil Mag*, **332**, Feb. 1915.

Comment on *"Operationalism Revisited: Measurement Accuracy, Scale Invariance and the Combinatorial Hierarchy"*
H. Pierre Noyes (2000)

In this paper I attempted to revive Bridgman's operationalism as (a) a useful way to remind physicists and philosophers of how far relativity and quantum mechanics have strayed from their grounding in experimental physics and (b) to try to arouse interest in the ANPA research program as a possible framework in which that criticism might have useful consequences. I mention in this comment two ways in which some of the specific physics attempted here has become obsolete.

The discussions of EPR and the quantum measurement problem from the discrete point of view given in Ref.'s 19-22 are inadequate, and in some respects incorrect or misleading. They are superceded by Etter's "link theory". The first paper on this theory is published here as Chapter 15, and more recent work is discussed in Etter's "Comment" on this paper. Etter provides a "covering" theory within which a finite and discrete version of quantum and "classical" statistical mechanics can peacefully coexist. The more recent work has already begun to prove its worth in the current lively discussion about "quantum computers".

The work on trying to tame the Feynman-Dyson-Tanimura derivations of electromagnetism and gravitation in this paper reflects my then recently started collaboration with Lou Kauffman. Naturally this is better presented in our published papers reprinted here as Chapter 11 and Chapter 12. More recent work by us in these two directions is discussed in Lou's comments on our joint efforts in Ch. 12.

Discrete physics and the derivation of electromagnetism from the formalism of quantum mechanics

By Louis H. Kauffman[1] and H. Pierre Noyes[2]

Department of Mathematics, Statistics and Computer Science, University of Illinois at Chicago, 851 South Morgan Street, Chicago, IL 60607-7045, USA
Stanford Linear Accelerator Center, Stanford University, Stanford, CA 94309, USA

Freeman Dyson has recently focused attention on a remarkable derivation of electromagnetism from apparently quantum mechanical assumptions about the motion of a single particle. We present a new version of the Feynman–Dyson derivation in a discrete context. In the course of our derivation, we have uncovered a useful and elegant reformulation of the calculus of finite differences.

1. Introduction

In unpublished work circa 1948, Richard Feynman discovered that a quantum mechanical particle whose coordinates and momenta obeyed the simplest non-relativistic commutation relations will admit a description of acceleration that is compatible with Newton's second law and with the action of a classical electromagnetic field. This remarkable derivation was recently brought to the attention of the scientific community by the elegant paper of Freeman Dyson (1990). In his editorial comment on the reconstructed proof, Dyson remarks, '... here we find Galilean mechanics and Maxwell equations coexisting peacefully. Perhaps it was lucky that Einstein had not seen Feynman's proof when he started to think about relativity.' The proof has been generalized by Tanimura (1992) in a paper that embeds the Feynman argument into the contexts of gravity and gauge theories.

There are many themes to consider in the project of understanding the Feynman–Dyson derivation. In this paper, we concentrate on the following consideration: Feynman and Dyson assume commuting spatial coordinates $X_1(t), X_2(t), X_3(t)$, each a differentiable function of the time t. This occurs in the context of commutation relations of the form $[X_i, \dot{X}_j] = \kappa \delta_{ij}$ (κ a constant) giving the formalism the outward appearance of quantum mechanics. In the usual approaches to quantum mechanics, one has the corresponding equation $[q_i, p_j] = i\hbar \delta_{ij}$, where q_i is the position operator and p_j is the momentum operator. These operators are not themselves functions of time in the Schrödinger representation of quantum mechanics, but they are functions of time in the Heisenberg formulation. As a consequence, the Feynman–Dyson derivation does apply directly to quantum mechanics in the Heisenberg formulation.

The derivation is *not* classical mechanics with the commutator interpreted as a Poisson bracket. As noted by Tanimura (1992), the Leibnitz rule needed in the proof holds for Poisson brackets only if the dynamical variables are derived from a

Proc. R. Soc. Lond. A (1996) **452**, 81–95
Printed in Great Britain

TEX Paper

L. H. Kauffman and H. P. Noyes

Hamiltonian or a Lagrangian. One major reason for being interested in the proof stems from the fact that this assumption is *not* made. We wish to point out that in a context of *discrete physics* the derivation can still be carried out, and that in this context there need not be any demand for simultaneous values of position and momentum operators. In fact, this idea is simply meaningless in our discrete context.

Because the variables and fields in the Feynman–Dyson derivation are non-commutative, the question of Lorentz invariance requires a special analysis that we shall not attempt in this paper, but comment on briefly in the Appendix. There is nothing paradoxical about the Feynman–Dyson derivation as it stands: it is a piece of mathematical physics asking for a good interpretation.

The purpose of this paper is to analyse the Feynman–Dyson derivation in a context of discrete physics. In this context, a spatial variable X_i has values $X_i, X_i', X_i'', \ldots$ at successive values of discrete time. A measurement of velocity depends upon the difference of position values at two different (neighbouring) values of discrete time. Thus, we may (by convention) identify the value of $\dot{X}_i$ with $X_i' - X_i$ and write $\dot{X}_i := X_i' - X_i$. Since velocity depends upon two times and position on only one time, the idea of simultaneous determination of position and velocity is meaningless in the discrete context.

In order to achieve our aims, we have had to go to the roots of the calculus of discrete differences and discover an ordered version of this calculus that just fits the desired application. In this discrete ordered calculus (described in §2 and 3 of this paper), the operation of differentiation acts also to shift a product to its left by one time step. Thus, $X\dot{X} := X'(X' - X)$, while $\dot{X}X := (X' - X)X$. In the discrete ordered calculus, $\dot{X}$ and X do not commute and a specific commutation relation such as $\dot{X}X - X\dot{X} = \kappa$ is regarded as a hypothesis about the structure of their non-commutativity.

Furthermore, the discrete ordered calculus (DOC) obeys the rule for the differentiation of the product, $(AB)\dot{} = A\dot{B} + \dot{A}B$, precisely without any time shifting (see §2). This makes DOC an appropriate vehicle to support the calculus and non-commutative algebra that we need for our work.

In §4 we work out the derivation of electromagnetism in this discrete context. We begin with the assumption of the commutation relations for X_i ($i = 1, 2, 3$):

1. $[X_i, X_j] = 0$
2. $[X_i, \dot{X}_j] = \kappa \delta_{ij}$.

Here the dot ($\cdot$) is the discrete derivative and κ is a commuting scalar in DOC. We discuss reformulations of these equations in §4.

With $F_i = \ddot{X}_i$ and

$$H_\ell = \frac{1}{2\kappa} \epsilon_{jk\ell} [\dot{X}_j, \dot{X}_k], \quad F_j = E_j + \epsilon_{jk\ell} \dot{X}_k H_\ell, \quad (F = E + v \times H),$$

we show that

(i) div $H = 0$,
(ii) $\partial H / \partial t + \nabla \times E = 0$.

This is the desired result. Note that ϵ_{ijk} is the alternating symbol, and that $F = E + v \times H$ defines E.

In order to interpret these equations as electromagnetism, we need the other two Maxwell equations:

$$\text{div } E = 4\pi\rho, \quad \frac{\partial E}{\partial t} - \nabla \times H = 4\pi j$$

In our context, following Dyson, we take these equations as *definitions* of ρ and j. With these conventions we have a non-commutative electromagnetic formalism. It remains to be understood how this formalism is related to standard electromagnetism, and how the considerations of special relativity enter into this non-commutative context. It is our purpose, in this first paper, to put the derivation on a firm footing in order to provide a platform for consideration of these problems in subsequent work.

We have taken great care to perform this derivation in the discrete ordered calculus. This involves taking the following *definitions* for partial derivatives of a function $f(X)$:

$$\frac{\partial f}{\partial X_i} = \frac{1}{\kappa}[f, \dot{X}_i], \quad \dot{f} = \frac{\partial f}{\partial t} + \dot{X}_j \frac{\partial f}{\partial X_j}.$$

(The Einstein summation convention is in effect.) These definitions are discussed in §4.

We wish to close this introduction with a remark about the commutativity of X and X'. X' is regarded as the indicator of X after one discrete time step. Formally, we can write both XX' and $X'X$. However, in our convention, XX' means [measure X', *then* measure X] and this would require the observer to step backwards in time! For this reason we do not assume that $XX' = X'X$, and this gives us the *formal freedom* to postulate (in §4) a set of commutation relations among $\{X_i, X_j'\}$ that can be regarded as the basis of our derivations. In a sequel to this paper, we shall discuss actual numerical solutions to these relationships.

Obviously, much more work remains to be done in this domain. We shall discuss gravity/quantum formalism in a sequel to this paper.

2. Motivating a discrete calculus

In one-dimensional standard quantum mechanics in the Heisenberg formulation (Dirac 1947) the uncertainty principle takes the form of a commutation relation

$$QP - PQ = \hbar i$$

where Q and P denote, respectively, the position and momentum operators for the quantum mechanical particle, and $\hbar$ is Planck's constant divided by 2π ($i^2 = -1$).

There are many interpretations of this formalism. In the Schrödinger picture of quantum mechanics, the system is represented by a wavefunction $\psi = \psi(x, t)$, where x denotes the spatial coordinate and t denotes the temporal coordinate. The operators Q and P are defined by the equations

$$Q\psi = x\psi, \quad P\psi = \frac{\hbar}{i}\frac{\partial \psi}{\partial x}.$$

Thus,

$$(QP - PQ)\psi = x\frac{\hbar}{i}\frac{\partial \psi}{\partial x} - \frac{\hbar}{i}\frac{\partial}{\partial x}(x\psi) = (\hbar i)\psi.$$

Hence, $QP - PQ = \hbar i$.

The Heisenberg picture is not tied to this particular interpretation. It simply asserts that the order of application of the position and momentum operators matters—and that the difference of these orders is described by the commutation relations.

We can find an almost identical commutation relation by thinking about position and momentum in a classical but discrete context. In a discrete universe, time goes forward in measured ticks, and space occurs only in discrete intervals. We can imagine position determined at an instant, but to find velocity or momentum the clock must advance one tick to allow computation of the ratio of change of position to change of time. In measuring position first and then momentum, we advance the clock *after* determining position. If momentum is measured before position, the clock advances *before* the measurement of position and the position is determined at a later time. In this way, PQ and QP differ due to the intervening time step.

Let us quantify these last remarks by working with discrete position X and discrete velocity $\dot{X}$. Let $X, X', X'', \ldots$ denote the sequence of values for X at successive times $t_0, t_1, t_2, \ldots$. Define the value of $\dot{X}$ to be $X' - X$ and write $\dot{X} := X' - X$ t   ʰicate this evaluation. We regard $\dot{X}$ as a discrete velocity with the time step normaʌₑd to unity by convention.

Let $X\dot{X}$ denote the *process*—measure $\dot{X}$ then measure X. Thus, on evaluating, we find

$$X\dot{X} := X'(X' - X)$$

since measuring $\dot{X}$ requires stepping forward in time to the position X'.

On the other hand, $\dot{X}X$ denotes the process—measure X then measure $\dot{X}$. Thus, $\dot{X}X := (X' - X)X$.

We conclude that

$$X\dot{X} - \dot{X}X := X'(X' - X) - (X' - X)X.$$

This difference is not zero, and if it turns out to be a constant (κ) then we have the equation $X\dot{X} - \dot{X}X := \kappa$: a discrete analogue to the Heisenberg commutation relation.

In order to take the derivations of Dyson (1990) and Tanimura (1992) and place them on a discrete foundation, we shall develop a time-ordered calculus that generalizes the ideas that have been presented in this section. We end this section with an informal discussion of some of the issues that are involved.

One issue that must be faced is the question of the commutativity of X and X'. We can formally write both $X'X$ and XX'. The first $(X'X)$ means—measure X, take a time step, measure X after the time step. However, XX' does not have operational meaning in this same sense, since X' demands a time step while X asks for the value at a previous time. *We therefore assume that $X'X$ and XX' are distinct* without yet making any explicit assumption about the value of their difference.

The second issue involves evaluation. We have been careful to write $\dot{X} := X' - X$ rather than $\dot{X} = X' - X$, since the dot in $\dot{X}$ is a special instruction to shift time to its left in the ordered calculus. The directed equals sign ($:=$) is used to indicate evaluation. Thus, we can write $A\dot{B} := A'(B' - B)$ and

$$A\dot{B}\dot{C} := (A\dot{B})'(C' - C) \tag{2.1}$$

$$:= A'\dot{B}'(C' - C) \tag{2.2}$$

$$:= A''(B' - B)'(C' - C) \tag{2.3}$$

$$:= A''(B'' - B')(C' - C). \tag{2.4}$$

(We assume that $(XY)' = X'Y'$.) Each step in evaluation must perform all the time shifts for any dot that is eliminated. We shall return to this issue in the next section.

Returning to $X\dot{X}$ and $\dot{X}X$, we evaluate and find

$$X\dot{X} - \dot{X}X := X'(X' - X) - (X' - X)X \tag{2.5}$$

$$= X'(X' - X) - X(X' - X) + X(X' - X) - (X' - X)X \quad (2.6)$$
$$= (X' - X)(X' - X) + XX' - X^2 - X'X + X^2 \quad (2.7)$$
$$= (X' - X)^2 + [X, X']. \quad (2.8)$$

Thus,

$$X\dot{X} - \dot{X}X := (X' - X)^2 + [X, X'],$$

where $[A, B] = AB - BA$. (In general, we will not assume that $[X, X'] = 0$.)

If X and X' commute, then $[X, \dot{X}] := (X' - X)^2$. In this context one might assume that $(X' - X)^2 = \kappa$ is constant and declare that $X\dot{X} - \dot{X}X = \kappa$.

If X and X' do not commute then the formula above shows how their commutator is related to $[X, \dot{X}]$.

To summarize, *the dot in $\dot{X}$ is an instruction to take a time step*. A product AB means *do B, then do A*. Therefore, $X\dot{X} := X'(X' - X)$, since measuring X after one time step yields X'.

3. A discrete ordered calculus—DOC

By a variable X we mean a collection of algebraic entities $X, X', X'', X''', \ldots$ called 'the values of X at successive steps of discrete time'. No assumptions of commutativity are made for these variables, but we do assume that multiplication is associative and that multiplication distributes over addition and that there is a unit element, 1, such that $1X = X1$ for all X. Furthermore, we assume that $1' = 1$. Similarly, there is a 0 such that $0 + X = X$ for all X and $0' = 0$.

At this point the reader will see that we are assuming that a non-commutative ring R has been given, and that $X, X', \ldots, Y, Y', \ldots$ belong to R. (Note that this means that we assume that $X' + Y = Y + X'$.) Thus, we can speak concisely by saying that we assume as a given a (non-commutative) ring R with unit (1) equipped with a unary operator $' : R \to R$, such that $1' = 1$ and $0' = 0$ *and* $(a + b)' = a' + b'$ for all a and b in R, and $(ab)' = a'b'$ for all a, b in R. In the context of the ring R, we shall define a discrete ordered calculus by first adjoining to R a special element J the sole purpose of which is to keep track of the time shifting. We assume that J has the properties:

1. $J' = J$;
2. $AJ = JA'$ for all $A \in R$; (of course $JJ' = JJ$ so this works for J as well).

We let $\widehat{R}$ be the ring obtained from R by formally adjoining J to R with these properties. Since $\widehat{R}$ is, by definition, a ring with unit, this means that

$$(X + Y)J = XJ + YJ, \quad J(X + Y) = JX + JY, \quad J0 = 0, \quad J1 = 1, \quad \text{etc.}$$

Now note that any expression in $\widehat{R}$ can be rewritten (using $AJ = JA'$) in the form of a sum of elements of the form $J^k Z$, where there is no appearance of J in Z. We can define an *evaluation map* $E : \widehat{R} \to R$ by the following equations:

(i) $E(A + B) = E(A) + E(B)$ for any $A, B \in R$;
(ii) $E(J^k Z) = Z$ whenever $Z \in R$.

E is defined on $\widehat{R}$ by writing $A \in \widehat{R}$ as a sum of elements of the form $J^k Z$ and then applying (i) and (ii) above. For example,

$$E(AJ + BJ(CJ)) = E(JA' + J^2 B''C') = A' + B''C'$$

(assuming that $A, B, C \in R$). It follows from our assumptions that $E : \widehat{R} \to R$ is well defined. Note that, by definition, $E(E(X)) = E(X)$, where we regard $R \subset \widehat{R}$ as the set of expressions in $\widehat{R}$ without any J. In fact, we note that $\widehat{R} \cong \bigoplus_{n=0}^{\infty} J^n R$, where

$$J^n R = \{J^n r \mid r \in R\}, \quad \text{and} \quad (J^n r)(J^m s) = J^{n+m} r^{(m)} s,$$

where $r, s, \in R$ and $r^{(m)} = r''^{...\prime}$, with m 'primes'. With this reformulation, the evaluation map is obviously well defined. Now we are prepared to define differentiation in $\widehat{R}$ and therefore initiate the discrete ordered calculus (DOC).

Definition 3.1. Define $D : \widehat{R} \to \widehat{R}$ by the equation $D(X) = J(X' - X)$. The presence of J in DX makes it a time shifter for expressions on its left. (Compare this approach with (Etter & Kauffman 1994). That approach arose from discussions about an early version of the present paper.)

Proposition 3.2. *Let $A, B \in \widehat{R}$. Then*

$$D(AB) = D(A)B + AD(B).$$

Proof.

$$\begin{aligned}
D(AB) &= J((AB)' - AB) \\
&= J(A'B' - A'B + A'B - AB) \\
&= J(A'(B' - B) + (A' - A)B) \\
&= JA'(B' - B) + J(A' - A)B \\
&= AJ(B' - B) + J(A' - A)B \\
&= AD(B) + D(A)B.
\end{aligned}$$

∎

We see from the proof of this proposition how the ordering convention in the discrete calculus has saved the product rule for differentiation.

In a standard commutative time-discrete calculus, one of the terms in the expansion of the derivative of a product must be time shifted. The same phenomenon occurs in the infinitesimal calculus, but there an infinitesimal shift is neglected in the limit:

$$\begin{aligned}
\frac{\mathrm{d}}{\mathrm{d}t}(fg) &= \lim_{h \to 0} \frac{f(t+h)g(t+h) - f(t)g(t)}{h} \\
&= \lim_{h \to 0} \frac{f(t+h)g(t+h) - f(t+h)g(t) + f(t+h)g(t) - f(t)g(t)}{h} \\
&= \lim_{h \to 0} f(t+h) \left(\frac{g(t+h) - g(t)}{h} \right) + \left(\frac{f(t+h) - f(t)}{h} \right) g(t) \\
&= \left[\lim_{h \to 0} f(t+h) \right] \frac{\mathrm{d}g}{\mathrm{d}t} + \left(\frac{\mathrm{d}f}{\mathrm{d}t} \right) g(t).
\end{aligned}$$

It is interesting to see how the evaluations work in specific examples. In writing examples it is convenient to write $\dot{A}$ for $D(A)$. Thus, $\dot{A} = J(A' - A)$. For example,

$$\begin{aligned}
(X\dot{Y})\dot{} &= J((X\dot{Y})' - X\dot{Y}) \\
&= J((XJ(Y' - Y))' - XJ(Y' - Y)) \\
&= J(X'J(Y'' - Y') - XJ(Y' - Y)) \\
&= J(JX''(Y'' - Y') - JX'(Y' - Y)) \\
&= J^2(X''Y'' - X''Y' - X'Y' + X'Y).
\end{aligned}$$

Thus, $E((X\dot{Y})) = X''Y'' - X''Y' - X'Y' + X'Y$. On the other hand,

$$\dot{X}\dot{Y} = J(X' - X)J(Y' - Y)$$
$$= J^2(X'' - X')(Y' - Y).$$

$$X\ddot{Y} = XJ(\dot{Y}' - \dot{Y})$$
$$= XJ(J(Y' - Y)' - J(Y' - Y))$$
$$= XJ^2(Y'' - Y' - Y' + Y))$$
$$= J^2 X''(Y'' - 2Y' + Y).$$

$$\dot{X}\dot{Y} + X\ddot{Y} = J^2[(X'' - X')(Y' - Y) + X''(Y'' - 2Y' + Y)]$$
$$= J^2[X''Y' - X''Y - X'Y' + X'Y + X''Y'' - 2X''Y' + X''Y]$$
$$= J^2[-X''Y' - X'Y' + X'Y + X''Y'']$$
$$= (X\dot{Y})\dot{}.$$

This is a working instance of our formula $D(AB) = D(A)B + AD(B)$. In the remainder of the paper it will be useful to write $A := B$ to mean that $E(A) = E(B)$. In particular, we will often use this to mean that B has been obtained from A by expanding some derivatives and throwing away some or all of the left-most J. This means that while it is true that $E(A) = E(B)$, A and B cannot be substituted for one another in larger expressions, since they contain different time-shifting instructions.

An example of this usage is

$$\dot{X}\dot{Y} := \dot{X}'(Y' - Y).$$

Note that

$$\dot{X}\dot{Y} = \dot{X}J(Y' - Y) = J\dot{X}'(Y' - Y).$$

Thus,

$$E(\dot{X}\dot{Y}) = E(\dot{X}'(Y' - Y)).$$

In calculating, the $:=$ notation allows us to 'do the Js in our heads'.

Discussion. With the DOC formalized we can return to the structure of the commutator $[X, \dot{Y}] = X\dot{Y} - \dot{Y}X$. We have

$$[X, \dot{Y}] := X'(Y' - Y) - (Y' - Y)X$$
$$:= X'(Y' - Y) - X(Y' - Y) + X(Y' - Y) - (Y' - Y)X$$
$$:= (X' - X)(Y' - Y) + XY' - XY - Y'X + YX$$
$$:= (X' - X)(Y' - Y) + (XY' - Y'X) - (XY - YX)$$

$$[X, \dot{Y}] := (X' - X)(Y' - Y) + [X, Y'] - [X, Y]$$

This formula will be of use to us in the next section.

Note how, in this formalism, we cannot arbitrarily substitute $\dot{X}$ for $X' - X$ since the definition of the dot ('·') as a time shifter can change the value of an expression. Thus, $(X' - X)(Y' - Y) \neq \dot{X}\dot{Y}$. It may be useful to write $X' - X = \|\dot{X}\|$, where $\|\dot{X}\|$ is, by definition, the difference, stripped of its time-shifting properties. Then, $(X' - X)(Y' - Y) = \|\dot{X}\| \|\dot{Y}\|$ and we can write

$$[X, \dot{Y}] := \|\dot{X}\| \|\dot{Y}\| + [X, Y'] - [X, Y].$$

88   *L. H. Kauffman and H. P. Noyes*

Since $[A, B] = AB - BA$, these commutators satisfy the Jacobi identity. That is, we have

$$[[A, B], C] + [[C, A], B] + [[B, C], A] = 0.$$

The proof is by direct calculation.

4. Electromagnetism

In this section we give a discrete version of the Feynman–Dyson (Dyson 1990) derivation of the source-free Maxwell equations from a quantum mechanical formalism. We shall work in the discrete ordered calculus (DOC) of § 3. We assume time-series variables X_1, X_2 and X_3 and the commutation relations

(i) $[X_i, X_j] = 0, \quad \forall_{ij},$

(ii) $[X_i, \dot{X}_j] = \kappa \delta_{ij},$

where κ is a constant and κ commutes with all expressions in DOC.

We further assume that there are functions $F_i(X\ \dot{X})$ $(i = 1, 2, 3)$ such that $\ddot{X}_i = F_i(X, \dot{X})$. (Here writing $F(X)$ means that F is a function of X_1, X_2, X_3.) It is the purpose of this section to show that F_i takes on the pattern of the electromagnetic field in vacuum. Our first task will be to rewrite the above relations in terms of the discrete ordered calculus.

Proposition 4.1. *Given that $[X_i, X_j] = 0$ for all $i, j = 1, 2, 3$, and letting $\Delta_i = X_i' - X_i$, the equations $[X_i, \dot{X}_j] = \kappa \delta_{ij}$ imply the equations*

$$E([X_i, X_j'] + \Delta_i \Delta_j) = E(\kappa \delta_{ij}).$$

Proof. First assume $[X_i, \dot{X}_j] = \kappa \delta_{ij}$. Then $[X_i, \dot{X}_j] := \|\dot{X}_i\| \|\dot{X}_j\| + [X_i, X_j'] - [X_i, X_j]$ by the calculation at the end of § 2. Here, $\|\dot{X}_i\| = X_i' - X_i = \Delta_i$ and $[X_i, X_j] = 0$. Thus,

$$[X_i, \dot{X}_j] := \Delta_i \Delta_j + [X_i, X_j'].$$

∎

Discussion. This proposition shows how the Heisenberg-type relations $X_i \dot{X}_j - \dot{X}_j X_i = \kappa \delta_{ij}$ translate into the time-series commutation relations $X_i X_j' - X_j' X_i := \kappa \delta_{ij} - \Delta_i \Delta_j$. From the point of view of discrete physics, it is these relations that will implicate electromagnetism. Since there is no *a priori* reason for the elements of time series to commute with one another, we can regard the equations

$$X_i X_j = X_j X_i, \quad X_i X_j' - X_j' X_i = \kappa \delta_{ij} - \Delta_i \Delta_j, \quad (\Delta_i = X_i' - X_i)$$

as setting the *context* for the discussion of the physics of a discrete particle. It is in this context that the patterns of electromagnetism will appear.

Derivation. We shall need to interpret certain derivatives in terms of our discrete formalism. First of all, we have

$$\frac{\partial X_i}{\partial X_j} = \delta_{ij}.$$

Therefore,

$$[X_i, \dot{X}_j] = \kappa \, \frac{\partial X_i}{\partial X_j}.$$

Consequently, we make the following definition.

Definition 4.2. Let G be a function of X, then we define $\partial G/\partial X_i$ by the equation

$$\frac{\partial G}{\partial X_i} = \kappa^{-1}\left[G, \dot{X}_i\right].$$

We also wish to *define $\partial G/\partial t$*. This time derivative is distinct from $\dot{G}$. It should satisfy the usual relationship for multivariable calculus:

$$\dot{G} = \frac{\partial G}{\partial t} + \dot{X}_j\frac{\partial G}{\partial X_j}, \quad \text{(summed over } j = 1,2,3\text{)}.$$

Therefore, we define $\partial G/\partial t$ by the equation

$$\frac{\partial G}{\partial t} = \dot{G} - \dot{X}_j\kappa^{-1}\left[G, \dot{X}_j\right], \quad \text{(summed over } j = 1,2,3\text{)},$$

when $[X_m, G] = 0$ for $m = 1,2,3$.

The condition $[X_i, G] = 0$ for $i = 1,2,3$ implies that G has no dependence on $\dot{X}_j$ $(j = 1,2,3)$ under mild hypotheses on G. For, if we assume that G is either a polynomial or a (non-commutative) power series in X_j and $\dot{X}_j$, then the equations $[X_i, \dot{X}_j] = \kappa\delta_{ij}$ and $[X_i, X_j] = 0$ show that (under these assumptions) G has no occurrence of $\dot{X}_j$. It is necessary to define $\partial/\partial t$ since our discrete theory does not carry the conventional time variable t.

With these definitions in hand, we can proceed to the consequences of the commutation relations (i) and (ii).

Lemma 4.3.
$$[\dot{X}_i, X_j] = [\dot{X}_j, X_i].$$

Proof.

$$X_iX_j = X_jX_i$$
$$\Rightarrow (X_iX_j)^{\cdot} = (X_jX_i)^{\cdot}$$
$$\Rightarrow \dot{X}_iX_j + X_i\dot{X}_j = \dot{X}_jX_i + X_j\dot{X}_i$$
$$\Rightarrow \dot{X}_iX_j - X_j\dot{X}_i = \dot{X}_jX_i - X_i\dot{X}_j$$
$$\Rightarrow [\dot{X}_i, X_j] = [\dot{X}_j, X_i]$$

∎

Lemma 4.4.
$$[\dot{X}_j, \dot{X}_k] + [X_j, \ddot{X}_k] = 0.$$

Proof.

$$X_j\dot{X}_k - \dot{X}_kX_j = \kappa\delta_{jk} \tag{4.1}$$
$$\Rightarrow \dot{X}_j\dot{X}_k + X_j\ddot{X}_k - \ddot{X}_kX_j - \dot{X}_k\dot{X}_j = 0 \tag{4.2}$$
$$\Rightarrow [\dot{X}_j, \dot{X}_k] + [X_j, \ddot{X}_k] = 0 \tag{4.3}$$

∎

400

L. H. Kauffman and H. P. Noyes

Since $\ddot{X}_j = F_j(X, \dot{X})$ we have the following lemma.

Lemma 4.5.
$$[\dot{X}_j, \dot{X}_k] + [X_j, F_k] = 0.$$

Thus,

$$\begin{aligned}
[X_\ell, [X_j, F_k]] &= -[X_\ell, [\dot{X}_j, \dot{X}_k]] \\
&= [\dot{X}_j, [\dot{X}_k, X_\ell]] + [\dot{X}_k, [X_\ell, \dot{X}_j]], \quad \text{(by the Jacobi identity)} \\
&= [\dot{X}_j, -\delta_{k\ell}\kappa] + [\dot{X}_k, \kappa\delta_{\ell j}] \\
&= 0 + 0 = 0.
\end{aligned}$$

Note that

$$[X_j, F_k] = -[\dot{X}_j, \dot{X}_k] = +[\dot{X}_k, \dot{X}_j] = -[X_k, F_j].$$

Thus,

$$[X_j, F_k] = -[X_k, F_j].$$

We now define the field H by the equation

$$-\kappa\epsilon_{jk\ell}H_\ell = [X_j, F_k],$$

where $\epsilon_{jk\ell}$ is the alternating symbol for 123. That is, $\epsilon_{123} = +1$ and $\epsilon_{abc} = \text{sgn}(abc)$ if abc is a permutation of 123, where $\text{sgn}(abc)$ is the sign of the permutation. Otherwise, $\epsilon_{abc} = 0$.

Note that $[X_\ell, [X_j, F_k]] = 0$ implies that $[X_\ell, -\epsilon_{jk\ell}H_\ell] = 0$, which in turn implies that $[X_\ell, H_s] = 0$. This implies that H_s has no dependence upon $\dot{X}$ since $\dot{X}$ has a non-trivial commutator with X. Under these circumstances we will regard H as a function of X and compute $\partial H/\partial t$ according to the formula

$$\frac{\partial H}{\partial t} = \dot{H} - \dot{X}_j\kappa^{-1}[H, \dot{X}_j]$$

as discussed above.

Definition 4.6.

$$E_j = F_j - \epsilon_{jk\ell}\dot{X}_k H_\ell.$$

With this definition of E, we have $F = E + v \times H$, where $v = \dot{X}$.

Lemma 4.7.

$$[X_m, E_j] = 0.$$

Proof.

$$\begin{aligned}
[X_m, E_j] &= [X_m, F_j - \epsilon_{jk\ell}\dot{X}_k H_\ell] \\
&= [X_m, F_j] - [X_m, \epsilon_{jk\ell}\dot{X}_k H_\ell] \\
&= -\kappa\epsilon_{mj\ell}H_\ell - \epsilon_{jk\ell}X_m\dot{X}_k H_\ell + \epsilon_{jk\ell}\dot{X}_k H_\ell X_m \\
&= -\kappa\epsilon_{mj\ell}H_\ell - \epsilon_{jk\ell}X_m\dot{X}_k H_\ell + \epsilon_{jk\ell}\dot{X}_k X_m H_\ell \\
&= -\kappa\epsilon_{mj\ell}H_\ell - \epsilon_{jk\ell}[X_m, \dot{X}_k]H_\ell \\
&= -\kappa\epsilon_{mj\ell}H_\ell - \epsilon_{jk\ell}\kappa\delta_{mk}H_\ell \\
&= (-\kappa\epsilon_{mj\ell} - \kappa\epsilon_{jm\ell})H_\ell \\
&= 0.
\end{aligned}$$

Thus, E also has no dependence upon $\dot{X}$.

Remark 4.8.

$$H_\ell = \tfrac{1}{2}\kappa^{-1}\epsilon_{jk\ell}[\dot{X}_j, \dot{X}_k].$$

Proof.

$$-\kappa\epsilon_{jk\ell}H_\ell = [X_j, F_k]$$

and by lemma4.5,

$$[X_j, F_k] = -[\dot{X}_j, \dot{X}_k].$$

Thus,

$$\epsilon_{jk\ell}H_\ell = \kappa^{-1}[\dot{X}_j, \dot{X}_k].$$

From this it follows that

$$H_\ell = \tfrac{1}{2}\kappa^{-1}\epsilon_{jk\ell}[\dot{X}_j, \dot{X}_k].$$

∎

Lemma 4.9.

$$\operatorname{div} H = \sum_{i=1}^{3}\frac{\partial H_i}{\partial X_i} = 0.$$

Proof.

$$\sum_\ell \frac{\partial H_\ell}{\partial X_\ell} = \kappa^{-1}\sum_\ell [H_\ell, \dot{X}_\ell]$$
$$= \tfrac{1}{2}\kappa^{-2}\epsilon_{jk\ell}[[\dot{X}_j, \dot{X}_k], \dot{X}_\ell]$$
$$= 0, \quad \text{(by the Jacobi identity).}$$

Thus, $\operatorname{div} H = 0$. ∎

Lemma 4.10.

$$\frac{\partial H_\ell}{\partial t} = \epsilon_{jk\ell}\frac{\partial E_j}{\partial X_k}.$$

Proof.

$$H_\ell = \tfrac{1}{2}\kappa^{-1}\epsilon_{jk\ell}[\dot{X}_j, \dot{X}_k]$$
$$\frac{\partial H_\ell}{\partial t} = \dot{H}_\ell - \dot{X}_j\kappa^{-1}[H_\ell, \dot{X}_j].$$

(See the discussion of $\partial/\partial t$ given earlier in this section.)

Now

$$\dot{H}_\ell = \tfrac{1}{2}\kappa^{-1}\epsilon_{jk\ell}[\dot{X}_j, \dot{X}_k]\dot{}$$
$$= \tfrac{1}{2}\kappa^{-1}\epsilon_{jk\ell}([\ddot{X}_j, \dot{X}_k] + [\dot{X}_j, \ddot{X}_k])$$
$$= \kappa^{-1}\epsilon_{jk\ell}[\ddot{X}_j, \dot{X}_k]$$
$$= \kappa^{-1}\epsilon_{jk\ell}[F_j, \dot{X}_k]$$
$$= \kappa^{-1}\epsilon_{jk\ell}[E_j + \epsilon_{jrs}\dot{X}_r H_s, \dot{X}_k]$$
$$= \kappa^{-1}\epsilon_{jk\ell}[E_j, \dot{X}_k] + \kappa^{-1}[\dot{X}_k H_\ell, \dot{X}_k] - \kappa^{-1}[\dot{X}_\ell H_k, \dot{X}_k].$$

And

$$[\dot{X}_k H_\ell, \dot{X}_k] - [\dot{X}_\ell H_k, \dot{X}_k] = \dot{X}_k H_\ell \dot{X}_k - \dot{X}_k \dot{X}_k H_\ell - \dot{X}_\ell H_k \dot{X}_k + \dot{X}_k \dot{X}_\ell H_k$$
$$= \dot{X}_k[H_\ell, \dot{X}_k] - \dot{X}_\ell[H_k, \dot{X}_k] + [\dot{X}_k, \dot{X}_\ell]H_k$$
$$= \kappa \dot{X}_k \frac{\partial H_\ell}{\partial X_k} - \kappa \dot{X}_\ell \frac{\partial H_k}{\partial X_k} + [\dot{X}_k, \dot{X}_\ell]H_k.$$

Thus,

$$\dot{H}_\ell = \kappa^{-1}\epsilon_{jk\ell}[E_j, \dot{X}_k] + \dot{X}_k \frac{\partial H_\ell}{\partial X_k} - \dot{X}_\ell \frac{\partial H_k}{\partial X_k} + \kappa^{-1}[\dot{X}_k, \dot{X}_\ell]H_k.$$

Hence,

$$\frac{\partial H_\ell}{\partial t} = \dot{H}_\ell - \dot{X}_j \frac{\partial H_\ell}{\partial X_j} = \epsilon_{jk\ell} \frac{\partial E_j}{\partial X_k} - \dot{X}_\ell \frac{\partial H_k}{\partial X_k} + \kappa^{-1}[\dot{X}_k, \dot{X}_\ell]H_k.$$

However, the second term on the right-hand side vanishes because div $H = 0$, and the third term vanishes by symmetry. To see this, note that

$$H_1 = \kappa^{-1}[\dot{X}_2, \dot{X}_2]$$
$$H_2 = -\kappa^{-1}[\dot{X}_1, \dot{X}_3]$$
$$H_3 = \kappa^{-1}[\dot{X}_1, \dot{X}_2].$$

Thus,

$$\kappa[\dot{X}_k, \dot{X}_\ell]H_k = [\dot{X}_1, X_\ell][\dot{X}_2, \dot{X}_3] - [\dot{X}_2, \dot{X}_\ell][\dot{X}_1, \dot{X}_3] + [\dot{X}_3, \dot{X}_\ell][\dot{X}_2, \dot{X}_2].$$

This vanishes for $\ell = 1, 2, 3$. Therefore,

$$\frac{\partial H_\ell}{\partial t} = \epsilon_{jk\ell} \frac{\partial E_j}{\partial X_k}.$$

∎

This lemma completes the derivation of Maxwell's equations. We have shown that

$$\text{div } H = 0$$

and

$$\frac{\partial H}{\partial t} + \nabla \times E = 0.$$

As Dyson (1990) remarks, the other two Maxwell equations

$$\text{div } E = 4\pi\rho, \qquad \frac{\partial E}{\partial t} - \nabla \times H = 4\pi j$$

can be taken to define the external charge and current densities ρ and j. *However*, it is important to realize that our entire theory has applied only to a single trajectory. We can regard this trajectory (and its 'particle') as defining an electromagnetic field, *or* we can regard this particle as moving in an external field with these properties. We cannot have it both ways. The analysis so far in no way takes into account the self-interaction of this particle or its interactions with other particles and fields. Of course, our talk at this stage about the 'trajectory' of a particle is an *analogue* of a physical trajectory. The trajectory we talk about is in the space of $A \times A \times A$, where A denotes the non-commutative operator algebra that underlies the theory. An eventual interpretation of this theory in terms of trajectories in physical space is a possible consequence of further analysis of our formalism. It is beyond the scope of this preliminary paper.

We feel that the foregoing analysis of the Feynman–Dyson derivation in a discrete context lays bare much of the beautiful structure of the electromagnetic formalism and its relation to a condition of discrete time. We hope to probe this structure more deeply in subsequent papers.

L.H.K. was partially supported by the National Science Foundation under NSF Grant no. DMS-9205277. H.P.N. was supported by the Department of Energy, contract DE-AC03-76SF00515.

Appendix A. Historical remarks

One of us (HPN) has already claimed that the Feynman proof is *not* paradoxical (Noyes 1991) in the context of the finite and discrete reconciliation between quantum mechanics and relativity (Noyes 1987) achieved by a new fundamental theory (Noyes 1989, 1992, 1994a; McGoveran 1989, 1991). Noting that the Feynman postulates,

$$\mathcal{F}_k(x, \dot{x}; t) = m\ddot{x}_k, \quad [x_i, x_j] = 0, \quad m[x_i, \dot{x}_j] = i\hbar\delta_{ij}$$

are independent of or linear in m, we can replace them by the *scale-invariant* postulates

$$f_k(x, \dot{x}; t) = \ddot{x}_k, \quad [x_i, x_j] = 0, \quad [x_i, \dot{x}_j] = \kappa\delta_{ij},$$

where κ is any fixed constant with dimensions of area over time $[L^2/T]$ and f_k has the dimensions of acceleration $[L/T^2]$. This step is suggested by Mach's conclusion (Mach 1875) that it is Newton's third law which allows mass *ratios* to be measured, while Newton's second law is simply a *definition* of force. Hence, in a theory which contains only 'mass points', the Newtonian scale invariance of classical MLT physics reduces to the Galilean scale invariance of a purely kinematical LT theory. Breaking scale invariance in such a theory requires not only some unique specification of a particulate mass standard, but also the requirement that this particle have some *absolute* significance.

As has been remarked recently (Noyes 1994b), this aspect of scale invariance had already been introduced into the subject by Bohr & Rosenfeld (1933). In their classic paper, they point out that because QED depends only on the universal constants $\hbar$ and c, the discussion of the measurability of the fields can to a large extent be separated from any discussion of the atomic structure of matter (involving the mass and charge of the electron). Consequently, they are able to derive from the *non-relativistic* uncertainty relations the same restrictions on measurability (over finite space-time volumes) of the electromagnetic fields that one obtains directly from the second-quantized commutation relations of the fields themselves. Hence, to the extent that one could 'reverse engineer' their argument, one might be able to get back to the classical field equations and provide an alternative to the Feynman derivation based on the same *physical* ideas.

Turning to the commutation relations themselves, we note that a velocity measurement requires a knowledge of the space interval *and* the time interval between two events in two well separated spacetime volumes. Further, to embed these two positions in laboratory space, we must (in a relativistic theory) know the time it takes a light signal to go to one of these two positions and back to the other via a third reference position with a standard clock. Thus, we need *three* rather than two reference events to discuss the *connection* between position and velocity measurements.

We can then distinguish a measurement of position followed by a measurement of velocity from a measurement of velocity followed by a measurement of position. The minimum value of the difference between the product of position and velocity for measurements performed in the two distinct orders then specifies the constant in the basic 'commutation relation' needed in the Feynman derivation. So long as this value is finite and *fixed*, we need not know its metric value. This specifies what we mean by *discrete physics* in the main body of the text.

Relativity need not change this situation. Specify c in a scale-invariant way as both the maximum speed at which information can be transferred (limiting group velocity) and the maximum distance for supraluminal correlation *without* information transfer (maximum coherence length). If the unit of length is ΔL and the unit of time is ΔT, then the equation $(\Delta L/c\Delta T) = 1$ has a *scale-invariant* significance. Further, the interval I, specified by the equation $c^2\Delta T^2 - \Delta L^2 = I^2$, can be given a *Lorentz-invariant* significance. We can extend this analysis to includes the scale-invariant definition $\Delta E/c\Delta P = 1$ and the Lorentz-invariant interval in energy-momentum space $(\Delta E^2/c^2) - \Delta P^2 = \Delta m^2$ provided we require that $\Delta P\Delta L/\Delta m = \Delta E\Delta T/\Delta m$. Then, given any arbitrary particulate mass standard Δm, mass ratios can be measured using a Lorentz-invariant *and* scale-invariant LT theory. We trust that this dimensional analysis of the postulates used in the Feynman proof already removes part of the mystery about why it works, and suggests how it can be made 'Lorentz invariant' in a finite and discrete sense.

Appendix B. On the form of the derivative

In our discrete ordered calculus (see §3) we have defined the derivative $\dot{X}$ by the formula $\dot{X} = J(X' - X)$, where J is a formal element satisfying $J' = J$ and $XJ = JX'$ for all X. Thus, we have the equation

$$\dot{X} = JX' - JX = XJ - JX = [X, J].$$

This suggests that the equation $\dot{X} = [X, J]$ is an analogue of the corresponding equation in the Heisenberg formulation of quantum mechanics:

$$\dot{X} = [X, U],$$

where U is the time-evolution operator.

Michael Peskin has pointed out to us (personal communication) that we could accomplish the discretization of time in our theory by taking $U = e^{-iH\Delta t}$ (formally), where Δt denotes a discrete timestep. Then $\dot{X} = [X, U]$ and $X' = U^{-1}XU$ serves to define the time step from X to X'. Our approach and Peskin's meet if we identify J and U! The physical interpretation of this identification deserves further investigation.

In any case, it is interesting to note that the differentiation formula $(XY)' = \dot{X}Y + X\dot{Y}$ follows directly from the formula $\dot{X} = [X, J]$ without the necessity of introducing the step X'. This relationship between the discrete ordered calculus and the algebra of commutators will be used in the next installment of our work.

References

Bohr, N. & Rosenfeld, L. 1933 *Det. Kgl. Danske Videnskabernes Selskab., Mat. -fys. Med.* **XII**, (8).

Dirac, P. A. M. 1947 *The principles of quantum mechanics*, 3rd edn, p. 117. Oxford University Press.

Dyson, F. J. 1990 *Am. J. Phys.* **58**, 209.

Etter, T. & Kauffman, L. H. 1994 Discrete ordered calculus. *ANPA WEST Jl* (in preparation).

Mach, E. 1919 *Science of mechanics*, 4th edn. Chicago, IL: Open Court Publishing Co.

McGoveran, D. O. & Noyes, H. P. 1989 Foundations for a discrete physics. SLAC-PUB-4526.

McGoveran, D. O. & Noyes, H. P. 1991 *Physics Essays* **4**, 115–120.

Noyes, H. P. 1987 On the discrete reconciliation between quantum mechanics and relativity in Schladming proceedings. *Recent developments in mathematical physics*, pp 287–290.

Noyes, H. P. & McGoveran, D. O. 1989 *Physics Essays* **2**, 76–100.

Noyes, H. P. 1991 Comment on Feynmans proof. . . . SLAC-PUB-5588.

Noyes, H. P. 1992 *Physics Today*, March, pp. 99–100.

Noyes, H.P. 1994a Bit string physics: a novel theory of everything. In *Proc. Workshop on Physics and Computation (PhysComp 94)* (ed. D. Matzke), pp. 88–94. Los Alamitos, CA: IEEE Computer Society Press.

Noyes, H. P. 1994b On the measurability of electromagnetic fields: a new approach. In *A gift of prophesy; essays in celebration of the life of Robert E. Marshak* (ed. E. G. Sudarshan), pp. 365–372. Singapore: World Scientific (incomplete, see also SLAC-PUB-5445).

Tanimura, S. 1992 *Annls Phys.* **220**, 229–247.

Received 28 November 1994; accepted 10 March 1995

N·H

ELSEVIER

5 August 1996

PHYSICS LETTERS A

Physics Letters A 218 (1996) 139–146

Discrete physics and the Dirac equation [*]

Louis H. Kauffman [a,1], H. Pierre Noyes [b,2]

[a] *Department of Mathematics, Statistics and Computer Science, University of Illinois at Chicago,*
851 South Morgan Street, Chicago, IL 60607-7045, USA
[b] *Stanford Linear Accelerator Center, Stanford University, Stanford, CA 94309, USA*

Received 1 April 1996; accepted for publication 13 May 1996
Communicated by P.R. Holland

Abstract

We rewrite the 1+1 Dirac equation in light cone coordinates in two significant forms, and solve them exactly using the classical calculus of finite differences. The complex form yields "Feynman's checkerboard" – a weighted sum over lattice paths. The rational, real form can also be interpreted in terms of bit-strings.

PACS: 03.65.Pm; 02.70.Bf
Keywords: Discrete physics; Choice sequences; Dirac equation; Feynman checkerboard; Calculus of finite differences; Rational versus complex quantum mechanics

1. Introduction

In this paper we give explicit solutions to the Dirac equation for 1+1 space–time. These solutions are valid for discrete physics [1] using the calculus of finite differences, and they have as limiting values solutions to the Dirac equation using infinitesimal calculus. We find that the discrete solutions can be directly interpreted in terms of sums over lattice paths in discrete space–time. We document the relationship of this lattice-path with the checkerboard model of Richard Feynman [2]. Here we see how his model leads directly to an *exact* solution to the Dirac equation in discrete physics and thence to an exact continuum solution by taking a limit. This simplifies previous approaches to the Feynman checkerboard [3,4].

We also interpret these solutions in terms of choice sequences (bit-strings) and we show how the elementary combinatorics of $i = \sqrt{-1}$ as an operator on ordered pairs ($i[a, b] = [-b, a]$) informs the discrete physics. In this way we see how solutions to the Dirac equation can be built using only bit-strings, and no complex numbers. Nevertheless the patterns of composition of i inform the inevitable structure of negative case counting [5,6] needed to build these solutions.

The paper is organized as follows. Section 2 reviews the Dirac equation and expresses two versions (denoted RI, RII) in light cone coordinates. The two versions depend upon two distinct representations of the Dirac algebra. Section 3 reviews basic facts about the discrete calculus and gives the promised solutions to the Dirac equation. Section 4 interprets these solutions in terms of lattice paths, Feynman checkerboard

[*] Work partially supported by Department of Energy contract DE–AC03–76SF00515 and by the National Science Foundation under NSF Grant Number DMS-9295277.
[1] E-mail: kauffman@uic.edu.
[2] E-mail: noyes@slac.stanford.edu.

and bit-strings. Section 5 discusses the meaning of these results in the light of the relationship between continuum and discrete physics.

2. The 1+1 Dirac equation in light cone coordinates

We begin by recalling the usual form of the Dirac equation for one dimension of space and one dimension of time. This is

$$i\hbar \frac{\partial \psi}{\partial t} = E\psi, \tag{1}$$

where the energy operator E satisfies the dictates of special relativity and obeys the equation

$$E = c\sqrt{p^2 + m^2 c^2}, \tag{2}$$

where m is the mass, c the speed of light and p the momentum. Dirac linearized this equation by setting $E = c\alpha p + \beta mc^2$ where α and β are elements of an associative algebra (commuting with p, c, m). It then follows that

$$\begin{aligned} c^2(p^2 + m^2 c^2) &= (c\alpha p + \beta mc^2)^2 \\ &= c^2 p^2 \alpha^2 + m^2 c^4 \beta^2 + c^3 pm(\alpha\beta + \beta\alpha). \end{aligned} \tag{3}$$

Thus whenever $\alpha^2 = \beta^2 = 1$ and $\alpha\beta + \beta\alpha = 0$, these conditions will be satisfied. Thus we have Dirac's equation in the form $i\hbar\partial\psi/\partial t = (c\alpha p + \beta mc^2)\psi$. For our purposes it is most convenient to work in units where $c = 1$ and $\hbar/m = 1$. Then $i\partial\psi/\partial t = (\alpha p/\hbar + \beta)\psi$ and we can take $p = (\hbar/i)\partial/\partial x$ so that the equation is

$$i\frac{\partial \psi}{\partial t} = \left(-\alpha i \frac{\partial}{\partial x} + \beta\right)\psi. \tag{4}$$

We shall be interested in 2×2 matrix representations of the Dirac algebra $\alpha^2 = \beta^2 = 1$, $\alpha\beta + \beta\alpha = 0$. In fact we shall study two specific representations of the algebra. We shall call these representations RI and RII. They are specified by the equations below

$$\text{RI}: \quad \alpha = \begin{pmatrix} -1 & 0 \\ 0 & 1 \end{pmatrix}, \quad \beta = \begin{pmatrix} 0 & -i \\ i & 0 \end{pmatrix}, \tag{5}$$

$$\text{RII}: \quad \alpha = \begin{pmatrix} -1 & 0 \\ 0 & 1 \end{pmatrix}, \quad \beta = \begin{pmatrix} 0 & 1 \\ 1 & 0 \end{pmatrix}. \tag{6}$$

As we shall see, each of these representations leads to an elegant (but different) rewrite in the 1+1 light cone coordinates for space–time. RI leads to an equation with real-valued solutions. RII leads to an equation that corresponds directly to Feynman's checkerboard model for the 1+1 Dirac equation (Ref. [2]). The lattice paths of Feynman's model are the key to finding solutions to both versions of the equation. We shall see that these paths lead to exact solutions to natural discretizations of the equations.

We now make the translation to light cone coordinates. First consider RI. Essentially this trick for replacing the complex Dirac equation by a real equation was suggested to one of us by Karmanov [7]. Using this representation, the Dirac equation is

$$i\frac{\partial \psi}{\partial t} = \left(\begin{pmatrix} i & 0 \\ 0 & -i \end{pmatrix} \frac{\partial}{\partial x} + \begin{pmatrix} 0 & -i \\ i & 0 \end{pmatrix} \right)\psi, \tag{7}$$

whence

$$\frac{\partial \psi}{\partial t} = \left(\begin{pmatrix} 1 & 0 \\ 0 & -1 \end{pmatrix} \frac{\partial}{\partial x} + \begin{pmatrix} 0 & -1 \\ 1 & 0 \end{pmatrix} \right)\psi. \tag{8}$$

If $\psi = \binom{\psi_1}{\psi_2}$ where ψ_1 and ψ_2 are real-valued functions of x and t, then we have

$$\begin{pmatrix} -\psi_2 \\ \psi_1 \end{pmatrix} = \begin{pmatrix} \partial\psi_1/\partial t - \partial\psi_1/\partial x \\ \partial\psi_2/\partial t + \partial\psi_2/\partial x \end{pmatrix}. \tag{9}$$

Now the light cone coordinates of a point (x, t) of space–time are given by $[r, l] = [\frac{1}{2}(t+x), \frac{1}{2}(t-x)]$ and hence the Dirac equation becomes

$$\begin{pmatrix} -\psi_2 \\ \psi_1 \end{pmatrix} = \begin{pmatrix} \partial\psi_1/\partial l \\ \partial\psi_2/\partial r \end{pmatrix}. \tag{10}$$

Remark. It is of interest to note that if we were to write $\psi = \psi_1 + i\psi_2$, then the Dirac equation in light cone coordinates takes the form $D\psi = i\psi$ where $D(\psi_1 + i\psi_2) = \partial\psi_1/\partial l + i\partial\psi_2/\partial r$. In any case, we shall refer to Eq. (9) as the RI Dirac equation.

Now, let us apply the same consideration to the second representation RII. The Dirac equation becomes

$$i\frac{\partial \psi}{\partial t} = \left(\begin{pmatrix} i & 0 \\ 0 & -i \end{pmatrix} \frac{\partial}{\partial x} + \begin{pmatrix} 0 & 1 \\ 1 & 0 \end{pmatrix} \right)\psi. \tag{11}$$

Thus

$$\frac{\partial \psi}{\partial t} = \left(\begin{pmatrix} 1 & 0 \\ 0 & -1 \end{pmatrix} \frac{\partial}{\partial x} + \begin{pmatrix} 0 & -i \\ -i & 0 \end{pmatrix} \right)\psi. \tag{12}$$

408

Hence

$$\begin{pmatrix} -i\psi_2 \\ -i\psi_1 \end{pmatrix} = \begin{pmatrix} \partial\psi_1/\partial l \\ \partial\psi_2/\partial r \end{pmatrix}. \qquad (13)$$

We shall call Eq. (13) the RII Dirac equation.

3. Discrete calculus and solutions to the Dirac equation

Suppose that $f = f(x)$ is a function of a variable x. Let Δ be a fixed nonzero constant. The discrete derivative of f with respect to Δ is then defined by the equation

$$D_\Delta f(x) = \frac{f(x+\Delta) - f(x)}{\Delta}. \qquad (14)$$

Consider the function

$$x^{(n)} = x(x-\Delta)(x-2\Delta)\ldots(x-(n-1)\Delta). \qquad (15)$$

Lemma.

$$D_\Delta x^{(n)} = nx^{(n-1)}. \qquad (16)$$

Proof.

$$(x+\Delta)^{(n)} - x^{(n)}$$
$$= (x+\Delta)(x)(x-\Delta)\ldots(x-(n-2)\Delta)$$
$$- (x)(x-\Delta)\ldots(x-(n-2)\Delta)(x-(n-1)\Delta)$$
$$= [(x+\Delta)-(x-(n-1)\Delta)]x^{(n-1)} = n\Delta x^{(n-1)}.$$

Thus

$$D_\Delta x^{(n)} = \frac{n\Delta x^{(n-1)}}{\Delta} = nx^{(n-1)}. \qquad (17)$$

We are indebted to Eddie Grey for reminding us of this fact [8].

Note that as Δ approaches zero $x^{(n)}$ approaches x^n, the usual nth power of x. Note also that

$$\frac{x^{(n)}}{\Delta^n n!} = C_n^{x/\Delta} \qquad (18)$$

where

$$C_n^z = \frac{z(z-1)\ldots(z-n+1)}{n!} \qquad (19)$$

is a (generalized) binomial coefficient. Thus

$$\frac{x^{(n)}}{n!} = \Delta^n C_n^{x/\Delta}. \qquad (20)$$

With this formalism in hand, we can express functions whose combination will yield solutions to discrete versions of the RI and RII Dirac equations described in the previous section. After describing these solutions, we shall interpret them as sums over lattice paths.

To this end, let $\partial_\Delta/\partial r$ and $\partial_\Delta/\partial l$ denote discrete partial derivatives with respect to variables r and l. Thus

$$\frac{\partial_\Delta f}{\partial r} = \frac{f(r+\Delta, l) - f(r, l)}{\Delta},$$
$$\frac{\partial_\Delta f}{\partial l} = \frac{f(r, l+\Delta) - f(r, l)}{\Delta}. \qquad (21)$$

Define the following functions of r and l

$$\psi_R^\Delta(r, l) = \sum_{k=0}^{\infty} (-1)^k \frac{r^{(k+1)}}{(k+1)!} \frac{l^{(k)}}{k!},$$

$$\psi_L^\Delta(r, l) = \sum_{k=0}^{\infty} (-1)^k \frac{r^{(k)}}{k!} \frac{l^{(k+1)}}{(k+1)!},$$

$$\psi_0^\Delta(r, l) = \sum_{k=0}^{\infty} (-1)^k \frac{r^{(k)}}{k!} \frac{l^{(k)}}{k!}. \qquad (22)$$

Note that as $\Delta \to 0$, these functions approach the limits:

$$\psi_R(r, l) = \sum_{k=0}^{\infty} (-1)^k \frac{r^{k+1}}{(k+1)!} \frac{l^k}{k!},$$

$$\psi_L(r, l) = \sum_{k=0}^{\infty} (-1)^k \frac{r^k}{k!} \frac{l^{(k+1)}}{(k+1)!},$$

$$\psi_0(r, l) = \sum_{k=0}^{\infty} (-1)^k \frac{r^k}{k!} \frac{l^k}{k!}. \qquad (23)$$

Note also, that if r/Δ and l/Δ are *positive integers*, then ψ_R^Δ, ψ_L^Δ and ψ_0^Δ are *finite sums* since $x^{(n)}/n! = \Delta^n C_n^{x/\Delta}$ will vanish for sufficiently large n when x/Δ is a sufficiently large integer.

Now note the following identities about the derivatives of these functions

$$\frac{\partial_\Delta \psi_R^\Delta}{\partial r} = \psi_0^\Delta, \qquad \frac{\partial_\Delta \psi_0^\Delta}{\partial r} = -\psi_L^\Delta,$$

$$\frac{\partial_\Delta \psi_L^\Delta}{\partial l} = \psi_0^\Delta, \qquad \frac{\partial_\Delta \psi_0^\Delta}{\partial l} = -\psi_R^\Delta. \tag{24}$$

In the limit $\Delta \searrow 0$, these can be regarded as continuum derivatives.

We can now produce solutions to both the RI and the RII Dirac equations. For RI, we shall require

$$\frac{\partial \psi_1}{\partial l} = -\psi_2, \qquad \frac{\partial \psi_2}{\partial r} = \psi_1. \tag{25}$$

We shall omit writing the Δ's in those equations, since all these calculations take the same form independent of the choice of Δ. Of course for finite Δ and integral r/Δ, l/Δ these series produce discrete calculus solutions to the equations.

Let

$$\psi_1 = \psi_0 - \psi_L, \qquad \psi_2 = \psi_0 + \psi_R. \tag{26}$$

It follows immediately that this gives a solution to the RI Dirac equation.

Similarly, if we let

$$\psi_1 = \psi_0 - i\psi_L, \qquad \psi_2 = \psi_0 - i\psi_R \tag{27}$$

then

$$\frac{\partial \psi_1}{\partial l} = -i\psi_2, \qquad \frac{\partial \psi_2}{\partial r} = -i\psi_1. \tag{28}$$

This gives a solution to the RII Dirac equation.

In the next section we consider the lattice path interpretations of these solutions.

4. Lattice paths

In this section we interpret the discrete solutions of the Dirac equation given in the previous section in terms of counting lattice paths. As we have remarked in the previous section, the solutions are built from the functions ψ_0, ψ_R and ψ_L. These functions are finite sums when r/Δ and l/Δ are positive integers, and we can rewrite them in the form

$$\psi_R(r, l) = \sum_{k=0}^{\infty} (-1)^k \Delta^{2k+1} C_{k+1}^{r/\Delta} C_k^{l/\Delta},$$

$$\psi_L(r, l) = \sum_{k=0}^{\infty} (-1)^k \Delta^{2k+1} C_k^{r/\Delta} C_{k+1}^{l/\Delta},$$

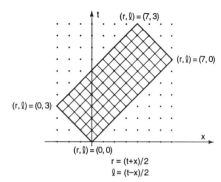

Fig. 1. Rectangular lattice in Minkowski space–time.

$$r = (t+x)/2$$
$$\ell = (t-x)/2$$

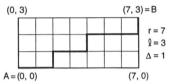

Fig. 2. An example of a path in the light-cone rectangle.

$$\psi_0(r, l) = \sum_{k=0}^{\infty} (-1)^k \Delta^{2k} C_k^{r/\Delta} C_k^{l/\Delta}, \tag{29}$$

where C_n^z is given in (19).

We are thinking of r and l as the light cone coordinates $r = \frac{1}{2}(t+x)$, $l = \frac{1}{2}(t-x)$. Hence, in a standard diagram for Minkowski space–time, a pair of values $[r, l]$ determines a rectangle with sides of length l and r on the left and right pointing light cones. (We take the speed of light $c = 1$.) This is shown in Fig. 1.

Clearly, the simplest way to think about this combinatorics is to take $\Delta = 1$. If we wish to think about the usual continuum limit, then we shall fix values of r and l and choose Δ small but such that r/Δ and l/Δ are integers. The combinatorics of an $r \times l$ rectangle with integers r and l is no different in principle than the combinatorics of an $(r/\Delta) \times (l/\Delta)$ rectangle with integers r/Δ and l/Δ. Accordingly, we shall take $\Delta = 1$ for the rest of this discussion, and then make occasional comments to connect this with the general case.

410

Finally, for thinking about the combinatorics of the $r \times l$ rectangle, it is useful to view it turned by $45°$ from its light-cone configuration. This is shown in Fig. 2. We shall consider lattice paths on the $r \times l$ rectangle from $A = [0,0]$ to $B = [r,l]$. Each step in such a path consists in an increment of either the first or the second light cone coordinate. The "particle" makes a series of "left or right" choices to get from A to B. In counting the lattice paths we shall represent *left* and *right* by

(left is vertical in the rotated representation). Now notice that a lattice path has two types of corners:

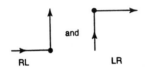

We can count RL corners by the point on the L axis where the path increments. We can count LR corners by the point on the R axis where the path increments. A lattice path is then determined by a choice of points from the L and R axes. More specifically, there are paths that begin in R (go right first) and end in L, begin in L and end in R, begin in L and end in L, begin in R and end in R. We call these paths of type RL, LR, LL and RR respectively. (Note that a RL corner is a two-step path of type RL and that an LR corner is a two step path of type LR.) It is easy to see that an RL path involves k points from the R axis and $k+1$ points from the L axis, an LR path involves $k+1$ points from the R axis and k points from the L axis, while an LL or RR path involves the choice of k points from each axis. See Fig. 3 for examples. As a consequence, we see that if $\|XY\|$ denotes the number of paths from A to B of type XY, then

$$\|RL\| = \sum_k C_k^r C_{k+1}^l,$$

$$\|LR\| = \sum_k C_{k+1}^r C_k^l, \qquad (30)$$

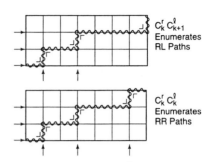

Fig. 3. Showing by example that $C_k^r C_{k+1}^l$ enumerates RL paths and $C_k^r C_k^l$ enumerates RR paths.

$$\|RR\| = \|LL\| = \sum_k C_k^r C_k^l.$$

We see, therefore, that our functions ψ_0, ψ_R and ψ_L can be regarded as weighted sums over these different types of lattice path. In fact, we can re-interpret $(-)^k$ in terms of the number of corners (choices) in the paths:

RR $\Rightarrow 2k$ corners,

LR $\Rightarrow 2k + 1$ corners,

RL $\Rightarrow 2k + 1$ corners,

LL $\Rightarrow 2k$ corners.

Hence if $N_c(XY)$ denotes the number of paths with c corners of type XY then

$$\psi_0 = \sum_c (-1)^{c/2} N_c(LL) = \sum_c (-1)^{c/2} N_c(RR),$$

$$\psi_R = \sum_c (-1)^{(c-1)/2} N_c(LR),$$

$$\psi_L = \sum_c (-1)^{(c-1)/2} N_c(RL). \qquad (31)$$

From the point of view of the solution to the RI Dirac equation ($\psi_1 = \psi_0 - \psi_L$, $\psi_2 = \psi_0 + \psi_R$) it is an interesting puzzle in discrete physics to understand the nature of the negative case counting that is entailed in the solution. (An attempt has been made by one of us to interpret this in terms of spin or particle number conservation in the presence of random electromagnetic fluctuations producing the paths [9].) The signs do not appear to come from local considerations along the path.

The RII Dirac solution gives a different point of view. Here $\psi_1 = \psi_0 - i\psi_L$, $\psi_2 = \psi_0 - i\psi_R$. Taken the hint given by the appearance of i, we note that $i^{2k} = (-)^k$ while $i^{2k+1} = (-1)^k i$. Thus

$$\psi_1 = \sum_c (-i)^c N_c(R), \quad \psi_2 = \sum_c (-i)^c N_c(L),$$

where $N_c(R)$ denotes the number of paths that start to the right and have c crossings, while $N_c(L)$ denotes the number of paths that start to the left and have c crossings. This shows that our solution in the RII case is precisely in line with the amplitudes described by Feynman and Hibbs (Ref. [2]) for their checkerboard model of the Dirac propagator. See also Refs. [10,3] for the relationship of the Feynman model to the combinatorics of the Ising model in statistical mechanics.

Returning now to the RI equation, we see that $(-i)^c N_c$ gives the clue to the combinatorics of the signs. In our RI formulation, no complex numbers appear and none are needed if we take a combinatorial interpretation of i as an operator on ordered pairs: $i[a, b] = [-b, a]$. Then we can think of a "pre-spinor" in the form of a labeled $\frac{1}{2}\pi$ angle associated to each corner:

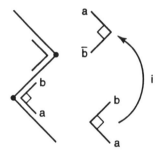

As the particle moves from corner to corner its pre-spinor is operated on by i. There is a combination of one sign change and one change in order. The total sign change from the beginning of the path to the end documents the positivity or negativity of the count.

As the referee kindly remarks to us, in the path-integral approach to nonrelativistic quantum mechanics it is well known that (for quadratic systems or

within a semi-classical approximation) each of these turning points contributes a phase $\frac{1}{2}\pi$ to the propagator. This is due to the Fresnel type of the propagator. Hence what we call a "pre-spinor" arises naturally in that approach.

5. Epilogue

If we had started by saying (in the RI case) we had a simple solution for the Dirac equation (discretized) using nothing but *bit-strings* (L,R choice sequences) and appropriate signs, then it would have been natural to ask: How are these signs justified on the basis of a philosophy of bit-strings? In retrospect we can answer: This pattern of signs is very simple, but not (yet) to be deduced from the notion of a distinction alone. Nevertheless, it does arise naturally from the simple structures that are available at that primitive level. The i operator ($i[a, b] = [-b, a]$) does not involve anything more sophisticated that the idea of exchanging the labels on the two sides of a distinction followed by the flipping of a label on a given side:

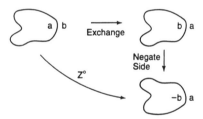

as is discussed elsewhere [11,12]. A choice sequence such as

$$\overset{\uparrow}{R}\ \overset{\uparrow}{L}\ RR\ \overset{\uparrow}{L}\ \overset{\uparrow}{R}RRR\ \overset{\uparrow}{L}$$

has "corners" wherever R meets L or L meets R. We have characterized these corners into two types RL and LR:

412

We then enumerate the choice sequences in terms of lattice paths in Minkowski space and the solutions to the Dirac equation emerge, along with a precursor to spin and the role of $i = \sqrt{-1}$ in quantum mechanics. We have shown exactly how this point of view interfaces with Feynman's checkerboard.

Corners in the bit-string sequence alternate from RL to LR and from LR to RL. The moral of Feynman's $(-i)^c$ where c is the number of corners is that this alteration should be regarded as an elementary rotation:

$$L \rightarrow \quad \rightarrow \quad \rightarrow \quad $$

One may wonder, why does this simple combinatorics occur in a level so close to the making of one distinction, and yet implicate fully the solutions to the Dirac equation in continuum 1+1 physics?! We cannot begin to answer such a question except with another question: If you believe that simple combinatorial principles underlie not only physics and physical law, but the generation of space–time herself, then these principles remain to be discovered. What are they? What are these principles? It is no surprise to the mathematician that i ends up as central to the quest. For i is a strange amphibian not only neither 1 nor -1, i is neither discrete nor continuous, not algebra, not geometry, but a communicator of both. In this essay we have seen the beginning of a true connection of discrete and continuum physics.

The continuum version of our theory merges the paths on the lattice to a sum over all possible paths on an infinitely divided rectangle in Minkowski space–time. The individual paths disappear into the values of the series

$$\psi_0 = \sum_{k=0}^{\infty} (-1)^k \frac{r^k}{k!} \frac{l^k}{k!},$$

$$\psi_L = \sum_{k=0}^{\infty} (-1)^k \frac{r^k}{k!} \frac{l^{k+1}}{(k+1)!},$$

$$\psi_R = \sum_{k=0}^{\infty} (-1)^k \frac{r^{k+1}}{(k+1)!} \frac{l^k}{k!}.$$

Here we have a glimpse of the possibilities inherent in a complete story of discrete physics *and* its continuum limit. The continuum limit will be seen as a *summary* of the real physics. It is a way to view, through the glass darkly, the crystalline reality of simple quantum choice.

Acknowledgement

As is discussed more fully in Ref. [9], this line of investigation started thanks to correspondence between V.A. Karmanov and I. Stein about the possibility of relating the Feynman–Hibbs suggestion to the Stein model [13–16], and a comment by D.O. McGoveran that an approximation suggested by Karmanov was already the exact result. Unfortunately these three authors could not come to consensus with each other and/or HPN as to how to present the work. Several drafts were also criticized by C.W. Kilmister and J.C. van den Berg. The work presented here follows a somewhat different approach, but has drawn heavily on the experience gained in collaboration and discussion with all five of these scientists.

References

[1] L.H. Kauffman and H.P. Noyes, Proc. R. Soc. A 452 (1996) 81, and references therein.
[2] R.P. Feynman and A.R. Hibbs, Quantum mechanics and path integrals (McGraw-Hill, New York, 1965), Problem 2-6, pp. 34–36.
[3] T. Jacobson and L.S. Schulman, J. Phys. A 17 (1984) 375.
[4] V.A. Karmanov, Phys. Lett. A 174 (1993) 371.
[5] T. Etter, Process, system, causality and quantum mechanics, in: Int. J. Gen. Syst., General systems and the emergence of physical structure from information theory, ed. K. Bowden, in press.
[6] T. Etter, The final collapse of the wave function, in: Proc. ANPA WEST 12, ed. F. Young (ANPA WEST, 112 Blackburn Avenue, Menlo Park, CA 94025, 1996).
[7] V.A. Karmanov, private communication to HPN (August 11, 1989).

L.H. Kauffman, H.P. Noyes / Physics Letters A 218 (1996) 139–146

[8] E. Grey, in: Proc. ANPA 17, ed. K. Bowden (September 1996) [available from C.W. Kilmister, Red Tiles Cottage, High Street, Barcombe, Lewes BN8 5DH, UK].

[9] H.P. Noyes, Phys. Essays 8 (1995) 434.

[10] H.A. Gersch, Int. J. Theor. Phys. 20 (1981) 491.

[11] L.H. Kauffman, Knots and physics (World Scientific, Singapore, 1991/1994).

[12] L.H. Kauffman, Special relativity and a calculus of distinctions, in: Discrete and combinatorial physics, Proc. ANPA 9, ed. H.P. Noyes (ANPA WEST, 112 Blackburn Avenue, Menlo Park, CA 94025, 1987) pp. 290–311.

[13] I. Stein, seminars at Stanford (1978, 1979).

[14] I. Stein, papers at the Second and Third Ann. Int. Conf. of the Alternative Natural Philosophy Association, King's College, Cambridge (1980, 1981).

[15] I. Stein, Phys. Essays 1 (1988) 155; 3 (1990) 66.

[16] I. Stein, The concept of object as the foundation of physics (Peter Lang, New York, 1996).

Comments on Joint Work with Pierre Noyes

Louis H. Kauffman
Department of Mathematics, Statistics and Computer Science
University of Illinois at Chicago
851 South Morgan Street
Chicago, IL, 60607-7045

1 Introduction

This paper is a comment on joint work with Pierre Noyes. We will discuss contents of the papers [11] and [12], here referred to respectively as paper number one and paper number two: (reprinted in this volume as Chapter 11 and Chapter 12).

1. L. H. Kauffman and H. P. Noyes, Discrete Physics and the Derivation of Electromagnetism from the Formalism of Quantum Mechanics , Proceedings of the Royal Soc. London A, Vol. 452 (1996), pp. 81-95.

2. L. H. Kauffman and H. P. Noyes, Discrete physics and the Dirac equation (with H. P. Noyes), Physics Lett. A, No. 218 (1996), pp. 139-146.

2 Discrete Ordered Calculus and Non-Commutative Electromagnetism

In paper number one, we construct a discrete background for the Feynman-Dyson derivation of non-commmutative electromagnetism from a set of commutation relations that are reminiscent of the Heisenberg approach to quantum mechanics. In fact this set of commutation relations is not quantum mechanics! In particular there is no assumption that the momentum operators in the different coordinate directions commute with one another. In fact, the construction of an analog of the electromagnetic field for this theory depends precisely on the non-commutativity of these momentum (or velocity) operators. (The "magnetic field" in the theory is essentially the non-commutative vector cross product of the velocity operator with itself.) For this reason, a conceptual underpinning of the original Feynmann-Dyson derivation is actually absent or unexplained in the original version. It was our intent to bring forth a conceptual domain in which this formalism could make sense.

Our approach is to use the notion of discrete measurement and to show that in this context the appropriate commutation relations are natural. In this realm non-commutativity of the operations of observing postion and observing momentum (or velocity) follow from the simple fact that a momentum observation requires two times, while a position observation requires only a single time. In a discrete model the clock must tick to observe momentum. The postion observed after that tick may be different than the postion observed before the tick. We construct a calculus (the discrete ordered calculus or DOC) based on this notion. It turns out that DOC is itself a non-commutative version of the classical calculus of finite differences, with the crucial property that the Leibniz rule (for the derviative of a product) is satisfied in this calculus. The Leibniz rule does not hold in the classical calculus of finite differences. As a result the DOC calculus behaves formally just as ordinary calculus (although non-commutative) and so can be used as the background calculus for our treatment of the Feynman Dyson derivation.

In the DOC caculus we write the derivative

$$DX = J(X' - X)$$

where the element J is a special discrete time-shift operator satisfying

$$ZJ = JZ'$$

for any Z in the ring R of (non-commutative) positions being acted upon. Here the terms X, X', X'', ... are a time series of such positions. The time-shifter, J, acts to automatically evaluate expressions in the resulting noncommutative calculus of finite differences. Note that J formalizes the operational ordering inherent in our initial discussion of velocity and position measurements. An operator containing J causes a time shift in the variables or operators to the left of J in the sequence order. On the other hand, it is significant to note that

$$DX = JX' - JX = XJ - JX = [X, J].$$

Thus the DOC derivative is a commutator and we may also write

$$X' = J^{-1}XJ,$$

exhibiting J as a discrete time evolution operator, analogous to the time evolution of states in quantum mechanics.

In DOC, X and DX have no reason to commute:

$$[X, DX] = XJ(X' - X) - J(X' - X)X = J(X'(X' - X) - (X' - X)X)$$

Hence

$$[X, DX] = J(X'X' - 2X'X + XX).$$

This commutator is non-zero even in the case where X and X' commute with one another. Consequently, we can consider physical laws in the form

$$[X_i, DX_j] = g_{ij}$$

where g_{ij} is a function that is suitable to the given application. In [11] (Chapter 11 in this book) we show how the formalism of electromagnetism arises when g^{ij} is δ^{ij}, the Kronecker delta. In [13] we will show how the general case corresponds to a "particle" moving in a noncommutative gauge field coupled with geodesic motion relative to the Levi-Civita connection associated with the g_{ij}. This result can be used to place the work of Tanimura [17] in a discrete context.

It should be emphasized that all physics that we derive in this way is formulated in a context of noncommutative operators and variables. We do not derive electromagnetism, but rather a noncommutative analog. It is not yet clear just what these noncommutative physical theories really mean. Our initial idealisation of measurement is not the only model for measurement that corresponds to actual observations. Certainly the idea that we can measure time in a way that has "steps between the steps of time" is an idealisation. It happens to be an idealisation that fits a model of the universe as a cellular automaton. In a cellular automaton an observation is what an operator of the automaton might be able to do. It is not necessarily what the "inhabitants" of the automaton can perform. Here is the crux of the matter. The inhabitants can have only limited observations of the running of the automaton, due to the fact that they themselves are processes running on the automaton. The theories we build on the basis of DOC can be theories *about* the structure of these automata. They will eventually lead to theories of what can be observed by the processes that run on such automata. It is possible that the well known phenomena of quantum mechanics will arise naturally in such a context. These points of view should be compared with [8].

In the case of electromagnetism, and our paper [11] we showed how to derive the same formal result as Feynman and Dyson and thus the electromagnetic equations appear in just the same form (disregarding the non-commutativity of the field coordinates) as classical electromagnetism. In fact a closer look as in [14] shows that in the discrete context there can be extra terms due to the possible non-vanishing of the non-commutative vector cross product of the H field (the magnetic field) with itself. This point is discussed in [13].

We are in the course of preparing a paper on the actual properties of discrete solutions to non-commutative electromagnetism. One way to approach this problem is to take scalar values in a finite field so that there is a possibility to solve the commutator equations in finite dimensional matrices. This is work in progress. To see the flavor of it, consider the commutator equation

$$[X, DX] = Jk$$

for a single variable X. Written out, this equation becomes

$$Jk = [X, J(X'-X)] = XJ(X'-X) - J(X'-X)X = J(X'(X'-X) - (X'-X)X).$$

If k and the elements of the time series $\{X, X', X'', ...\}$ are all commuting scalars then this equation reduces to

$$k = (X - X')^2.$$

Thus

$$X' = X \pm k^{1/2},$$

a Brownian random walk, is a solution to the simplest one-dimensional commutator equation.

2.1 Gauge Fields and Differential Geometry

Letting X_i $(i = 1, 2, ..., d)$ denote a set of spatial variables (non-commutative time series in the sense of our discrete ordered calculus), we will look at a collection of basic assumptions about the commutation of these variables and of their derivatives. It is natural from the point of view of the discrete ordered calculus to have

$$[X_i, X_j] = 0$$

for all i and j. There are no other natural commutations from the point of view of this calculus.

We shall define g_{ij} by the equation

$$[X_i, \dot{X}_j] = g_{ij}.$$

Here $\dot{X}_j$ is shorthand for DX_j and

$$[A, B] = AB - BA.$$

Along with this commutator equation, we will assume that

$$[X_i, X_j] = 0,$$

$$[X_i, g_{jk}] = 0$$

and

$$[g_{rs}, g_{jk}] = 0.$$

Here it is assumed that g_{ij} is non-degenerate in the sense that there exists g^{ij} so that

$$g^{ij} g_{jk} = \delta^i_k$$

and that

$$g_{ij}g^{jk} = \delta_i^k.$$

Here we are using the Einstein summation convention that implicitly assumes that we sum over repeated indices in an expression. Symbol δ_j^i is a Kronecker delta, equal to 1 when i equals j and 0 otherwise.

The first result that is a direct consequence of these assumptions is the symmetry of the "metric" coefficients g^{ij}. That is, we shall show that

$$g^{ij} = g^{ji}.$$

Lemma 3. $g_{ij} = g_{ji}.$

Proof.

$$g_{ij} - g_{ji}$$
$$= [X_i, \dot{X}_j] - [X_j, \dot{X}_i]$$
$$= [X_i, \dot{X}_j] + [\dot{X}_i, X_j]$$
$$= D[X_i, X_j]$$
$$= 0.$$

For the purpose of doing calculus in this situation we define $\dot{X}^i$ by the equation

$$\dot{X}^i = g^{ik}\dot{X}_k.$$

The operator $\dot{X}^i$ is simply the index shift of the corresponding $\dot{X}_i$. We do not define a corresponding X^i. It is easy to check the equation

$$[X_i, \dot{X}^j] = \delta_i^j.$$

Consequently, we define the derivative of an operator F with respect to X_i by the equation

$$\partial^i F = [F, \dot{X}^i]$$

420

and the corresponding lowered derivative by the formula

$$\partial_i F = [F, \dot{X}_i].$$

Note that we have

$$\partial_i X_j = g_{ij}.$$

With these partial derivatives in hand, we define $\dot{F}$ by the formula

$$\dot{F} = \partial^k F \dot{X}_k.$$

If F commutes with g^{ij} then it is easy to see that

$$\dot{F} = \partial_k F \dot{X}^k.$$

These formulas extend (implicitly) the definition of the time series to entities other than the operators X_i since

$$\dot{F} = DF = J(F' - F).$$

A stream of consequences then follows by differentiating both sides of the equation

$$g_{ij} = [X_i, \dot{X}_j].$$

Note that

$$\dot{g}_{ij} = [\dot{X}_i, \dot{X}_j] + [X_i, D^2 X_j]$$

by the Leibniz rule

$$D[A, B] = [DA, B] + [A, DB].$$

Note also that we can freely use the Jacobi identity

$$[A, [B, C]] + [C, [A, B]] + [B, [C, A]] = 0.$$

In particular, the Levi-Civita connection

$$\Gamma_{ijk} = (1/2)(\partial_i g_{jk} + \partial_j g_{ik} - \partial_k g_{ij})$$

associated with the g_{ij} comes up almost at once from the differentiation process described above. To see how this happens, view the following calculation where

$$\nabla_{ij} F = [X_i, [X_j, F]].$$

We apply the operator ∇_{ij} to the second DOC derivative of X_k.

Lemma 4. $\Gamma_{ijk} = (1/2)\nabla_{ij}D^2 X_k$

Proof.

$$\nabla_{ij}D^2 X_k = [X_i, [X_j, D^2 X_k]]$$

$$= [X_i, \dot{g}_{jk} - [\dot{X}_j, \dot{X}_k]]$$

$$= [X_i, \dot{g}_{jk}] - [X_i, [\dot{X}_j, \dot{X}_k]]$$

$$= [X_i, \dot{g}_{jk}] + [\dot{X}_k, [X_i, \dot{X}_j]] + [\dot{X}_j, [\dot{X}_k, X_i]$$

$$= [X_i, \partial_s g_{jk} \dot{X}^s] + [\dot{X}_k, g_{ij}] + [\dot{X}_j, -g_{ik}]$$

$$= \partial_i g_{jk} - \partial_k g_{ij} + \partial_j g_{ik}$$

$$= 2\Gamma_{ijk}.$$

It is remarkable that the form of the Levi-Civita connection comes up directly from this non-commutative calculus without any apriori geometric interpretation.

One finds that

$$D^2 X_i = G_i + g_{ir}g_{js}F^{rs}\dot{X}^j + \Gamma_{ijk}\dot{X}^j \dot{X}^k$$

where

$$F^{rs} = [\dot{X}^r, \dot{X}^s].$$

It follows from the Jacobi identity that

$$F_{ij} = g_{ir}g_{js}F^{rs}$$

satisfies the equation

$$\partial_i F_{jk} + \partial_j F_{ki} + \partial_k F_{ij} = 0,$$

identifying F_{ij} as a noncommutative analog of a gauge field. G_i is a non-commutative analog of a scalar field. The details of these calculations will be found in [13].

This description of the equations for a noncommutative particle in a metric field illustrates the role of the background discrete time in this theory. In terms of the background time the metric coefficients are not constant. It is through this variation that the spacetime derivatives of the theory are articulated. The background is a process with its own form of discrete time, but no spacetime structure as we know and observe it. Our observation of spacetime structure appears as a rough (commutative) approximation to the processes described as consequences of the basic noncommutative equations of the discrete ordered calculus.

2.2 Poisson Brackets and Commutator Brackets

Dirac [6] introduced a fundamental relationship between quantum mechanics and classical mechanics that is summarized by the maxim *replace Poisson brackets by commutator brackets*. Recall that the Poisson bracket $\{A, B\}$ is defined by the formula

$$\{A, B\} = (\partial A/\partial q)(\partial B/\partial p) - (\partial A/\partial p)(\partial B/\partial q),$$

where q and p denote classical position and momentum variables respectively.

In our version of discrete physics the noncommuting variables are functions of discrete time, with a *DOC* derivative D as described in the first section. Since $DX = XJ - JX = [X, J]$ is itself a commutator, it follows that

$$D([A, B]) = [DA, B] + [A, DB]$$

for any expressions A, B in our ring R. A corresponding Leibniz rule for Poisson brackets would read

$$(d/dt)\{A, B\} = \{dA/dt, B\} + \{A, dB/dt\}.$$

However, here there is an easily verified exact formula:

$$(d/dt)\{A, B\} = \{dA/dt, B\} + \{A, dB/dt\} - \{A, B\}(\partial\dot{q}/\partial q + \partial\dot{p}/\partial p).$$

This means that the Leibniz formula will hold for the Poisson bracket exactly when

$$(\partial\dot{q}/\partial q + \partial\dot{p}/\partial p) = 0.$$

This is an integrability condition that will be satisfied if p and q satisfy Hamilton's equations

$$\dot{q} = \partial H/\partial p,$$
$$\dot{p} = -\partial H/\partial q.$$

This, of course, means that q and p are following a principle of least action with respect to the Hamiltonian H. Thus we can interpret the *fact* $D([A, B]) = [DA, B] + [A, DB]$ in the discrete context as an analog of the principle of least action. Taking the discrete context as fundamental, we say that Hamilton's equations are *motivated* by the presence of the Leibniz rule for the discrete derivative of a commutator. The classical laws are obtained by following Dirac's maxim in the opposite direction! Classical physics is produced by following the correspondence principle upwards from the discrete.

Taking the last paragraph seriously, we must reevaluate the meaning of Dirac's maxim. The meaning of quantization has long been a basic mystery of quantum mechanics. By traversing this territory in reverse, starting from the noncommutative world, we begin these questions anew.

In making this backwards journey to classical physics we see that it is necessary to make at least one further restriction on the commutation relations. For in Poisson brackets it is the case that

$$\{q_i, q_j\} = 0$$

$$\{p_i, p_j\} = 0$$

$$\{q_i, p_j\} = \delta_{ij}.$$

In our formalism, we would identify X_i as the correspondent with q_i and $\dot{X}^j$ as the correspondent of p_j. Thus to match the Poisson algebra we would need to demand that

$$F^{ij} = [\dot{X}^i, \dot{X}^j] = 0.$$

Under these circumstances we would have

$$D^2 X_i = G_i + \Gamma_{ijk}\dot{X}^j\dot{X}^k,$$

the usual form of the description of geodesic motion with respect to the Levi-Connection. (See the previous section for the discussion of these equations.) The non-commutative calculus structure that lies below this has a non-trival gauge field that arises from the extra commutation relation. This raises further questions about the nature of the generalization that we have made. Originally Hermann Weyl [18] generalized classical differential geometry and discovered gauge theory by allowing changes of length as well as changes of angle to appear in the holonomy. Here we arrive at a very similar situation via the properties of a non-commutative discrete calculus of observations. A closer comparison with the geometry of gauge theories is called for.

3 Paper Number Two and the Dirac Equation

The Dirac notation $\langle A|B\rangle$ [6] denotes the probability amplitude for a transition from A to B. Here A and B could be points in space (for the path of a particle), fields (for quantum field theory), or geometries on spacetime (for quantum gravity). The probability amplitude is a complex number. The actual probability of an event is the absolute square of the amplitude. If a complete set of intermediate states $C_1, C_2, ...C_n$ is known, then the amplitude can be expanded to a summation

$$\langle A|B\rangle = \Sigma_{i=1}^{n}\langle A|C_i\rangle\langle C_i|B\rangle.$$

This formula follows the formalism of the usual rules for probability, and it allows for the constructive and destructive interference of the amplitudes. It is the simplest case of a quantum network of the form

$$A---*---C---*---B$$

where the colors at A and B are fixed and we run through all choices of colors for the middle edge. The vertex weights at the vertices labelled $*$ are $\langle A|C\rangle$ and $\langle C|B\rangle$ respectively. A measurement at the C edge reduces the big summation to a single value.

Consider the generalization of the previous example to the graph

$$A---*---C^1---*---C^2---*---...---*---C^m---B$$

With A and B fixed the amplitude for the net is

$$< A|B >= \Sigma_{1\leq i_1\leq...\leq i_m\leq n} < A|C_{i_1}^1 >< C_{i_2}^2|C_{i_3}^3 > ... < C_{i_m}^m|B >$$

One can think of this as the sum over all the possible paths from A to B. In fact in the case of a "particle" travelling between two points in space, this is exactly what must be done to compute an amplitude - integrate over all the paths between the two points with appropriate weightings. In the discrete case this sort of summation makes perfect sense.

Now consider the summation discussed above in the case where $n = 2$. That is, we shall assume that each C^k can take two values, call these values L and R. Furthermore let us suppose that $< L|R >=< R|L >= \sqrt{-1}$ while $< L|L >=< R|R >= 1$. The amplitudes that one computes in this case correspond to solutions to the Dirac equation [6] in one space variable and one time variable. This example is related to an observation of Richard Feynman [9]. In [12] we give a very elementary derivation of this result and we show how these amplitudes give *exact* solutions to the discretized Dirac equation. Everything is quite exact and one can understand just what happens in taking the limit to the continuum. In this example a state of the network consists in a sequence of choices of L or R. These can be interpreted as choices to move left or right along the light-cone in a Minkowski plane. It is in summing over such paths in spacetime that the solution to the Dirac equation appears. In this case, time has been introduced into the net by interpreting the sequence of nodes in the network as a temporal direction.

More specifically, let (a, b) denote a point in discrete Minkowski spacetime in lightcone coordinates. This means that a denotes the number of steps taken to the left and b denotes the number of steps taken to the right. We let $\psi_L(a, b)$ denote the sum over the paths that enter the point (a, b) from the left and $\psi_R(a, b)$ the sum over the paths that enter (a, b) from the right. Each path P contributes $i^{c(P)}$ where $c(P)$ denotes the number of corners in the path. View the diagram below.

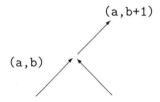

It is clear from the diagram that

$$\psi_L(a, b + 1) = \psi_L(a, b) + i\psi_R(a, b).$$

Thus we have that

$$\partial\psi_L/\partial R = i\psi_R$$

and similarly

$$\partial\psi_R/\partial L = i\psi_L.$$

This pair of equations is the Dirac equation in light cone coordinates.

This discrete derivation of the Dirac equation is simpler than the method used in [12]. I am indebted to Charles Bloom [1] for pointing this out to us. In fact, this form of the discretization is essentially Feynman's original method as is evident from the reproduction of Feynman's handwritten notes in Figure 8 of the review paper [16] by Schweber. It is still an open problem to generalize this exercise of Feynman to four dimensional discrete spacetime.

As in the Dirac equation example, one way to incorporate spacetime is to introduce a temporal direction into the net. At a vertex, one must specify labels of *before* and *after* to each edge of the net that is incident to that vertex. If there is a sufficiently coherent assignment of such local times, then a global time direction can emerge for the entire network. Networks endowed with temporal directions have the structure of morphisms in a category where each morphism points from past to future. A category of quantum networks emerges equipped with a functor (via the algebra of the vertex weights) to morphisms of vector spaces and representations of generalized symmetry groups. Appropriate traces of these morphisms produce the amplitudes.

Quantum non-locality is built into the network picture. Any observer taking a measurement in the net has an effect on the global set of states available for summation and hence affects the possibilities of observations at all other nodes in the network. By replacing space with a network we obtain a precursor to spacetime in which quantum mechanics is built into the initial structure.

Remark. A striking parallel to the views expressed in this section can be found in [7]. Concepts of time and category are discussed by Louis Crane [4], [5] in relation to topological quantum field theory. In the case of Crane's work there is a deeper connection with the methods of our paper, as I shall explain below.

3.1 Temporality and the Crane Model for Quantum Gravity

Crane uses a partition function defined for a triangulated four-manifold. Let us denote the partition function by $Z(M^4, A, B) = < A|B >_M$ where M^4 is a four-manifold and A and B are (colored - see the next sentence) three dimensional submanifolds in the boundary of M. The partition function is constructed by summing over all colorings of the edges of a dual complex to this triangulation from a finite set of colors that correspond to certain representations of the the quantum group $U_q(SU(2))$ where q is a root of unity. The sum is over products of $15J_q$ symbols (natural generalizations of the $6J$ symbols in angular momentum theory) evaluated with respect to the colorings. The specific form of the partition function (here written in the case where A and B are empty) is

$$Z(M^4) = N^{v-e}\Sigma_\lambda\Pi_\sigma dim_q(\lambda(\sigma))\Pi_\tau dim_q^{-1}(\lambda(\tau))\Pi_\zeta 15J_q(\lambda(\zeta)).$$

Here λ denotes the labelling function, assigning colors to the faces and tetrahedra of M^4 and $v - e$ is the difference of the number of vertices and the number of edges in M^4. Faces are denoted by σ, tetrahedra by τ and 4-simplices by ζ. We refer the reader to [2] for further details.

In computing $Z(M^4, A, B) = < A|B >_M$ one fixes the choice of coloration on the boundary parts A and B. The analog with quantum gravity is that a colored three manifold A can be regarded as a three manifold with a choice of (combinatorial) metric. The coloring is the combinatorial substitute for the metric. In the three manifold case this is quite specifically so, since the colors can be regarded as affixed to the edges of the simplices. The color on a given edge is interpreted as the generalized distance between the endpoints of the edge. Thus $< A|B >_M$ is a summation over "all possible metrics" on M^4 that can extend the given metrics on A and B. $< A|B >_M$ is an amplitude for the metric (coloring) on A to evolve in the spacetime M^4 to the metric (coloring) on B.

The partition function $Z(M^4, A, B) = < A|B >_M$ is a topological invariant of the four manifold M^4. In particular, if A and B are empty (a vacuum-vacuum amplitude), then the Crane-Yetter invariant, $Z(M^4)$, is a function of the signature and Euler characteristic of the four-manifold [2]. On the mathematical side of the picture this is already significant since it provides a new way to express the signature of a four-manifold in terms of local combinatorial data.

From the point of view of a theory of quantum gravity, $Z(M^4, A, B) = < A|B >_M$, as we have described it so far, is lacking in a notion of time and dynamical evolution on the four manifold M^4. One can think of A and B

as manifolds at the initial and final times, but we have not yet described a notion of time within M^4 itself.

Crane proposes to introduce time into M^4 and into the partition function $< A|B >_M$ by labelling certain three dimensional submanifolds of M^4 with special grouplike elements from the quantum group $U_q(SU(2))$ and extending the partition function to include this labelling. Movement across such a labelled hypersurface is regarded as one tick of the clock. The special grouplike elements act on the representations in such a way that the partition function can be extended to include the extra labels. Then one has the project to understand the new partition function and its relationship with discrete dynamics for this model of quantum gravity.

Lets denote the special grouplike element in the Hopf algebra $G = U_q(SU(2))$ by the symbol J. Then, as discussed at the end of the previous section, one has that the square of the antipode $S : G \longrightarrow G$ is given by the formula $S^2(x) = J^{-1}xJ$. This is the tick of the clock. The DOC derivative in the quantum group is given by the formula $DX = [X, J] = J(S^2(X) - X)$. I propose to generalize the discrete ordered calculus on the quantum group to a discrete ordered calculus on the four manifold M^4 with its hyperthreespaces labelled with special grouplikes. This generalised calculus will be a useful tool in elucidating the dynamics of the Crane model. Much more work needs to be done in this domain.

References

[1] Bloom, Charles [1998], (private communication).

[2] Crane, Louis, Kauffman, Louis H., Yetter, David N. [1997], State sum invariants of 4-manifolds, *Journal of Knot Theory and Its Ramifications*.

[3] Connes, Alain [1990], *Noncommutative Geometry*, Academic Press.

[4] Crane, Louis [1996], Clock and category: Is quantum gravity algebraic?, *J. Math. Phys.* **36** (11), November (1996), pp. 6180-6193.

[5] Crane, Louis [1997], A proposal for the quantum theory of gravity, (preprint).

[6] Dirac, P.A.M. [1968], *Principles of Quantum Mechanics,* Oxford University Press.

[7] T. Etter and H. Pierre Noyes, Process, System, Causality and Quantum Mechanics', *Physics Essays,* **12**, No. 4, Dec. 1999, (in press) and Ch. 15 in this book.

[8] E. Fredkin [1990], Digital Mechanics, *Physica D* **45**, pp. 254-270.

[9] R.P. Feynman and A.R. Hibbs [1965], *Quantum Mechanics and Path Integrals,* McGraw Hill Book Company.

[10] Kauffman, Louis H. [1991, 1994], *Knots and Physics,* World Scientific Pub.

[11] Kauffman, Louis H. and Noyes, H. Pierre [1996], Discrete Physics and the Derivation of Electromagnetism from the formalism of Quantum Mechanics, *Proc. of the Royal Soc. Lond. A,* **452**, pp. 81-95, and Chapter 11 in this book.

[12] Kauffman, Louis H. and Noyes, H. Pierre [1996], Discrete Physics and the Dirac Equation, *Physics Letters A,* 218 ,pp. 139-146, and Chapter 12 in this book.

[13] Kauffman, Louis H. and Noyes, H. Pierre (In preparation).

[14] Kauffman, Louis H. [1996], Quantum electrodynamic birdtracks, *Twistor Newsletter Number 41.*

[15] Kauffman, Louis H. [1998], Noncommutativity and discrete physics, *Physica D* 120 (1998), 125-138.

[16] Schweber, Silvan S. [1986], Feynman and the visualization of space-time processes, *Rev. Mod. Phys.* Vol. 58, No. 2, April 1986, 449 - 508.

[17] Tanimura, Shogo [1992], Relativistic generalization and extension to the non-Abelian gauge theory of Feynman's proof of the Maxwell equations, *Annals of Physics, vol. 220,* pp. 229-247.

[18] Weyl, Hermann [1922], *Space–Time–Matter,* Methuen, London (1922).

ARE PARTONS CONFINED TACHYONS?[*]

H. Pierre Noyes

Stanford Linear Accelerator Center
Stanford University, Stanford, California 94309

ABSTRACT

We note that if hadrons are gravitationally stabilized "black holes", as discrete physics suggests, it is possible that PARTONS, and in particular quarks, could be modeled as tachyons, i.e. particles having $v^2 > c^2$, without conflict with the observational fact that neither quarks nor tachyons have appeared as "free particles". Some consequences of this model are explored.

Invited paper presented at the
Twelfth Annual International Meeting of the **Western Chapter** of the
ALTERNATIVE NATURAL PHILOSOPHY ASSOCIATION
Cordura Hall, Stanford University, February 17-19, 1996

[*] Work supported by the Department of Energy, contract DE–AC03–76SF00515.

1. INTRODUCTION

Kendall, Friedman and Taylor received the Nobel prize for their discovery of hard scattering centers inside the proton when it is bombarded with high energy electrons. These centers are reminiscent of the nuclei discovered by Rutherford when he bombarded atoms with α—particles. Unfortunately for simplicity they do not have as transparent a geometrical interpretation. These "deep inelastic" electron scattering experiments were analyzed by Bjorken and shown to follow the relativistic kinematics of scattering from electrically charged objects "within" the proton, so the analogy to Rutherford scattering is indeed close, once one recognizes that, up to a point, the enormously different energy scale makes little difference. Feynman arrived at similar conclusions and dubbed these scattering centers "partons", but was careful to point out that all the essential features of his own work were already present in Bjorken's analysis.

It was soon recognized that these scattering centers might indeed be the "quarks" postulated by Gell-Mann or the "Aces, Deuces and Treys" of Zweig. These theoretical entities had been invented to account for a number of regularities in the hadronic mass spectrum and strong interaction dynamics of elementary particles. So the deep inelastic scattering analysis soon came to be called the "quark-parton" model. "Free quarks" were eagerly sought in a number of ingenious ways, but so far have made themselves conspicuous by their absence. This embarrassment was eventually removed by the quantum-chromodynamic strand of what is now called the "standard model of quarks and leptons". In contrast to leptons whose "running" electromagnetic (and, in principle, weak) coupling constants become ever stronger as the energy of the bombarding particles increases, the quantum chromodynamic or strong coupling constant becomes ever weaker as the energy increases leading to what is called "asymptotic freedom". But the same model implies that the quark coupling constant becomes ever stronger as the energy becomes smaller leading to "infrared slavery" and quark confinement within the hadrons. Unfortunately making an actual calculation of the "confining potential" from first principles has as yet proved to lie beyond the ingenuity of theorists; direct attack on the problem using super-computers is only now beginning to yield believable — and not very accurate — results. Here we use some clues from our new fundamental theory[1,2] to construct a different *dynamical* model for quark confinement which might be easier to compute.

That quarks are confined in our theory was argued some time ago. According to McGoveran's Theorem,[3,4] in any finite and discrete theory with a universal ordering operator there can be no more than 3 homogeneous and isotropic macroscopic ("spacial") dimensions. These are exhausted in labeling particles by the three absolutely conserved quantum numbers allowed in current particle physics.

There are four candidates: lepton number, baryon number, charge and either the "z-component of weak isospin" or "weak hypercharge". The extended GellMann-Nishijima relation leaves only three of these to specify independently and measure at macroscopic distances. Any other conserved, discrete quantum number must "compactify" or be "confined". Since quarks and gluons carry the conserved quantum number of color, we conclude that this quantum number cannot appear as a label for free particles and hence that gluons and quarks are confined. But this general argument does not lead to a computational scheme.

There is another theoretical entity which has never been observed, namely any massive particle traveling faster than the speed of light. The generic name for such particles is "tachyon", perhaps more familiar to some of my readers as part of the technobable in the STAR TREK series than in the scientific literature. The possibility of their existence was explored extensively by Gary Feinberg,[5] and others.[6,7] There is no direct experimental evidence that such particles can be made to appear in the laboratory. Since neither quarks nor tachyons have been observed directly we adopt here the speculative stance that both might turn out to be the same type of entity, and try to construct a model that achieves this in a natural way. In Newtonian physics, if an attracting center is massive enough so that circular velocity is c at some radius R, neither light nor particles limited to that velocity at R can ever escape this "black hole", as was first pointed out by Laplace in 1795. In the Newtonian framework, particles in elliptical orbits tangent to this confining circle at aphelion would necessarily have still larger velocities as they plunged toward and swung around the attracting center, and hence could be called "confined tachyons". This, in short, is our fanciful dynamical parton-tachyon model.

Conventional physicists will reject this idea out of hand because I have taken a Newtonian picture and extended it into the supraluminal region where it is wildly inappropriate. If I ended my remarks here, I would have to agree with them. But in discrete physics we have found that apparently "non-relativistic" ideas can find strange connections with relativistic physics, for instance in the successful derivation of Maxwell's Equations from superficially "non-relativistic" commutation relations and Newton's second law of motion.[8] So I am prepared to pursue this wild idea here and see what comes of it.

2. THE CLASSICAL MODEL

We start by assuming that we know nothing about relativity and quantum mechanics other than that the velocity of light c has a special significance in both theories. Then, following Laplace, we note that there are can be interactions (in his case, gravitational) strong enough so that particles with velocity c cannot escape the neighborhood of the source of this interaction. Our approach is kinematic, rather than dynamic, in that we take as paradigmatic a conic section trajectory. For trajectories obeying Kepler's Laws, it can be shown that these (a) conserve *specific angular momentum* or angular momentum per unit mass about the attracting center given by $rv_\perp = r^2\dot\theta$ where r is the distance from that center, $v_\perp$ is the component of velocity perpendicular to the line from the center to the point on the trajectory, and θ is the angle that line (in the plane of the trajectory) makes with a line from the center to the perihelion point and (b) that the acceleration of the point is toward the center and varies inversely with the square of the distance. These *kinematic* statements are derived from Kepler's Laws in the Appendix.

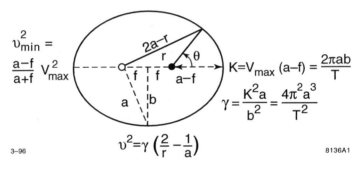

$$v^2 = \gamma\left(\frac{2}{r} - \frac{1}{a}\right)$$

Figure 1. Radial coordinates and algebraic parameters for elliptical motion.

This derivation also provides us with the generalized equation for conic section motion with velocity v along such a trajectory whose square is $v^2 = \dot r^2 + (r\dot\theta)^2$

$$v^2 = \gamma\left(\frac{2}{r} - \frac{1}{a}\right) \quad elliptical$$

$$v^2 = \gamma\left(\frac{2}{r}\right) \qquad parabolic \tag{2.1}$$

$$v^2 = \gamma\left(\frac{2}{r} + \frac{1}{a}\right) \quad hyperbolic$$

We illustrate the radial coordinates for elliptical motion in Fig. 1 and for hyperbolic motion in Fig. 2. To calculate the constant γ from space-time measurements on

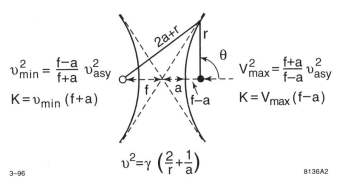

$$v_{min}^2 = \frac{f-a}{f+a} v_{asy}^2$$

$$K = v_{min}(f+a)$$

$$V_{max}^2 = \frac{f+a}{f-a} v_{asy}^2$$

$$K = V_{max}(f-a)$$

$$v^2 = \gamma \left(\frac{2}{r} + \frac{1}{a}\right)$$

3-96   8136A2

Figure 2. Radial coordinates and algebraic parameters for hyperbolic motion.

elliptical orbits rather than dynamically (i.e. by first introducing the concepts of *force* and *inertial mass*), we symbolize the specific angular momentum by $K \equiv rv_\perp = r^2\dot\theta$ and use Kepler's Second and Third Laws (cf. Appendix) to obtain the relations

$$\gamma = \frac{K^2 a}{b^2} = \frac{4\pi^2 a^3}{T^2} \tag{2.2}$$

where T is the *period*, that is the time it takes a point moving on an elliptical trajectory to return to the same position. If the acceleration is centrifugal (away from the center) we call it repulsive rather than attractive. Such motion is necessarily hyperbolic, and motion takes place on the opposite branch of the hyperbola (cf. Fig. 2). For this case (eg Rutherford scattering), the trajectory equation is given by

$$\frac{1}{r} = -\frac{a + f \cos\theta}{f^2 - a^2} \tag{2.3}$$

but the equation for the square of the velocity remains unchanged.

These results, being classical, are all *scale invariant*. We now analyze how separate them into two types of motion, in the first of which (*particulate motion*), the velocity is always less than c, and the second of which (*confined tachyonic motion*) always has velocities greater than c. To do this we first note that, for attractive interactions, the velocity is always maximum at the distance of closest approach to the attracting center $r_{min} = |a - f|$. If we have velocity close to but below c at this point, any case of particulate motion will, necessarily, lie outside the circle of radius

$$c^2 r_{min}^{(c)} = \gamma_c \tag{2.4}$$

For elliptical ($a > f$) motion close to the limiting case, the velocity will fall to $\frac{a-f}{a+f}c$ at r_{max} (aphelion). For hyperbolic motion ($f > a$), the velocity will fall to $\frac{f-a}{f+a}c$

along the asymptote; the particle escapes. At the distance of closest approach the motion is always perpendicular to the line from the object to the center, so in all cases, the specific angular momentum is $K_c = r_{min}c = |a - f|c$. If the distance of closest approach $r_{min}^{(v)}$ is greater than $r_{min}^{(c)}$, then the velocity v_{max} is always less than c, and we have that

$$\frac{|a - f|}{a + f}v_{max}^{(v)} \leq v \leq v_{max}^{(v)} \tag{2.5}$$

where the lower limit on the velocity is either the apehelion velocity for elliptical motion, or the asymptotic velocity for hyperbolic motion. Clearly the specific angular momentum is always given by $K_v = |a - f|v_{max}$

If the *maximum* distance from the attractive center is less than $r_{min}^{(c)}$, the trajectory is always elliptical and always tachyonic.

For repulsive interactions the asymptotic velocity is the maximum velocity and is always less than c, the motion is always hyperbolic and the velocity falls to its minimum value at the distance of closest approach, which is $a + f$. Clearly for repulsive interactions $K_v = (a + f)v_{min}$ and is zero for the straight line trajectory where the asymptotic velocity is pointed straight at the center.

We trust the classical kinematics of our confined tachyon motion inside the "black hole radius" $r_{min}^{(c)}$ and the allowed particulate motion outside with $v < c$ is now clear, and that for a given interaction strength γ_c the two regions are cleanly separated by that radius. Before showing how the two regions can communicate, and why we have excluded the (classically allowed) cases of an object with $v < c$ at large distances but $v > c$ at short distances, we must first examine the modifications that relativistic quantum mechanics imposes on our classical model.

3. THE QUANTUM MODEL

Our quantum model is basically that discussed in the paper on the fine structure spectrum of hydrogen[9] in which David McGoveran and I showed that the Sommerfeld formula and an improvement in accuracy by four orders of magnitude for the calculation of the fine structure constant could be obtained combinatorially. There we derived the formula

$$\left(\frac{E_n}{m_{ep}c^2}\right)^2\left[1 + \left(\frac{1}{137n}\right)^2\right] = 1 \tag{3.1}$$

where $m_{ep} = m_e m_p/(m_e + m_p)$, m_e, m_p are the masses of the electron and proton respectively, and $E_n = (m_p + m_e)c^2 - \epsilon_n$ is the mass-energy of their n^{th} bound state,

ϵ_n is the binding energy of that state, n is Bohr's principle quantum number for hydrogen and $1/137$ is the combinatorial calculation of the fine structure constant $e^2/\hbar c$. This formula (using empirical values) is equivalent to Bohr's relativistic correction to his non-relativistic model for the hydrogen atom, and was published by him in 1915.[10]

Using the geometrical-kinematic approach we employed above, we could model a quantized circular orbit for the same problem[11] as a regular polygon with n sides of length $137nh/m_e c$. Each side is traversed with velocity $c/137n$, with the direction but not the magnitude changing at each corner of the polygon. This model leads to the same formula for the energy and period of the motion.

In the fine structure paper we also point out that the general formula for a (relativistic) bound state of mass μ composed of exactly two particles with masses m_1, m_2 due to an interaction measured by the coupling constraint $g^2/\hbar c = 1/137n$ in a S-matrix theory give the equivalent formula (with $\hbar = 1 = c$)

$$(g^2_{\mu m_1 m_2}\mu)^2 = (m_1 + m_2)^2 - \mu^2 \tag{3.2}$$

We also pointed out that this formula is *non-perturbative* and gives a good value for the basic (phenomenological) strong interaction coupling constant $G^2_{\pi N \overline{N}}$ between pions and nucleons.

In a subsequent paper[12] we noted that the same formula could also be derived by assuming that the relativistic energy allowed to a "vacuum fluctuation" in a finite and discrete "second quantized field theory" which has at most N particles of mass m is $M_x = Nm$. Then a bound state of mass μ "at rest" can virtually dissociate into a system with that energy balanced by the momenta of N quanta of momentum μc. This allows us to write that

$$\mu^2 = (Nm)^2 - (N\mu)^2 = E^2 - p^2 \tag{3.3}$$

The handy-dandy formula (3.2) is recovered by taking $g^2 = 1/N$ and $m = m_1 + m_2$. We conclude that this is formula is the proper starting point for the relativistic quantum mechanical connection between the coupling constant between a particle of mass μ and its two constituents and the mass values themselves.

We have seen that "classical" tachyons are obtained by replacing the invariant relation $E^2 - p^2 = m^2$ by the relation $p^2 - E^2 = \mu^2$ thus guaranteeing that $v = p/E \geq 1$ in units of the limiting velocity c. It thus seems that, so far as the two-body problem goes, the handy-dandy formula, and hence the relativistic Rutherford formula, should hold just as well for confined tachyons as for confined quarks. This immediately explains why the parton Bohr formula (and the Bohr-Sommerfeld generalization) proved so immediately useful as a guide to the discover and exploration of the hydrogen-like energy levels of "quarkonium", i.e. the bound states of a quark-antiquark pair.

This is fully compatible with the derivation of the handy-dandy formula from the S-matrix bound state wave function normalization condition $\int \phi = 1$ where $\phi(s) = \frac{g^2 \mu}{s-\mu^2}$ and s is square of the invariant four momentum. The form follows from the definition of a bound state as a pole in the S-matrix, and gives us explicitly

$$\int_{(m_1+m_2)^2}^{\infty} \phi^2(s)ds = 1 \qquad (3.4)$$

leading to the handy dandy formula, Eq.3.2. In the usual interpretation the "off mass shell" values of run from the rest-energy (momentum equal to zero) value of $(m_1 + m_2)c^2$ to infinity. But for tachyons, the momentum runs from the minimum value $(m_1 + m_2)c$ (energy equal to zero) to infinity. The algebraic consequence is the same!

4. Conclusion

We conclude that, carefully interpreted, from an S-matrix point of view, once one accepts that quarks, or partons, or tachyons are confined, the paradigmatic aspects of the quark-parton model can be extracted just as well from a confined tachyon model. It then becomes a question of which is most useful as an aid to the imagination in formulating the interactions. For some who like to think classically, tachyons orbiting inside a black hole may seem a better place to start than either quantum field theory or discrete physics. This little exercise is offered to them as a starting point which, like discrete physics but unlike quantum field theory, has a well defined "correspondence limit", even though that limit has the usual problem of any black hole model of whether or not information can be successfully extracted once particles have passed the event horizon.

5. Appendix: KEPLER'S LAWS AND INVERSE SQUARE LAW ACCELERATIONS

5.1 KEPLER,S LAWS

Since my starting point for discussing tachyons is a kinematic rather than a dynamical model for confined motion in circular and elliptical orbits, I give here a derivation of this model from Kepler's Laws taken from draft chapter of my book on discrete physics.[13] I state Kepler's Laws as follows:

I. The paths of the planets are ellipses with the sun at one, single focus.

II. The line from sun to planet sweeps out equal areas in equal times.

III. The ratio of the square of the periodic time for the planet to complete its orbit to the cube of the length of the major axis is independent of the length of the minor axis and the same for all planets.

Starting with the First Law, we draw the ellipse (cf. Fig. 1) with the line between the sun and the planet of length r, and θ the angle between this line and the line from the sun to the perihelion point (distance of closest approach to the sun). Since an ellipse is defined to be the locus of all points such that the lengths of the lines from the planet to the two foci add to some constant value which we call $2a$, if the distance between the two foci is called $2f$, the two foci and the planet define a triangle with sides r, $2f$, $2a - r$. Using the fact that for any triangle with sides a, b, c, we know that $2ab \cos \theta_{ab} = a^2 + b^2 - c^2$, we have that

$$2rf \, \cos \, (\pi - \theta) = r^2 + 4f^2 - (2a - r)^2$$

from which follows the orbit equation connecting r and θ, namely

$$\frac{1}{r} = \frac{a + f \, \cos \, \theta}{a^2 - f^2} = \frac{a + f \, \cos \, \theta}{b^2}$$

Here a is called the semi-major axis of the ellipse. As we can see from Fig. 1 perihelion distance is $a - f$, aphelion distance is $a + f$ and the square of the semi-minor axis, or half-width, of the ellipse is given by $b^2 = a^2 - f^2$.

Turning to the Second Law, we recall that for a triangle with sides a, b, c the square of the area is given by $A^2 = \frac{1}{16}[(a + b)^2 - c^2][c^2 - (a - b)^2]$. Hence the area swept out when we go from position r to $r + \Delta r$ due to velocity v in time Δt is given by

$$\Delta A^2 = \frac{1}{16}[(2r + \Delta r)^2 - v^2 \Delta t^2][v^2 \Delta t^2 - \Delta r^2]$$

We first note that the velocity component along r is $\dot{r} = \frac{\Delta r}{\Delta t}$ and that the square of the velocity component perpendicular to r, $v_\perp = r\dot\theta$, is

$$v_\perp^2 = v^2 - \dot r^2 = r^2\dot\theta^2$$

Hence, to the extent that we can neglect Δr and $v\Delta t$ compared to $2r$, we have that

$$\frac{\Delta A}{\Delta t} = \frac{1}{2}rv_\perp = \frac{1}{2}r^2\dot\theta$$

In Newtonian dynamics the quantity $mrv_\perp = mr^2\dot\theta$ is called the angular momentum (of a particle of mass m moving past a point at a distance r from it). Unfortunately there is no accepted term for the quantity $rv_\perp$. We choose to call it *specific angular momentum* and note that $rv_\perp = r^2\dot\theta$. It has dimensions of angular momentum per unit mass, or $\mathbf{L}^2/\mathbf{T}$, or area per unit time. When it is conserved in particle motion, we will call it Kepler's Constant and symbolize it by K. If the planet takes a time T (its period) to return to the same position in its orbit, the line from sun to planet sweeps out the area of the ellipse, which is πab. Hence

$$K = rv_\perp = r^2\dot\theta = \frac{2\pi ab}{T}$$

independent of time.

In order to relate this motion to Kepler's Third Law, namely that a^3/T^2 has the same value for all planets, we first differentiate the orbit equation to obtain a connection between the square of the velocity, the radial position, Kepler's constant and the parameters of the ellipse. Since

$$\dot r = \frac{f\,sin\,\theta\;r^2\dot\theta}{b^2} = \frac{f\,sin\,\theta}{b^2}K; \qquad r\,\dot\theta = \frac{K}{r}$$

we have that

$$v^2 = K^2[\frac{f^2 sin^2\theta}{b^4} + \frac{1}{r^2}] = \frac{K^2 a}{b^2}(\frac{2}{r} - \frac{1}{a}) = \gamma(\frac{2}{r} - \frac{1}{a})$$

where we have used Kepler's Third Law to define the constant

$$\gamma = \frac{K^2 a}{b^2} = \frac{4\pi^2 a^3}{T^2}$$

This shows that our velocity equation applies to any planet in the solar system. Note that it is independent of the focal distance f or the eccentricity of the orbit $\epsilon = f/a$ or the semi-major axis $b^2 = a^2 - f^2 = a^2(1 - \epsilon^2)$. The velocity equation for a circular orbit ($r = a = b$) defines a particular specific angular momentum, K_a, and velocity V_a which relate to the orbital parameters in the following way

$$V_a = \frac{K_a}{a} = \frac{2\pi a}{T} = \frac{\gamma}{K_a}$$

and allow us to bound the velocities in orbit as follows:

$$\frac{a-f}{a+f} \le \frac{v^2}{V_a^2} \le \frac{a+f}{a-f}$$

The velocity equation, which is the key to solar system dynamics and the technological requirements for interplanetary travel,[14] is often derived using the conservation of energy, and other concepts taken from Newtonian and post-Newtonian physics. We emphasize that the derivation given here is purely kinematic, given Kepler's three laws.

5.2 THE INVERSE SQUARE LAW

Our next step, drawing on Newton's analysis of these laws, is to emphasize is the key role played by the concept of *change in velocity* or *acceleration*. For the planetary problem, we start by differentiating the equation for radial velocity in an orbit to obtain

$$\ddot{r} = \frac{f \cos\theta \, \dot{\theta} K}{b^2} = \left(\frac{1}{r} - \frac{a}{b^2}\right)\frac{K^2}{r^2} = \frac{K^2}{r^3} - \frac{\gamma}{r^2}$$

The first term, called the *centrifugal acceleration*, arises because we have constrained the motion to move in an orbit obeying Kepler's Second Law and depends (through K) on the orbital parameters a, b, T for the individual planet. But the second applies to all the planets in the solar system, and varies with the inverse square of the distance from sun to planet.

One of Newton's great insights came when he realized that the motion of the Galilean satellites about Jupiter also obeyed Kepler's laws with Jupiter playing the role of the sun and that our own satellite (the moon) could be thought of as falling around the earth. This suggests that for each gravitating center there is some constant γ for centripetal acceleration toward that center given by γ/r^2. In the solar system, Newton could identify γ_{Sun} common to all the planets, $\gamma_{Jupiter}$ for Galilean satellites relative to Jupiter and γ_{Earth} for the motion of our own moon relative to the Earth. But then, for an object with $K = 0$ at the surface of the earth, $r = R_E$, the acceleration toward the center of the earth could be identified with the post-Galilean physical constant g used to describe the motion of freely falling objects close to the surface of the earth by taking

$$g = \frac{\gamma_E}{R_E^2} = \frac{4\pi^2 a_M^3}{T_M^2 R_E^2}$$

Note that the radius of the earth, R_E had been measured by Erosthenes in terms of local earth distances long ago, the length of the month, T_M, was known in seconds, and the semi-major axis of the moon's orbit, a_M was measured in terms of the earth's radius by the parallax of the moon sighted from well separated positions on the earth at the same time. Therefore the value of g could be calculated from known quantities and compared with local measurements of g in the *same* units of length and time.

Here I abandon this (hopefully well known) pseudo-historical account to emphasize that the mathematical analysis we have carried through will apply to any geometrically describable motion which follows elliptical (and more generally, conic section) paths conserving specific angular momentum (Kepler's Second Law), and ties these firmly to inverse square law accelerations. Thus this analysis applies to the motion of charges which obey Coulomb's Law (a fact which Rutherford exploited to arrive at his model of the nuclear atom), and could be applied to a quark model mediated by massless gluons if we can make sense out of such a bizzare "classical" model for strong interactions. Newton, of course, went on from his analysis to postulate the law of universal gravitation, which for the restricted case we have treated of a test object of mass small compared to the attracting center of mass M amounts to asserting that the constant γ we isolated above has the value GM where G is a universal (but in Newton's time unmeasurable) constant. Thus only relative planetary masses were known until Cavendish made a terrestrial measurement of G and hence was said to have "weighed the earth" (and the sun!). Coulomb's measurements provided (in principle) an absolute value for charges in an analagous way, but we will have to do a lot more work before our model can give an absolute measurement of the basic strong interaction.

5.3 THE VELOCITY OF LIGHT

One of Galileo's many inventions was a method of determining longitude by observing the local time at which a satellite of Jupiter was eclipsed by Jupiter, and comparing this with the time the same exclipse could be observed at some standard location. He did not publish it because he was negotiating with Spain to pay him for developing the necessary tables of eclipses, and the negotiations did not succeed before he was distracted by other events. A century later determination

of longitude was still a matter of great practical importance. Quantitative investigation of the variation of eclipse times compared to those predicted by Kepler's
Laws eventually revealed a puzzling discrepancy periodic in the synodic period
$1/T_{EJ} = (1/T_E - 1/T_J)$ with which the distance between Earth and Jupiter varies.
In 1727 Rømer noticed that if the light coming from Jupiter to Earth has a finite
velocity, then this could quantitatively account for the discrepancy and in so doing provide a *measurement* of the velocity of light, now usually symbolized by c.
His value is in fact within 30 % of more recent determinations. As we now know,
this physical constant has a much more general significance than this, and is now
recognized as a *universal* constant on a par with G.

In 1795 Laplace noted that this has a peculiar implication.[15] We have given
above the velocity equation for any material object subject to gravity; it was
extended by Newton to include the parabolic and hyperbolic conic sections as well:

$$v^2 = V_a^2(\frac{2}{r} - \frac{1}{a}) \quad elliptical$$

$$v^2 = V_a^2(\frac{2}{r}) \qquad parabolic$$

$$v^2 = V_a^2(\frac{2}{r} + \frac{1}{a}) \quad hyperbolic$$

Hence any object subject to gravity which has a squared velocity at perihelion
greater than $2V_a^2$ can escape on a hyperbolic trajectory, but if it has less, it will
remain trapped in orbit, or (if less than V_a^2) never even achieve orbit. Laplace
assumed that light acts in this way, and calculated that a massive object with
the density of the earth and 250 times the radius of the sun would have so strong
a surface gravity that light could never leave its surface. We now have much
stronger reasons than Laplace to believe that light *is* subject to gravity, and some
observational evidence for "black holes" which cannot (of course) be seen but
which cause what can be interpreted as gravitational phenomena that are visible
at sufficient distances from their centers.

Using our Newtonian formulae for the case of $V_a^2 = c^2$, the radius of such an
object is related to its mass by

$$R_{bh} = \frac{GM_{bh}}{c^2}$$

We see that neither G nor c by themselves break classical scale invariance, but if
we can find some natural phenomenon that sets the scale of masses, or lengths, or
times, or angular momenta, we will have ready to hand a way to set a scale which
takes us from familiar "classical" laws of physics into a realm of new phenomena
where those laws may not apply.

REFERENCES

1. H. P. Noyes, "Some Remarks on Discrete Physics as an Ultimate Dynamical Theory", *Proc. ANPA 17* (in press) and SLAC-PUB-95-7017, Oct. 1995.

2. H. P. Noyes, "A Bit-String 'Theory of Everything' ", in *Proc. Workshop on Physics and Computation (PhysComp '94)*, D. Matzke, ch., IEEE Comp. Soc. Press, Los Amitos, CA, 1994, pp. 88-94.

3. D. O. McGoveran and H. P. Noyes, "Foundations of a Discrete Physics" in *Discrete and Combinatorial Physics* (Proc. ANPA 9), H. P. Noyes, ed, ANPA WEST, 112 Blackburn Ave., Menlo Park, CA 94025, 1987, theorem 13, p.59; and SLAC-PUB-4526 (June, 1989), p.32.

4. H. P. Noyes and D. O. McGoveran, *Physics Essays* **2**, 76-100 (1989), p.91.

5. G. Feinberg, *Phys.Rev.* **159**, 1089 (1967), — **D17**, 1651 (1978).

6. D. K. Deshpande, E.C.G. Sudarshan and O.M.P. Bilaniuk, *Am.J.Phys.* **30**, 718 (1962).

7. S. Tanaka, *Prog. Theor. Phys. (Kyoto)* **24**, 171 (1960).

8. L. H. Kauffman and H. P. Noyes, "Discrete Physics and the Derivation of Electromagnetism from the Formalism of Quantum Mechanics", *Proc. Roy. Soc.* **A 452**, 81-95 (1996).

9. D. O. McGoveran and H. P. Noyes, *Physics Essays*, **4**, 115-120 (1991).

10. N. Bohr, *Phil.Mag.* 332, Feb. 1915.

11. H. P. Noyes, "Quantized Conic Sections; Quantum Gravity" in Proc. ANPA WEST 9 and SLAC-PUB-6057, March 15, 1993

12. H. P. Noyes, "Electromagnetism From Counting", in *Physical Interpretations of Relativity Theory . III.*, M.C.Duffy, ed., Imperial College, September 1992 and SLAC-PUB-5858 (December, 1992); see also H.P.Noyes *Proc. ANPA WEST 9* and SLAC-PUB-6057, p. 17.

13. H. P. Noyes *DISCRETE PHYSICS: A New Fundamental Theory*, J. C. van den Berg, ed. (in preparation).

14. A. C. Clarke, *Interplanetary Flight*, Temple Press, London, 1950.

15. P. S. Laplace, *Le System du Monde*, Paris, 1795; for quotation and discussion, see C. W. Misner, K. S. Thorne and J. A. Wheeler, *Gravitation*, W. H. Freeman and Co., 1973.

Comment on "*Are Partons Confined Tachyons?*"
H. Pierre Noyes (2000)

Thanks to a recent conversation with James Lindesay, my thinking about this somewhat whimsical idea has become considerably clarified. I quote freely from his remarks on this occasion. If one thinks about the event horizon of a black hole in the Schwarzschild metric, the coordinate singularity which occurs there is not a *physical* singuarity. For instance, if one has enough mass in the black hole so that its Schwarzschild radius is as large as the radius of the orbit of Pluto, the local acceleration toward the center is no more than g, and local phenomena are no more disrupted by gravitational effects than they are at the surface of the earth. Since, according to general relativity, a point particle cannot be given a velocity as large as c inside that radius, even though it can pass freely through that radius, it will have less than escape velocity and will inevitably fall back into the "black hole". A complex system can be disrupted passing through that radius, leaving part of its mass behind, which is one way of looking at Hawking radiation.

What does happen inside that radius, using the Schwarzschild metric, is that asymptotic "space" coordinates and "time" coordinates are interchanged, and the same thing happens in energy-momentum space. Therefore, if one takes the position (consistent with general relativity) that the invariant mass must remain positive definite, p^2 will be greater than E^2 for massive particles. In that sense, motion of particles inside the event horizon will be "tachonic", giving some support to the idea suggested in this paper. However, one remains as far as ever from being able to connect this idea to observable phenomena. I include the paper in this collection in the hope that it, like the work on "anti-gravity" presented in Ch.8, may free up the thinking of other physicists.

A Short Introduction to BIT-STRING PHYSICS * †

H. Pierre Noyes

Stanford Linear Accelerator Center

Stanford University, Stanford, CA 94309

Abstract

This paper starts with a personal memoir of how some significant ideas arose and events took place during the period from 1972, when I first encountered Ted Bastin, to 1979, when I proposed the foundation of ANPA. I then discuss program universe, the fine structure paper and its rejection, the quantitative results up to ANPA 17 and take a new look at the handy-dandy formula. Following this historical material is a first pass at establishing new foundations for bit-string physics. An abstract model for a laboratory notebook and an historical record are developed, culminating in the bit-string representation. I set up a tic-toc laboratory with two synchronized clocks and show how this can be used to analyze arbitrary incoming data. This allows me to discuss (briefly) finite and discrete Lorentz transformations, commutation relations, and scattering theory. Earlier work on conservation laws in 3- and 4- events and the free space Dirac and Maxwell equations is cited. The paper concludes with a discussion of the quantum gravity problem from our point of view and speculations about how a bit-string theory of strong, electromagnetic, weak and gravitational unification could take shape.

Revised and considerably extended version of

two invited lectures presented at the18^{th} annual international meeting of the

ALTRNATIVE NATURAL PHILOSOPHY ASSOCIATION

Wesley House, Cambridge, England, September 4-7, 1996

---

*Work supported by Department of Energy contract DE–AC03–76SF00515.

†Conference Proceedings, entitled *Merologies*, will be available from ANPA c/o Prof.C.W.Kilmister, Red Tiles Cottage, Hight Street, Bascombe, Lewes, BN8 5DH, United Kingdom.

1 Pre-ANPA IDEAS: A personal memoir

1.1 First Encounters

When I first met Ted Bastin in 1972 and heard of the Combinatorial Hierarchy (hereinafter CH), my immediate reaction was that it must be dangerous nonsense. Nonsense, because the two numbers computed to reasonable accuracy — $137 \approx \hbar c/e^2$ and $2^{127} + 136 \approx \hbar c/G m_p^2$ — are *empirically determined*, according to conventional wisdom. Dangerous, because the idea that one can gain insight into the physical world by "pure thought" without empirical input struck me then (and still strikes me) as subversive of the fundamental Enlightenment rationality which was so hard won, and which is proving to be all too fragile in the "new age" environment that the approach to the end of the millennium seems to encourage [84, 86].

Consequently when Ted came back to Stanford the next year (1973)[8], I made sure to be at his seminar so as to raise the point about empirical input with as much force as I could. Despite my bias, I was struck from the start of his talk by his obvious sanity, and by a remark he made early on (but has since forgotten) to the effect that *the basic quantization is the quantization of mass*. When his presentation came around to the two "empirical" numbers, I was struck by the thought that some time ago Dyson[19] had proved that if one calculates perturbative QED up to the approximation in which 137 electron-positron pairs can be present, the perturbation series in powers of $\alpha = e^2/\hbar c \approx 1/137$ is no longer uniformly convergent. Hence, the number 137 *as a counting number* already had a respectable place in the paradigm for relativistic quantum field theory known as renormalized quantum electrodynamics (QED). The problem for me became *why* should the arguments leading to CH produce a number which *also* supports this particular physical interpretation.

As to the CH itself, I refer you to Clive Kilmister's introductory talk in these proceedings[39], where he discusses an early version of the bit-string construction of the sequence of discriminately closed subsets with cardinals $2^2 - 1 = 3 \rightarrow 2^3 - 1 = 7 \rightarrow 2^7 - 1 = 127 \rightarrow 2^{127} - 1 \approx 1.7 \times 10^{38}$ based on bit-strings of length 2,4,16,256 respectively.. The first three terms can be mapped by square matrices of dimension $2^2 = 4 \rightarrow 4^2 = 16 \rightarrow 16^2 = 256$. The 256^2 discriminately independent matrices made available by squaring the dimension needed to map the third level are many two few to map the $2^{127} - 1$ discriminately closed subsets in the fourth level, terminating the construction. In the historical spirit of this memoir, I add that thanks to some archeological work John Amson and I did in John's attic in St. Andrews, the original paper on the hierarchy by Fredrick Parker-Rhodes, drafted late in 1961, is now available[79].

I now ask you to join with me here in my continuing investigation of how the CH can be connected to conventional physics. As you will see in due course, this research objective differs considerably from the aims of Ted Bastin and Clive Kilmister. They, in my view, are unnecessarily dismissive of the results obtained in particle physics and physical cosmology using the conventional (if mathematically inconsistent) relativistic quantum field theory, in particular quantum electrodynamics (QED), quantum chromodynamics (QCD) and weak-electromagnetic unification (WEU).

Before we embark on that journey, I think it useful to understand some of the physics background. Dyson's argument itself rests on one of the most profound and important papers in twentieth century physics. In 1937 Carl Anderson discovered in the cosmic radiation a charged particle he could show to be intermediate in mass between the proton and electron. This was the first to be discovered of the host of particles now called collectively "mesons". *One* such particle had already been postulated by Yukawa, a sort of "heavy photon" which he showed, using a "massive QED", gave rise to an exponentially bounded force of finite range. If the mass of the Yukawa particle was taken to be a few hundred electron masses, this could be the "nuclear force quantum". Anderson's discovery prompted Gian Carlo Wick to try to see if the existence of such a particle could be accounted for simply by invoking the basic principles of quantum mechanics and special relativity. He succeeded brilliantly, using only one column in *Nature*[96]. We summarize his argument here.

Consider two massive particles which are within a distance R of each other during a time Δt. If they are to act coherently, we must require $R \leq c\Delta t$. [Note that this postulate, in the context of my *neo-operationalist* approach[68] based on measurement accuracy[76], opens the door to *supraluminal* effects at short distance, which I am now starting to explore[69]]. Because of the uncertainty principle this short-range coherence tells us that the energy is uncertain by an amount $\Delta E \approx \hbar/\Delta t$. But then mass-energy equivalence allows a particle of mass μ or rest-energy $\mu c^2 \geq \Delta E$ to be present in the space time-volume of linear dimension $\approx R\Delta t$. Putting this together, we have the *Wick-Yukawa Principle*:

$$R \leq c\Delta t \approx \frac{c\hbar}{\Delta E} \leq \hbar/\mu c \tag{1}$$

Put succinctly, if we try to localize two massive particles within a distance $R \leq \hbar/\mu c$, then the uncertainty principle allows a particle of mass μ to be present. If this meson has the proper quantum numbers to allow it to transfer momentum between the two massive particles we brought together in this region, they will experience a force, and will emerge moving in different directions than those with which they entered the *scattering region*. Using estimates of the range of nuclear forces obtained from deviations from Rutherford scattering in the 1930's one can then estimate the mass of the "Yukawa particle" to be $\approx 200 - 300$ electron masses.

We are now ready to try to follow Dyson's argument. By 1952, one was used to picturing the result of Wick-Yukawa uncertainty at short distance as due to "vacuum fluctuations" which would allow N_e electron-positron pairs to be present at distances $r \leq \hbar/2Nm_e c$. This corresponds to taking $\mu = 2Nm_e$ in Eq. 1. Although you will not find it in the reference[19], in a seminar Dyson gave on this paper he presented what he called a crude way to understand his calculation making use of the non-relativistic coulomb potential. I construct here my own version of the argument.

Consider the case where there are N_e positive charges in one clump and N_e negative charges in the other, the two clumps being a distance $r = \hbar/m_e c$ apart. Then a single charge from one clump will have an electrostatic energy $N_e e^2/r = N_e [e^2/\hbar c] m_e c^2$ due to the other clump and visa versa. I do recall that Dyson said that the system is dilute enough so that non-relativistic electrostatic estimates of this type are a reasonable approximation. If under the force of this attraction, these two charges we are considering come together and scatter producing a Dalitz pair ($e^+ + e^- \rightarrow 2e^+ + 2e^-$) the energy from the fluctuation will add another pair to the system. Of course this process doesn't happen physically because like charges repel and the clumps never form in this way. However, in a theory in which like charges attract [which is equivalent to renormalized QED with $\alpha_e = [e^2/\hbar c] \rightarrow -\alpha_e$ in the renormalized perturbation series], once one goes beyond 137 terms such a process will result in the system *gaining* energy by producing another pair and the system collapses to negatively infinite energy. Dyson concluded that the renormalized perturbation series cannot be uniformly convergent, and hence that QED cannot be a fundamental theory, as I have subsequently learned from Schweber's history of those heroic years[85].

Returning to 1973, once I had understood that, thanks to Dyson's argument, 137 can be interpreted as a *counting number*, I saw immediately that $2^{127} + 136 \approx 1.7 \times 10^{38} \approx \hbar c/G m_p^2$ could *also* be interpreted as a counting number, namely the number of baryons of protonic mass which, if found within the Compton wavelength of any one of them, would form a black hole. These two observations removed my objection to the calculation of two pure numbers that, conventionally interpreted, depend on laboratory measurements using arbitrary units of mass, length and time. I could hardly restrain my enthusiasm long enough to allow Ted to finish his seminar before bursting out with this insight. If this cusp turns out to be the point at which a new fundamental theory takes off — as I had hoped to make plausible at ANPA 18 — then we can tie it firmly into the history of "normal science" as the point where a "paradigm shift", in Kuhn's sense of the word[40], became possible.

However, my problem with *why* the calculation made by Fredrick Parker-Rhodes[79, 80] lead to these numbers remained unresolved. Indeed, I do not find a satisfactory answer to that question even in Ted and Clive's book published last year[10]. [I had hoped to get further with that quest at the meeting (ANPA 18, Sept.5-8, 1996), but discussions during and subsequent to the meeting still leave many of my questions unanswered. I intend to review these discussions and draw my own

450

conclusions at ANPA 19 (August 14-17,1997).]

1.2   From "NON-LOCALITY" to "PITCH": 1974-1979

My interest in how to resolve this puzzle has obviously continued to this day. I was already impressed by the quality of Fredrick's results in 1973, and made a point of keeping in contact with Ted Bastin. This led to my meeting with Fredrick Parker-Rhodes, Clive Kilmister and several of the other Epiphany Philosophers at a retreat in the windmill at Kings Lynn, followed by discussions in Cambridge. At that point the group were trying to put together a volume on *Revisionary Philosophy and Science.* I agreed to contribute a chapter, and finished about half of the first draft of what was to become "Non-Locality in Particle Physics" [55, 56] on the plane going back to Stanford. In the course of finishing that article, I noted for the first time that the Dyson route to the hierarchy number places an energy cutoff on the validity of QED at $E_{max} = 2 \times 137 m_e c^2$, which is approximately equal to the pion (Yukawa particle) mass. I have subsequently realized that this explains a puzzle I had been carrying with me since I was a graduate student.

This puzzle, which I have sometimes called the *Joe Weinberg memnonic*, came from quite another direction[91]. An easy way to remember the hierarchy of nuclear, QED, and atomic dimensions expressed in terms of fundamental constants is the fact that

$$1.4 \; Fermi \approx \frac{e^2}{2m_e c^2} = \left[\frac{e^2}{\hbar c}\right] \frac{\hbar}{2m_e c} = \left[\frac{e^2}{\hbar c}\right]^2 \frac{\hbar^2}{2m_e e^2} \approx 0.265 \; Angstrom \qquad (2)$$

Why nuclear dimensions should be approximately half the "classical electron radius" (i.e. $\frac{e^2}{2m_e c^2} \approx$ $1.4 \times 10^{-15} meter$) and hence $[1/137]^2$ smaller than than the radius of the positronium atom (i.e. $\frac{\hbar^2}{2m_e e^2} \approx 2.65 \times 10^{-10} meter$) was almost completely mysterious in 1947. It was known that the mass of the electron attributed to it's electrostatic "self-energy" as due to its charge distributed over a spherical shell fixed at this radius would have the mass m_e, but the success of Einstein's relativity had shown that this electron model made no sense[78]. The square of this parameter was also known to be proportional to the cross section for scattering a low energy electromagnetic wave from this model electron (Thompson cross section $[8\pi/3](e^2/m_e c^2)^2$), but again why this should have anything to do with nuclear forces was completely mysterious.

As we have already seen, it *was* known that the Wick-Yukawa principle [96] accounted roughly for the range of nuclear forces if those forces were attributed to a strongly interacting particle intermediate in mass between proton and electron. However, the only known particle in that mass range (the muon) had been shown experimentally to interact with nuclei with an energy 10^{13} times smaller than the Yukawa theory of nuclear forces demanded[18]. The Yukawa particle (the pion) was indeed discovered later that year, but there was still no reason to connect it with the "classical electron radius". Joseph Weinberg left his students to ponder this puzzle.

The trail to the solution of this conundrum starts with a 1952 paper by Dyson[19], despite the fact that neither he nor I realized it at the time. Two decades later, when I first heard a detailed account of the *combinatorial hierarchy*[8], and was puzzled by the problem of how a counting number (i.e. 137) could approximate a combination of empirical constants (i.e. $\hbar c/e^2$), I realized that this number is both the number of terms in the perturbation series and the number of virtual electron-positron pairs where QED ceases to be self-contained. But, empirically, $m_\pi \approx 2 \times 137 m_e$. Of course, if neutral this system is highly unstable due to 2γ decay, but if we add an electron-antineutrino or a positron-neutrino pair to the system, and identify the system with π^- or π^+ respectively, the system *is* stable until we include weak decays in the model. This suggests that the QED theory of electrons, positrons and γ-rays breaks down at an energy of $[2(\hbar c/e^2)+1]m_e c^2$ due to the formation of charged pions, finally providing me with a tentative explanation for the Joe Weinberg memnonic. As noted above, I first presented this speculative idea some time ago [55, 56].

By the time I wrote "NON-LOCALITY", I was obviously committed to engaging in serious research on the combinatorial hierarchy as part of my professional activity. Ted was able to get a research contract to spend a month with me at Stanford. I had hoped that this extended period of interaction would give me a better understanding of what was going on; in the event little progress was made on my side. By 1978 I had met Irving Stein, and was also struggling to understand how he could get both special relativity and the quantum mechanical uncertainty principle from an elementary random walk. His work, after much subsequent development, is now available in final form [88].

Meanwhile Ted had attended the 1976 Tutzing Conference organized by Carl Friedrick von Weizsacker and presented a paper on the combinatorial hierarchy by John Amson. I agreed to accompany Ted to the 1978 meeting and present a joint paper. I arrived in England to learn of the startlingly successful successful calculation of the proton-electron mass ratio, which Ted and I had to discuss and digest in order to present Fredrick's result [11, 81] at the Tutzing meeting, which followed almost immediately thereafter. This formula has been extensively discussed at ANPA meetings. It was originally arrived at by assuming that the electron's charge could come apart, as a statistical fluctuation, in three steps with three degrees of freedom corresponding to the three dimensions of space and that the electrostatic energy corresponding to these pieces could be computed by taking the appropriate statistical average cut off at the proton Compton radius $\hbar/m_p c$. The only additional physical input is the CH value for the electronic charge $e^2 = \hbar c/137$. Take $0 \le x \le 1$ to be the fractional charge in these units and $x(1-x)$ the charge factor in Coulomb's

452

law. Take $0 \leq y \leq 1$ to be the inverse distance between the charge fractions in that law in units of the proton Compton radius. Then, averaging between these limits with the appropriate weighting factors of $x^2(1-x)^2$ and $1/y^3$ respectively, Fredrick's straightforward statistical calculation gives

$$\frac{m_p}{m_e} = \frac{137\pi}{<x(1-x)><\frac{1}{y}>} = \frac{137\pi}{(\frac{3}{14})[1+\frac{2}{7}+\frac{4}{49}](\frac{4}{5})} \tag{3}$$

At that time the result was within a tenth of a standard deviation of the accepted value. I knew this was much too good because, for example, the calculation does not include the effect of the weak interactions. I was therefore greatly relieved when a revision of the fit to the fundamental constants changed the empirical value by 20 standard deviations, giving us something to aim at when we know how to include additional effects.

I also learned from the group during those few days before Tutzing that up to that point no one had proved the *existence* of the combinatorial hierarchy in a mathematical sense! Subsequent to the Tutzing meeting, thanks to the kind hospitality of K.V.Laurikainen in Finland, I was able to devote considerable time to an empirical attack on that problem and get a start on actually *constructing* specific representations of both the *level* 2 $\longrightarrow$ *level* 3 and the *level* 3 $\longrightarrow$ *level* 4 mappings.

It turned out that neither John Amson's nor our contributions to the Tutzing conferences, despite promises, appeared in the conference proceedings. Fortunately we had had an inkling at the meeting that this contingency might arise. In the event we were able to turn to David Finkelstein and write a more careful presentation of the developments up to that point for publication in the *International Journal of Theoretical Physics*[11]. The first version, called "Physical Interpretation of the Combinatorial Hierarchy" (or PICH for short) still lacked a formal existence proof, but Clive came up with one; further, he and John Amson (whose unpublished 1976 Tutzing contribution had been extended and completed to serve as an Appendix) were able to say precisely in what sense the CH is *unique*. The final title was therefore changed to "Physical Interpretation and mathematical structure of The Combinatorial Hierarchy" affectionately known as PITCH. The finishing touches on this paper were completed at the first meeting of ANPA. This brings my informal history to the point at which Clive ended his historical sketch in his first lecture.

1.3   ANPA 1: The foundation of the organization

Although I was obviously putting considerable time into trying to understand the CH, and the Parker-Rhodes formula for m_p/m_e showed that there might be more to the physics behind it than the basic coupling constants, I was by no means convinced that the whole effort might not turn out in the long run to be an unjustifiable "numerology". I therefore, privately, took the attitude that my efforts should go into trying to derive a clear *contradiction* with empirical results which would prove the CH approach to be wrong. Then I could drop the whole enterprise and get back to my (continuing) conventional research, where I felt more at home. I was not the only one with doubts at this time. Clive told us, years later, that he had been somewhat afraid to examine the foundational arguments too closely for fear that the whole scheme would dissolve!

In the spring of 1979 I happened to make the acquaintance of an investment counselor named Dugal Thomas who was advising a large fraction of the private charitable foundations in the US. He offered to help me with fundraising if I could put together a viable organization for supporting Ted Bastin's type of research. I threw together a proposal very quickly. Dugal located a few prospective donors; like all subsequent efforts to raise substantial funds for ANPA this initial effort came a cropper. Soon after that effort started I also learned that I had received a Humboldt U.S.Senior Scientist award, giving me the prospect of a year in Germany and some extra cash. Consequently I felt encouraged to approach Clive to see if he would serve as treasurer for the proposed organization. Clive agreed to approach Fredrick to see if he would match the small amount of "seed money" I was prepared to invest in ANPA. [The name and original statement of purpose came from the proposal I had already written. I intended that the term "natural philosophy" in the name of the organization would hark back to the thinkers at the start of the scientific revolution who were trying to look at nature afresh and shake themselves loose from the endless debates of the "nominalist" and "realist" metaphysicists of the schools.] With Fredrick's promise in hand, Clive and I approached Ted Bastin with the invitation to be the Coordinator, and asked John Amson to join us as a founding member.

The result of all this was what can be properly called ANPA 1, which met in Clive's Red Tiles Cottage near Lewes in Sussex in the early fall of 1979. John Amson was unable to attend, but endorsed our statement of purpose (modified by Ted to include specific mention of the CH) and table of organization. Once these details were in hand we had a proper scientific meeting, including thrashing out an agreed manuscript for PITCH. I gave a paper on the quantum mechanical three and four body problem, which I was working on in Germany. I noted in particular that the three channel Faddeev equations go to the seven channel Faddeev-Yakubovsky equations when one goes from three to four independent particles, reminiscent of the CH $1 \longrightarrow 2$ level transition. It is taken a long time to see what the relationship is between these two facts, but now that I am developing a "bit-string scattering theory" with Ed Jones[75], this old insight is finding an appropriate home.

Selected Topics

All meetings subsequent to ANPA 1 have been held annually in Cambridge, England. Proceedings were prepared for ANPA 7[58], and some of the papers incorporated in a SLAC-PUB[59]. The ANPA 9 proceedings [60] are available from ANPA West. Proceedings ANPA's 10 to 17 are available from Clive Kilmister. This is obviously not the place to attempt the impossible task of summarizing 16 years of work by more than 20 dedicated people in a way that would do justice to their varied contributions. I have therefore chosen to pick a few topics where I still find continued discussion both interesting and important.

2 Program Universe

2.1 Origin of Program Universe

About a decade and a half ago, Clive attempted to improve the clarity of what Ted has called "the canonical approach" [9] by admitting into the scheme a second operation besides the *Discrimination* operation, which had been central to the project ever since John Amson introduced it [2] and related it to subsequent developments [4]. Clive called this second operation *Generation* because at that stage in his thinking he saw no way to get the construction off the ground without generating bit-strings as well as discriminating them. I think he had in mind at that time a random sequence of G and D operations, but did not quite know how to articulate it. Because Mike Manthey and I were unsure how to construct a specific theory from what we could understand of this new approach, we decided to make a simple-minded computer model of the process and see how far it would lead. The first version [42] turned out to be unnecessarily complicated, and was replaced [73] by the version described below in section 2.3.

One essential respect in which the construction Mike Manthey and I turned out differs from the canonical approach is that we explicitly introduced a random element into the generation of the bit-strings rather than leaving the background from which they arise vague. Some physicists, in particular Pauli, have seen in the random element that so far as proved to be inescapable in the discussion of quantum phenomena an entrance of the the "irrational" into physics. This seems to me to equate "rationality" with determinism. I think this is too narrow a view. Statistical theories of all sorts are used in many fields besides physics without such approaches having to suffer from being castigated as irrational. In particular, biology is now founded on the proposition that evolution is (mainly) explicable as the natural selection of heritable stability in the presence of a random background. The caveat "mainly" is inserted to allow for historical contingencies[33]. Even in physics, the idea of a random "least step" goes back at least to Epicurus, and of a least step to Aristotle. I would characterize Epicurus as an exemplary rationalist whose aim was to help mankind escape from the superstitious terrors generated by ancient religions. This random element enters program universe via the primitive function "flipbit" which Manthey uses to provide either a zero or a one by unsynchronized access to a closed circuit that flips these two bits back and forth between two memory locations. Before discussing how this routine is used, we need to know a bit more about bit-strings and the operations by which we combine them.

2.2 Bit-Strings

Define a bit-string $a(a;W)$ with length W and Hamming measure a by its W ordered elements $(\mathbf{a})_w \equiv a_w \in 0,1$ where $w \in 1,2,...,W$. Define the Dirac inner product, which reduces two bit-strings to a single positive integer, by $\mathbf{a} \cdot \mathbf{b} \equiv \Sigma_{w=1}^{W} a_w b_w$. Hence $\mathbf{a} \cdot \mathbf{a} = a$ and $\mathbf{b} \cdot \mathbf{b} = b$. Define *discrimination* between bit-strings of the same length, which yields a third string of the same length, by $(\mathbf{a} \oplus \mathbf{b})_w = (a_w - b_w)^2$. Clive and I arrived at this way of representing discrimination during a session in his office after ANPA 2 or 3. From this representation the *basic bit-string theorem* follows immediately:

$$(\mathbf{a} \oplus \mathbf{b}) \cdot (\mathbf{a} \oplus \mathbf{b}) = a + b - 2\mathbf{a} \cdot \mathbf{b} \tag{4}$$

This equation could provide the starting point for an alternative definition of "$\oplus$" which avoids invoking the explicit structure used above.

We also will need the *null string* $\Phi(W)$ which is simply a string of W zeros. Note that $\mathbf{a} \oplus \mathbf{a} = \Phi(W)$, that $(\mathbf{a} \oplus \mathbf{a}) \cdot (\mathbf{a} \oplus \mathbf{a}) = 0$ and that $\mathbf{a} \cdot \mathbf{\Phi} = 0$. The complement of the null string is the *anti-null string* $\mathbf{W}(W)$ which consists of W ones and has the property $\mathbf{W} \cdot \mathbf{W} = W$. Of course $\mathbf{W} \cdot \mathbf{\Phi} = 0$.

Define *concatenation*, symbolized by "$\|$", for two string $\mathbf{a}(a; S_a)$ and $\mathbf{b}(b; S_a)$ with Hamming measures a and b and respective lengths S_a and S_b and which produces a string of length $S_a + S_b$, by

$$\begin{aligned}(\mathbf{a}\|\mathbf{b})_s &\equiv a_s \ \ if \ \ s \in 1,2,...,S_a \\ &\equiv b_{S_a - s} \ \ if \ \ s \in S_a + 1, S_a + 2, ..., S_a + S_b\end{aligned} \tag{5}$$

For strings of equal length this doubles the length of the string and hence doubles the size of the bit-string space we are using. For strings of equal length it is sometimes useful to use the shorthand but somewhat ambiguous "product notation" $\mathbf{ab}$ for concatenation. Note that while "$\cdot$" and "$\oplus$" are, separately, both associative and commutative, in general concatenation is not commutative even for strings of equal length, although it is always, separately, associative.

2.3 Program Universe

To generate a growing universe of bit-strings which at each step contains $P(S)$ strings of length S, we use an algorithm known as *program universe* which was developed in collaboration with

M.J.Manthey [42, 73]. Since no one knows how to construct a "perfect" random number generator, we cannot start from Manthey's "flipbit", and must content ourselves with a pseudo-random number generator that, to some approximation which we will be wise to reconsider from time to time, will give us either a "0" or a "1" with equal probability. Using any available approximation to "flipbit" and assigning an order parameter $i \in 1, 2, ..., P(S)$ to each string in our array, Manthey[42] has given the coding for constructing a routine "PICK" which picks out some arbitrary string $\mathbf{P}_i(S)$ with probability $1/P(S)$. Then program universe amounts to the following simple algorithm:

> PICK any two strings $\mathbf{P}_i(S), \mathbf{P}_j(S)$, $i, j \in 1, 2, ..., P$ and compare $\mathbf{P}_{ij} = \mathbf{P_i} \oplus \mathbf{P_j}$ with $\Phi(S)$.
>
> If $\mathbf{P}_{ij} \neq \Phi$, adjoin $\mathbf{P}_{P+1} := \mathbf{P}_{ij}$ to the universe, set $P := P + 1$ and recurse to PICK. [This process is referred to as ADJOIN.]
>
> Else, for each $i \in 1, 2, ..., P$ pick an arbitrary bit $\mathbf{a}_i \in 0, 1$, replace $\mathbf{P}_i(S + 1) := \mathbf{P}_i(S) \| \mathbf{a}_i$, set $S := S + 1$ and recurse to PICK. [This process is referred to as TICK.]

It is important to realize that if we take a snapshot of the universe of bit-strings so constructed at any time, with the $\mathbf{P}_i$ written as rows of 0's and 1's in a rectangular array containing S columns, there is nothing in the *process* that generated them which distinguishes this universe from any of the $S!$ other universes of 0's and 1's of this height and width which could be obtained by using any of the $S!$ possible permutations of the columns. In this sense any run of program universe up to this point could just as well have produced any of these other universes. The point here is that, since the rows are produced by discrimination, and the order of the bits is the same for each row, the result is independent of the order of the bits. Similarly, since the column of bits which is adjoined to this block representation just before $S \rightarrow S + 1$ is some (hopefully good!) approximation to a Bernoulli sequence, the probability of it having k 1's and $P(S) - k$ 0's is simply $P(S)!/k!(P(S) - k)!$ independent of how the rows are ordered by the order parameter i. That is, even though we have introduced an order parameter for the rows in order to make it easy to code the program in a transparent way, this parameter in itself is not intended to play any role in the physical interpretation of the model. At this stage in our argument, this means that program universe can end up with any one of the $2^{P(S)}S!$ possible block rectangles containing only 0's and 1's of height $P(S)$ and width S with some probability which is presumably calculable. This probability is relevant when we come to discuss cosmology. Nevertheless, if we look at the *internal structure* of some fixed portion of *any one* of these universes, the way in which they are constructed will allow us to make some useful and general statements. Further, these rectangular blocks of "0" 's and "i" 's are *tables* and hence have *shapes* in the precise sense defined by Etter's *Link Theory* [22, 23, 24, 25, 26]. I hope to have time to discuss Etter's theory at ANPA 19.

I have called another symmetry of the universes so constructed *Amson invariance* in reference to his paper on the BI-OROBOUROS [5]. He notes that there is nothing in the discrimination operation which prevents us from using the alternative representation for discrimination given by

$$0 \oplus' 0 = 1; \ 0 \oplus' 1 = 1 = 1 \oplus' 0; \ 1 \oplus' 1 = 1 \tag{6}$$

This will produce a dual representation of the system in which the roles of the *bits* "0" and "1" (which obviously can no longer be thought of as integers in a normal notation) are interchanged. Then when the construction of the *combinatorial hierarchy* is completed at level 4, one will have the complete system and its dual. But then, one can answer the question which has been asked in these meetings: "Where do the bits in the CH come from?" in an interesting way. In John's construction the bits are simply the two dual representations of the CH! Consequently one has a nested sequence of CH's with no beginning and no end. The essential point for me here is not this nested sequence —- which will be difficult to put to empirical test —- but the emphasis it gives to the fact that the two symbols are *arbitrary* and hence that their interchange is a *symmetry operation*. This has helped me considerably in thinking about how the particle-antiparticle symmetry and CPT invariance come about in bit-string physics.

Note that in the version of program universe presented here the arbitrary bits are concatenated only at one growing end of the strings. Consequently, once the string length S passes any fixed length L the $P(L)$ strings present will consist of some number $n_L \leq L$ of strings which are discriminately independent. Further, once $S > L$, the portion of all string of length L changes only by discrimination between members of this collection. Consequently it can end up containing at most $2^{n_L} - 1$ types of distinct, non-null strings no matter how much longer program universe runs. Whether it ever even reaches this bound, and the value of n_L itself, are *historically contingent* on which run of program universe is considered. This observation provides a model for *context sensitivity*. One result of this feature of program universe is that at any later stage in the evolution of the universe we can always separate any string into two portions, a *label string* $\mathbf{N}_i(L)$ and a *content string* $\mathbf{C}_i(S - L)$ and write $P_i(S) = \mathbf{N}_i(L) \| \mathbf{C}_i(S - L)$ with $i \in 1, 2, ..., n_L$, making the context sensitivity explicit.. Once we separate labels from content, the permutation invariance we talked about above can only be applied to the columns in the label and/or to the columns in the content parts of the strings separately. Permutations which cross this divide will interfere with any physical interpretation of the formalism we have established up to that point.

In preparation for a more detailed discussion on the foundations of bit-string physics, we note here that the alternatives ADJOIN and TICK correspond *precisely* to the production of a virtual particle represented by a 3-leg "Feynman" diagram, or "3-event", and to the scattering process represented by a 4-leg "Feynman" diagram, or "4-event" respectively. We have to use quotes around

Feynman, because our diagrams obey finite and discrete conservation laws consistent with measurement accuracy. This whole subject will be more fully developed elsewhere, for instance in discussing bit-string scattering theory [75].

Another aspect of program universe is worth mentioning. We note that TICK has a *global* character since a 4-event anywhere in the bit-string universe will necessarily produce a "simultaneous" increase of string length in our space of description. This means that it will be a candidate for representing a coherent cosmological time in an expanding universe model. The time interval to which TICK refers is the shortest meaningful (i.e. finite and discrete) distance that the radius of the universe can advance in our theory divided by the velocity of light. We will return to this idea on another occasion when we discuss cosmology.

3 Lessons from the rejection of the Fine Structure paper

3.1 Background

In preparation for ANPA 9, Christoffer Gefwert[32], David McGoveran[49] and I[61] prepared three papers intended to present a common philosophical and methodological approach to discrete physics. Unfortunately, in order to get the first two papers typed and processed by SLAC, I had to put my name on them, but I want it in the record that my share in Gefwert's and McGoveran's papers amounted mainly to criticism; I made no substantial contribution to their work. We started the report on ANPA 9[60] with these three papers, followed by a paper on Combinatorial Physics by Ted Bastin, John Amson's Parker Rhodes Memorial Lecture (the first in this series), a second paper by John, and a number of first rate contributed papers. Clive Kilmister's concluding remarks closed the volume. I went to considerable trouble to get the whole thing into camera ready format and tried to get the volume into the Springer-Verlag lecture note series, but they were unwilling to accept such a mixed bag. They were interested in the first three papers and were willing to discuss what else to include, but I was unwilling to abandon my comrades at ANPA by dropping any of their contributions to ANPA 9. We ended up publishing the proceedings ourselves, with some much needed help on the physical production from Herb Doughty, which we gratefully acknowledge.

David and I did considerably more work on my paper, and I tried to get it into the mainstream literature, but to no avail. Our joint version ended up in *Physics Essays*[73]. In the interim David had seen how to calculate the fine structure of hydrogen using the discrete and combinatorial approach, and presented a preliminary version at ANPA 10[46]. I was so impressed by this result (see below) that I tried to get it published in *Physical Review Letters*. It was rejected even after we rewrote it in a vain attempt to meet the referee's objections. In order for the reader to form his own opinion about this rejection, I review the paper[51] here and quote extensively from it.

The first three pages of the paper reviewed the arguments leading to CH and the essential results already achieved. These will already be familiar to the careful reader of the material given above. With this as background we turned to the critical argument:

We consider a system composed of two masses, m_p and m_e — which we claim to have computed from first principles[62] in terms of $\hbar, c$ and $G_{[Newton]}$ — and identified by their labels using our quantum number mapping onto the combinatorial hierarchy [73]. In this framework, their mass ratio (to order α^3 and $(m_e/m_p)^2$) has also been computed using only $\hbar, c$ and 137. However, to put us in a situation more analagous to that of Bohr, we can take m_p and m_e from experiment, and treat 1/137 as a counting number representing the coulomb interaction; we recognize that corrections of the order of the square of this number *may* become important one we have to include degrees of freedom involving electron-positron pairs. We attribute the binding of m_e to m_p in the hydrogen atom to coulomb events, i.e. only to those events which involve a specific one of the 137 labels at level 3 and hence occur with probability 1/137; the changes due to other events average out (are *indistinguishable* in the absence of additional information). We can have any periodicity of the form 137j where j is any positive integer. So long as this is the only periodicity, we can write this restriction as 137j steps $= 1$ *coulomb event*. Since the internal frequency $1/137j$ is generated independently from the *zitterbewegung* frequency which specifies the mass scale, the normalization condition combining the two must be in quadrature. We meet the bound state requirement that the energy E be less than the system rest energy $m_{ep}c^2$ (where $m_{ep} = m_e m_p/(m_e + m_p)$ is used to take account of 3-momentum conservation) by requiring that $(E/\mu c^2)^2[1 + (1/137N_B)^2] = 1$. If we take $e^2/\hbar c = 1/137$, this is just the relativistic Bohr formula[14] with N_B the principle quantum number.

[Here I inserted into McGoveran's argument a discussion of the Bohr formula and how it might be derived from dispersion theory. This insertion was motivated by the vain hope that any referee would see that our reasoning was in fact closely related to standard physics. We will look at this result, called the handy-dandy formula, in a new way in the section of this paper carrying that title.]

The Sommerfeld model for the hydrogen atom (and, for superficially different but profoundly similar reasons [12], the Dirac model as well) requires two *independent* periodicities. If we take our reference period j to be integer and the second period s to differ from an integer by some rational fraction Δ, there will be two minimum values $s_0^{\pm} = 1 \pm \Delta$, and other values of s will differ from one or the other of these values by integers: $s_n = n + s_0$. This means that we can relate ("synchronize") the fundamental period j to this second period in two different ways, namely to

$$137j \frac{steps}{(coulomb\ event)} + 137s_0 \frac{steps}{(coulomb\ event)} = 1 + e = b_+ \tag{7}$$

or to

$$137j \frac{steps}{(coulomb\ event)} - 137s_0 \frac{steps}{(coulomb\ event)} = 1 - e = b_- \tag{8}$$

where e is an event probability. Hence we can form

$$a^2 = j^2 - s_0^2 = (b_+/137)(b_-/137) = (1 - e^2)/137^2 \tag{9}$$

Note that if we want a finite numerical value for a, we cannot simply take a square root, but must determine from context which of the symmetric factors [i.e. $(1 - e)$ or $(1 + e)$] we should take (c.f. the discussion about factoring a quadratic above). With this understood, we write $s_n = n + \sqrt{j^2 - a^2}$.

We must now compute the probability e that j and s are mapped to the same label, using a single basis representation constructed within the combinatorial hierarchy. We can consider the quantity a as an event probability corresponding to an event **A** generated by a global ordering operator which ultimately generates the entire structure under consideration. Each of the two events j and s can be thought of as derived by sampling from the same population. That population consists of 127 strings defined at level three of the hierarchy. In order that j and s be independent, at least the last of the 127 strings generated in the construction of s (thus completing level three for s) must not coincide with any string generated in the construction of j. There are 127 ways in which this can happen.

There is an additional constraint. Prior to the completion of level three for s, we have available the $m_2 = 16$ possible strings constructed as a level two representation basis to map (i.e. represent) level three. One of these is the null string and cannot be used, so there are 15 possibilities from which the actual construction of the label for s that completes level 3 are drawn. The level can be completed just before or just after some j cycle is completed. So, employing the usual frequency theory of probability, the expectation e that j and s as constructed will be indistinguishable is $e = 1/(30 \times 127)$.

In accordance with the symmetric factors $(1 - e)$ or $(1 + e)$ the value e can either subtract from or add to the probability of a coulomb event. These two cases correspond to two different combinatorial paths by which the independently generated sequences of events may close (the "relative phase" may be either positive or negative). However we require only the probability that all s_0 events be generated within one period of j, which is $1 - e$. Hence the difference between j^2 and s^2 is to be computed as the "square" of this "root", $j^2 - s_0^2 = (1 - e)^2$. Thus, for a system dynamically bound by the coulomb interaction with two internal periodicities, as in the Sommerfeld or Dirac models for the hydrogen atom, we conclude that the value of the fine structure constant to be used should be

$$\frac{1}{a} = \frac{137}{1 - \frac{1}{30 \times 127}} = 137.0359\ 674...$$ (10)

in comparison to the accepted empirical value of[1]

$$\frac{1}{\alpha} \simeq 137.0359\ 895(61)$$ (11)

Now that we have the relationship between s, j and a, we consider a quantity H' interpreted as the energy attribute expressed in dynamical variables at the $137j$ value of the system containing two periods. We represent H' in units of the invariant system energy μc^2. The independent additional energy due to the shift of s_n relative to j for a period can then be given as a fraction of this energy by $(a/s_n)H'$, and can be added or subtracted, giving us the two factors $(1 - (a/s_n)H')$ and $(1 + (a/s_n)H')$. These are to be multiplied just as we multiplied the factors of a above, giving the (elliptic) equation $(H')^2/(\mu 2c^4) + (a^2/s_n^2)(H')^2/\mu^2 c^4 = 1$, Thanks to the previously derived expression of $s = n + s_0$ this can be rearranged to give us the Sommerfeld formula[87]

$$H'/\mu c^2 = [1 + \frac{a^2}{(n + \sqrt{j^2 - a^2})^2}]^{-1/2}$$ (12)

Several corrections to our calculated value for α can be anticipated,....

3.2 The rejection

It is obvious to any physicist that if an *understandable* theory can be constructed which allows one to calculate the fine structure constant and Newton's gravitational constant to high accuracy, it should be possible to create a major paradigm shift in theoretical physics. But even though David McGoveran[46, 51] had showed us how to add four more significant figures to the calculation of the inverse fine structure constant, we were unable to make the chain of thought understandable to most of the physics community. To quote an anonymous referee for *Physical Review Letters*:

I recommend that this letter be rejected. How happy we should all be to publish a physical theory of the fine structure constant! Any such theory, right or wrong, would be worth publishing. But this letter does not contain a theory which might be proved right or wrong. The formula for the fine-structure constant comes out of a verbal discussion which seems to make up its own rules as it goes along. Somewhere underlying the discussion is a random process, but the process is never precisely defined, and its connection to the observed quantities is not explained. I see no way by which the argument of this letter could be proved wrong. Hence I conclude that the argument is not science.

It should be obvious already that, because of my professional background, I have some sympathy with this criticism. In fact, though I was careful not to discuss this with the people in ANPA, for a number of years my research into the meaning of the CH was, in a sense, aimed at giving the canonical ANPA arguments sufficient precision so that they *could* be proved wrong. Then, I could drop my involvement with these ideas and get back to doing more conventional physics. What convinced me that ANPA must be on the right track was, in fact, the McGoveran calculation and his later extension of the same ideas to yield several more mass ratios and coupling constants in better agreement with experiment[47]. Only this past year have we succeeded in getting two publications about discrete physics into leading mainstream physics journals[35, 36]. But the basic canonical calculations are, even today, not in the kind of shape to receive that blessing. This is not the place to review my disheartening attempts to get this and other ANPA calculations before a broader audience. As an illustration of this failed strategy of hoping that the quality of our results would do the job by itself I remind you in the next section what these results were.

4   Quantitative Results up to ANPA 17

We emphasize that the only experimental constants needed as input to obtain these results are $\hbar, c$ and m_p.

The bracketed rational fraction corrections given in bold face type are due to McGoveran[47]. The numerical digits given in bold face type emphasize remaining discrepancies between calculated and observed values.

Newton's gravitational constant (gravitation):

$$G_N^{-1}\frac{\hbar c}{m_p^2} = [2^{127} + 136] \times [\mathbf{1} - \frac{1}{\mathbf{3 \cdot 7 \cdot 10}}] = 1.693\ 31\ldots \times 10^{38}$$

$$experiment = 1.693\ \mathbf{58}(21) \times 10^{38}$$

The fine structure constant (quantum electrodynamics):

$$\alpha^{-1}(m_e) = 137 \times [1 - \frac{1}{30 \times 127}]^{-1} = 137.0359\ \mathbf{674}....$$

$$experiment = 137.0359\ 895(61)$$

The Fermi constant (weak interactions — β-decay):

$$G_F m_p^2 / \hbar c = [256^2 \sqrt{2}]^{-1} \times [1 - \frac{1}{3 \cdot 7}] = 1.02\ \mathbf{758} \ldots \times 10^{-5}$$

$$experiment = 1.02\ 682(2) \times 10^{-5}$$

The weak angle (gives weak electromagnetic unification, the Z_0 and $W^{\pm}$ masses).

$$sin^2 \theta_{Weak} = 0.25[1 - \frac{1}{3 \cdot 7}]^2 = 0.2267 \ldots$$

$$experiment = 0.2259(46)]$$

The proton-electron mass ratio (atomic physics):

$$\frac{m_p}{m_e} = \frac{137\pi}{< x(1-x) >< \frac{1}{y} >} = \frac{137\pi}{(\frac{3}{14})[1 + \frac{2}{7} + \frac{4}{49}](\frac{4}{5})} = 1836.15\ \mathbf{1497} \ldots \qquad (13)$$

$$experiment = 1836.15\ 2701(37)$$

The standard model of quarks and leptons (quantum chromodynamics):

The pion-electron mass ratios

$$m_\pi^{\pm} / m_e = 275[1 - \frac{2}{2 \cdot 3 \cdot 7 \cdot 7}] = 273.12\ \mathbf{92} \ldots$$

$$experiment = 273.12\ 67(4)$$

$$m_{\pi^0} / m_e = 274[1 - \frac{3}{2 \cdot 3 \cdot 7 \cdot 2}] = 264.2\ \mathbf{143} \ldots$$

$$experiment = 264.1\ 373(6)]$$

The muon-electron mass ratio:

$$m_\mu / m_e = 3 \cdot 7 \cdot 10[1 - \frac{3}{3 \cdot 7 \cdot 10}] = 207$$

$$experiment = 206.768\ 26(13)$$

The pion-nucleon coupling constant:

$$G^2_{\pi N \bar{N}} = [(\frac{2M_N}{m_\pi})^2 - 1]^{\frac{1}{2}} = [195]^{\frac{1}{2}} = 13.96....$$

$$experiment = 13.3(3), \ or \ greater \ than \ 13.9$$

I eventually came to the conclusion that the only way to get the attention of the establishment would be to show, in detail, that these results can be derived from a finite and discrete version of relativistic quantum mechanics (it turns out, in a finite and discrete Foch space) which is compatible with most of the conventional approach. The rest of the paper is devoted to a sketch of what I think are constructive accomplishments in that direction. The next section is a draft of the start of a paper illustrating the new strategy.

5 The Handy-Dandy Formula

One essential ingredient missing from current elementary particle physics is a *non-perturbative* connection between masses and coupling constants. We believe that one reason that contemporary conventional approaches to relativistic quantum mechanics fail to produce a *simple* connection between these two basic classes of parameters is that they start by quantizing classical, manifestly covariant continuous field theories. Then the uncertainty principle necessarily produces infinite energy and momentum at each space-time point. While the renormalization program initiated by Tomonoga, Schwinger, Feynman and Dyson succeeded in taming these infinities, this was only at the cost of relying on an expansion in powers of the coupling constant. Dyson[19] showed in 1952 that this series cannot be uniformly convergent, killing his hope that renormalized QED might prove to be a fundamental theory [85]. Despite the technical and phenomenological successes of non-Abelian gauge theories, this difficulty remains unresolved at a fundamental level. What we propose here as a replacement is an expansion in *particle number* rather than in coupling constant. The first step in this direction already yields a simple formula with suggestive phenomenological applications, as we now show.

We consider a two-particle system with energies e_a, e_b and masses m_a, m_b which interact via the exchange of a composite state of mass μ. We assume that the exterior scattering state is in a coordinate system in which the particles have momenta of equal magnitude p but opposite direction. The conventional S-Matrix approach starts on energy-shell and on 3-momentum shell with the algebraic connections[6]

$$\begin{aligned} e_a^2 - m_a^2 &= p^2 = e_b^2 - m_b^2 \\ M^2 &= (e_a + e_b)^2 - |\vec{p}_a + \vec{p}_b|^2 \end{aligned} \qquad (14)$$

$$|\vec{p}|(M; m_a, m_b) \;=\; \frac{[(M^2 - (m_a + m_b)^2)(M^2 - (m_a - m_b)^2)]^{\frac{1}{2}}}{2M}$$

but then requires an analytic continuation in M^2 off mass shell. Although this keeps the problem finite in a sense, it leads to a non-linear self-consistency or *bootstrap* problem from which a systematic development of *dynamical* equations has yet to emerge.

We take our clue instead from non-relativistic multi-particle scattering theory [27, 92, 93, 94, 28, 3, 97] in which once a two-particle bound state vertex opens up, at least one of the constituents must interact with a third particle in the system before the bound state can re-form. This eliminates the singular "self energy diagrams" of relativistic quantum field theory from the start. Further, the algebraic structure of the Faddeev equations automatically guarantees the unitarity of the three particle amplitudes calculated from them [31]. The proof only requires the unitarity of the two-body input[54, 57]. This suggests that it might be possible to develop an "on-shell" or "zero range" multi-particle scattering theory starting from some two-particle scattering amplitude formula which guarantees s-wave on-shell unitarity.

In order to implement our idea, rather than use Eq.15 we define the parameter k^2, which on shell is the momentum of either particle in the zero 3-momentum frame, in terms of the variable s which in the physical region runs from $(m_a + m_b)^2$ (i.e. elastic scattering threshold) to the highest energy we consider by

$$k^2(s; m_a + m_b) = s - (m_a + m_b)^2 \tag{15}$$

Then we can insure on-shell unitarity for the scattering amplitude $T(s)$ with the normalization $Im\, T(s) = \sqrt{s - (m_a + m_b)^2}|T|^2$ in the physical region by

$$
\begin{aligned}
T(s) \;&=\; \frac{e^{i\delta(s)} sin\, \delta(s)}{\sqrt{s - (m_a + m_b)^2}} \;=\; \frac{1}{k\, ctn\, \delta(s) - i\sqrt{s - (m_a + m_b)^2}} \\
&=\; \frac{1}{\pi} \int_{(m_a + m_b)^2}^{\infty} ds' \frac{\sqrt{s' - (m_a + m_b)^2}|T(s')|^2}{s' - s - i\epsilon} \;=\; \frac{2}{\pi} \int_0^{\infty} dk' \frac{sin^2 \delta(k')}{k^2 - (k')^2 - i\epsilon}
\end{aligned}
\tag{16}
$$

We arrived at this way of formulating the two-body input for multi-particle dynamical equations in a rather circuitous way. It turns out that this representation does indeed lead to well defined and soluble *zero range* three and four particle equations of the Faddeev-Yakubovsky type[57, 75], and that *primary singularities* corresponding to bound states and CDD poles [17] can be introduced and fitted to low energy two particle parameters without destroying the unitarity of the three and four particle equations. However, if we adopt the S-Matrix point of view which suggests that elementary particle exchanges should appear in this non-relativistic model as "left hand cuts" starting at $k^2 = -\mu_x^2/4$, where μ_x is the mass of the exchanged quantum [72], then we discovered[57] that the unitarity of the 3-body equations can no longer be maintained; our attempt to use this model as a starting point for doing elementary particle physics was frustrated.

We concluded that a more fundamental approach was required, in the pursuit of which[11, 73] the non-perturbative formula which is the subject of this paper was discovered[51]. However the reasoning was considered so bizarre as, according to one referee, not even to qualify as science. This paper aims to rectify that deficiency by carrying through the derivation in the context of a relativistic scattering theory, which we will call *T-Matrix theory* in order to keep it distinct from the more familiar S-Matrix theory from which it evolved. Thanks to a comment by Castillejo[16] in the context of our treatment of the fine structure of the spectrum of hydrogen[51], we finally realized that the success of our new approach required us from the start to view our *T-Matrix* as embedded in a multi-particle space. This can be accomplished using the relativistic kinematics of Eq.16 rather than of Eq.15 for the off-shell extension which leads to dynamical equations.

As we know from earlier work on partial wave dispersion relations[72], if we know that the scattering amplitude has a pole at $s_\mu = \mu^2$, or equivalently at $k^2 + \gamma^2 = 0$ where $\gamma = +\sqrt{(m_a + m_b)^2 - \mu^2}$ then a subtraction in the partial wave dispersion relation given by Eq.17 easily accommodates the constraint while preserving on-shell unitarity in the physical region. This allows us to define the dimensionless *coupling constant* g^2 as the "residue at the bound state pole" with appropriate normalization. We choose to do this by the alternative definition of $T(s)$ given below:

$$\begin{aligned} T(s; g^2, \mu^2) &= \frac{g^2 \mu}{s - \mu^2} = \frac{g^2 \mu}{k^2(s) + \gamma^2} \\ &= \frac{1}{k \ ctn \ \delta(s) + ik(s)} \end{aligned} \tag{17}$$

Consistency with the dispersion relation, assuming a constant value for g^2, then requires that at $k^2 = 0$

$$\begin{aligned} T((m_a + m_b)^2); g^2, \mu^2) &= \frac{1}{\gamma} = \frac{g^2 \mu}{\gamma^2} \\ k \ ctn \ \delta((m_a + m_b)^2) &= \gamma \end{aligned} \tag{18}$$

Consequently $g^2 \mu = \gamma$ and by taking γ^2 also from Eq. 190 we obtain our desired result, the *handy-dandy formula* connecting masses and coupling constants:

$$(g^2 \mu)^2 = (m_a + m_b)^2 - \mu^2 \tag{19}$$

In the non-relativistic context where $\gamma_{NR}^2 = 2m_{ab}\epsilon_{ab}$, $m_{ab} = m_a m_b/(m_a + m_b)$, $\epsilon_{ab} = m_a + m_b - \mu$, this evaluation of the value of $k \ ctn \ \delta$ at low energy is equivalent to assuming that the phase shift is given by the *mixed effective range expansion*[53]:

$$k \ ctn \ \delta = \gamma + k^2/\gamma = -\gamma + (k^2 + \gamma^2)/\gamma \tag{20}$$

corresponding to the *zero range* bound state wave function $r\psi(r) = e^{-\gamma r}$ which assumes its **asymptotic** form very close to point where the positions of the two particles coincide. As Weinberg discusses in considerable detail in his papers on the quasi-particle approach [92, 93, 94], this constraint requires the bound state to be purely composite — i.e. to contain precisely two particles with no admixture of effects due to other degrees of freedom. We believe that his analysis supports our contention that we can claim the same interpretation for our relativistic model of a bound state, and hence that we have derived the proper two-particle input for relativistic dynamical n-particle equations of the Faddeev-Yakubovsky type. These equations, which are readily solved for three and four particle systems, will be presented on another occasion[75].

What follows next is an unsystematic presentation of results, some of which were initially obtained using the *combinatorial hierarchy*[11, 73], but which we now claim to have placed firmly within at least the phenomenology of standard elementary particle physics.....

THE TIC-TOC LABORATORY: A Paradigm for Bit-String Physics

Just prior to ANPA 19 I will be attending a conference organized by Professor Zimmermann entitled *NATURA NATURANS: Topoi of Emergence*. The following notes are intended to serve as raw material for my presentation there. Some of these ideas came out of extensive correspondence I have had with Ted and Clive following ANPA 18, and owe much to their comments. In particular the section 6 should be compared to Clive's discussion of a scientific investigation in his paper in these proceedings[39].

I would also like to remind you before we start of Eddington's parable that if we set out to measure the length of the fish in the sea, and we find that they are all greater than one inch long we have the option of concluding (a) that all the fish in the sea are greater than one inch long or (b) that we are using a net with a one inch mesh. Thinking of my approach in this way, I seem to be finding out that *because* I insist on finite and discrete measurement accuracy together with standard methodological principles. I am bound to end up with something that looks like a finite and discrete relativistic quantum mechanics that has the "universal constants" we observe in the laboratory. Whether the cosmology we observe is also constrained to the same extent is an interesting question. My guess is that we will find that historical contingency plays a significant role.

6 A Model for Scientific Investigation

I restrict our formalism so that it can serve as an abstract model for physical measurement in the following way.

We assume that we encounter entities one at a time, save an entity so encountered, compare it with the next entity encountered, decide whether they are the same or different, and record the result. If they are the same, we record a "0" and if they are different we record a "1". The first of the two entities encountered is then discarded and the second saved, ready to be compared to the next entity encountered. The recursive pursuit of this investigation will clearly produce an ordered string of "0" 's and "1" 's, which we can treat as a bit-string. We further assume that this record — which is our abstract version of a *laboratory notebook* — can be duplicated, communicated other investigators, treated as the input tape for a Turing machine, cut into segments which can be duplicated, combined and compared using our bit-string operations, the results recorded, and so on.

Our second assumption is that if we cut this tape into segments of length N and determine how many such segments have the Hamming measure a, the probability we will find the integer a, given N will approach $2^{-N} \frac{N!}{a!(N-a)!}$ in the sense of the *law of large numbers*. Without further tests all such strings characterized by the two integers $a \leq N$ will be called *indistinguishable*. It should be obvious that I make this postulate in order to be able to, eventually, derive the Planck black body spectrum from my theory. Remember that Planck's formula has stood up to all experimental tests for 97 years, a remarkable achievement in twentieth century physics! We have recently learned that in fact it also represents to remarkable precision the cosmic background radiation at $2.73°K$. For those who want to know how and why the quantum revolution started with the discovery of Planck's formula, rather than just myths about what happened, I strongly recommend Kuhn's last major work[41].

Any further structure coming out of our investigation is to be found using the familiar operation of *discrimination* $\mathbf{a} \oplus \mathbf{b}$ between two strings $\mathbf{a}$, $\mathbf{b}$ of equal length, by *concatenation* $\mathbf{a}\|\mathbf{b}$ (which doubles the string length for equal length strings), and by taking the *Dirac inner product* $\mathbf{a} \cdot \mathbf{b}$ which takes two strings out of the category of *bit-strings* and replaces them by a positive integer. This third operation is *also* how we determine the Hamming measure of a single string: $\mathbf{a} \cdot \mathbf{a} \equiv a$. It will become our abstract version of *quantum measurement*, which we interpret as the determination of a *cardinal*.

Clearly the category change between "bit-string" and "integer" is needed if we are to have a theory of *quantitative measurement*. I take this to be the hallmark of *physics* as a science.

The category change produced by taking the inner product allows us to relate two strings which combine by discrimination to the integer equation:

$$2\mathbf{a} \cdot \mathbf{b} = a + b - (\mathbf{a} \oplus \mathbf{b}) \cdot (\mathbf{a} \oplus \mathbf{b}) \tag{21}$$

If it is taken as axiomatic that (a) we can know the Hamming measure of a bit string and (b) that this implies that we can know the Hamming measure of the discriminant between two bit-strings, then this *basic bit-string theorem* seems very natural.

Once we start combining bit-strings and recording their Hamming measures, and in particular writing down sequential records of these integers, the analysis clearly becomes *context sensitive*. It is our abstract model for a *historical record*.

The underlying philosophy is the assumption that in appropriate units *any* physical measurement can be abstractly represented by a positive integer with an uncertainty of $\pm\frac{1}{2}$. If we were using real numbers, this would be expressed by saying that the value of the physical quantity represented by $n \pm \frac{1}{2}$ has a 50% chance of lying in the interval between $n - \frac{1}{2}$ and $n + \frac{1}{2}$. But in discrete physics, such a statement is *meaningless* in the sense used by operationalists. Clearly, part of our conceptual problem is to develop a language describing the uncertainty in the measurement of integers which does *not* require us to construct the real numbers.

7   Remark on Integers

It is clear from our comment on measurement accuracy that we will find it useful to talk about *half-integers* as well as integers. This will also be useful when we come to talk about angular momenta and other "non-commuting observables" in our language. But how much farther must we go beyond the positive integers? It was Kronecker who said "God gave us the integers. All else is the work of man." One of our objectives is to keep this extra work to a minimum.

I am certain that the largest string length segment we will need to construct the quantum numbers needed to analyze currently available data about the observable universe of physical cosmology and particle physics is 256, and that all we need do with such segments is to combine or compare or reduce them by the operations listed above, i.e. discrimination, concatenation and inner product. Using as a basis bit-strings of length $16W$, I also see how to represent negative integers, positive and negative imaginary integers, and complex integer quaternions. Discussion of how far we need go in that direction, or into using rational fractions other than $\frac{1}{2}$ is, in my opinion, best left until we find a crying need to do so. In any case, we have to lay considerable groundwork before we do.

For the moment we assume all we need know about the integers is that

$$1 + 0 = 1 = 0 + 1; \quad 1 \times 0 = 0 = 0 \times 1; \quad 1 + 1 = 2 \tag{22}$$

that we can iterate the third equation to obtain the counting numbers up to some largest integer N that we pick in advance as adequate for the purpose at hand, and that given any integer n so generated other than 0 or N, any second integer n' will be greater than, equal to, or less than the first. **That is, we** assume that the three cases

$$n' > n \quad or \quad n' = n \quad or \quad n' < n \tag{23}$$

are *disjoint*. This already implies that we can talk about larger integers[34].

I have found that McGoveran's phrase "naming a largest integer in advance", used above, needs be give more structure in my theory. I assume that all the quantum numbers I need consider can be obtained using strings of length 256 or less. If we have 2^{256} such strings, we have more than enough to *count* — in the sense of Archimedes *Sand Reckoner* — the number electrons and nuclei ("visible matter") in our universe. The mass of the number of nucleons of protonic mass needed to form these nuclei is considerably less than current estimates of the "closure mass" of our universe, leaving plenty of room for the observed "dark matter". I also believe that considerably less than the 256! orders in which we could combine the 2^{256} distinct strings of length 256 will suffice to provide the raw material for a reasonable model of a historical record of both cosmic evolution and terrestrial biological, social and cultural evolution. Such a model can be correct without being complete.

8 From the Tic-Toc Lab to the Digital Lab

Our model for an observatory is a number of *input* devices which, relying on our general model, produce bit-strings of arbitrary length which we can segment, compare, duplicate and operate on using the three operations $\mathbf{a} \oplus \mathbf{b}$, $\mathbf{a} \| \mathbf{b}$, and $\mathbf{a} \cdot \mathbf{b}$, and record the results. Our model for a laboratory adds to these observatory facilities, *output* devices which convert bit-strings into signals we can *calibrate* in the laboratory by disconnecting our input devices from the (unknown) signals coming from outside the laboratory and connecting them to our locally constructed output devices. We require that the results correspond to the predictions of the theory which led to their construction. We then take the critical step from being observers to being *participant observers* by connecting our output devices to the outside of the laboratory and seeing if the input signals from the outside into the laboratory change in a correlated way. Just how we construct input and output devices is a matter of *experimental protocol*, which we must test by having other observer-participants construct similar devices and assuring ourselves that they achieve comparable results to our own. *All* of this "background" is presupposed in what I mean by the phrase "the practice of physics" which Gefwert, McGoveran and I employed in our discussion of methodology at ANPA 9[32, 49, 61]. It will be seen that from the point of view of *theoretical physics* I am claiming that all of our operations can, in principle, be reduced to bit-string operations looking at input tapes to the laboratory, preparing output tapes connected to the outside world, and comparing the new inputs which result from these outputs. It will sometimes be convenient to refer to these tapes by more familiar names, such as "clocks", "accumulating counters", etc., without going through the detailed translation into laboratory protocol that is required for the actual practice of physics.

The most important device with which to start either an astronomical observatory or a physical laboratory is a reliable clock. For us a *standard clock* will simply consist of an input device which produces a tape with an alternating sequence of "1" 's and "0" 's, which we will also respectively refer to as *tic*'s and *toc*'s. It may be a device we construct ourselves or something that occurs without our intervention other than what is involved in producing the tape. Note that the fact that we have constructed it still leaves the physical clock *outside* our (theoretical) bit-string world; it remains essentially just as mysterious as, for example, the pulsations of a pulsar, as recorded in our observatory.

We have two types of clock with period $2W$: a *tic-toc* clock in which the sequence of $2W$ alternating symbols starts with a "1", and a *toc-tic* clock in which the sequence starts with a "0". We will represent the first by a bit-string which we call $\mathbf{L}(W; 2W)$ and the second by a bit-string $\mathbf{R}(W; 2W)$. The arbitrariness of the designation R or L corresponds the fact that our choice of the symbols on the bit-strings is also arbitrary, reflecting what we called Amson invariance in our discussion of program universe above. Independent of the specific symbolization, these two bit-strings have the following properties in comparison with each other and with the (unique) anti-null string $\mathbf{I}(2W)$ of length $2W$ (which could be called a "tic-tic clock"):

$$
\begin{aligned}
\mathbf{R} \cdot \mathbf{R} &= W &= \mathbf{L} \cdot \mathbf{L} \\
\mathbf{R} \cdot \mathbf{I} &= W &= \mathbf{L} \cdot \mathbf{I} \\
\mathbf{R} \cdot \mathbf{L} &= 0 \\
\mathbf{I} \cdot \mathbf{I} &= 2W \\
\mathbf{R} \oplus \mathbf{L} &= \mathbf{I}
\end{aligned}
\tag{24}
$$

We now have two calibrated clocks one of which we can use to make measurements, and the second to obtain the *redundant* data which is so useful in checking for experimental error — a matter of laboratory protocol which I could expound on at some length, but will refrain from so doing. We now consider an *arbitrary* signal $\mathbf{a}(a; 2W)$, and compare it with the anti-null string. We must have either that (1) $\mathbf{a} \cdot \mathbf{I} < W$ or that (2) $\mathbf{a} \cdot \mathbf{I} = W$ or that (3) $\mathbf{a} \cdot \mathbf{I} > W$. Actually, we need consider only the first two cases, because if we define $\bar{\mathbf{a}}$ by the equation $\bar{\mathbf{a}} \equiv \mathbf{a} \oplus \mathbf{I}$, we can reduce the third case to the first simply by replacing $\mathbf{a}$ by $\bar{\mathbf{a}}$. If we know the Hamming measure a, for instance by running the string through an accumulating counter which simply records the number of "1" 's in the string, we do not have to make this test because, independent of the order of the bits in the string the definitions of the inner product, the anti-null string $\mathbf{I}$ and the conjugate string $\bar{\mathbf{a}}$ guarantee that

$$
\begin{aligned}
\mathbf{a} \cdot \mathbf{a} &= a = a \cdot \mathbf{I} \\
\bar{\mathbf{a}} \cdot \bar{\mathbf{a}} &= 2W - a = \bar{\mathbf{a}} \cdot \mathbf{I} \\
\mathbf{a} \cdot \bar{\mathbf{a}} &= 0
\end{aligned}
\tag{25}
$$

Hence an accumulating counter, which throws away most of the information contained in the bit-string $\mathbf{a}$, still gives us useful structural information if we know the context in which it is employed to produce its single integer result a.

If we compare the bit-string $\mathbf{a}$ with our standard clock *before* throwing away the string, we get two additional integers, only one of which is independent. Define these by

$$
a_R \equiv \mathbf{a} \cdot \mathbf{R}; \qquad a_L \equiv \mathbf{a} \cdot \mathbf{L}
\tag{26}
$$

Without actually constructing them, we now know that there exist in our space of length $2W$ two bit-strings $\mathbf{a}_R$ $\mathbf{a}_L$ with the following properties

$$
\begin{aligned}
\mathbf{a}_R \oplus \mathbf{a}_L &= \mathbf{a} \\
\mathbf{a}_R \cdot \mathbf{a}_R &= a_R = \mathbf{a} \cdot \mathbf{R} \\
\mathbf{a}_L \cdot \mathbf{a}_L &= a_L = \mathbf{a} \cdot \mathbf{L} \\
\mathbf{a}_R \cdot \mathbf{L} = 0 = \mathbf{a}_R \cdot \mathbf{a}_L &= 0 = \mathbf{a}_L \cdot \mathbf{R} \\
a_R + a_L &= a
\end{aligned}
\tag{27}
$$

Suppose we have a second arbitrary string $\mathbf{b}$ coming from some independent input device. Clearly we can get some structural information in the same way as before, succinctly summarized by the three integers b, b_R and b_L and the constraint $b = b_L + b_R$. If the two sources are *uncorrelated*, these amount to a pair of *classical measurements*, which we can, given enough data of the same type, analyze statistically by the methods developed in classical statistical mechanics. But if the two sources are correlated *and* we construct the string $\mathbf{a} \oplus \mathbf{b}$ and take its inner product both with itself and with our standard clock before throwing it away, we will have the starting point for a model of *quantum measurement*. This is a deep subject, on which much light has been shed by Etter's recent papers on Link Theory [22, 23, 24, 25, 26]. We have started to investigate the connection to bit-string physics[70], but have only scratched the surface. We trust that the more systematic analysis started in this paper will, eventually, help in bringing the two together. Here, we will, instead, show how our tic-toc laboratory can give us useful information about the world in which it is embedded.

9 Finite and Discrete Lorentz Transformations

We now consider a situation in which our laboratory is receiving two independent input signals one of which, for segments of length $2W$, repeatedly gives Hamming measure a and the other $2W - a$. Because of our experience with the Doppler shift, we leap to the conclusion that our laboratory is situated between two standard clocks similar to our own which are sending output signals to us. We assume that they are at relative rest but that our own lab is moving toward the one for which the recorded Hamming measure is larger than W and away from the second one for which the recorded Hamming measure is smaller than W. We calculate our velocity relative to these two stationary, signalling tic-toc clocks as $v_{lab} = (a - W)/W$ measured relative to the velocity of light. If our lab is in fact a rocket ship and we have any fuel left, we can immediately test this hypothesis by turning on the motors and seeing if, after they have been on long enough to give us a known velocity increment Δv, our velocity measured relative to these external clocks changes to

$$v' = \frac{v + \Delta v}{1 + v\Delta v} \tag{28}$$

If so, we have established our motion relative to a given, external framework. Rather than go on to develop the bit-string version of finite and discrete Lorentz boosts, which is obviously already implicitly available, I defer that development until we have discussed the more general bit-string transformations developed in the section below on commutation relations. For an earlier approach, see[64].

This situation is not so far fetched as might seem at first glance. Basically, this is how the motion of the earth, and of the solar system as a whole, have been determined relative to the $2.73°K$ cosmic background radiation in calibrating the COBE satellite measurements that give us such interesting information about the early universe.

To extend this "calibration" of our laboratory relative to the universe to three dimensions, we need only find much simpler pairs of signals than those corresponding to the background radiation, namely pairs, which for the moment we will call U and D which have the properties

$$\begin{aligned}
\mathbf{U} \cdot \mathbf{U} &= W = \mathbf{D} \cdot \mathbf{D} \\
\mathbf{U} \cdot \mathbf{I} &= W = \mathbf{D} \cdot \mathbf{I} \\
\mathbf{U} \cdot \mathbf{D} &= 0 \\
\mathbf{U} \oplus \mathbf{D} &= \mathbf{I}
\end{aligned} \tag{29}$$

These look just like our standard clock, but compared to it we find that

$$\begin{aligned}
\mathbf{U} \cdot \mathbf{R} &= W + \Delta = \mathbf{D} \cdot \mathbf{L} \\
\mathbf{U} \cdot \mathbf{L} &= W - \Delta = \mathbf{D} \cdot \mathbf{R}
\end{aligned} \tag{30}$$

where (for $\mathbf{U}, \mathbf{D}$ distinct from $\mathbf{R}, \mathbf{L}$) we have that $\Delta \in 1, 2, ..., W - 1$. These form the starting point for defining directions and finite and discrete rotations. As has been proved by McGoveran[45], using a statistical result obtained by Feller[29], at most three independent sequences which repeatedly have the same number of tic's in synchrony can be expected to produce such recurrences often enough to serve as a "homogeneous and isotropic" basis for describing independent dimensions, showing that our tic-toc lab *necessarily* resides in a three-dimensional space.

It is well known that finite and discrete rotations of any macroscopic object of sufficient complexity, such as our laboratory, *do not commute*. It is therefore useful to develop the non-commutative bit-string transformations before we construct the formalism for finite and discrete Lorentz boosts *and* rotations as a unified theory. Once we have done so, we expect to understand better why finite and discrete commutation relations imply the finite and discrete version of the free space Maxwell[35] and Dirac [36]equations, which we developed in order to answer some of the conceptual questions raised by Dyson's report[20] and analysis[21] of Feynman's 1948 derivation[30] of the Maxwell equations from Newton's second law and the non-relativistic quantum mechanical commutation relations; we intend to extend our analysis to gravitation because we feel that Tanimura's extension of the Feynman derivation in this direction[89] raises more questions than it answers.

10 Commutation Relations

If we consider three bit-strings which discriminate to the null string

$$\mathbf{a} \oplus \mathbf{b} \oplus \mathbf{h}_{ab} = \mathbf{\Phi} \tag{31}$$

they can always be represented by three *orthogonal* (and therefore *discriminately independent*) strings[66, 70]

$$
\begin{aligned}
\left(\mathbf{n}_a \oplus \mathbf{n}_b \oplus \mathbf{n}_{ab}\right) \cdot \left(\mathbf{n}_a \oplus \mathbf{n}_b \oplus \mathbf{n}_{ab}\right) &= n_a + n_b + n_{ab} \\
\mathbf{n}_a \cdot \mathbf{n}_b &= 0 \\
\mathbf{n}_a \cdot \mathbf{n}_{ab} &= 0 \\
\mathbf{n}_b \cdot \mathbf{n}_{ab} &= 0
\end{aligned}
\tag{32}
$$

as follows

$$
\begin{aligned}
\mathbf{a} &= \mathbf{n}_a \oplus \mathbf{n}_{ab} \Rightarrow a = n_a + n_{ab} \\
\mathbf{b} &= \mathbf{n}_b \oplus \mathbf{n}_{ab} \Rightarrow b = n_b + n_{ab} \\
\mathbf{h}_{ab} &= \mathbf{n}_a \oplus \mathbf{n}_b \Rightarrow h_{ab} = n_a + n_b
\end{aligned}
\tag{33}
$$

It is then easy to see that the Hamming measures a, b, h_{ab} satisfy the triangle inequalities, and hence that this configuration of bit-strings can be interpreted as representing and integer-sided triangle. However, if we are given only the three Hamming measures, and invert Eq.35 to obtain the three numbers n_a, n_b, n_{ab}, we find that

$$
\begin{aligned}
n_{ab} &= \frac{1}{2}[+a+b-h_{ab}] \\
n_a &= \frac{1}{2}[+a-b+h_{ab}] \\
n_b &= \frac{1}{2}[-a+b+h_{ab}]
\end{aligned}
\tag{34}
$$

Hence, if either one (or three) of the integers a, b, h_{ab} is (are) *odd*, then n_a, n_b, n_{ab} are *half-integers* rather than integers, and we *cannot* represent them by bit-strings. In order to interpret the angles in the triangle as *rotations*, it is important to start with orthogonal bit-strings rather than strings with arbitrary Hamming measures.

In the argument above, we relied on the theorem that

if $\mathbf{n_i} \cdot \mathbf{n}_j = n_i \delta_{ij}$ when $i, j \in 1, 2, ..., N$,

then

$$
(\Sigma_{\oplus, i=1}^N \mathbf{n}_i) \cdot (\Sigma_{\oplus, i=1}^N \mathbf{n}_i) = \Sigma_{i=1}^N n_i
\tag{35}
$$

which is easily proved [70]. Thus in the case of two discriminately independent strings, under the even-odd constraint derived above, we can always construct a representation of them simply by concatenating three strings with Hamming measures n_a, n_b, n_{ab}. This is clear from a second easily proved theorem:

$$
(\mathbf{a}\|\mathbf{b}\|\mathbf{c}\|....) \cdot (\mathbf{a}\|\mathbf{b}\|\mathbf{c}....) = a + b + c + ...
\tag{36}
$$

Note that because we are relying on concatenation, in order to represent two discriminately independent strings $\mathbf{a}, \mathbf{b}$ in this way we must go to strings of length $W \geq a + b + h_{ab}$ rather than simply $W \geq a + b$, as one might have guessed simply from knowing the Hamming measures and the Dirac inner product.

If we go to *three* discriminately independent strings, the situation is considerably more complicated. We now need to know the *seven* integers $a, b, c, h_{ab}, h_{bc}, h_{ca}, h_{abc}$, invert a 7×7 matrix, and put further restrictions on the initial choice in order to avoid quarter-integers as well as half-integers if we wish to construct an orthogonal representation with strings of minimum length $W \geq a + b + c + h_{ab} + h_{bc} + h_{ca} + h_{abc}$. We have explored this situation to some extent in the references cited, but a systematic treatment using the reference system provided by tic-toc clocks remains to be worked out in detail.

The problem with non-commutation now arises if we try to get away with the scalars a, a_R, Δ, W arrived at in the last section when we ask for a transformation either of the basis $\mathbf{R}, \mathbf{L} \to \mathbf{U}, \mathbf{D}$ or the rotation of the string **a** under the constraint $a_R + a_L = a = a_U + a_D$ while keeping the two sets of basis reference strings fixed. This changes $a_R - a_L$ to a different number $a_U - a_D$, or visa versa. If one examines this situation in detail, this is exactly analagous to raising or lowering j_z while keeping j fixed in the ordinary quantum mechanical theory of angular momentum. Consequently, if one wants to discuss a system in which both j and j_z are conserved, one has to make a *second* rotation restoring j_z to its initial value. It turns out that, representing rotations by $\oplus$ and bit-strings then gives different results depending on whether j_z is first raised and then lowered or visa versa; finite and discrete commutation relations of the standard form result. We will present the details of this analysis on another occasion. In effect what it accomplishes is a mapping of conventional quantum mechanics onto bit-strings in such a way as to get rid of the need for continuum representations (eg. Lie groups) while retaining *finite and discrete* commutation relations. Then a new look at our recent results on the Maxwell[35] and Dirac[36] equations should become fruitful.

11 Scattering

We ask the reader at this point to refer back to our section on the Handy Dandy Formula when needed. There we saw (Eq. 17) that the unitary scattering amplitude $T(s)$ for systems of angular momentum zero can be computed at a single energy if we know the phase shift $\delta(s)$ at that energy. For the following analysis, it is more convenient to work with the dimensionless amplitude $a(s) \equiv \sqrt{s - (m_a + m_b)^2} T(s)$, which is related to the tangent of the phase shift by

$$a(s) = e^{i\delta(s)} \tan \delta(s) = \frac{\tan \delta(s)}{1 + i \tan \delta(s)} \equiv \frac{t(s)}{1 + it(s)} \qquad (37)$$

In a conventional treatment, given a real interaction potential $V(s, s') = V(s', s)$, $T(s)$ can be obtained by solving the Lippmann Schwinger equation $T(s, s'; z) = V(s, s') + \int ds'' T(s, s''; z) R(s'', s'; z) T(s'', s'; z)$ with a singular resolvent R and taking the limit $T(s) = \lim_{z \to s + i0+, s' \to s} T(s, s'; z)$. Here we replace this integral equation by an *algebraic* equation for $t(s)$:

$$t(s) = g(s) + g(s)t(s) = \frac{g(s)}{1 - g(s)} \qquad (38)$$

One can think of this equation as a sequence of scatterings each with probability $g(s)$ which is summed by solving the equation. Here $g(s)$ will be our model of a *running coupling constant*, which we assume known as a function of energy. We see that if $g(s_0) = 0$ there is no scattering at the energy corresponding to s_0, while if $g(s_0) = +1$, the phase shift is $\frac{\pi}{2}$ at the corresponding energy and $a(s_0) = -i$; otherwise the scattering is finite.

The above remarks apply in the physical region $s > (m_a + m_b)^2$, where in the singular case a phase shift of $\frac{\pi}{2}$ causes the cross section $4\pi \sin^2 \delta / k^2$ to reach the unitarity limit $4\pi \lambda_0^2$ where $\lambda_0 = \hbar/p_0$

is the de Broglie wavelength at that energy; this is called a resonance and the cross section goes through a maximum value at that energy. If, as in S-matrix theory, we analyticly continue our equation below elastic scattering threshold, the scattering amplitude is real and the singular case corresponds to a bound state pole in which the two particles are replaced by a single coherent particle of mass μ, within which the particles keep on scattering until some third interaction supplies the energy and momentum needed to liberate them. There can also be a singularity corresponding to a repulsion rather than attraction, which is called a "CDD pole" in S-matrix dispersion theory[17]. The corresponding situation in the physical region is a cross section which never reaches the unitarity limit. To cut a long story short, these four cases correspond to the four roots of the quartic equation (Eq. 19) called the handy-dandy formula, which we repeat here, replacing the running coupling constant by its value at the singularity which we call $g_0 = g(s_0)$

$$(g_0)^4 \mu^2 = (m_a + m_b)^2 - \mu^2 \tag{39}$$

Again to cut a long story short, the model for a running coupling constant which Ed Jones and I are exploring[75] is simply

$$g_{m_a,m_b;\mu}(s) = \sqrt{\frac{\pm[(m_a + m_b)^2 - \mu^2](m_a + m_b)^2}{[k^2(s) - (m_a + m_b)^2]s}} g_{m_a m_b;\mu}(0) \tag{40}$$

The singularity at $s = 0$ is included only when m_a and m_b have a bound state of zero mass, usually called a quantum.

We have seen that when $m_a = m_e$, $m_b = m_p$ and $\mu = m_H$ the handy-dandy formula gives the relativistic Bohr formula for the hydrogen spectrum. Replacing m_p by m_e in the formula gives the corresponding formula for positronium (i.e the bound state of an electron-positron pair). But for that system, one can think of the photons produced in electron-positron annihilation as bound states of the pair with zero rest mass. This interaction is important in high energy electron-positron scattering, where it is called "Bhabha scattering". Introducing the $s^{-\frac{1}{2}}$ in this way is supposed to insure that our theory gives the correct Feynman diagram (and hence cross section) for this effect, but until we have checked the detailed derivation and predictions I warn the reader to treat this formula (Eq.42) as a guess rather than as a result actually derived from the theory.

In my paper at ANPA WEST 13[70], I started to explore the connections of this type of scattering theory to bit-strings *and* to Etter's Link Theory by making the hypothesis that

$$tan\delta_{ab} = \frac{\pm(\mathbf{a} \oplus \mathbf{b}) \cdot (\mathbf{a} \oplus \mathbf{b})}{\mathbf{a} \cdot \mathbf{b}} \tag{41}$$

Unfortunately the details are about a sketchy as presented here, but at least should provide insight into where I am headed.

12 Quantum Gravity

The initial intent of Eddington, and following him of Bastin and Kilmister, was to achieve the reconciliation of quantum mechanics with general relativity. I emphasize here that they were aware of the problem, and thought they had a research program which might solve it, long before the buzz-words "Grand Unified Theory", "Theory of Everything", "Final Theory" or "Ultimate Dynamical Theory" [90, 67] became popular. In a sense they achieved the first major step with the publication of Ted Bastin's paper in 1966[7]. According to John Amson, that paper came about after many attempts to understand Fredrick's breakthrough[79] had led John to the discovery of discriminate closure[2] which gave more mathematical coherence to the scheme, and also convinced Clive and Fredrick that it was time to publish. In the event, the four authors could not agree on a text in time to meet a deadline, and authorized Bastin to go ahead with his version as sole author. This paper really does unify electromagnetism with gravitation (which was also Einstein's long sought and unachieved goal) in the sense that both coupling constants are derived from a common theory. Ted also correctly identified $(256)^4$ with the weak interactions, but missed the $\sqrt{2}$ needed to connect it numerically with the Fermi constant because he was unfamiliar with the difference between 3-vertices (Yukawa-type couplings) and 4-vertices (Fermi-type couplings) in the quantum theory of fields. As we all know, this paper was met by resounding silence. I am optimistic, in spite of my past failures, that the bit-string theory now has enough points of contact with more conventional approaches to fundamental physics to get us all into court.

Since the critical problem for many physicists is how we deal with "quantum gravity", I start there. For weak gravitational fields it makes sense to start in a flat space-time[95]. Then it can be shown that spin 2 gravitons of zero mass lead to the Einstein field equations, but as Meisner, Thorne and Wheeler note[52], the "Resulting theory eradicates original flat geometry from all equations, showing it to be unobservable." Consequently they feel that this approach says nothing about "... the greatest single crisis of physics to emerge from these equations: complete gravitational collapse." My qualitative answer is that this crisis arises from using a continuum theory at short distance where only a quantum theory makes sense. I now try to make the alternative presented here plausible.

My first step is to establish the existence of quantum gravitational effects for *neutral* particles, namely neutrons. That neutrons are gravitating objects in the classical sense was proved at Brookhaven soon after the physicists there learned how to extract epithermal neutrons from their high flux reactor and send them down an evacuated pipe a quarter of a mile long. The neutrons fell (within experimental error) by just the amount that Galileo would have predicted. That they are quantum mechanical objects was proved by Overhauser[77] by cutting a single silicon crystal

10 centimeters long into three connected planes and using critical reflection, both *calculable* from measured $n - Si$ cross sections and demonstrable *experimentally*, to form in effect a two-slit apparatus for neutrons with the positions known to atomic precision over a distance of ten centimeters. Then the shift in the interference pattern between the case when two beams were both horizontal to the case when they were in the vertical plane with one higher than the other for part of its path was proved to be precisely that predicted by non-relativistic quantum mechanics using the Newtonian gravitational potential in the Schroedinger equation. It was this brilliant experiment which convinced me that quantum mechanics is a general theory and not just a peculiarity in the behavior of electrically charged particles at short distance. In my opinion, Overhauser deserves the Nobel prize for this work, which opened up the study of the foundations of quantum mechanics to high precision experimental investigation. In the hands of Rausch[82] and others this technique has led to many tests of the model of the neutron as a quantum mechanical particle acting coherently with a precisely known mass and magnetic dipole.

Having established that neutral particles react gravitationally to the Newtonian gravitational potential $V_N(m_1, m_2; r) = G_N \frac{m_1 m_2}{r}$ as expected, it makes sense to extend our relativistic bit-string model for the Coulomb potential $V_C(m_1, m_2; r) = \frac{Z_1 Z_2 e^2}{r}$ to the gravitational case. Here Z_1, Z_2 are the electric charges expressed in units of the electronic charge e. If we are guided by Bastin's remark quoted above to the effect that the basic quantization is the quantization of mass, the analogy suggests that there is a (currently unknown) unit of mass, which we will call Δm. Then, to complete the analogy with the Coulomb case, we can replace $\alpha_C = e^2/\hbar c \approx 1/137$ with a much smaller constant $\alpha_N = G_N \Delta m^2/\hbar c$. If we also define $N_i = m_i/\Delta m$ for any particle i with *gravitational mass* m_i, the quantized version of the Coulomb and Newtonian interactions become formally equivalent, differing only by two dimensionless constants and two *quantum numbers*, independent of the units of charge or mass:

$$V_C(Z_1, Z_2; r) = Z_1 Z_2 \alpha_C \frac{\hbar c}{r}; \quad V_N(N_1, N_2; r) = N_1 N_2 \alpha_N \frac{\hbar c}{r} \tag{42}$$

We can now apply the Dyson argument to gravitation with more precision. We have seen that renormalized QED extended to enough precision to generate $N_e = 137$ electron-positron pairs, becomes unstable because of (statistically rare) clumping of clusters with enough electrostatic energy to form another pair. We interpret the fact that this disaster does not occur to the formation of a pion with mass $m_\pi \approx 2 \times 137 m_e$. In the neutral particle case, if one assumes CPT invariance, one can still distinguish fermions from anti-fermions by their spin even if they have no other quantum numbers. Hence, independent of whether or how neutral fermions and anti-fermions interact, we can expect gravitational clumping to occur for each type separately. Recall Dyson's remark that the system is dilute enough so that the non-relativistic potential can be used reliably to estimate the interaction energy of the clump. We know experimentally that Δm is much smaller than the electron mass so that the critical radius of the clump is $\hbar/\Delta mc >> \hbar/m_e c$, so Dyson's comment still applies..

In contrast to the electromagnetic case where the cutoff mass-energy of the pion requires us to go outside QED for the physics, in the gravitational case we have a cutoff energy ready to hand, namely the Planck mass $M_{Pk} \equiv [\hbar c/G_N]^{\frac{1}{2}}$. If we assemble a Planck's mass worth of neutral particles of mass Δm at rest within their own Compton wavelength, i.e. $N_G \Delta m = M_{Pk}$, and nothing else intervenes, they will fall together until they are all within a distance of $\hbar/M_{Pk}c$. At that point they will have a gravitostatic energy $N_G G_N \Delta m M_{Pk}/[\hbar/c\Delta M] = M_{Pk}c^2$, which is just sufficient to contain the kinetic energy they acquired in reaching this concentration. Inserting the definition of the Planck Mass into this gravitational energy equation we find that it is algebraically equivalent to the boundary condition with which we started: $N_G \Delta m = M_{Pk}$. They will form a black hole with the Planck radius. We conclude that $N_G = M_{Pk}/\Delta m$ neutral, gravitating objects of mass Δm at rest within their own Compton wavelength will collapse to a black hole with the Planck radius, a quantum version of the disaster that Wheeler is concerned about.

If there are, in fact, neutral fermions with no other properties than their mass, they would form such such black holes and might serve as a model for the dark matter which we know to be at least ten times as prevalent in the universe as ordinary matter. It then becomes a question in big-bang cosmology whether or not they contribute the needed effects to correlate additional observations. We defer that question to another occasion.

If we assemble enough particles of the types we know about to add up to a Planck mass, they can start collapsing and will radiate much of their energy on the way down to higher concentrations. If attractions are balanced by repulsions they could end up close enough together to form a black hole. However, as Hawking showed, small black holes interact with the "vacuum" outside the event horizon and radiate electromagnetically with the consequence that they are "white hot" and soon evaporate; the calculation was extended to rotating, charged black holes by Zurek and Thorne[98]. At the quantum scale a new possibility enters, namely that a quantum number may be possessed by the system which *cannot* be radiated away by emitting a single particle with that quantum number while conserving energy, momentum and spin. The obvious candidates for such conserved quantum numbers are baryon number, charge and lepton number, suggesting that the lightest baryon (the proton), the lightest charged lepton (the electron) and the lightest neutral lepton (the electron-type neutrino) are *gravitationally stabilized* black holes with spin $\frac{1}{2}\hbar$. This idea did not make it into the mainstream literature[63]. We use it freely in what follows. The problem then is to explain why $(M_{Pk}/m_p)^2 \approx 2^{127}$, why $m_p/m_e \approx 1836$ and why $m_{\nu_e}/m_e \leq 5 \times 10^{-5} m_e$.

Before we leave gravitation, however, we need to show within bit-string physics that the graviton has spin 2. We know from our discussion of the handy-dandy formula that we can account for spin $\frac{1}{2}$ electrons, positrons and protons and their interactions with the appropriate spin 1 photons. We have shown elsewhere[65] that we can construct the quantum numbers of the standard model of quarks and leptons. In particular, this will include the electron, muon and tau neutrinos and their anti-particles. Extending this approach to a string of length 10 we can have 6 spin $\frac{1}{2}$ fermions and 2 spin 1 bosons. On another occasion I will show how these construct 5 gravitons and 5 anti-gravitons represented in terms of strings of length 10, and go on from that to make a model for dark matter that can be expected to be approximately 12.7 times as prevalent in the universe as electrons and nucleons.

It remains to show that we can meet the three classical tests of general relativity, a problem met on another occasion [71]. Briefly, any relativistic theory gives the solar red shift (Test 1), the factor of 2 compared to special relativity in the bending of light by the sun comes from the spin 1 of the photon (Test 2), and the factor 6 compared to special relativity for the precession of the perihelion of Mercury[13] from the spin 2 of the graviton (Test 3). The calculation by Sommerfeld on which the third argument partly depends comes from simply replacing the factors $N_i = m_i/\Delta m$ in the Coulomb potential by $E_i/\Delta m$, where E_i includes the changing velocity of the orbiting particle in elliptical orbits, and hence is natural in our theory.

13 STRONG, ELECTROMAGNETIC, WEAK, GRAVITATIONAL UNIFICATION (SEWGU): A look ahead

I now show, very briefly, how SEWGU *might* take shape, if all goes well, giving the zeroth approximation to some old results such as $[\alpha_e^{-1}(0)]_0 = 137$, some new results I believe such as $[m_\tau/m_\mu]_0 = 16$, and some guesses such as $[m_t/m_Z]_0 = 2$ which I have little confidence in. The notation makes no distinction between fact and speculation, so *CAVEAT LECTOR!*

Start with strings of length 8 to label the 6 L,R *bare* states of the three types of neutrinos ν_e, ν_μ, ν_τ and their anti-particles, together with two slots for three generations, $g_1 = (10), g_2 = (11), g_3 = (01)$. To get masses and energies we have to add content strings and do a detailed analysis using bit-string scattering theory. We expect the following results to emerge

$$[\frac{m_\mu}{m_e}]_0 = 210 = [\frac{m_{\nu_\mu}}{m_{\nu_e}}]_0 \tag{43}$$

$$[\frac{m_\tau}{m_\mu}]_0 = 16 = [\frac{m_{\nu_\tau}}{m_{\nu_\mu}}]_0 \tag{44}$$

together with the massless bosons $\gamma_L, \gamma_R, \gamma_C$ (two states of spin 1 photons with the coulomb interaction as a third state) and $g_{2L}, g_{1L}, g_0, g_{1R}, g_{2R}, g_N$ (five states of the graviton plus the Newtonian interaction).

The electromagnetic coupling at the mass of the Z_0, $[\alpha_e^{-1}(m_Z)]_0 = 128$.

The masses of the charged and neutral pion compared to the electron:

$$[\frac{m_{\pi^\pm}}{m_e}]_0 = 275 = 2[\alpha_e^{-1}(0)]_0 + 1 \tag{45}$$

$$[\frac{m_{\pi^0}}{m_e}]_0 = 274 = 2[\alpha_e^{-1}(0)]_0 \tag{46}$$

We can also use label strings of length eight to get (bare) quarks with eight colors (red,orange,yellow,green,blue,purple,black,white) which can form the colorless pion triplet (π^+, π^0, π^-) and the nucleon-antinucleon doublet $(n, p, \bar{n}, \bar{p})$. To identify these as first generation hadrons, we neead to extend the string length from 8 to 10. This allows us to go on to strings of length 16 and include the neutrinos with the two generation slots accounting for the shared coupling. After a few years of effort, we expect parameters of the full Cabbibo-Kobayashi-Maskawa coupling scheme to emerge. If this fails, we may have to abandon bit-string physics!

Calling the strong coupling constant α_π rather than α_s to emphasize the conceptual difference, we are confident that at low energy

$$[\alpha_\pi(m_\pi^2)]_0 = 1 \tag{47}$$

$$[\alpha_\pi^{-1}(4m_p^2)]_0 = 7 \tag{48}$$

At high energy we expect to show that

$$[\frac{m_Z}{m_p}]_0 = 2 \times 7^2 \tag{49}$$

$$[\frac{m_t}{m_Z}]_0 = 2 \tag{50}$$

where m_t is the top quark mass. If this works out we may be able to predict a new coupling between quarks and leptons that goes beyond the standard model which *might* explain the anomalous results recently obtained at ZEUS and HERA.

I expect to be able to derive the m_p/m_e formula in a way consistent with McGoveran's last paper on that problem[48], but now directly using bit-string dynamics.

I expect to be able to understand the mapping between Foch space labels and bit-string geometry in terms of the Eulerean rectangular block Kilmister told me about with edges of length 44, 117, 240 using strings of length 256 divided into a label of length 16 and content string of length 240.

Finally, I expect to recover the old result

$$[\frac{M_{Pk}^2}{m_p^2}]_0 = 2^{127} \tag{51}$$

in terms of a running coupling constant for gravitation normalized by $\alpha_G(M_{Pk}^2) = 1$ using bit-string scattering theory.

14 Epilogue

I realize all too well how sketchy these notes are and apologize for that. I hope to get a systematic and reasonably complete outline of the full theory hammered out in a year or two. "But always at my back I hear time's winged chariot hurrying near. And yonder all before us lie deserts of vast eternity.... The grave's a fine and private place, but none I think do there embrace..." theoretical physics! So I decided to rough out this paper to insure that it is part of the Fixed Past before some singular event in the Uncertain Future terminates my activities.

References

[1] M.Aguilar-Benitz, et. al., *Particle Properties Data Booklet*, North Holland, Amsterdam, 1988, pp. 2, 4.

[2] J.Amson, "A Discrimination System", unpublished workpaper dated 31 August 1965; presented with comments in *International Journal of General Systems*, (1996)[15].

[3] E.O.Alt,P.Grassberger and W.Sandhas, *Nucl.Phys.*, **B 2**, 167 (1967).

[4] J.Amson, Appendix to [11].

[5] J.Amson, " 'BI-OROBOUROS' — A Recursive Hierarchy Construction", in [58, 59] (unpublished).

[6] R.M.Barnett, et. al. *Phys.Rev.* **D**, No. 1, Part I, p.175 (1996).

[7] T.Bastin, "On the Scale Constants of Physics", *Studia Philosophica Gandensia*, **4**, 77 (1966).

[8] Seminar on [7] given at Stanford in 1973.

[9] T.Bastin, remark in private conversation with Clive Kilmister and HPN, May 9, 1996.

[10] T.Bastin and C.W.Kilmister, *Combinatorial Physics*, World Scientific, Singapore, 1995.

[11] T.Bastin, H.P.Noyes, J. Amson and C.W.Kilmister, *Int'l J. Theor. Phys.*, **18**, 445-488 (1979).

[12] L.C.Biedenharn, *Found. of Phys.*, **13**, 13 (1983).

[13] P.G.Bergmann, *Introduction to the Theory of Relativity*, Prentice-Hall, New York, 1942, pp 212-218 provides a clear summary of Sommerfeld's calculation.

[14] N.Bohr, *Phil. Mag.*, 332-335 (February 1915).

[15] K.Bowden, ed. " General Systems and the Emergence of Physical Structure from Information Theory", special issue of the *International Journal of General Systems* (in press).

[16] L.Castillejo, private comment to HPN c.1990.

[17] L.Castillejo, R.H.Dalitz and F.J.Dyson, *Phys. Rev.* **101**, 453 (1956).

[18] E.Conversi, E.Pancini and O Piccioni, *Phys. Rev.* **71**, 209 (1947).

[19] F.J.Dyson, *Phys Rev*, **85**, 631 (1952).

[20] F.J.Dyson, *Physics Today*, No.2, 32 (1989).

[21] F.J.Dyson, *Am. J. Phys*, **58**, 209 (1990).

[22] T.Etter, "Boolean Geometry", *ANPA West Journal* 6, No. 1, 15-33 (1996).

[23] T.Etter, "How to Take Apart a Wire", *ANPA West Journal* 6, No. 2, 16-28 (1996).

[24] T.Etter, "Quantum Mechanics as a Branch of Merology, *Proc. PhysComp96*.

[25] T.Etter, "Process, System, Causality, and Quantum Mechanics: A Psychoanalysis of Animal Faith", in [15].

[26] T.Etter, "Quantum Mechanics as Meta-Law", submitted to *Physica D*.

[27] L.D.Faddeev, *JETP*, **39**, 1459 (1960).

[28] L.F.Faddeev, *Mathematical Aspects of the Quantum Mechanical Three-body Problem*, Davey, New York, 1965.

[29] W.Feller, *An Introduction to Probability Theory and its Applications*, Vol.I, 3rd Edition, p. 316; For more detail see the first edition, p. 247 et. seq, Wiley, New York, 1950.

[30] R.P.Feynman,c.1948. Conversation with F.J.Dyson, discussed by Dyson [20].

[31] D.Z.Freedman, C.Lovelace and J.M.Namyslowski, *Nuovo Cimento* **43 A**, 258 (1966).

[32] C.Gefwert, D.O.McGoveran and H.P.Noyes, "Prephysics", in [60] pp 1-36.

[33] S.J.Gould, "Darwinian Fundamentalism", *N.Y.Review of Books*, June 12, 1997, pp 34-37.

[34] L.Kauffman, phone conversations with HPN, spring, 1997.

[35] L.H.Kauffman and H.P.Noyes, *Proc. Roy. Soc. (London)*, **A 452**, 81-95 (1996).

[36] L.H.Kauffman and H.P.Noyes, "Discrete Physics and the Dirac Equation", *Physics Letters* **A 218**, 139-146 (1996).

[37] C.W.Kilmister, ed. *Discrete Physics and Beyond: Proc.ANPA 10*, ANPA, c/o C.W.Kilmister, Red Tiles Cottage, High Street, Barcombe, Lewes, BN8 5DH, UK.

[38] C.W.Kilmister, *Eddington's search for a Fundamental Theory: A key to the universe*, Cambridge, 1994.

[39] C.W.Kilmister, "The Hierarchy Reviewed", in *Proc. ANPA 18*, T. Etter, ed., ANPA, c/o C.W.Kilmister, Red Tiles Cottage, High Street, Barcombe, Lewes, BN8 5DH, UK.

[40] T.S.Kuhn, *The Structure of Scientific Revolutions*, University of Chicago Press, 1962.

[41] T.S.Kuhn, *Black Body Theory and the Quantum Discontinuity, 1894-1912*, Oxford, 1978; afterword in University of Chicago reprint, 1989.

[42] M.J.Manthey, "Program Universe" in [58, 59], pp 101-110.

[43] M.J.Manthey, ed., *Objects in Discrete Physics: Proc. ANPA 11 (1989)*; ANPA, c/o C.W.Kilmister, Red Tiles Cottage, High Street, Barcombe, Lewes, BN8 5DH, UK.

[44] M.J.Manthey, ed., *Alternatives in Physics and Biology:Proc. ANPA 12 (1990)*; ANPA, c/o C.W.Kilmister, Red Tiles Cottage, High Street, Barcombe, Lewes, BN8 5DH, UK.

[45] Discussed as Theorem 13, [49] pp 59-60.

[46] D.O.McGoveran, "The Fine Structure of Hydrogen", in [37], pp. 28-52.

[47] D.O.McGoveran, "Advances in the Foundations",.in[43], pp 82-100.

[48] D.O.McGoveran, "Motivations for using the Combinatorial Hierarchy", in[44], pp 7-17.

[49] D.O.McGoveran and H.P.Noyes, "Foundations of a Discrete Physics" in [60], pp 37-104.

[50] D.O.McGoveran and H.P.Noyes, "Foundations of a Discrete Physics" SLAC-PUB-4526, June, 1989.

[51] D.O.McGoveran and H.P.Noyes, *Physics Essays*, **4**, 115-120 (1991).

[52] C.W.Misner, K.S.Thorne, and J.A.Wheeler, *Gravitation*, Freeman, New York, 1973, Box 18.1, p. 437..

[53] H.P.Noyes, *Ann.Rev.Nuc.Sci.*, **22**, 465-484 (1972).

[54] H.P.Noyes, *Czech.J.Phys.*, **B24**, 1205-1214 (1974).

[55] H.P.Noyes, "Non-locality in Particle Physics", SLAC-PUB-1405 (1974).

[56] H.P.Noyes, "Non-Locality in Particle Physics" prepared for an unpublished volume entitled *Revisionary Philosophy and Science*, R.Sheldrake and D.Emmet, eds; available as SLAC-PUB-1405 (Revised Nov. 1975).

[57] H.P.Noyes, *Phys.Rev* **C 26**, 1858-1877 (1982).

[58] H.P.Noyes, ed. *Proc. ANPA 7*, 1985; I have what may be the only copy.

[59] H.P.Noyes, "ON THE CONSTRUCTION OF RELATIVISTIC QUANTUM MECHANICS: A Progress Report, SLAC-PUB-4008, June 1986 (Part of Proc. ANPA 7).

[60] H..P.Noyes, ed. *'DISCRETE AND COMBINATORIAL PHYSICS: Proc. ANPA 9 (1987)*, ANPA West, 112 Blackburn Ave., Menlo Park, CA 94025.

[61] H.P.Noyes, "DISCRETE PHYSICS: Practice, Representation and rules of correspondence" in [60], pp 105-137.

[62] H.P.Noyes, in *Physical Interpretations of Relativity Theory.II.*, M.C.Duffy, ed. (Conf. at Imperial College, London, 1990), p. 196 and SLAC-PUB-5218, July, 1990.

[63] H.P.Noyes, "Comment on 'Statistical Mechanical Origin of the Entropy of a Rotating, Charged Black Hole", SLAC-PUB-5693, November 1991 (unpublished).

[64] H.P.Noyes, "The RQM Triangle", in *Reconstructing the Object* (Proc. ANPA WEST 8), F.Young, ed. 1992; ANPA West, 112 Blackburn Ave, Menlo Park, CA 94025.

[65] H.P.Noyes, "Bit-String Physics, a Novel 'Theory of Everything' ", in *Proc. Workshop on Physics and Computation (PhysComp '94)*, D. Matzke, ed.,94, Los Amitos, CA: IEEE Computer Society Press, 1994, pp. 88-94.

[66] H.P.Noyes, "On the Lorentz Invariance of Bit-String Geometry", SLAC-PUB-95-6760 (September 1995), a slightly revised version of the presentation at ANPA WEST 8, Feb. 18-20, 1995.

[67] H.P.Noyes, "Some Remarks on Discrete Physics as an Ultimate Dynamical Theory", in *Philosophies: Proc. ANPA 17 (1995)*, K.G.Bowden, ed., pp 26-37.

[68] H.P.Noyes, "Operationalism Revisited", *Science Philosophy Interface*, **1**, 54-79 (1996).

[69] H.P.Noyes, "Are Partons Confined Tachyons?", in *Proc. ANPA WEST 12 (1996)*, F.Young, ed., ANPA West, 112 Blackburn Ave., Menlo Park, CA 94025.

[70] H.P.Noyes, "On the Distinction Between Classical and Quantum Probabilities", in *Proc. ANPA WEST 13 (1997)*, T.Etter ed., ANPA West, 112 Blackburn Ave., Menlo Park, CA 94025.

[71] H.P.Noyes and D.O,McGoveran, "Observable Gravitational and Electromagnetic Orbits and Trajectories in Discrete Physics", in *Physical Interpretations of Relativity Theory.I*, (Proc. II), M.C.Duffy, ed, British Society for the Philosophy of Science, 1988, p.42.

[72] H.P.Noyes and D.Y.Wong, *Phys.Rev.Letters*, **3**, 191(1959).

[73] H.P.Noyes and D.O.McGoveran, *Physics Essays*, **2**, 76 (1989).

[74] H.P.Noyes, C.Gefwert and M.J.Manthey, "Toward a Constructive Physics", SLAC-PUB-3116 Rev., September, 1983.

[75] H.P.Noyes and E.D.Jones, "Zero Range Scattering Theory.II.A minimal model for relativistic quantum mechanical 2, 3, and 4 particle scattering" (in preparation).

[76] H.P.Noyes, *DISCRETE PHYSICS: A New Fundamental Theory*, J.C. van den Berg, ed. (in preparation).

[77] R.Collela, A.W.Overhauser and S.A.Werner, *Phys.Rev.Lett.*, **34**, 1472-4 (1975).

[78] A.Pais, *Subtle is the Lord*, Oxford, 1982, pp 155-159.

[79] A.F.Parker-Rhodes, "Hierarchies of Descriptive Levels in Physical Theory", Cambridge Language Research Unit, internal document I.S.U.7, Paper I, 15 January 1962. Included in [15].

[80] A.F.Parker-Rhodes, *The Theory of Indistinguishables, Synthese*, **150**, Reidel, Dordrecht, Holland (1981).

[81] In [80], p 184, Fredrick gives a different version of the m_p/m_e calculation.

[82] H.Rausch, "Quantum Measurements in Neutron Interferometry", in *Proc. 2nd Int. Symp. Foundations of Quantum Mechanics*, Tokyo, 1986, pp 3-17.

[83] "Review of Particle Properties", *Physical Review* **D**, Part I, 1 Aug. 1994, p. 1233.

[84] C.Sagan *The Demon-Haunted World*, Random House, New York, 1995.

[85] S.Schweber, *QED and the Men Who Made It*, Princeton, 1994, p.565.

[86] A.Sokal, "Transgressing the Boundaries: Toward a transformative hermeneutics of Quantum Gravity", *Social Text* 46/47, 1996; — , "A Physicist Experiments with Cultural Studies", *Lingua Franca*, May/June 1996.

[87] A.Sommerfeld, 1916 *Ann. d. Physik*, **51**, 1.

[88] I.Stein, *The Concept of Object as the Foundation of Physics*, Fritz Lang, New York, 1996.

[89] S.Tanimura, *Annals of Physics*, **220**, 229-247 (1992)

[90] G.'t Hooft, "Questioning the answers, or stumbling on good and bad Theories of Everything", in *Physics and our View of the World*, J. Hilgewoord, ed., Cambridge University Press, 1994.

[91] J.Weinberg, in graduate course on classical electrodynamics, UCBerkeley 1947.

[92] S.Weinberg, *Phys.Rev.*, **130**, 776 (1963).

[93] S.Weinberg, *Phys.Rev.*, **131**, 440 (1963).

[94] S.Weinberg, *Phys.Rev.*, **133 B**, 232 (1964).

[95] S.Weinberg, *Phys. Rev.* **B 138**, 988-1002 (1965).

[96] G.C.Wick, *Nature*, **142**, 993 (1938).

[97] O.A.Yakubovsky, *Sov.J.Nucl.Phys.*, **5**, 937 (1967).

[98] W.H.Zurek and K.S.Thorne, *Phys. Rev. Lett.*, **54**, 2171-2175 (1985).

PROCESS, SYSTEM, CAUSALITY, AND QUANTUM MECHANICS
A Psychoanalysis of Animal Faith[*][**]

Tom Etter

112 Blackburn Avenue
Menlo Park, California 94025-2704

and

H. Pierre Noyes

Stanford Linear Accelerator Center
Stanford University, Stanford, CA 94309

Abstract. We shall argue in this paper that a central piece of modern physics does not really belong to physics at all but to elementary probability theory. Given a joint probability distribution J on a set of random variables containing x and y, define a link between x and y to be the condition x=y on J. Define the state D of a link x=y as the joint probability distribution matrix on x and y without the link. The two core laws of quantum mechanics are the Born probability rule and the unitary dynamical law whose best known form is Schroedinger's equation. Von Neumann formulated these two laws in the language of Hilbert space as prob(P) = trace(PD) and D'T = TD respectively, where P is a projection, D and D' are (von Neumann) density matrices, and T is a unitary transformation. We'll see that if we regard link states as density matrices, the algebraic forms of these two core laws occur are completely general theorems about links. When we extend probability theory by allowing cases to count negatively, we find that the Hilbert space framework of quantum mechanics proper emerges from the assumption that all D's are symmetrical in rows and columns. On the other hand, Markovian systems emerge when we assume that one of every linked variable pair has a uniform probability distribution. By representing quantum and Markovian structure in this way, we see clearly both how they differ and also how they can coexist in natural harmony with each other, as they must in quantum measurement, which we will examine. Looking beyond quantum mechanics, we see how both structures have their special places in a much larger continuum of formal systems that we have yet to look for in nature.

* Work supported in part by Department of Energy contract DE—AC03--76SF00515.
** *Physics Essays*, in press, **No.4**, Dec.1999

The problem of how to connect "classical" to "quantum" physics is of current interest, partly because the study of mesoscopic systems with a large number of denumerable quantized degrees of freedom has become technically feasible. For example, it has proved possible to make a coherent superposition of an n_{Bohr}=50 and an n_{Bohr}=51 Rydberg atom which retains its coherence long enough after they have separated by several atomic diameters to allow us to study in some detail how the coherence decays as a function of time[1]. This mesoscopic "Schroedinger's Cat" behaves in the way von Neumann expected it would, and poses no apparent barrier to invoking the Copenhagen "collapse of the wave function" prescription[1]. But it remains true that the border between a system described by standard quantum mechanics and the "same" system described by standard classical physics remains infinitely fragile and unexplained unless all three (classical-border-quantum) are accepted as brute fact. One way out of dilemmas of this type which has been used throughout the history of science is to create a "covering theory" within which the two theoretical descriptions become special cases, and mixtures of systems which represent the two extreme cases are logically compatible. Then future research on mixed systems has at least the possibility of uncovering new, observable parameters which can clarify the connection. In our opinion the covering theory approach to the foundations of quantum mechanics has been insufficiently explored compared to the approach most physicists use of treating quantum mechanics as the covering theory which contains classical physics, or the assumption made by some ("hidden variable") philosophers that classical physics must contain quantum mechanics as a special case. We present this paper as a possible, and we believe plausible, alternative.

The main goal of this paper is to show that basic quantum mechanics can best be understood as a branch of the theory of probability. For this to happen, probability theory must be expanded in two ways: first and foremost, probabilities must be allowed to go negative. Second, and this is more a matter of technique than a change in our assumptions, the composition of transition matrices must be subsumed into a more general method of composition based on the linking of variables. We'll see that *in this expanded theory*, probability amplitude is not some new kind of "fluid" or physical field, but simply (extended) probability itself. By regarding it as such, we turn some of the most striking basic discoveries of modern physics into pure mathematics, thereby freeing them from their purely physical context to become tools available to every branch of science.It is important to realize that we assume from the start that there is an uncontrollably *arbitrary* or "random" element in nature. Some writers on the foundations of quantum mechanics — for example Stapp[2] (following Pauli) — call such an arbitrary element *irrational*. In our view[3], this equates *determinism* to *rationality*, and inevitably leads to a *dualistic* metaphysics. But the metaphysical stance we adopt also has a respectable tradition, going back at least to Aristotle's discussion of the paradoxes of Zeno, followed by the "least swerve" of Epicurus. We believe that calling such an approach "irrational" is unduly pejorative. We therefore believe it legitimate to take as our exemplar of "classical physics" (and classical statistics) Markov processes. As with probability theory, the concept of "Markov process" has to be extended beyond the familiar "Markov chains" by allowing boundary conditions to be stated in the future, the past or both together. This allows us to talk about "backward time" in an interesting way.

Before we confront you with an appropriate mathematical formalism in which to present our novel approach to the foundations of quantum mechanics, we present an introductory parable, followed by a discursive overview of our theory (Ch.1) and a look at some of the philosophical problems raised by trying to modify the conventional approach to physics in order to meet quantum phenomena (Ch. 2). Some readers may prefer to skip directly to Chapter 3 (PROCESS

AND SYSTEM: THE MATHEMATICS); however, if they do so they are warned that the reasons for many of the subtleties introduced there are likely to remain hard to grasp without the preparation we have provided in the earlier sections.

INTRODUCTION: COUNTING SHEEP

Once upon a time there was a sheep farmer who had ten small barns, in each of which he kept five sheep. When asked how many sheep he had altogether, he replied "many", for people in those days counted on their fingers, and no one had ever thought of counting beyond ten.

Every morning he would drive his sheep over the hill and through the woods to their pasture, where they assembled in five fields, ten sheep to a field. The farmer, who was of a reflective bent, saw here a curious and beautiful law of nature: "Ten barns each with five sheep, and then five fields each with ten sheep!" Unfortunately this law did not always hold, and when the wolves howled on the hill at night, it failed quite often. The farmer had an explanation for this: "The howling of the wolves greatly upsets my sheep, and the laws of nature, like the laws of man, are often disobeyed when agitated spirits prevail". The farmer realized that to make his law universal he would have to modify it thus: "When tranquillity reigns, ten of five turn into five of ten."

We today who know arithmetic would say that the farmer's law, though true enough in his particular situation, isn't a very good law by scientific standards. It needs to be "factored" into two laws, the first being the simple and very general law that $xy = yx$ and the second a more complicated and specialized law having to do with sheep and wolves. The farmer was indeed aware that $xy = yx$, at least in the case of 5 and 10, but what he could not see is that the essential condition for xy to be yx has nothing to do with sheep or wolves or tranquillity but is simply that the total number of sheep remain constant. One reason he couldn't see this is that he lacked any conception of the total number of his sheep; that's because in those days there were no numbers beyond ten, just "many".

There are three morals to this tale. The first is that it's not enough just to ask whether a law is right or wrong—we should also ask whether it gets to the point. The second is more subtle: if the point escapes us, maybe it's because we lack the raw materials of thought needed to even conceive of it. The third is not subtle at all: learn to count!

We have learned to count beyond ten sheep and even beyond three dimensions, but we still are under a very stifling conceptual limitation in not being able to count beyond the two types of phenomena that we call *classical* and *quantum*. This paper will set these two among many more. It will do this by teaching us some new ways to count cases, such as how to keep counting when the count goes below zero! This will provide us with the raw materials for thought we need to clearly see some crucial points that quantum philosophy has so far missed, notably the significance, or rather the insignificance, of the wave function, and the essentially acausal nature of quantum processes.

Quantum mechanics has revealed many puzzling patterns in nature, perhaps the most puzzling being the EPR correlation. The explanations we hear for this phenomenon all too often resemble the farmer's spirit of tranquillity. We'll see that we can "factor" a simple piece of probability theory out of physics that makes sense of things like EPR in much the same way that $xy = yx$ makes sense of the farmer's sheep counts. What's basically new here is that quantum phenomena in general can be represented as simple large number phenomena whose laws belong to the arithmetic of case counting. But what about non-locality? The collapse of the wave front? Quantum measurement? How does the present paper fit in with the more familiar ways of interpreting the formalism of quantum mechanics? Let me briefly address this question.

It was recognized quite early that quantum mechanics bears the earmarks of a purely statistical theory. The Schroedinger equation looks very much like the equation governing a Markov process, and we actually get the Schroedinger equation if we multiply the generator of a self-adjoint continuous Markov chain by i. Now Markov processes belong entirely to the theory of probability—there's no physics in them at all. Could it be that quantum mechanics does for mechanics what statistical mechanics did for the theory of heat? Can mechanics, and space and time along with it, be reduced to the statistical behavior of something simpler and more fundamental, or perhaps even to theorems in the bare science of probability itself?

Two obstacles have stood in the way of such a simplification.

The first is that probabilities can't be imaginary. It turns out that we can define imaginary probabilities in terms of negative probabilities, but that wouldn't seem to be of much help—have you ever counted fewer than 0 cases? But then again, come to think of it, have you ever seen a pile of gold bricks with fewer than 0 bricks in it? And yet the mathematics of negative piles of gold bricks has become indispensable for keeping accounts, especially government accounts. So why not "keep accounts" with negative probabilities? This first obstacle doesn't look like it ought to stop us in our tracks for long, and if it were the only obstacle, it probably wouldn't.

The second has proved more obstinate. We saw that the Schroedinger and Markov equations have the same form, but we must next ask, do they govern the same quantities? The answer appears to be no. The numbers that in the Markov equation are probabilities, are the square roots of probabilities in the Schroedinger equation. This little disparity has for sixty years kept alive the notion of the "wave function" whose "amplitude" squared is probability.

And this little disparity, this seeming technicality, has been the logjam that has kept quantum philosophy circling in the same stagnant pool of inadequate ideas for the last sixty years. The present paper aims to break up that logjam.

Let's assume we have overcome obstacle 1 and can now work within an extended theory of probability where cases count negatively as well as positively. As we'll see, most of the math in this extended probability theory is the same as in the all-positive theory. Now suppose we write down the Schroedinger equation as a formal Markov chain, in which the differential transition matrix operates on a state vector. The numbers in this vector are the amplitudes of the "wave function", which we would now like to think of as probabilities. The well-known problem with this construction is that the basic rule for quantum probability, the Born rule, says that these amplitudes must be squared to give the probabilities that are actually observed in quantum measurement.

However, and now we are coming to the key idea, this chain of transition matrices we call a Markov chain is only one of many ways to represent the joint probability distribution that is the Markov process itself. Another is by a chain of joint probability matrices linked by conditioning events of the form x=y. We define the state S of such a link as the probability matrix on x and y with the link removed. A state is called quantum if the distributions on x and y are identical, and a chain is called quantum if its states are quantum and its matrices are unitary. Now here is the punch-line: in a quantum chain, the unlinked probabilities on x and y behave like quantum amplitudes, while the linked probabilities, which are the diagonal entries of S, are the squares of these amplitudes, and hence behave like the probabilities they are supposed to be. There is no wave function, only probabilities, positive and negative.

The link method embeds quantum mechanics in the mathematics of Markov processes in such a way that quantum amplitudes are represented by unlinked probabilities and quantum measurement probabilities by linked probabilities. As we'll see, the same embedding maps classical Markov chains into isomorphic representations of themselves. In linked chains of any kind, linked probabilities are quadratic in unlinked probabilities, but in classical chains one of the factors is constant and factors out, so probability is in effect linear in amplitude, i.e. the classical "state vector" is the usual probability vector. If this sounds a bit cryptic, don't worry, it will all be spelled out in detail.

Most discussions of the meaning of quantum mechanics these days seem to be about the problem of the "collapse of the wave function." In link theory this problem simply vanishes, since there is no wave function to collapse. Imagine if the Eighteenth Century caloric were still hanging around as the official theory of heat: we'd be chronically plagued by ever more complicated theories explaining the collapse of the "caloric field" when you measure an atom's energy. What a relief to get away from the spell of such nonsense!

This large-number explanation of quantum mechanics raises two basic questions: Large numbers of what? And must we buy it?

The answer to the first question is implicit in the above discussion, but needs to be said simply: The things we count large numbers of are cases. Simple arithmetic reveals that the core quantum laws, in a generalized form, are features of any probabilistic system whatsoever. Von Neumann's formulation of the Born probability rule prob(P) = trace(PS) holds at every connection between the parts of such a system, and the dynamical rule S'T = TS governs every part that is connected at two places.

We brought up caloric to draw a parallel between our present situation and the situation in physics when it was discovered that the laws governing heat could be interpreted as statistical laws of atomic motion. However, there is a big difference. In the case of heat, the statistical theory sat on top of the Newtonian theory of motion, whereas in our case there is no underlying empirical theory at all. Probability theory is just the arithmetic of case counting, so the generalized quantum laws are like xy = yx in that their truth is assured, the only empirical issue being where and when they apply.

The answer to the second question is no, we don't have to. However, the same can be said about the arithmetical explanation of five fields with ten sheep each. It's logically possible that when true tranquillity reigns, the gods always make sure that every field contains ten sheep (presumably the age of true tranquillity is long since past). It's also logically possible that the non-local "guide wave" explanation of quantum phenomena is the right one. With both sheep and quantum, the arithmetical explanation makes so much more sense that it would be most malicious of the gods to reject it just to save our old habits of thought.

We'll see that there is another reason to prefer the arithmetical explanation, which is that, as our discussion of Markov processes suggests, it also applies to classical things like computers. This at last enables us to make sense of quantum measurement, which has always been a great mystery. Quantum and classical now stand revealed as two "shapes" made of the same stuff, so there is nothing more mysterious about their both being parts of the same process than there is about round wheels and square windows both being parts of the same car. The radical path also leads to a good Kantian solution of Hume's problem, which is that of finding causality in the order of succession, and we'll see that the choice between acausal and causal/classical thinking is to some extent a choice of analytical method, like the choice between polar and rectilinear coordinates.

But then comes the big question: What about the other shapes? The ones other than quantum or classical that we have never before imagined, and therefore never thought to look for in nature? We'll briefly touch on the big question, but it calls for a much bigger answer than we can give here, or now.

1 PROCESS AND SYSTEM: AN OVERVIEW

The term 'animal faith' in our subtitle is taken from the title of Santayana's book "Skepticism and Animal Faith" [4]; it refers to what gets us through the day and keeps our thought processes going even through our spells of radical doubt. Santayana was concerned to delimit animal faith and to contrast it with other, presumably higher, things. Our agenda here is quite different and is closer to Kant's: it is to articulate and transform into explicit principles the animal faith implicit in certain of our concepts that play a key role in science. Once this is done, once these principles become explicit, we'll find that they take on a new life of their own, and are full of surprises. First, they reveal the simple mathematical structure that unifies quantum and classical. But then they strangely turn against themselves, revealing their own limitations, and even rather impolitely suggesting that perhaps we should find better ways to get through the day. This paper, however, is about their more cheerful messages.

Like Freud's "talking cure", our analysis will start by paying attention to how we talk about everyday things, and then go on to explore subterranean labyrinths in search of hidden meanings. But it is not a search for what has been repressed, or for what has fallen into the unconsciousness of habit, but for that ancient animal heritage of "know how" that we, as thinking people, unconsciously draw upon in formulating our most sophisticated thoughts.

Lest this sound too ambitious, let me add that we are confining our analysis to what underlies certain commonplace scientific words, notably: *information, part, place, event, variable, process, procedure, system, input, output* and *cause*. The unconscious beliefs we are searching for are those that belong to the smoothly working hidden machinery behind the easy flow of thoughts in which these words occur. To put it another way, without certain implicitly held principles these commonplace words, which are indispensable in any discussion of scientific matters, would be quite meaningless, and these are the principles we are trying to capture and articulate.

We shall see that, when these principles are precisely articulated, they become a tightly knit system that has some surprising consequences, including, as mentioned, the two 'core laws' of quantum mechanics. More generally, the consequences of an analysis of animal faith are of two kinds—call them Kantian and Freudian; the Kantian kind reinforce our animal faith, while the Freudian kind force us to question it.

Let's begin our analysis with information. Suppose you see something and write down what you see; this is a *datum*. *Data* is what has been seen and noted. To qualify as *information*, however, data must have some element of surprise. If you know that A is going to happen, seeing it happen may be gratifying or reassuring, but it is not *informative*. Thus *information* is what you have seen, contrasted with what you *might* have seen. When you *describe* something, you normally supply a number of connected pieces of information, so *description* is information broken up into *connected parts*.

These very simple and ordinary observations are beginning to give hints of a pattern, but they come to a halt with the difficult word ' *parts*'. It's curious how in the long and contentious history of speculations about what the world is made of, almost all the debate has been about what are its *ingredients*, and almost none about the more basic problem of what it means for these proposed ingredients to fit together as parts. Taking ordinary material things apart and putting them together is such a natural activity that we blithely extend it to entities of every kind, without ever imagining that this might lead to problems.

And this is certainly the case with information, as when we speak confidently of "partial information", "the whole story", etc. No doubt we are at ease with such talk because the parts and wholes of information so often coincide with the parts and wholes of the language that conveys it. But underlying this linguistic idea of part and whole there is a deeper level of meaning, and excavating that deeper level will be the main task of our analysis.

First of all, we must carefully distinguish between two kinds of part- whole relationship: that which we find in space and space-time, and that which we find in material structures like buildings and computers and molecules. Let's call the first relationship *extension* and the second composition. The parts of an extension are its *regions*, and the identity of a region, i.e. that which makes it different from other regions, is its *place*. The parts of a composition, on the other hand, are its *components*, and the identity of a component is not its place but its *type*. Components are *interchangeable* parts; you can duplicate them, remove them from their places and use them elsewhere, etc. But a region, far from being an interchangeable part, is by its nature unique; the people of Palo Alto can leave California, even the buildings of Palo Alto can leave California, but Palo Alto itself can't leave California since it's a *place* in California.

Matter, as we currently conceive of it, is composed of elementary particles, the most perfectly interchangeable of parts imaginable, which makes matter a composition par excellence, which is to say, its essential self is nowhere. There are those today who speak of making matter out of space or space out of matter, but this is nonsense; you can't make extension out of composition nor composition out of extension. Those who claim to have a theory of *everything*, even if they are right, have only found half of the Holy Grail; the other half is a theory of *everywhere*!

The material objects we encounter in everyday life are always both regions and components; they have both places and types. Thought is a constant dialogue between extension and composition. Practically speaking, the important distinction to keep in mind is between parts that don't keep their identity when you remove them from their context, like random variables in a joint probability distribution, and those that do, like logic gates in a computer. Only the latter can be used to *construct* things.

In mathematics, the dialogue is between geometry and algebra. Think of analytic geometry, which is just such a dialogue. In pure Euclidean metric geometry, we start out with a homogeneous space whose places are *points*, i.e. we can only identify them by *pointing* to them as "here" or "there", or "there",... etc. In order to keep track of these points, we move into algebra. We do this by giving each point a unique identity as a composition of an x-vector, a y-vector and a z-vector; this identification scheme is known as a coordinate system. Notice that vectors themselves are not regions or places; you specify a vector as a certain *type* of thing that can be found anywhere. Notice also that we must carefully specify just what it means for one vector to be a part, i.e. a component, of another; this particular kind of part-whole relation is what defines linear algebra as opposed to other kinds of algebra. Finally, notice that giving coordinates to points does not entirely get rid of *here* and *there*; thus the dialogue continues since we must locate the 0 vector *here*, and we must then point "there", "there", "there" in the directions of x, y and z. Nor can we completely get rid of *this kind* and *that kind* in geometry, since geometric structure involves numerical ratios of distance, so we have *this* and *that* number, and onward to *this* and *that* figure, etc.

How does the distinction between extension and composition apply to informational structures?

Information accumulates as a progressive *extension* of the "body of knowledge". Incoming items of information build on each other, qualify each other, and in general, can only be understood in relationship to their "neighbors". We say, "*this* happened, and then *this* happened and then *this* happened ..." etc., where in each case to know *what* happened we must know something about what happened previously. From time to time, though, we *abstract* an item of information from its place and give it a kind of autonomy by turning it into a story, a design, a warning, an example, a rule, a law, a procedure etc. that can be retold or reused in other contexts; in short, we turn it into a component. One way to analyze a process is to systematically reconstruct it from components that have been so abstracted; we'll call this a *reductive* analysis and the resulting composition a *system*. Let's reflect a bit on the distinction between process and system, which will play an important role in interpreting our mathematical results.

The word "process", in its most general sense, means something that "proceeds or moves along". But the word also has the connotation of *procedure* as in *due process*. Thus a process of a certain kind refers to that which is allowed to happen, or which can happen, under certain specified conditions. This is how we shall construe the present technical meaning of the word. If the possibilities under the specified conditions are assigned probabilities we'll call the process a *stochastic* process. More exactly, a stochastic process is a joint probability distribution on a set of so-called stochastic variables. In Chapter 3, we'll focus on Markov processes, which are the simplest and best known stochastic processes, and we'll see that the core quantum laws as they occur in Markov processes take the form most familiar to physicists.

A system is a process analyzed into interchangeable parts. A good system brings order to a complex whole by portraying it as a regular arrangement of a small variety of such parts. For stochastic systems the paradigm case is a Markov chain, which is a "chain" of connected copies of a single part called a *transition matrix* T_{ij} We can think of T_{ij} as a representation of a stochastic variable j (the column variable) whose probability distribution p(j) is a function of a free parameter i (the row variable). Connecting means assigning the free parameter i to the stochastic variable j of the prior transition matrix, thereby turning the numbers T_{ij} into conditional probabilities. If we assign the free parameter of the first component to an *unconditioned* variable, then chaining transmits the definiteness of the first variable down the line to create a joint probability distribution on all the succeeding variables, and the resulting process is a Markov process.

The contrast between process and system is roughly that between extension and composition. A process is extended, its regions being *events*. Like many English nouns ('noun', for instance), the word "event" can be taken either in the definite or the indefinite sense; an event can be something in particular that happened, in which case it has its unique place in space-time, or it can be a particular *kind* of happening, as when we speak of "the event heads" in probability theory. One way to describe a process is as an arrangement of events in the second sense, which we can think of as labeling its extended parts by certain of their qualities; another word for this is *map-making*. A map is a composition of essences, or predicates, to use the modern term, but the parts of the process that these essences identify need not be removable components, so a map does not in general represent the process itself as a composition.

Thus we see that there are two ways to analyze a process, or any other extended whole: We can *map* it; this is the way of the naturalist, and also of the explorer, the historian and the astronomer. Or we can *reduce* it to a heap of autonomous parts, which we then reassemble into a composition that has the same map the naturalist would draw. Nowadays we often hear about the naturalist as the good guy and the reductionist as the bad guy. But building things and taking things apart belong to life as much as exploring and drawing pictures. When the reductionist goes bad, it's usually because the naturalist hasn't given him a good enough map; what the reductionist then recreates from his storehouse of interchangeable parts may resemble the naturalist's whole, but the essence of the original is missing.

To give to a part the autonomy of a component, we must usually do more than just copy it as it appears in place. Rather, we must "de-install" it from its original context, which is to say, we must transform it into a new entity whose features are no longer conditioned by place. To reductively analyze a process is to de-install its partial regions, at least in our imagination, turning them into items that can stand alone, and that can be duplicated in such a way that the duplicates can be reassembled into a duplicate of the original whole. Reductive analysis gives the observer a more penetrating gaze, which sees not only what and where things are but where and how they come apart.

The metaphor of de-installing and re-installing components should be a vivid one for anyone who has had to wrestle with the problem of de-installing an APP from Windows. In fact, de-installing a computer program is literally a special case of the operation we shall describe here of *disconnecting* a component. We'll go into all this in detail later, but for now let it just serve as a reminder of how much hard work goes into seeing our ordinary world as full of *things*. This hard work is largely unconscious, and is based on "know-how" dating from our dim animal past. Our problem today is that, with the progress of scientific thought, we have wandered into domains where this unconscious skill no longer serves all our needs; for instance, it falls far short of telling us how to de-install a quark from a nucleus, or an event from a time loop, or for that matter, how to de-install a thought from the stream of consciousness. Thus it has become essential to dredge up and articulate the principles we unconsciously rely on in this work of de-installing things, and consciously learn to do it better, for otherwise we risk disastrous encounters with things we can't imagine and therefore can't see.

The naturalist, in contrast to the reductionist, sees things in place; he sees what and where they are in the context of the process as a whole. This involves seeing extensional separations and boundaries, but it also involves seeing how things *function* within the whole. Actually, as we'll see, functional structure is best described in terms of the following three-place relation:

Separability: We say that b *separates* A and C, or that A and C are *separable* at b, if fixing b makes the uncertainty or indefiniteness as to what is the case with A independent of the uncertainty or indefiniteness as to what is the case with C.

The most familiar example of this notion is the separability of the future from the past by the present: If we know everything about the present, then getting more information about the past does not give us any more information about the future, and vice versa, or so we suppose. (Remember, we are now examining our presuppositions, not looking for objective truths). Note that this says nothing about the determinism of the future by the present—all it says is that the past and future, however else they may be related, only "communicate" with each other via the present. Which brings up another familiar example: the separability of two communicating parties by their line of communication. If A and C are talking by phone, tapping their phone line (b) can tell you everything that's going on between them; whatever else is going on with the two of them at the time is going on with each of them independently.

Another term for separability is conditional independence. In the theory of stochastic processes the two terms have essentially the same meaning, though we'll keep them both to refer to slightly different mathematical formulations (see Chapter 2.)

Independence, unconditional and conditional: Events A and C are called *independent* if the probability of A&C is the probability of A times the probability of C. A and C are called *conditionally independent* given condition B if they are independent in the probability distribution conditioned by B, i.e. if $p(A\&C|B) = p(A|B)p(C|B)$. If for some random variable x, A and C are independent given condition x=k, where k is any value of x, then we say that x *separates* A and C. In the above telephone example, if we regard b as a random variable, then b separates A and C.

Since every probability distribution is conditioned by something, there is a sense in which all independence is conditional. The relation of independence, like every notion of being different or separate, is really a three-term relation: A and C are independent in a certain context B. We have just considered the case where that context is the result of placing a certain condition on a probability distribution. But the more familiar case is that in which the context is the result of *removing* a certain condition on a probability distribution. Notice that although there is only one way to impose a condition, there are many ways to remove a condition; this is one reason why there are so many different ways to represent a given process as a system. The converse of "deconditioned" independence is:

Conditional dependence: We say that A and C are *conditionally dependent* given B if $p(A\&C|B)$ is unequal to $p(A|B)p(C|B)$.

A very important kind of conditional dependence is that in which the condition B is of the form x=y, where, without B, A&x is independent of y&C. This is called linking, and we'll return to it shortly. There is a fundamental theorem that relates linking to separability:

Disconnection theorem: Process A&x&C is separable at x if and only if there exists a process in which A&x is independent of y&C which reduces to A&x&C under the condition x=y. This, and other theorems stated in this chapter, are proved in Chapter 3.

Speaking more generally, what makes separability so important is that it always coincides with the possibility of breaking a process into two de-installed components of some kind. We separate A and C by listening in on their phone line b, which is at the same time their connection and their common functional boundary, at least as far as their conversation goes. On the other hand we *disconnect* A and C, turning them into autonomous components, by *cutting* their phone line; this ends their conversation because their phone line is their separation boundary. When we mark the separation boundaries in a process, we may be functioning as naturalists, but we are also like the butcher with his blue pen marking the chops in a carcass.

To reconnect A and C so as to restore the whole process is to equate what is the case with b1 and what is the case with b2, thus restoring the original situation of A&b&C. Just what it means to equate b1 with b2 turns out to be less obvious than it sounds, and in fact can be understood in two very different ways; we'll call these two ways *i-o connection* and *linking*. Understanding exactly how an i-o connection differs from a link is the key to understanding how quantum processes differ from classical processes such as phone conversations. Let's start with i-o connection.

We'll simplify slightly by supposing that A is talking to C on a one-way line b. If that line is cut, what happens to the signal voltage at the two cut ends b1 and b2?

The cut line b1 from A is an *output*. This means, ideally, that what comes down the wire depends only on A, so the signal at b1 is unchanged by the cut. The cut line b2 from C is an *input*, however, and our description of the situation doesn't specify how inputs behave when they are disconnected. An open line will normally produce white noise, sometimes mixed with hum and faint radio signals. However, and this is a crucial point, the designer of a system usually doesn't have to take into account disconnected inputs; he only needs to think about how the system behaves when inputs are "driven" by outputs, whether from the user or from other components. To put it another way, the designer can regard inputs as *free variables*; his components are *open processes*, analogous to open formulae in the predicate calculus.

Open processes are interconnected by equating inputs to outputs, which means *assigning* free variables to non-free or *bound* variables (assigning variables is a familiar concept for computer programmers.) A composition created in this way will be called an *i-o system*. If all the free variables of the components are assigned, we call the system *closed*; otherwise it's an *open* system.

We are on familiar ground here; this is how engineers and programmers are taught to think these days. Once again, here are the main ideas:

Input: A free variable in an open process.

I-o connection: The assignment of a free variable to a bound variable.

Output: The bound variable to which a free variable is assigned.

Disconnecting is just the reverse of connecting. More exactly:

I-o disconnection: A and C are disconnected at variable b means we are given two processes A&b1 and b2&C with b2 free such that when b2 is assigned to b1, the result is the original process A&b&C.

I-o system: A composition of connected open processes. Every i-o system represents a process, but we must be careful to distinguish this process from the system, just as we must be careful to distinguish a vector from its representation as a sum in a particular basis.

There's something not quite right about all this. After all, a free variable is only an abstract linguistic entity, whereas our cut wire, lying there on the ground and picking up white noise, remains very much a real material object. The parts of a telephone system are not "indefinite processes"—they are just as definite as houses and trees and stones. Or so says animal faith. There is in fact a way to describe the situation that's more in keeping with this bit of animal faith, which is to replace the concept of assigning a free variable by the concept of *linking* a so-called *white* variable.

White variable: A random variable on which there is a uniform probability distribution. In the case of a discrete variable, whiteness means that every value is equally probable, and hence the variable has a finite number of values, since their probabilities must add up to 1. A white real variable has a uniform probability density which must integrate to 1, so it's range is bounded.

Theorem: If x and y are independent random variables with the same range, and y is white, then the condition x=y on their joint probability distribution does not change the distribution on x.

In our telephone example, let's now forget about free variables and think of the components A&b1 and b2&C as closed processes. Taken together, the two can be regarded as independent regions in the process A&b1&b2&C. When we connect b1 to b2, the result is a new process A&b&C. How does this new process differ from the old? The obvious answer is that it is just like the old one except for its behavior being restricted by the condition that b1 equals b2. Connecting two wire ends means making their signals equal.

It is not immediately clear how this answer, which is couched in the physical language of broken wires, signal voltages etc., applies to processes as such. (Remember, a process, as we are now defining it, is simply a joint probability distribution on a set of variables.). However, the phrase "… restricted by the condition…" gives a strong hint. To say "the probability distribution D1 is just like the probability distribution D2 restricted by the condition C" means that you get D2 from D1 by conditioning all event probabilities in D1 by event C. Thus our answer is basically this: To connect any two variables x and y, condition all of the probabilities of the disconnected system by the event x=y. This particular bit of animal faith brought to light will turn out to be the key to making sense out of quantum mechanics, so let's forthwith coin words that nail it down:

Links: Place a condition of the form x=y on two variables x and y of a stochastic process, thereby creating a new process in which the unconditional probability p(E) of any event E is p(E |x=y) in the old process. This condition is called a link.

Link system: A stochastic process plus a set of links. We'll go over all of this in considerably more detail in chapters 2 and 3. A summary of the relevant concepts, terminology and notation of probability theory will be found in Chapter 3, Section 1. Let's now briefly look at some high spots.

First, if x and y are independent and y is white, then the link between them behaves exactly like an i-o connection, with y acting like the input; this follows immediately from the above theorem. We'll show in Chapter 3 that we can neatly translate any i-o system into a link system which represents the same process and has essentially the same formal description. Engineers should have no problem adapting to link descriptions.

Second, and this is extremely important, we'll prove that this natural mapping of i-o connections to links does not go backward! Though every i-o system is equivalent to a link system, most link systems have no i-o equivalents, and even when one can make an i-o model of a link process, that model is usually exponentially more complicated than some link representation of the same process. Quantum systems are among those having no i-o counterparts, though quantum measurement systems do, as we'll see in Section 3.7. If technology can learn to really deal with this larger class of non i-o systems, it will turn into a fundamentally new kind of enterprise.

Third, a crucial definition, that of a *link state*:

Link state: *Given* any link x=y, the *state* of that link is defined as the matrix representing the joint probability distribution on x and y without the link (actually, it's this matrix normalized by dividing by its trace, but that's a small detail.) This is the concept that replaces the quantum wave function, and does away with the problem of the collapse of the wave function and other related quantum nuisances, as we discuss later.

In the introduction we briefly mentioned caloric. Heat, in the Eighteenth Century, was regarded as a fluid called caloric which was somehow related to the mechanical behavior of objects, but subject to its own non-mechanical laws. A series of experiments during the first half of the nineteenth century, starting with Count Rumford's crude qualitative observations, and ending with the definitive work of Joule and Helmholtz, convinced physicists that they didn't actually need to postulate a new substance called caloric, since heat is simply energy. When it was further realized that this energy is the random kinetic energy of molecular motion, thermodynamics became statistical mechanics.

History to some extent repeated itself in the early twentieth century with de Broglie's wave mechanics. His "matter waves" were at first thought to be waves in some new kind of fluid, the quantum analogue of caloric, but after Born discovered his quantum probability rule, they lost most of their substantiality, and physicists began to call them "probability waves".

Quantum mechanics, however, didn't go the way of thermodynamics. Because the so-called probability wave is really a wave of the square root of probability, "quantum caloric" didn't seem to be reducible to anything already known. And, alas, it is still with us, ignored by most working physicists, but hanging around the edge of physics in a kind of limbo where from time to time it is reworked and touted as a wonderful new discovery. But fortunately this unhappy state of affairs is almost over.

Here, in a nutshell, is how link states reduce "quantum caloric" to something already known, namely probability:

Suppose x and y are independent random variables. Then the probability that x and y will both have some value k is the product of the probabilities that they will have k separately, i.e. $p(x=k \ \& \ y=k) = p(x=k)p(y=k)$. Let's call $p(x=k)$ and $p(y=k)$ the unlinked probabilities of x being k and y being k. Now suppose we impose the link condition x=y. The linked probability of x=k is then the probability, conditioned by x=y, that both x=k and y=k. The probability that x is k, as a function of k, is proportional to $p(x=k)p(y=k)$.

In short, linked probabilities are always quadratic in unlinked probabilities. If the distributions on x and y are identical, which is the quantum situation, then linked probabilities are the squares of unlinked probabilities. That's essentially all there is to it! Quantum amplitudes are not the intensities of some mysterious new fluid, but are simply unlinked probabilities. One might ask "Probabilities of what?", but we'll see that we don't have to answer this question, just as we don't have to answer the question of what sets of things have numbers x and y in order to understand the meaning and truth of $xy = yx$.

A generalized form of quantum amplitudes will be found in any statistical situation whatsoever if we subject that situation to link analysis. Quantum amplitudes proper will show up if probabilities can go negative (we'll return to this in a minute) and we de-install components in such a way that the link states possess quantum symmetries; as we'll see in Chapter 3, this is always possible, though not always advisable.

Link states are quantum states represented as von Neumann density matrices. However, we can start in the more familiar way with quantum states as vectors over the amplitude field, and when we apply link analysis to the Schroedinger equation, amplitudes still turn out to be unlinked probabilities. Here, very roughly, is why this is so; the details are in Chapter 3.

In a continuous Markov chain, regarded as an i-o system, the probability distribution on the state variable can be represented as a vector which evolves in time according to the law $v' = T(v)$, where v is the distribution vector at time t, v' the vector at t+dt, and T the differential transition matrix. The Schroedinger equation, expressed in the language of Hilbert space, has exactly the same form, the difference being that the components of v are amplitudes rather than probabilities, and T is a unitary matrix rather than a transition matrix.

Let's now represent the Markov chain as a link system rather than an i-o system. The changing state vector is now the diagonal of a link matrix S that changes according to the law $S' = TST^{-1}$. If we think of S and S' as density operators, this is the von Neumann generalization of $v' = T(v)$. An immediate corollary is the Born probability rule in the generalized von Neumann form $prob(P) = trace(PS)$. The rule for turning an i-o system into a link system gives S the form $|v><w|$, where w is a "white" vector, i.e. it represents a uniform probability distribution, so it is only v that varies with time. The diagonal entries are quadratic in the entries of v and w, but since the constant entries of w are equal they factor out, so the probability vector is just v, the same probability vector that occurs in the i-o system.

Next, let's do the same with the quantum chain. Again we have S' = TST⁻¹, but now **T** is unitary. And again we have S = |v >< w|, but now instead of w being white, we have w=v, which makes the diagonal entries into the squares of the entries in v, i.e. state probabilities are the squares of "amplitudes"! The amplitudes are no longer the values of some mysterious wave function but are simply the probabilities in the joint distribution on a pair of unlinked variables. Quantum caloric is nowhere in sight!

That takes care of the main obstacle to a statistical interpretation of quantum mechanics. The square law is no longer a reason not to regard probability waves as true waves of probability. It's true, the wave amplitudes aren't the probabilities that we measure; rather, they are the probabilities that we would measure if the quantum process were disconnected at a separation point. But the same is true in a Markov chain, analyzed as a link process, the only difference being that in the Markov chain the output probabilities don't change when the output is linked to an input. In a quantum chain, "input" and "output" have identical distributions, and since unlinked state variables are independent, linking yields the unlinked probabilities squared.

There is still the problem of amplitudes going negative. They can also go imaginary, but as Mackey observed, the scalar i can be interpreted as a unitary symmetry in real quantum mechanics)[5]. For the more technically minded reader who has never encountered Mackey's real quantum mechanics, we mention an example[6] of how a real, positive definite probability distribution can first interpreted as positive and negative case counts and then reinterpreted as a familiar complex amplitude system. The case considered is a problem proposed by Feynman[7]: interpreting a random sequence as the *Zitterbewegung* of a relativistic quantum particle executed with velocity ± c and step-length *h/mc*, how does the "sum over histories" yield the single particle Dirac equation in 1+1 dimensions? Any random sequence with N symbols 0 or 1 having N_1 ones and N_0 zeros specifies a standard "case count" probability of finding a 1 as $p_1 = N_1/N$, and of finding a 0 as $p_0 = N_0/N = 1-p_1 = (N-N_1)/N$. If we interpreted the ones as "steps to the right" and the zeros as "steps to the left", this sequence can be interpreted as the trajectory of a particle with average velocity $v=(N_1-N_0)c/N$, and a combinatorial analysis[6] shows that the sum over such trajectories with this parameter fixed is indeed the exact solution of a finite difference equation which becomes the looked for Dirac equation in the continuum limit. The only subtle point is that one must use a real representation of the Dirac (Pauli) matrices, which is always allowed in this context.

So real, positive case counts do become standard quantum mechanics in an appropriate context. To get complex quantum mechanics (and a complex representation of the Dirac matrices) only requires a matrix transformation and still represents *identical* physics. In that representation it turns out that $\pm i$ represents when a sequence of steps in the same direction changes to a sequence of steps in the opposite direction. So the choice between real and complex representations offers different *conceptual* advantages without changing the physics. We could go on to discuss CPT invariance and the connection to the finite and discrete Maxwell equations[8], but will leave that for another occasion.

What could possibly be the meaning of negative probabilities? How can there be fewer than 0 things having a certain property P? How can an event E occur less than 0 times?

These are not questions that can be answered by constructing some ingenious mechanism, since they involve logic itself. Earlier we compared negative probabilities to negative piles of gold, our point being that there is no problem in constructing a useful mathematics that incorporates them. But that doesn't tell us what they really are, and we now doubt very much that their deeper significance can be understood at all within the scope of science as we know it.

This could be enough to drive many people back to quantum caloric. But before you join the stampede, be aware that the logical mysteries in quantum mechanics don't come from the link state interpretation. They come from the bare fact of amplitude cancellation, and practically stare you in the face with EPR. How can it be that A is possible, and also that B is possible, but that (A OR B) is impossible? —that's the mystery of the two-slit experiment. Here "OR" is the logician's "or" which allows A and B both to be true, A true and B false, A false and B true, but requires (A OR B) to be false if both are false. The double slit example we have is mind is: Case A: slit A open leading to a detection; Case B: slit B open leading to a detection; case (A OR B): both slits open but there is no detection. Von Neumann was well aware of the logical strangeness of quantum phenomena, and tried to cope with it by means of a non-Boolean construction that he called quantum logic. This turned out to be a dead end, but other strange logics dating from the period of quantum logic show more promise. Enough of this for now, however. We'll return briefly to the logic of negative case-counting in Chapter 2, but it's too big a subject to do justice to here; suffice to say that the logical anomalies don't vitiate our simple mathematics. So where do we stand with quantum mechanics?

We've gotten rid of "matter waves" but the mystery of quantum interference remains, since to derive the Schroedinger equation in link theory does require negative probabilities, even in the process itself. But now, at least, the mystery of interference has become a *simple* mystery, and there is no longer any reason to think of it as belonging to physics in particular. Once one has really grasped how universal the two generalized quantum laws are at the case-counting level, it's as hard to take seriously the arcane "models" of quantum mechanics that abound today as it is to take seriously the ancient sheep farmer's "spirit of tranquillity" after one has understood that xy=yx.

But we're afraid we are getting somewhat ahead of our story. We started out to psychoanalyze animal faith, to dredge up some of the unconscious beliefs that underlie our scientific thinking. "Wait a minute!", you protest, "I don't know about you, but I can't find any of this fancy mathematics in my unconscious!" Fair enough, neither of the authors can either. It's the analyst's job to propose, not to pronounce, and his proposals must always be confirmed by the analysand, who in this case is all of us. However, what we're looking for in our unconscious is certainly not fancy mathematics. Animal faith in itself, i.e. in place, can't even be put into words. To "find" it means to articulate something that does the same job, namely that of supporting our habitual modes of conscious thinking. In our current jargon, to articulate an item of animal faith means to de-install it, i.e. to transform it into a proposition that can stand alone. However, once articulated and understood, this new creation can then be put to the test by re-installing it in our intuition: Does it ring a bell? Do we think, "Ah, yes, that's what I've always thought, even if perhaps I never quite said it that way."? If the items we have articulated pass this kind of test, then we can confidently move on to the interesting surprises.

Quantum mechanics is one such interesting surprise; another, which is the topic of Chapter 2, is a very Kantian answer to Hume's skepticism about the existence of causes. Kant's own answer was that we necessarily see causes everywhere because, to paraphrase the way Bertrand Russell put it, we wear "causal-colored glasses." [9] By combining the philosopher von Wright's insight into how we operationally define causality with our analysis of process and system, we'll bring this statement up to date: We necessarily see input-output systems everywhere. Since input-output systems are closely related to computer models, it could be said that we necessarily see computers everywhere - we all wear computer-colored glasses!

If all this sounds like the grand march of progress, be aware that the two surprises above are not entirely harmonious. What makes our Kantian analysis of causality possible is the fact that the separability structure of a process only partly determines the form of the link matrix S, and we can show that it is always possible to choose a causal S (i.e., an S with an input). However, quantum S's are not causal! Furthermore, and this is crucial, the choice of S for a given disconnection will, in general, affect the separability structure of the disconnected parts. Thus things that come apart neatly with one type of S may exhibit bad "non-localities" for another type of S. EPR comes apart very neatly with a quantum analysis, but any causal analysis compatible with the data will exhibit non-locality, as Bell's theorem shows. Thus, though we are always perfectly free to choose causal S's and thus turn the universe into a giant computer, this computer may have a gosh-awful tangle of long distance wiring that could be eliminated by using other S forms.

Kant has had his say—now it's Freud's turn. Freud's project was not to justify our beliefs but to change our minds. He was, of course, working with people whose minds were very much in need of changing (which of course doesn't include you and me). Still, he may be able to give us a few useful tips.

As long as our assumptions remain unconscious, we have no choice but to believe them, or rather, we have no choice but to act as if we believed them. The question of belief doesn't actually arise until we are able to contemplate and weigh alternatives. For this to happen with some article of animal faith, we must first of all raise it up from the unconscious depths in such a way that it becomes a proposition, i.e. we must de-install it. Then one of two things may happen. We may take a clear look at this proposition and recognize its absurdity: "Wait a minute, my mother isn't really threatening to lock me in my crib if I try to go outdoors!", or whatnot. Or, we may see that the proposition is obviously true. If this happens, then another question arises: is this truth "objective", or does it depend on how you look at it?

How do we answer this last question? How do we find out whether something depends on how we look at it? One way is to see whether we can look at it differently. If we believe that all print is too fuzzy to read, we might try on glasses. Of course we may have to experiment to find the right glasses. Kant's error was to confuse our glasses with our eyes, so-to-speak.

The Kantian conclusion of Chapters 2 and 3 is that we all wear causal glasses, which turn out to be the same as computer-colored glasses. Does this mean that we are forever fated to see everything as a computer, or can we take these glasses off? As mentioned, "causal glasses" correspond to a particular kind of link matrix. We can work mathematically with very different kinds of link matrices, and indeed we must do so in order to practice quantum physics. Chapter 3 will make this very clear, and will point us in the direction of an expanded kind of science that works with the full range of acausal state matrices. But the question still remains: Can we, as human beings living our daily lives, ever take off our causal glasses?

For mathematicians, all things are possible. The experimenter who is willing to push hard enough may occasionally force nature to reveal its unnatural proclivities. But you and I, when we are just trying to get on with it, must constantly ask questions whose answers begin with "Because". That's our human nature. But what will be the effect on how we live and act if we can learn from the coming science how to go beyond "Why" and "Because"? What sort of beings might we someday become?

2 HUME'S PROBLEM

All modern discussions of causality begin with Hume. It was Hume who first clearly pointed out that merely knowing the regular succession of events does not tell us what causes what. The fact that B always follows A gives us no grounds for concluding that A "forces" B to happen. Why speak of causes at all, then? One school of thought says we should stop doing so; here is the young Bertrand Russell on the subject.[10]

"... the reason why physics has ceased to look for causes is that, in fact, there are no such things. The law of causality, I believe, like much that passes muster among philosophers, is a relic of a bygone age, surviving, like the monarchy, only because it is erroneously supposed to do no harm."

The mature Bertrand Russell, however, found out that like the rest of us he was unable to say very much outside of logic and pure mathematics without bringing in causes. Cause and effect pervades everyday life; the solution to Hume's problem can't be confined to the ivory tower of philosophy. A recent news item reported the discovery of a correlation in Mexico City between heavy chili pepper consumption and stomach cancer. For us chili pepper lovers, this was most unwelcome news: Do chilies cause cancer, or might there be some other explanation? What's at stake here is no mere philosophical abstraction; it's whether we have to stop eating chilies!

If there is no causal link between A and B, then, even if there is a high correlation between A and B, our wanting to bring about or prevent B is no reason to do anything about A. But if we learn that A causes B, or that it significantly increases the probability of B, then whatever power we may have over A becomes power over B too. The understanding of cause and effect is essential for practical life, indeed it is essential for our very survival, because it is the knowledge of means and ends.

This obvious fact of life has been curiously neglected by philosophers. The logician G. H. von Wright, however, was an exception; he actually went so far as to propose that the ends-means relation be used as the defining criterion of causality.[11] That is, he would have us say that, by definition, A causes B if and only if A can be used as a means to B, and more generally, that there is a causal connection from A to B if and only if we can change B by changing A. Notice that this is an operational definition: it says that the presence or absence of a causal connection between A and B can be detected by performing a certain operation, which is to change A and notice whether there is a concomitant change in B. That is, of course, just how diet scientists would go about seeing whether chilies cause cancer.

Does von Wright's definition solve Hume's problem? It certainly gives causality an empirical anchor that cannot be neglected in any proposed solution. However, a very important question still remains: is it possible to find formal patterns in the data itself that correspond to the causal connections revealed by the dispositional relation of means-to-ends? This is a difficult and subtle question, and the search for its answer ultimately takes us into territory where the question itself essentially disappears. First, though, there is a lot of work to be done closer to home.

The people who really live with causality are not philosophers but engineers, and curiously enough, they rarely use the word "cause". That's not because, as Russell said, "there is no such thing", but because they have developed a much richer and more discriminating vocabulary for the various causal structures and relationships they deal with. "Variable", "input", "output", "process", and "function" are standard causal terms. Computer science has added some important newcomers such as, "assignment", "call", and "subroutine". Indeed, the concept of subroutine neatly revives the whole Aristotelian foursome of causes: The material cause of a subroutine is the computer it runs in, the formal cause is its source code, the efficient cause is its call, and the final cause is the purpose it serves in the calling program.

It's clear from this brief survey of terms that, for the engineer, causality is an aspect of how something is composed, not of how it is extended. The engineer's job is to put together certain components into a system that realizes a causal process, i.e. a process having inputs that produce certain desired outputs. Furthermore, the components themselves are causal processes, and putting components together means using the outputs of certain components to causally control the inputs of others. As remarked in Chapter 1, the information engineer does not normally concern himself with the behavior of unconnected inputs. For him, and more generally for low power transfer, the behavior of input B connected to output A is usually the behavior of A in the absence of the connection, so B can be regarded as an indeterminate or *free* parameter. Thus, what the engineer designs, in our current jargon, is a functional i-o system; its components are what we called *open processes*. The functional dependence of outputs on inputs is called a *transfer function*. The concept of a transfer function clearly matches von Wright's concept of causality as the potential for the relationship of means to ends, so it makes sense to refer to functional i-o systems as *causal systems*.

In a causal system, so defined, the outputs of each component are functions of its inputs, and their composition by assignment of inputs to outputs is functional composition. However, we now know that, in the case of macroscopic components, the functional dependence of output on input is a large-number effect resting on the statistical behavior of atoms. A more accurate, and more general, account of the engineer's systems makes the output of a component into a *probabilistic* function of the input, with the causal transfer function replaced by a transition matrix of conditional probabilities; these are the i-o systems of Chapter 1. The composition of transfer "functions" is no longer functional composition but matrix multiplication, which reduces to functional composition in the special cases where the transition matrices contain only 0's and 1's. Having carefully noted the distinction between the mathematician's function and the engineer's transfer function, we'll now relax our language a bit and use the words "cause" and "causal" in connection with both. Thus we'll often refer to i-o systems as causal systems and to i-o states as causal states, especially in a context where we are distinguishing these from quantum systems and states, which are essentially a-causal. A more accurate qualifier would be 'statistico-causal', but it's too much of a mouthful. Since "classical physics" is often taken to imply strict determinism, we emphasize that the "classical" systems we consider in this paper *always* contain stochastic elements.

Suppose an input is connected to an output—how do you know which is which? To find out you disconnect them; the output is the one that continues to do by itself what they did together when connected - another operational test of causality. But suppose you can't disconnect them, and suppose also that you can't manipulate the inputs to the system; all you can do is watch. Then how do you identify causality? Is there nothing we can learn about what causes what by just watching? Hume's problem is still with us.

Here's another take on Hume's problem: Are there patterns in the joint probability distribution of an observed process that mark the causal relations we would see if we were to experiment with its disconnected parts? To put it another way, do cause and effect belong to the process itself, or are they artifacts that result from taking the process apart?

The theory of link systems throws new light on this question. Recall the disconnection theorem, which says that a process can be disconnected or "de-linked" at x if and only if it is separable at **x**. This shows us a way to transform a process into a link system: First disconnect x into x1 and x2, where x2 is "white" (corresponding to an input), which produces two independent processes, call them A and B. Then link A and B by imposing the condition x1=x2; this gives us a two-component link system exhibiting our original process. Now do the same for separating variables in A and B etc. until there are no separation points left. The result is a system which "factors" the original process into "prime" components that cannot be further subdivided. Recall that separability belongs to the process itself, i.e. it is intrinsic to the probability distribution. Thus the disconnection theorem reveals severe limitations on how we can construe the correlations in a process as causal. A highly correlated joint probability distribution need not have any separating variables at all, so causality requires something besides mere correlation in the data itself. We have at least a partial answer to Hume.

We'll see in Chapter 3 that the earmarks of causality in the data itself can actually be quite distinctive. Given the temporal order of the events in the process, it turns out that there is a unique prime factorization into components with white inputs. We learned in Chapter 1, and it will be proved in Chapter 3, that such systems are in natural 1-1 correspondence with i-o systems representing the same processes. Thus from a knowledge of the given alone, i.e. of the statistical behavior of the system without our interference, we can completely describe the i-o components we would get if we took the process apart at its (intrinsic) separation boundaries. The cause and effect relation, which we first located in the transfer functions and interconnections of the parts, is already *there* in the process itself!

OK, Mr. Hume, what do you say to that?

"Not so fast, sir; you were too glib with your prime factorization. Why should the x2's be white? And, if you regard what is given to our understanding to be only the probability of certain events, then on what basis do you bring in temporal order? If you must call upon such invisible help to reveal causality in the bare order of correlation, you do but reinforce my conviction that causes are phantoms of the imagination."

Good points. Let's translate them into the modern idiom. It will help to focus our ideas if we confine ourselves to processes in which the separating boundaries are all in a line; these are called Markov processes. Their definition can be stated in terms of a three-term relation among events called the Markov property, which is defined as follows:

Markov property: To say that the triple of events (A,B,C) have the Markov property means that they satisfy the equation $p(C|A\&B) = p(C|B)$, which we'll refer to as the *Markov equation*.

Markov process: A process in which the variables are indexed by a time parameter such that, at any time, if A is any event not involving future variables, and B is the state of the present, and C is any event not involving past variables, then A, B and C satisfy the Markov equation.

Recall that the probability $p(X|Y)$ of X conditioned by Y is defined as $p(X\&Y)/p(Y)$. Bearing this in mind, notice that if we multiply both sides of the Markov equation by $p(A\&B)/p(B)$ we get the equation $p(A\&C|B) = p(A|B)p(C|B)$. This relationship among A, B and C is known as *conditional independence*; in words, A is independent of C, given B. Recall that in Chapter 1 we referred to separability as conditional independence; this equation defines what we were referring to. Event B separates events A and C means that, given B, the probability of A&C is the product of the probabilities of A and C. It turns out to be more convenient to define separability by a slightly different equation, which is gotten by multiplying both sides of the Markov equation by $p(A\&B)p(B)$.

Separability: To say that B separates A and C means that $p(A\&B\&C)p(B) = p(A\&B)p(B\&C)$.

If none of the joint probabilities of A, B and C are zero, then the Markov property, conditional independence, and separability are all logically equivalent. However, for conditional independence to make sense, p(B) cannot be zero, while for the Markov equation to make sense p(A&B) must also not be zero. Since separability always makes sense, whatever the probabilities, it's the most convenient of the three to work with. Such fine points needn't concern us now, however; the really important point is this:

Conditional independence and separability are symmetrical in A and C, which means that the defining property of a Markov process is symmetrical in time! Think about that for a minute.

The Markov equation seems to have a temporal arrow, since it's about conditioning of the future by the present and past, but we now see that this is an illusion. Our argument for causality being in the process itself rested on a theorem that assumes a given temporal order of events, and in the usual discussions of Markov processes, temporal order is taken for granted. But Hume's challenge was to find causality in the bare order of correlation. We now see that this bare order does not distinguish past from future, so it would seem that Mr. Hume is right: Cause and effect really are in the eye of the beholder.

Kant, who has been listening to this discussion with growing impatience, can restrain himself no longer. Since his style is notoriously obscure, we will freely translate his remarks, with a little help from the eminent historian of philosophy Harald Hoffding.[12]

"We gather that by 'process' you mean something that is accessible to experience. Now experience, in contrast to mere imagination, is a complex composed of elements some of which are due to the faculty of knowledge itself, while others are the result of the way in which this faculty is determined to activity from without. As I have shown, space and time are forms of our perception. For whatever the nature of our sensations, and however much they may change, the spatial and temporal relations in which their content is presented to us remain the same; a space or a time does not change, whatever be its filling out. It follows that the temporal order of the process is given, since it belongs to the form whereby the outer determinant of our experience of the process can manifest itself in experience as a process. Hume is correct that cause and effect are in the 'eye of the beholder', but wrong to imagine that we could see otherwise than as we do."

Kant's ambition was to describe once and for all the "forms" that belong to the "faculty of knowledge", leaving the "outer determinants" of experience in a limbo where they have become known as "things in themselves". Today we are more modest. For one thing, we no longer make his absolute distinction between the subjective and the objective; rather we see objectivity as a matter of degree. Also, instead of trying to directly analyze the faculty of knowledge, we often rest content with characterizing the contribution of the subject to an experience in terms of a "choice of viewpoint", and of the object as a particular feature of the experience that is invariant under the choice of viewpoint. The subjective-objective boundary is of course quite mobile, and depends on the range of possible viewpoints we are considering.

Even transposed into a modern setting, Kant's observations about time direction are telling. So far we have been taking the notion of process as simply given. But we must ask: given to what? If given merely to imagination, then of course processes can be anything we want them to be, and they can go backwards or forwards or any which way in time. But if we are empirical scientists, then processes are something given to *experience*, at least potentially, which means that in thinking about processes going backwards we must consider whether we could actually observe such a process. And here we must heartily agree with Kant that the answer is no, since experience itself has an inexorable temporal arrow, whatever its content. Indeed the process as something given to experience is given one piece at a time, and the order in which these pieces are given is precisely the temporal order that we find in the process itself. This being the case, what could it possibly mean that time goes backwards?

We've been describing the process itself as what is given, but here's something else to keep in mind: The time reversal of the *given* is the *taken away!*

Let's be more concrete. Suppose we are observing a process and writing down what we observe. Our observations accumulate in a diary which starts with 100 blank pages and progressively fills with writing. Now each letter that we write reduces the nominal possibilities for what the diary *could* contain by a factor of 26; this is the Shannonian sense in which it constitutes information. Thus the process itself proceeds step-by-step alongside of a process of information gathering, which progressively reduces the vast initial range of possible states of the diary to one state, the chronicle of *what actually happened.*

We decided in Chapter 1 that it is not a particular chronology that we will call a process, but rather the conditions and rules that govern it. Let us then imagine an enormous series of repetitions governed by the same conditions and rules, each producing a diary of what actually happened. We now are given an enormous heap of diaries from whose statistics we abstract the joint probability distribution that we having been calling a stochastic process. The question is whether the stochastic process so abstracted exists as a thing-in itself apart from its potentiality of "determining our faculty of knowledge to activity", namely, the activity that produces this heap of diaries.

The Kantian answer is no. The very *idea* of process is inseparable from the idea of a progressive accumulation of observed information in a diary. Such a progression has of course a time arrow, that of the progressive narrowing of a range of possibilities. Thus a Markov process backwards in time is nonsense. To be given a process without a time arrow is inconceivable, since the given-ness of a process is conceptually tied to the given-ness of successive stages of its presentation to a potential observer.

It took a lot of fancy footwork, but time and with it causality seem at last to have returned to their familiar shapes. Finally we are back safe in the bosom of animal faith. Or are we?

Kant says it's OK to believe what our animal faith tells us, because we can't help it —such beliefs belong to the very "faculty of knowledge". Darwin's theory of evolution also suggests that it's OK, but for a somewhat different reason, namely that our faculty of knowledge evolved to enable us to survive in an all- too-real world of hostile and dangerous "things-in-themselves". The two messages, despite their opposite philosophical starting points, are actually not all that far apart. Darwin's message applies an important corrective to Kant's, however, which is that in analyzing our faculty of knowledge we must take into account the special conditions that obtained in the environment where that faculty evolved. Thus, the belief that the world is flat is certainly "true enough" for creatures who never travel more than a few miles from where they are born, but is badly wrong for long-distance navigators.

Homo sapiens is perhaps alone among the animals in being able to modify his animal beliefs in response to drastic changes in the environment, including those of his own making. Furthermore, and in this we humans are probably unique, our faculty of knowledge includes a faculty of abstract reasoning that can extrapolate to conditions far beyond our present environment. Sometimes this precedes a big move: think how we used every branch of modern science to prepare ourselves for exploring and colonizing outer space.

Kant lived in the Age of Enlightenment whose idols were Euclid and Newton, and rather naively incorporated their systems into his "intrinsic" faculty of knowledge, thus confusing a brief phase of human culture with something final and absolute. One wonders what he would have said about Einstein's theory of relativity. Kant thought it was impossible to experience space as non-Euclidean, and yet today we have good reason to believe that certain regions of space are quite non-Euclidean, and we know at least some of the ways in which this impinges on experience. The most drastic departures from Euclid involve warps not only in space but in time, producing loops where time closes back on itself. For the inhabitants of such a loop, if there be such, the future is also the past!

Needless to say, time loops pose a serious problem for our neat Kantian explanation of time-direction and causality. If the future is the past, which way does our experience go? How can there be a diary today about the future? In brief, *what happens*? If what happens is what could be experienced by someone involved in the happening, what would such experiences in a time loop be like? And what would it mean for experience itself to happen?

We do not explore this question here, even though it has become a respectable subject that Hawking and others discuss freely in technical language.[13,14,15]. We have discussed these questions in more detail elsewhere.[16] There we conclude that the generalized concept of happening we have been looking for, whatever else it may require, does require a generalized theory of probability in which we can countenance debts as well as surpluses of cases. We've come a long way from our starting place, which was to respond to Hume's challenge to find causality in the "bare order of correlation". In effect, Hume's problem effectively disappeared, giving way to others more puzzling and serious. The problem with Hume's problem is that it assumes we are simply given a world where things happen. This is indeed true in everyday life, but when we start to think about what happens in time loops or atomic nuclei, we find that it's the concept of happening itself that is the problem. Causality, time direction, and history all seem to be bound up together in a way that makes it hard to say which is the more fundamental. This tightly knit complex of ideas is a familiar one, and we have made some progress in analyzing it. But we have also discovered that it doesn't transfer very well to either the domain of the very large or of the very small. To make sense of modern physics seems to require that we give up not only causality and history but even logic. The next chapter will lay out in a more systematic way the kind of mathematical theory that must take their place.

3 PROCESS AND SYSTEM: THE MATHEMATICS

The main goal of this paper is to show that basic quantum mechanics can best be understood as a branch of the theory of probability. For this to happen, probability theory must be expanded in two ways: First and foremost, probabilities must be allowed to go negative. Second, and this is more a matter of technique than a change in our assumptions, the composition of transition matrices must be subsumed into a more general method of composition based on the linking of variables. We'll see that *in this expanded theory*, probability amplitude is not some new kind of "fluid" or physical field, but simply probability itself. By regarding it as such, we turn some of the most striking basic discoveries of modern physics into pure mathematics, thereby freeing them from their purely physical context to become tools available to every branch of science. Just how much more of physics as we know it will follow in this course remains to be seen.

510

3.1 Some Basic Concepts of Standard Probability Theory:

Event, sample space, case, variable, range, joint variable, logic, atom, probability measure, probability distribution, random variable, summation theorem, marginal, stochastic process, independence, conditional probability, conditional independence, separability, Markov property, Markov process.

Event: Following Feller[17] we'll take *event* to be our basic undefined concept. We'll use the word in pretty much the ordinary sense. Like most English nouns, it can be either definite or indefinite.

Sample space, case: A *sample space* is a set of mutually exclusive possibilities called *cases*. We say that an event E *belongs* to a sample space L if for every case of L, it makes sense to ask whether E is true or false. Another word for event is *proposition*. Examples: The event *Heads* belongs to the sample space whose cases are the possibilities Heads and Tails. The event *Seven* belongs to the sample space whose cases are the 36 possible throws of a pair of dice.

Variable: The term will be used in the scientist's sense, meaning something that can vary, rather than in the logician's sense of a place-holder for a term. A variable x is said to belong to a sample space L if for any value k of x, the event x=k belongs to L. For example, let L be the set of all cases for ten coin tosses, and let x be the number of heads; then x=1, x=2, x=3 ... x=10, are events in L and x is a variable in L. More abstractly, a variable is simply a function on L; when we regard x in this way, the proper notation is x(L). The set of all possible values of a variable is called its range.

Joint variable x℈y: Though variables are often numerical, we'll place no restrictions on what kind of things their values can be. For instance, a variable can range over all of the joint values of several other variables. The notation x℈y will refer to a variable that ranges over all joint values of x and y.

There is an ambiguity in the term "joint value", which is whether or not it involves the order of things joined. Is x℈y the same variable as y℈x? And is (x℈y)℈z the same variable as x℈(y℈z)? Actually, there are two concepts of joining, which in everyday language we would render as "x and y" and "x, then y". The first and simplest, which is unstructured joining, is commutative and associative, so we can write x℈y℈z℈...etc...without parentheses and with the variables in any order. The second, which is the one we'll use here, takes joint values to be *ordered* n-tuples; it has the decisive advantage over the first that variables of the same type are in natural 1-1 correspondence. There's actually a third concept called *list structured* joining that is very useful in constructing complicated link diagrams, but it belongs to a more advanced and specialized chapter of our theory.

Logic, 1 and 0, generating: Events can be combined with AND, OR and NOT, and a set of events closed under these operations will be called a *logic* (this use of the word 'logic' comes from von Neumann). We'll abbreviate AND by "& " and NOT by "~". A logic is of course a Boolean algebra. The *null* element of a logic, abbreviated 0, is the element A&~A, which is the same for any A. The *universal* element, abbreviated 1, is the element ~0 = A OR ~A. Given any set K of events, the set of all events that result from applying AND, OR and NOT to the members of K is a logic, and it will be called the logic *generated* by K. Given a variable x, the logic generated by all events of the form x=k, where k is a value of x, is called the logic of x, and the members of that logic are called events in x. More generally, the logic of x, y, z ... is the logic generated by all events of the form x=i, y=j, z=k etc., and the members of that logic are called events *in* x, y, z etc...

Atom: The atoms of a logic are the cases of that logic. To put it another way, an *atom* of a logic is defined as an event that cannot be further refined by any other event of the logic, i.e. A is an atom if, for any E in the logic, either A&E = A or A&E = 0. Atoms are mutually exclusive possibilities, and the set of all atoms is exhaustive within the logic. In the logic of the variable x, the atoms are the events of the form x=k for all k, while in the logic of x, y, z... the atoms are the events of the form (x=i)&(y=j)&(z=k)... for all i, j, k... If the number of atoms is finite, then the subsets of atoms uniquely correspond to the events in the logic. The cases of the sample space are the atoms of its logic, so atom and case are basically synonymous terms. However, we'll generally reserve the word 'case' for atoms that we count, such as the atoms of the sample space.

Probability: Underlying all our more sophisticated ideas about probability is Pascal's classic definition of the probability p(E) of an event E as the number of cases favorable to E divided by the total number of cases. If the cases of a sample space can be regarded as equally probable, then this definition defines a so-called *probability measure* on the events of that sample space, satisfying the following three axioms:

1) It's additive: If E1 contradicts E2 then p(E1 OR E2) = p(E1)+p(E2)
2) It's non-negative; for all E, p(E)≥0.
3) It's normal, i.e. p(1) = 1.

For now we are dealing with *classical* probabilities, which are never negative. For these we can always imagine an underlying sample space with equiprobable cases that define p(E) according to Pascal's definition. In dealing with negative probabilities, the notion of an underlying case set must be generalized so that it can contain both positive and negative cases. (The notion of negative set membership has been carefully investigated by mathematicians, and Blizard has shown that the Zermello-Fraenkel axioms can be generalized to allow for it.[18]) In this more general form, the case set has a very important role to play in link theory, since it enables us to distinguish among several quite different conceptions of independence; the details here will be found in future papers.

Probability distribution on x: A function defined on the range of x which for each k in that range has the value p(x=k). The standard notation for this function is p(x), which can be very confusing, since it conflates the scientist's and logician's two very different notions of variable. However, it's such a well-established tradition that we'll have to live with it; when you get confused, just remember to replace x by something that looks like a sentence.

Stochastic, or random variable: A variable on which is given a probability distribution. The qualifiers 'stochastic' and 'random' are only needed when a variable must be distinguished from variables of other kinds, like free variables. Since our results have nothing to do with randomness, "stochastic" is the preferred qualifier.

Joint probability distribution: on variables x, y, z..., written p(x,y,z...), is defined as the probability distribution on the variable x&y&z...

Summation theorem and marginals: The summation theorem says that given a joint probability distribution on variables x and y, we can find the probability distribution on x alone by summing over all values of y; i.e.. $p(x=i) = \sum_{j \in y} p((x=i)\&(y=j))$, or, in our abbreviated notation, $p(x) = \sum_y p(x,y)$. This is a theorem we'll use repeatedly. A corollary is that a joint probability distribution uniquely defines the joint probability distribution on any subset of its variables; such partial joint probability distributions are called *marginals*.

Stochastic process: Define a *stochastic process*, or *process* for short, as a set of variables on which is given a joint probability distribution. In the usual definition it is also assumed that these *stochastic variables* involve events which can be ordered in time. This won't do for our present purposes, though, since, we need a new concept of process general enough to apply to time loops. We can start out by thinking of process variables as having at least approximate locations in the space-time of current physics. However, since we aim to eventually construct space-time out of probabilities, this is only a temporary expedient.

Independence: Events E, F, G,... are called independent if $p(E\&F\&G...) = p(E)p(F)p(G)...$ Note that for a set of events to be independent, it's not enough that every pair of events in that set be independent; independence is a joint condition of all the events. Two variables are called independent if events in the first are independent of events in the second. A set of variables are called independent if events in different variables are independent. This definition pretty much captures our intuitive concept of independence as lack of correlation, although there is one thing about it that is not so intuitive: Suppose that either $p(E)=1$ or $p(E)=0$. Then for any F, $p(E\&F)$ is either $p(F)$ or 0; in either case it is $p(E)p(F)$. Such a "definite" E is independent of all events, including itself! This oddity will turn out to be important for the concept of probability space (see Section 3.8).

Given a finite sample space of equiprobable cases, there is another way to define the independence of E and F, which is that the sample space can be arranged as a rectangular set in which the cases of E are a vertical stripe and the cases of F are a horizontal stripe. For ordinary probability theory in which cases always count positively, the two definitions are equivalent. However, with both negative and positive cases, the two diverge, since $p(E)$ can be 0 even though its cases set is not "rectangular" with respect to that of F. It turns out that there is even a third definition of independence which applies to imaginary and complex probabilities (see Section 3.6) when these are defined in terms of real probabilities. The above product rule is the definition we'll adopt here, however.

Conditional probability: The conditional probability of E given F, written $p(E|F)$, is defined as $p(E\&F)/p(F)$.

Since a conditional probability results from dividing by $p(F)$, it is only well-defined if $p(F)$ is not 0. Negative probabilities are constantly producing 0's in weird places, so it's best to avoid conditional probabilities if there's an alternative; fortunately, there usually is.

Condition: An event C that is assumed to be true, thereby reducing the sample spaces to the set of those cases favorable to it. Axioms for a process are often given as conditions on a "free" process whose variables are independent. A very important kind of condition for our present work is the *link*, which is an event of the form x=y. The probability of E in the reduced sample space conditioned by C is $p(E|C)$ in the unconditioned sample space.

Conditional independence: Events E and F are called *conditionally independent* given C if they are independent in the sample space conditioned by C. The usual algebraic formulation of this relationship is: $p(E\&F|C) = p(E|C)p(F|C)$. More generally, a set of events E1, E2, E3... is conditionally independent given C if $p(E1\&E2\&E3...|C) = p(E1|C)p(E2|C)p(E3|C)...$

As we noted in Chapter 2, these equations becomes meaningless when the probability of the condition is 0. The following definition is equivalent to conditional independence when $p(B)$ is not 0, but also makes sense when it is 0:

Separability: Events A and C are called separable by event B if $p(A\&B\&C)p(B) = p(A\&B)p(B\&C)$. This definition comes from noting that it can be obtained by multiplying the conditional independence condition, $p(A\&C|B)= p(A|B)p(C|B)$, on both sides by $p(B)^2$. The equation defining the separability of n events by B is derived by multiplying both sides of the appropriate equation for conditional independence by $p(B)^n$.

We say that variable y separates variables x and z if $p(x,y,z)p(y) = p(x,y)p(y,z)$. which is an elliptical way of saying that if for any values i, j, and k of variables x, y and z we have $p((x=i)\&(y=j)\&(z=k))p(y=j) = p((x=i)\&(y=j))p((y=j)\&(z=k))$.

Much of our work will be with Markov processes. Feller[17] begins his chapter on Markov processes with the concept of Markov chain, which he defines as a series of trials in which the probabilities at each trial depend on the outcome of the previous trial, the rule of dependence being the same for all trials. He then goes on to say that a Markov chain is characterized by a matrix of conditional probabilities, called the *transition matrix*, together with an initial "state vector" giving the probability distribution at the initial trial.

There are problems with this definition. Part of what motivates it is the need to separate out the *law* of the process, given by the transition matrix, from the *boundary condition* on the process given by the initial state vector. And yet there is no reason why the initial vector should not contain 0's, for which conditional probabilities are meaningless. Thus the Markovian law of the process, which makes perfect intuitive sense whatever the initial state, cannot properly be described by a matrix of conditional probabilities. The so-called transition matrix must be defined in some other way.

The remedy for this and related confusions is one we have discussed at length in Chapters 1 and 2; this remedy is to clearly distinguish between the process itself and its representation as a *system*. Feller began by describing a process and then switched over to describing it as a system, without noticing the change. Section 3.6 will be devoted to Markov systems, so here we'll define only the process. First a more general concept, which we've already met in Chapter 2:

Markov property: Three events A, B and C are said to have the Markov property if $p(C|A\&B) = p(C|B)$. Multiplying both sides by $p(A\&B)p(B)$, we see that this is simply separation, i.e. it says that B separates A from C.

Markov process: A stochastic process whose variables can be arranged in a sequence such that each variable separates the variables before it from those after it.

Markov transition matrix: Abstractly, a matrix in which the sum of every column is 1 (just to confuse matters, such matrices are sometimes called *stochastic matrices*). More concretely, the matrix of conditional probabilities from variable $x(t)$ to variable $x(t+1)$ in a Markov process.

3.2 Processes

Primary variables, secondary variables, join, global variable, independent parts

Here are some useful technical terms:

Primary variables: In Section 3.1, a process was defined as a set of random variables on which we are given a joint probability distribution; these variables will be called the *primary* variables of the process. We will assume that the primary variables are given in a certain linear order, which may or may not be related to the probability distribution.

Secondary variables: A secondary variable is a joint variable in a set of primary variables, called its members. We will assume that the members always occur in their natural order, so there is one and only one secondary variable for every subset of primary variables. We'll speak of secondary variables as also belonging to the process. Both primary and secondary variables will be denoted by small letters.

Variables: Unless otherwise stated, the word 'variable' will refer to either a primary or secondary variable of a process.

White variable: A variable on which the probability distribution is uniform.

Global variable: The secondary variable whose membership includes all primary variables. The process itself will be denoted by capitalizing the global variable, e.g., W is the process whose global variable is w.

Independent part, component: A sub-process U of W is called an independent part of W, or a *component* of W, if the primary variables of U are independent of all other primary variables of W. The primary variables of U are also called its *members*.

Prime: A process having no independent parts is called *prime*.

Prime factorization theorem: The set of prime components of a process is unique.

One often hears that the whole is more than the sum of its parts. What is generally meant is that the whole is more than the mere heap of its parts, which is true unless the whole happens to be a heap of parts. A process is indeed the heap of its prime parts, which are in fact components, in the sense that the description of each does not require that we take into account the individual peculiarities of the others. However, we saw that one of the surprises of negative and imaginary probability theory is that there are several quite different definitions of independence, and thus the members of a heap share, as it were, the "nationality" of their heap, which involves the particular kind of independence that its members have from each other.

3.3 Links and Cuts

If you have been dozing through these definitions, that's OK—you can always go back to them. But now is the time to wake up. Pay attention! Here comes the key idea of the whole paper:

Link: Given a process W containing variables x and y, we *link* x to y in two steps.

Step 1: Apply the condition x=y to W to obtain a new *conditioned* process W' with the same variables as W but with a new joint probability distribution W' in which the probability p'(E) of an event E is p(E|x=y) in W.

Step 2: Then modify W' by dropping the duplicate variable y.

We can think of a link as an operator L(W,x,y) on W, x and y that produces a new process W', in which the probability of an event E is its probability in W conditioned by the event x=y, and in which the variable y is dropped. We'll write W|(x=y) for this operation, i.e. W' = W|(x=y); more generally, W|LMN will mean applying links L, M and N to process W.

There is nothing in the definition of link that says x and y must be different variables; W|x=x) is a perfectly legitimate operator. What, then, does it do? It does nothing to the probabilities, since the condition x=x is already in effect. It does, however, drop the second variable, namely x. To link a variable to itself simply means to drop it from W.

x and y can be either primary or secondary variables. If they are secondary, however, they must contain corresponding members that are also implicitly linked. Thus if x = x1&x2 then y must be of the form y1&y2, and the condition x = y implies also that x1 = y1 and x2 = y2. This means that when y is dropped, we must also drop all of its primary members, and with them all other secondary variables containing any of these members, which are no longer needed, since they are duplicated by secondary variables with corresponding members in x. Dropping a variable is a bit more complicated than the name suggests.

The converse of linking could be called "de-linking", but the term doesn't really fit, since it implies that there are two things to be de-linked, whereas in fact there is only one. To restore the original W, we must first turn x in two equal variables, and then remove the condition that equates the two. We'll call this operation:

Cutting: To *cut* the process W' at x means to produce a new process W containing a new variable y having the same range as x such that W' = W$|$(x=y). Unlike linking, cutting is not an operator, since there are an infinite number of ways to cut W' at x. Unlike linking, cutting is not an operator, ince there are an infinite number of ways to cut W' at x.

3.4 States

Link state, density matrix, state of a variable, trace, propositions, projections, Born's rule, pure states, mixed states

We now come to another key concept.

State of a link: Given a link W$|$x=y, the state of that link is defined as the joint probability distribution on x and y in W.

Equivalent states: States that result from cuts on the same variable in the same process.

Density matrix: The joint probabilities on x and y arranged in a matrix with x as the horizontal index and y as the vertical. Following *one* quantum tradition, we'll use the terms 'state' and "density matrix" interchangeably.

State of a variable x: The density matrix of the link x=x. Notice that the off-diagonal elements of this matrix are all 0, since it's impossible for its two indices to differ, and impossibility has probability 0! A more rigorous definition of the state of a variable is the diagonal matrix whose indices range over the range of that variable, and whose diagonal is the probability distribution on that variable. For instance, if x ranges from 1 to 3, then its state is the 3 by 3 matrix M_{ij} such that M_{11} is the probability that x=1, M_{22} the probability that x=2, M_{33} the probability that x=3, and all other entries are 0.

We'll find that the state of x encompasses all of the classical meanings of the word "state", including Markov states and deterministic states like those of a computer. A Markov state is usually represented by a vector; this traditional vector can now be defined as the diagonal of the state matrix of the time-dependent random variable of the process. (We'll see the details of this later).

Trace: The trace of a matrix S, written tr(S), is defined as the sum of its diagonal entries.

Important trace theorem: tr(AB) = tr(BA). Important corollary: tr(A) = tr(TAT^{-1}), i.e. the trace of a matrix is invariant under all linear transformations.

Proposition: Literally, a synonym for event. (We are borrowing the term from von Neumann, who applied it to quantum variables.) There is a difference in connotation, though. Events are often described by noun phrases, whereas one always represents propositions by sentences. We can think of propositions as sentences in the predicate calculus whose free variables are assigned to process variables.

Predicate: A "propositional function" whose free variables are unassigned. A predicate becomes a proposition when all of its variables are either assigned or quantified. A link, for instance, is a proposition resulting from the assignment of the equality predicate to a pair of variables.

Projection: In linear algebra, a projection P is defined as a linear operator satisfying the idempotent law PP = P. It can be shown that we can always find a basis in which such a P is represented by a diagonal bit matrix. For now we'll confine ourselves to diagonal projections, i.e.

a *projection* will be defined as a matrix in which all off-diagonal elements are 0 and the diagonal elements are either 0 or 1.

Von Neumann showed that propositions about the measurement of a quantum observable x can be represented by projections that commute with the eigenvalues of that observable. This led him to an elegant generalization of the Born probability rule: If P is a projection representing a certain proposition, and D is a density matrix representing a quantum state, then, given D, the Born probability of that proposition is tr(PD).

Representing predicates by projections: If $P(x)$ is a proposition about x, define the *characteristic function* of $P(x)$ to be the bit-valued function $B(x)$ which is 1 or 0 corresponding to whether $P(x)$ is true or false for x. The diagonal matrix P having $B(x)$ as its diagonal is the projection that we'll use to represent the proposition $P(x)$. We'll normally use the same letter P for both the predicate and the projection, and when the predicate is assigned, for the resulting proposition; it's usually clear from the context which of these things we mean.

If we multiply the projection P by the state matrix D of x, we clearly knock out all cases for which the predicate P is false, so the sum of the remaining entries in D is the probability of proposition P, i.e. $p(P) = tr(PD)$. This makes Born's probability rule start to look a bit less mysterious. However, what really demystifies Born's rule is that it holds not just for the state of a variable but for any link state. More exactly:

Generalized Born's rule: Let S be the link state of x=y in process W, and let P be any proposition about x. Then the probability of P in the conditioned process $W|x=y$ is tr(PS)/tr(S).
Proof:
First note that the state S' of x in $W|x=y$ is S with all off-diagonal elements set to 0, normalized by dividing by $p(x=y)$. Since P is diagonal, PS is the result of multiplying the rows of S by corresponding diagonal elements of P. Thus the diagonal elements of S' are simply those of S divided by $p(x=y) = tr(S)$. Since we just saw that $p(P) = tr(PS')$, we conclude that $p(P) = tr(PS)$. QED.

Our result differs from von Neumann's by the factor tr(S). Had we defined a link state as S/tr(S), thus making tr(S) always 1, we would have gotten von Neumann's result exactly. This was our original definition of state, and it works fine for classical and quantum states. However, we would like our definition of state to still make sense even as we push our theory into new domains where, because of the possible cancellation of plus and minus cases, the trace of S can be 0. For this reason, We believe that our present unnormalized state is preferable in the long run.

Pure states: Given a link $W|x=y$, if x and y are independent in W, the state of the link is called *pure*.

It follows directly from the definition of independence that the density matrix of a pure state has the form $|v><u|$ in Dirac notation. If u is white, the state will be called *causal*. On the other hand, if u=v, the state is quantum. Causal and quantum overlap if u and v are both white., but in no other case.

Pure causal state: The state of a link between independent variables, of which the second is white.

Causal trace theorem: If S is pure causal, then tr(S) = 1/n, where n is the dimension of S.
Proof:
To say that w is white means that for any i, $p(w=i) = 1/n$. Let S be the state of link $W|x=w$. Since S is pure, $p((x=i)\&(w=i)) = p(x=i)p(w=i) = (1/n)p(x=i)$. But $p((x=i)\&(w=i))$ is S_{ii}, the i'th diagonal term of S. Thus $tr(S) = \sum_i S_{ii} = (1/n\sum_i p(x=i) = 1/n$.

White connection theorem: Roughly: Pure causal links don't affect the component on the "output" side. More exactly, if x is in a component A that is independent of y, and y is white, then the link $W \mid x=y$ doesn't change the joint probability distribution on A.

Proof:

Let z be the join of the variables of A other than x, i.e. A = X&Z. Then p(x,z,y) = p(x,z)p(y) = (1/n)p(x,z) in W. By the above trace theorem, p(x=y) = 1/n. Thus p(x,z,y | x=y), which, ignoring y, is the joint probability distribution on x and z in $W \mid x=y$, is equal to (1/n)p(x,z)/p(x=y) = p(x,z), which is the joint probability distribution on x and z in W. QED.

This theorem is the key to mapping the engineer's input-output systems into link theory. Remember, we asked the question: When an input is wired to an output, how do you know which is which? The answer is: cut the wire and see which loose end is unaffected by the cut. This answer can be amplified a bit. If the wire is from box A to box B which is otherwise unconnected to A, then cutting it not only leaves the output of A unaffected, but all of A unaffected. This is essentially the content of the white connection theorem, if we regard the causal link as a connecting wire. The word "causal" means what it says!

Mixed states: Needless to say, pure states play a very important role in our theory. In quantum mechanics, impure states are called *mixed* states; this is because they can always be written as linear combinations of pure states. This turns out to be true quite generally of link states, so we'll borrow the quantum term *mixed state* for any state that is not pure. Let us point out, though, that the decomposition into pure states of a mixed classical state, such as one might find in a computer program, is generally quite artificial, since the pure components will seldom be classical and will often involve negative probabilities.

3.5 Link Systems

Disconnection, fundamental theorem of disconnection, link systems, proper systems, predicates as components, relational composition, abstract links and their states, contraction, product theorem, tensors, public variables, systems, transition matrices, causal boxes and systems

We now come to a simple but profound connection between purity and separability which could be the key to unifying space and matter. First a definition:

Disconnection: To *disconnect* x from z at y means to cut their process at y in such a way that x&y becomes independent of y'&z , where y' is the new variable produced by the cut.

Fundamental theorem of disconnection: Let W be the join of three variables x, y and z, i.e. w=x&y&z,, where p(y) is not 0 for any y. Then x can be disconnected from z at y if and only if y separates x and z. (Recall that separation means p(x,y,z)p(y) = p(x,y)p(y,z).)

Proof:

Suppose there exists a disconnection of W that produces a new process W'. Let y' be the new variable of W', i.e. W = W' | (y=y'). To avoid confusion, let's use p for probabilities in W and p' for probabilities in W'. Then in W' we have p'(x,y,y',z) = p'(x,y)p'(y',z). This implies that, in W', x is independent of z for any values of y and y'. In particular, x is independent of z if y=k and y'=k. Since this is true for any k, it is consistent with the linking condition y=y', which is what turns W' into W. This shows that y separates x and z in W. Notice that going from disconnection to separation doesn't require the assumption that p(y) is never 0.

Going backwards is a bit harder. Suppose y separates x and z in W. We need to construct a new W' containing x, y and z plus a new variable y' such that x&y is independent of y'&z in W', and such that W = W'|(y=y'). Separability in W means that $p(x,y,z)p(y) = p(x,y)p(y,z)$. Let W' be a process consisting of two independent parts W1' and W2', where W1' contains x and y, and W2' contains y' and z. Let $p'(x,y) = p(x,y)$. We construct $p'(y',z)$ as $p(z|y')/n$ where n is the number of cases of z for which $p(z|y')$ is non-null. Then $p(z|y')=p(z\&y)/p(y')$ where $p(y') = \Sigma_z$ $p(z|y')$. Note that this is where the requirement $p(y') \neq 0$ is needed.

We must now show that under the condition y'=y, W' is W, which is to say, $p'(x,y,z) = p(x,y,z)$. First, we will use separability to separate out z in the joint probability distribution of W.

Lemma 1: $p(x,y,z) = p(x,y)p(z|y)$. Proof: Because $p(y)$ is never 0, we can freely speak of probabilities conditioned by y. Thus we can write $p(x,y,z)$ as $p(x,z|y)p(y)$. We can also write separability as conditional independence, i.e. $p(x,z|y) = p(x|y)p(z|y)$. Combining the two we get: $p(x,y,z) = p(x,y)p(y)p(z,y) = p(x,y)p(z|y)$. QED.

Lemma 2: The joint probability distribution on x, y, y' and z in W' is $p(x,y)p(z|y')$.
Proof:
By the independence of W1' and W2', we have $p'(x,y,y',z) = p'(x,y)p'(y',z)$. Since $p'(x,y)$ was defined as $p(x,y)$ this becomes $p(x,y)p'(y',z). = p(x,y)p(z|y')/n$, where n is the number of rows in the transition matrix $p(y'|z)$. QED.

Now let's link y and y'. This involves two steps: first we identify the variables y and y', and then we normalize the resulting distribution by dividing by $p'(x=y)$. We see by lemma 2 that the first step turns the joint distribution on W' into $p(x,y)p(z|y)/n$. Now $p'(x=y)$ is the trace of the state matrix of the link, which we can easily see is 1/n (the steps showing this are spelled out in the proof of the causal link chain theorem in the next section.) Thus dividing by $p(y=y')$ yields $p(x,y)p(z|y)$, which by lemma 1 is $p(x,y,z)$. Thus $p'(x,y,z) = p(x,y,z)$. QED.

The above disconnection theorem is what tells us how to mark a process for dissection, so-to-speak. First, we find a variable x, primary or secondary, which separates the process into two conditionally independent parts. We then disconnect the process at x into two unconditionally independent parts, and then repeat for these two parts, etc. until we can repeat no more, at which point we have arrived at a prime factorization. These prime factors, as such, constitute a mere heap of independent parts. The disconnection procedure, however, leaves a record of past connections, a kind of wiring diagram, that can be used to reconstitute the original process.

The question arises whether such a prime factorization is unique. There are actually two questions here: first, whether the factors themselves are unique as processes, to which the answer is definitely no (except in the trivial case where the process has no factors except itself) and second, whether the prime "boundaries" are unique, or to put it another way, whether every factoring procedure ends up with the same sets of primary variables in its factors. The second question has no simple yes or no answer. Often it is yes, but there are important cases (EPR, for instance) where a seemingly arbitrary choice in making the first cut will determine whether or not any further cuts are even possible.

Link system: A pair of processes W and W' together with a sequence of links on W that produce W'.

Proper link system: A link system in which all links are between different independent parts of W, and no two links involve the same variable.

Propriety theorem: In a proper system, the links commute, i.e. you can get from W to W' by applying the links in any order.
Proof:

Divide the linked variables into two sets X and Y, where X contains the first variable of each link, and Y contains the second. Let x be the join of variables in X, y the join of variables in Y. Then the link x=y clearly encompasses the links of all the members of X and Y, whose order is immaterial as long as they are properly matched.

The linking order can indeed make a difference if the linked pairs have variables in common. If you link y to z and then x to y, all three of x, y and z will be tied together and called x. However, if you first link x to y, y disappears, so you when then you try to link y to z, there is no process variable to link. This sort of thing cannot happen in a proper system, of course.

But suppose you need to link all three of x, y and z together? Indeed, this is exactly what an engineer does all the time when he "branches" an output to go to several inputs. Must the engineer then work with an improper system?

The answer is no. There are in fact several ways to branch properly. One rather interesting way is to extend link theory by treating predicates as "abstract" components which, when linked to random variables, become conditions on the process. Thus to connect x, y and z, we introduce the abstract component x'=y'=z' plus the three proper links. The resulting more general theory merges stochastic composition with the predicate calculus, and may be of mathematical interest. However, much of what it accomplishes can be done in other ways; for instance, we can regard x'=y'=z' as a stochastic process with three equal white variables, and the links x=x', y=y' and z=z' also connect x, y and z. At any rate, abstract components won't concern us here. With one exception, that is; links themselves should definitely be regarded as abstract components. But they are the only ones we need for now.

Not having to worry about the order and grouping of links is such a plus that we'll assume that all our systems are proper. Under this assumption, there is a very readable and natural diagrammatic notation for link systems, which uses icons to represent component processes, and arrows between them going from x's to y's to represent links. Indeed, if we draw and label the icons right, we can often avoid variable names altogether, since the variables will be identified by their places.

Unfortunately, the practicalities of word processing, e-mail, etc. make it expedient to have a linear backup notation. Boxes are written thus: $A[y_1,y_2, .. ; x_1, x_2, ..]$, where the variables before the semicolon are the ingoing arrows, those after, the outgoing. A link is shown by an equation, so A(y ; x) (x=x') B(x' ; z) links x in A to x' in B. If there is no need to reconstruct the particular lost variables in W, we can omit the equation and simply use the same letter in A and B, e.g. A(y ; x) B(x ; z). Needless to say, diagrams are much easier to understand, and it's often best to turn this linear notation into actual diagrams rather than trying to decipher it as it is. There are more powerful diagrammatic notations than this for dealing with complex structures, but they aren't needed for the simple examples we'll be studying here. It's worth introducing one new concept from this extended notation, though:

Box: An icon that represents a component of a system, but that may also be a container, implicit or explicit, for a more detailed system. From the outside a box is to be regarded simply as a process, but it's permissible to look inside it for smaller components (think of the transistors inside an integrated circuit or the electrons inside an atom.) In the next two sections we'll show all component icons as boxes, and sometimes refer to them as such.

Transformation: A two-variable sub-process T of a link system in which both variables are linked to others. If the variables of T are x' and y, in that order, and if x=x' is the link to x and y=y' is the link to y, then the link state (density matrix) of x=x' is called the state *before* T, while the link state of y=y' is called the state *after* T. T is called *causal* if x' is white.

Transformation product theorem: Roughly, to combine two successive transformations, multiply their matrices. More exactly, if T and U are transformations with variables x',y and y',z, then linking y and y' and then ignoring y produces the transformation nUT, where n is a normalizer.

Proof:

Combine the summation theorem with the definitions of linking and matrix multiplication.

Corollary: Equivalent cut theorem. Suppose that cut C1 in W' produces the process W1 with state variables x and y. Call the state of that cut S1. Define W2 as the process that results from multiplying x by the invertible diagonal matrix D and y by D^{-1}. Then the resulting state matrix S2 is equivalent to S.

Corollary: A pure state with no 0's in its diagonal is equivalent to a causal state.

Corollary. Born's rule gives the same probabilities for equivalent states.

3.6 Markov Chains and Schroedinger's Equation

The rule of law, boundary conditions, Markov chains, generators, time-reversed Markov chains, link chains, the link dynamical rule, differential Markov chains, inverse Markov chains, double boundary conditions, chains, sharp states, Schroedinger's equation, complex amplitudes

The Ancient and Medieval worlds observed nature as carefully as we do, hoping as we do to discover what remains fixed in the midst of change. But science, as we know it, began when our focus shifted from changeless *things* to the changeless laws of *change itself*. The goal of "natural philosophy" then became to find those universal "laws of causality" that determine the future from the present. The amazing success of Newtonian mechanics convinced many people that this goal had actually been attained, and so went the scientific consensus for 200 years. But then came quantum mechanics.

What most shocked people about quantum mechanics was its seeming break with Newtonian determinism. Chance, seemingly exiled to the limbo of ancient superstition, had, like Napoleon returning from Elba, once again forced itself upon the civilized world. Could Chaos be far behind? The ideal goal of scientific theory had to retrench from predicting certainties to predicting probabilities. Our ultimate "law of causality" was an illusion; the best we could hope for was to find the ultimate rule giving the best *probability distribution* on the state variable x(t+1) of the world at time t+1 as a function of the state variable x(t) of the world at time t. Which brings us to *Markov chains*.

A Markov chain is a Markov process whose transition matrix is the same at all times t (see Section 3.1 for the definition of a Markov transition matrix.) Call this unchanging matrix G the generator of the process. Given G plus the state at time 0, we can calculate p(x(1)), p(x(2)) .. etc., and also p(x1,x2) and p(x2,x3) ...etc. It follows from the Markov property that we can calculate the whole joint distribution (this is a well-known theorem.)

A generator, in contrast to an initial state, is the *law* of a process. It's important to realize that not every Markov process has such a law. Given the Markov property alone, the matrix T of conditional probabilities between successive states can be an arbitrary function of time, in which case the distinction between law and initial condition, between "essence" and "accident", is artificial at best.

What happens to the generator if we reverse time? Except in very special cases, it does not exist. The transition matrices of a time-reversed Markov chain are almost always time-dependent. This makes the chain property a good indicator of time direction—score one more point against Hume! It's far from being a definitive indicator, though, since this reversed time-dependence is always lawful, and the reversed chain can always be embedded in a larger process with the chain property.

We have not yet mentioned systems; the concept of generator applies only to the process itself. However, it follows quickly from our results in Section 3.5 that we can always represent a Markov chain by a link system whose components, except for the first one, look and behave just like G. More exactly:

Causal link chain theorem: Given a Markov chain C with generator G and initial vector V, there is a link system whose first component is V and whose successive components all have the matrix $(1/n)$G, where n is the *dimension* of V etc., i.e. the number of values in the range of the state variables.

Proof:

Let x be the initial chain variable, and x',y be the variables (indices) of G. Then we must show that the joint distribution on x, y that results from assigning x', the conditioning index in G, to x, is the same as that resulting from linking the joint distribution in x', y having the joint distribution matrix $(1/n)$G to the independent variable x with the link (x=x').

Recall that the sum of every column of a transition matrix is 1, so $(1/n)$G is a joint distribution matrix in which the variable x' is white. Thus, by the white connection theorem, linking x and x' does not change the distribution on x, and so multiplies the i'th column of $(1/n)$G by $p(x)/p(x=y) = np(x)$. But this is just what you get when you multiply the initial vector by the transition matrix G.

QED.

Causal link chain: The link system derived from a Markov chain as above. When we don't need to carefully distinguish system from process, we'll usually call such a system a Markov chain too.

Causal transformation: A transformation T in which the probability distribution on the row index is white, i.e. all columns have the same sum. The transformations in a causal link chain are of course causal. A transition matrix as ordinarily defined is a causal transformation for which all elements are non-negative.

Theorem: The product TU of causal transformations T and U is causal.

Theorem: The link states of a causal link chain are of the form $|v><w|$, where w is white.

We have now mapped Markov chains onto link chains in such a way that their generators turn into causal transformations with the same matrices, except for a normalizing factor n. It was important to keep track of such *normalizers* in proving our basic theorems, but in the future it will make life easier to ignore them until it comes time to compute actual probabilities. This is almost always possible because, if we know any multiple of a probability distribution on x, we can divide by its sum to get the probabilities themselves. When it comes to dealing with negative probabilities, there may be no sensible normalizer at all. For instance, it can happen that even though all the case counts we need are well defined, the total case count is 0.

The passage of time in a Markov chain can be represented by a time-dependent transition matrix T(t), where $T(t+1) = T(t)$G. By the product theorem in Section 3.5, this same "transition law" (modulo a normalizer) is true in a causal link chain. The two concepts differ radically, however, when it comes to the meaning of the word 'state'. In a Markov chain, the states are what we have called the state vectors of the variables, which directly represent their probability distributions. These state vectors are indeed still well-defined in link theory, and have the same meaning. However, the word 'state' itself has been given a very new meaning, and now refers to the density matrix of a link. Since the state vector is the diagonal of the state matrix, probabilities are gotten by the rule $p(P) = tr(PS)$, as we saw in the last section.

In Markov chains, the law governing change of state, or as we'll call it, the *dynamical* rule, is $V' = T(V)$, where T is the time-dependent transition matrix, V the initial state, and V' the final state. Things are very different when we come to link states.

Link dynamical rule. Let T be a causal transformation. If the matrix of T has an inverse, then $S' = TST^{-1}$, where S is the state before T, S' the state after. If T does not have an inverse, then the most that can be said is that $S'T = TS$.

Proof:

Let x' and y be the variables of T, where the links to T are $x=x'$ and $y=y'$ for external variables x and y'. Now consider the join of the components other than T, and abstract from this the sub-process with variables x and y'. Cut the link $x=x'$. By the product theorem, the matrix S in the unlinked variables x and x' is UT. Similarly, the matrix S' in the unlinked variables y and y' is TU. We now have $S = UT$ and $S' = TU$. Thus $S'T = TUT = TS$. If T has an invertible matrix, then we can multiply on the right by T^{-1} giving the stronger rule $S' = TST^{-1}$

The classical dynamical rule is the offspring of Newtonian determinism, and is based on a certain concept of explanation, which equates our understanding of a phenomenon with our ability to reliably predict it. With the return of chance, reliability becomes a matter of degree. The best laws are now those that give us the best estimates of probability. Nevertheless, the aim of classical explanation remains the same, namely reducing information about the future to information about the past. Thus the form of an explanatory law is that of a function that yields a probability distribution on the future as a function of a probability distribution on the present.

Notice that all this changes with the link dynamical rule. S' is only a function of S if T is invertible. Otherwise, the dynamical rule only specifies a certain relationship between future and past, in which the two enter symmetrically (note that if T is invertible, we can also write $S = TS'T^{-1}$.) With this shift come new questions about the very meaning of dynamical explanation. The traditional scientist asks *how* the past dominates the future, but never *whether*. But link dynamics reveals many other kinds of relational structure besides domination. We are beginning to see hints of how link theory could enable us to remove our "causal colored glasses".

Differential Markov chain: One whose generator makes a very small change in the state.

It greatly simplifies many calculations to write the generator of a differential Markov chain in the form $G = 1+Hdt$, where 1 is the identity matrix. Ignoring second-order effects, we then have $G^2 = 1+2Hdt$ etc., so we can calculate T(t) by integration rather than by repeated matrix multiplication. In differential Markov chains, it's customary to speak of H rather than G as the generator of the process. Let $\psi(t)$ be the state vector of the chain at t, and define $d\psi(t)$ as $\psi(t+dt)-\psi(t)$; then for any t we have $d\psi = G(\psi) - \psi = H(\psi)dt$. Many people have remarked on the resemblance between the equation $d\psi = H(\psi)dt$ and Schroedinger's equation. We'll now see that this resemblance is no accident.

First, note that $1+Hdt$ is always invertible, its inverse being $1-Hdt$ (we are ignoring second-order terms, so $(1+Hdt)(1-Hdt) = 1-H^2dt^2 = 1$). Treating $1-Hdt$ as a generator would give us our original Markov chain reversed in time!

But doesn't this clash with our earlier observation that the time-reversal of a Markov chain isn't a Markov chain? No, because closer inspection reveals that $1-Hdt$ always contains negative "probabilities". Of course if we allow negative probabilities, then the answer is yes, and we have a useful tool for retrodicting state vectors. However, be careful here. Though the state evolution is reversed by the inverse process, the forward process itself, which includes the joint distribution on all variables, is not, since the inverse process always has negative joint probabilities.

Inverse Markov chain: A formal chain with a generator that is the inverse of a causal transformation.

If we encounter a time-reversed Markov chain, there are two things we can do in order to preserve the chain property: We can turn ourselves around in time, in effect putting the boundary condition on the future, or we can allow negative probabilities and work with the inverse chain, which means giving up studying the transitions and contenting ourselves with state vector dynamics. Suppose, though, that we encountered a chain C made up of two chains C1 and C2, the one forward and the other reversed. Then it would seem that we only have the second option. However, link theory gives us a third, which is to put a separate condition on both the past and future of C, where the initial condition is white for C1, and the final condition is white for C2. This is illustrated in Fig. 1.

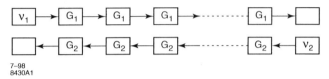

7–98
8430A1

Fig. 1. A Markov process in time can be generated either from a vector given when the process starts and propagate forward in time or from a vector given when the process ends and propagate backward in time (see text).

The "generator" G = G1&G2 of C is of course not a causal transformation, so C is not a Markov chain. But G persists throughout the time of the process and thus should qualify as a law. What is essentially new is that in order to define the process as satisfying this law we must now be given two boundary conditions, one initial and one final, and that without this final condition, the process has no law! This observation leads us to a basic generalization of the concept of Markov chain:

Chain: A link system given by a repeated transformation G called a generator, an initial component V and a final component W. This is illustrated in Fig. 2.

7–98
8430A2

Fig. 2. A chain with double boundary conditions.

As we'll now see, quantum systems result from opening up our formalism to allow for *both* negative probabilities *and* chains with double boundary conditions.

We've been focusing on transformations; let's now turn to states. As we saw above, the link states of a causal chain are of the form $|v><w|$, where w is white. We know from quantum physics that quantum density matrices are self-adjoint, which means that pure quantum states are of the form $|v><v|$. What kind of link chains have quantum states?

Suppose the initial link state of a quantum chain is $|v><v|$. By the dynamical rule, the second link state will then be $G|v><v|G|^{-1}$. For the second state to be quantum, i.e. to be of the form $|v'><v'|$, we must have $<v|G^{-1}=(G|v>)^{\dagger}=<v|G^{\dagger}$. (Recall that $T^{\dagger}$ is the adjoint of T, which

for real matrices is just T with rows and columns reversed.) If this equality is to be true for any initial state vector $|v>$, i.e. if we are to be able to separate the law of the process from its boundary conditions, then we must have $G^{-1} = G^{\dagger}$. But a transformation whose inverse is its adjoint is a unitary transformation.

Thus we see that the generator of a quantum chain is unitary. One question remains; Given the initial vector $|v>$, what must be the final state vector $<w|$? Let T be the product GGGG... of all the G's. Then $|w><w| = T|v><v|T^{-1}$, so $<w| = <v|T^{-1}$. We can write T^{-1} as $G^{-1}\,G^{-1}\,G^{-1}\,G^{-1}$... which leads to a way of diagramming quantum chains that is very useful in analyzing quantum measurements, as we'll see in the next section. First an important definition:

Sharp vector, state: A vector is called *sharp* if it contains a single non-zero entry. A quantum state matrix S is also called *sharp* if it contains a single non-zero entry.

Since S is self-adjoint, its non-zero entry must be in its diagonal, which means it is a pure state whose vector components are also sharp. This leads to the diagram of a quantum system which is prepared in a sharp state given in Fig. 3.

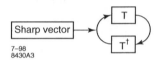

7–98
8430A3

Fig. 3. A sharp vector linked to a quantum system.

As with a differential Markov chain, we can write the equation of a quantum chain in the form d $\psi = H(\psi)dt$, where ψ is the state vector. But where is the i? you ask. We still haven't arrived at physics yet; what we have been studying is a more general mathematical structure known as *real* quantum mechanics. As mentioned, Mackey[5] showed how to get *complex* quantum mechanics as a special case of real quantum mechanics by introducing a (real) linear operator, which we'll call i, that commutes with every other operator. We'll take a slightly different route.

Instead of starting out with i as an operator, we'll start with a matrix representation of the complex numbers, where a+ib is represented as a 2x2 matrix C_{jk}, with $C_{11} = C_{22} = a$, and $C_{12} = -C_{21} = b$. One can quickly verify that these matrices behave like complex numbers under addition and multiplication. A complex "probability measure" is then simply an additive measure on a Boolean algebra having these matrices as its values. Let x be a variable on which there is such a complex probability distribution c(x). We can map this onto a real distribution on three variables x, j and k such that for any x, p(j,k) is the j,k'th entry of c(x). With this mapping, a complex probability acquires a definite meaning in terms of real probabilities: it is a joint distribution on two binary variables j and k satisfying p(j=1&k=1)= p(j=2&k=2) and p(j=2&k=1)= - p(j=1&k=2). A complex variable x is then a process in x, j and k such that, for any m, p(j,k|x=m) is a complex probability.

Suppose E and E' are independent events with complex probabilities c and c'. What, then, is the probability of E&E'? We would of course like it to be cc'. However, if c and c' are independent real 2x2 matrices, i.e. independent joint distributions on i,j and i',j', then combining E and E' produces a distribution p(j,k,j',k')=p(j,k)p(j',k') which does not even make sense as a complex probability. This presents us with a choice: We must either give up trying to reduce complex probabilities to real probabilities, or else give up our old notion of independence, putting the complex product rule c(E&E')=cc' in its place. Since quantum mechanics uses this complex product rule in defining the inner product of two quantum objects, we'll take the second course, and ask what it means when translated into real probability theory.

To multiply c and c' means to *contract* k and j', i.e., to link k to j' and then ignore k, or equivalently, since cc'= c'c, to contract j and k'. Thus to combine a complex process W with an independent complex process W' into a complex process W&W' we must always link a binary variable of one with a binary variable of the other. In quantum mechanics, an n-dimensional unitary transformation T can be thought of as operation on an object which is an entangled composite of an n-dimensional real object R and a two-dimensional real object J, where the entanglements satisfies the skew-symmetry rules above. The 2x2 pieces of T can be thought of as operators on J. Suppose T' operates on an independent R'&J'. To bring T and T' into the same universe of discourse, we must link j and j', thus multiplying the transformations on J times those on J'. But it only makes sense to multiply transformations if they operate on the same object, so we conclude that J=J'! In other words, there is a particular two-state particle that belongs to every quantum object.

As is well known, only the relative phase of quantum amplitude can be observed, the absolute phase being unobservable. This is equivalent to saying that the state of J is unobservable. Note that this means that "pure" quantum states, regarded as real states, are always mixtures with two-dimensional degeneracy. An interesting way to construct complex quantum mechanics is to start with real objects Ri, add J to each of them, and then confine ourselves of those states of Ri&J in which either J is independent of Ri (in the real sense), or is entangled in such a way that changing the state of J does not affect the probabilities in measurements of Ri. This allows us to make the following definition:

Complex quantum mechanics: Real quantum mechanics where everything intersects in a single quantum bit whose state doesn't matter.

Discussing why the real quantum mechanics constructed by Mackey[5] needs this additional symmetry would take us past the interface between mathematics and physics we have imposed on this paper into the domain of "the physical interpretation of quantum mechanics". For future reference we note that complex non-relativistic quantum mechanics allow time reversal invariance to be simply formulated, and in the context of relativistic quantum mechanics makes the formal definition of CPT invariance easier to express.

3.7 The Quantum Measurement Problem and its Solution

The measurement problem in link theory, the paradoxes of negative case counting, standard measurement theory, the projection postulate, the disturbance rule, the three axioms that define laboratory objects, functional objects, experiments, confined quantum objects, quantum time vs. laboratory time, prepared quantum chains, definite present and open future, the projection postulate in link theory, the collapse of the wavefront, the confined quantum object is a laboratory object, the white state, time loops revisited, macroscopic matter

The Born rule tells us that, for any proposition P and quantum variable x, the probability of a measurement of x satisfying P is tr(PS), where S is the link state of x. The dynamical rule tells how the state S changes with the passage of time. So doesn't that wrap it up? What more is there to be said about quantum observation? In the standard Hilbert space approach, there is the problem of how to construe our "classical" measuring instruments as quantum objects, which led von Neumann to his famous construction of the movable boundary between object and observer. But this problem doesn't arise in link theory, since quantum and classical can perfectly well coexist in the same joint distribution. Indeed, as we have seen, the very distinction between

526

classical and quantum rests in part on arbitrary decisions about how to analyze a process. So what, then, is left of the infamous quantum measurement problem? Have we in fact solved it?

Not quite. In fact, the measurement problem takes a particularly acute form in link theory, as we'll now see. Fig. 4a shows a single complete measurement of a variable x in a pure state $|v><v|$; we simply draw an arrow from x to a recorder, linking x to an input variable of the recorder which is assumed to be white. So far so good; there's no measurement problem yet.

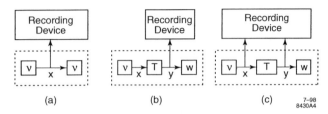

(a) (b) (c)

7–98
8430A4

Fig. 4. a) A single complete measurement.
b) A single complete measurement following a transition
c) Two complete measurements which cannot reproduce the results of quantum mechanics (see text).

If we wait until time t(y) to measure (Fig. 4b), then the state will have been transformed into $TST^\dagger$, and an arrow from y to the recorder will again give results in agreement with standard theory. Still no measurement problem.

But now let's consider two complete measurements, one before and one after transformation T (Fig. 4c). Each of these measurements will give the same probabilities that it did when it was the sole measurement. However, that's not what the standard theory predicts. What went wrong?

Here's one thing that went wrong. Consider a *selection* from among all the records of just those records for which the first measurement had the value k. Within this subset, every event in x will have probability 1 or 0, so x is independent of all other events, and v, which we'll assume is not sharp, is replaced by a sharp vector v'. The result will be in most cases that the state of y is no longer quantum. In other words, measuring the object throws it out of a quantum state.

But we're in worse trouble than that. In the system of Fig. 4c, it may well happen that some of the joint probabilities on x and y are negative, which means our recorder has recorded "negative" cases! So what's wrong with recording negative cases, you may ask ? Remember, recorded cases are not just alternatives, they are *instances*; they are actual things, marks on a tape, whatnot. Suppose your closet contains ten positive white shirts and ten negative green shirts. If you reach into your closet for a shirt, you will come out empty handed, since there are 0 shirts there. On the other hand, if you reach in for a white shirt, you'll have ten to choose from. That's quantum logic.

To see the problem of repeated measurement more sharply, consider a third measurement on z which comes after a second transformation $T^\dagger$ (remember, T is unitary, so $T^\dagger = T^{-1}$); this is shown in fig. 5.

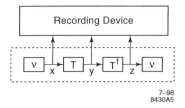

7–98
8430A5

Fig. 5. Illustration of the quantum measurement problem (see text).

As before, we'll run a long series of recorded experiments. Since $TT^\dagger = 1$, x and z are perfectly correlated in this process. Since the recorder is white and therefore doesn't effect the probabilities of x, y and z, this perfect correlation will show up in the total set of records, i.e. x and z are equal in every record. But now consider the subset S_k of all records for which y=k. Since y separates x and z, we know that x and z are conditionally independent given y, which means that they are independent within S_k. Thus the records in S_k will in most cases show x and z as having different values! S_k is like your white shirts in the closet—it's a non-null subset of the null set. But real shirts and real records just don't behave like that. Obviously we've got to take another tack.

Up to now in this paper we have managed to avoid the logical anomalies that result from negative case counting by carefully steering away from them, the trick here being to think only about the arithmetic of ratios. But this won't work forever. Sooner or later we have to face the paradox of the white shirts. This is the paradox of the two-slit experiment seen straight on. With both slits open, the case counts cancel at the null in the "wave interference pattern" where the detector lies, and no electrons are detected. But with one slit closed, the "white shirt" electrons alone are selected, and they make it through. We'll see that the solution of the measurement problem involves transforming this seeming paradox into a mathematical account of what it means for the future to be *open*.

Since the predictions of the standard theory of quantum measurement are in close agreement with experiment, we must represent measurement by link systems that make these same predictions. To see just what is required, here is a brief summary of the standard theory, which is based on an assumption that von Neumann called the *projection postulate*.

Projection postulate. Given a state S and a proposition (projection) P, there is an ideal measurement M to test for the truth of P that leaves those things that pass the test in the state PSP/p, and those things that fail in the state P'SP'/(1-p), where p = tr(PS) and P' is the negation 1-P of P.

It follows directly that M leaves the object in the state PSP+P'SP'. Let Q be any observable with eigenvalues $q_1, q_2, \dots q_n$, and let $P_1, P_2 \dots P_n$ be the "projectors" of these eigenvalues, i.e., the (ortho) projections whose ranges are the eigenspaces of $q_1, q_2 \dots q_n$. It is easy to show from the projection postulate that we can measure Q by performing a series of tests for, $P_1, P_2 \dots P_n$. The object will only pass one of these tests, P_i, which tells us that Q has the value q_i The ensemble of objects thus tested will be in the mixture $\Sigma_i P_i SP_i$, which gives us the general rule for the minimum "disturbance" of a state by a measurement.

Disturbance rule: Given an object in state S, the least disturbing measurement we can make of quantity Q is one which leaves the object in the state $\Sigma_i P_i SP_i$, where the P_i are the projectors of Q. The disturbance rule gives us a way to prepare a beam of objects in any state S, which is to start with a beam in any state and measure it in a way that yields the components of S, and then select out a subset of objects from each component to give the proper weight of that component in S.

Practically speaking, the standard theory of measurement works very well, at least for simple systems. But the projection postulate has caused a lot of trouble for people concerned with quantum foundations, since, by "collapsing the wavefront", it messes up what is otherwise a beautifully clean and simple unitary dynamics. Since link theory has no wavefront to collapse, it ought to be able to do better, and indeed it does; in fact, we count its success in this regard as its major accomplishment to date. We'll see in a while what the projection postulate means in terms of linking. But first we need a description in our current language of the kinds of things that go on in a physics laboratory.

The first thing to be said about our idealized and simplified laboratory is that there is a clock on the wall. Someone, or something, is always watching that clock and logging every event, so that at the end of every experiment there is a diary of the experimental procedures and the outcomes of measurement. The second thing to be said is that these experimental procedures can be repeated as often as necessary to obtain meaningful relative frequencies. Let's now characterize the *objects* in our lab.

Laboratory object: A process K in which we distinguish two kinds of *external* variables, called *input* and *output* variables. Each such variable exists for only one clock tick; for instance x could be the keyboard input to a computer at time t; we'll write the time at which x occurs as t(x). Input variables are those that we control, while output variables are those that we measure, or use to control inputs. K may have other variables besides its inputs and outputs, and it need not have inputs at all, though it must always have at least one output.

When we speak simply of *the input* to K, we'll mean all of its input variables together, or more exactly, the join of all these variables. Similarly for *the output* from K. The input (output) *before* or *after* time t_1 is the join of input (output) variables x_i such that $t_i(x_I) < t_1$, or $t(x_i) > t_1$.

We'll assume that laboratory objects satisfy three axioms.

1) Controllability: The input is white.

2) Causality: Given the input before t, the output at or before t is independent of the input at or after t (no influences backwards in time.)

3) Observability: The output never has negative probabilities.

These three axioms broadly capture our common-sense notion of processes that we can observe and manipulate. Notice that they have nothing whatsoever to do with classical mechanics. Bohr characterized the laboratory as the place where Newtonian mechanics reigns, but this was a mistake. Newton's laws neither imply nor are implied by the above axioms; indeed they belong to a wholly different universe of discourse (that's one reason why the young Russell, who admired physics more than common sense, was so eager to jettison causality.)

Functional object: One in which the output is a function of the input. The requirement of whiteness on the input imposes the condition on the joint distribution that all probabilities are either 0 or 1/n, where n is dimension (total number of cases.) It's convenient to ignore normalization and represent such a joint distribution by the bit array which is the characteristic function of the object's function regarded as a relation.

Experiment: A link system whose components are laboratory objects, one of which is a functional object known as a *recorder*. The output of the recorder is a variable at the end of the process whose value is a record of all inputs and outputs to the object on which the experiment is performed. It is assumed that every input variable in an experiment is linked to a prior output variable, and from this it follows that the process represented by the experiment is itself a laboratory object.

Let's now turn to the construction of quantum objects. We'll show these as enclosed in a dotted box, as in Fig. 4 and Fig. 5. Our aim now is to design dotted boxes that enclose their quantum weirdness, so-to-speak, i.e., boxes whose in-going and outgoing arrows satisfy our definition of a laboratory object. Such confined quantum objects are, properly speaking, the true objects of scrutiny in a quantum physics lab.

Notice that the dotted boxes in Fig. 4 are confined quantum objects; they satisfy 1) and 2) because they have no inputs, and 3) because their probabilities satisfy Born's rule and are therefore non-negative. The dotted boxes in Fig. 5 are not confined objects, however, since their outputs, i.e. the joint distributions on x, y and z, do have negative probabilities. Our challenge now is to redraw Fig. 5 as a confined object that gives the right numbers.

There's one thing doesn't look right about the box in Fig. 5, which is that to prepare it in state $|v><v|$ requires placing a boundary condition both before time t(x) and after time t(y). If we think of these conditions as inputs from the lab, the second condition violates 2), the causality principle. This is assuming that the time sequences inside and outside the box are the same. But there is another possibility, and a more likely one, which is that "time" in the quantum domain is somewhat autonomous from laboratory time, or to put it more abstractly, that the chains of quantum separability that we compared to Markov chains are in themselves not temporal sequences at all. Under this assumption, it is reasonable to define a *prepared* quantum chain as one whose beginning and end are both at the initial laboratory time of the experiment (Fig 6a).

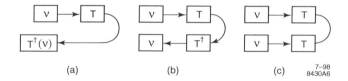

(a) (b) (c)

7–98
8430A6

Fig. 6. a) A prepared quantum chain.
b) A prepared quantum chain redrawn.
c) A prepared quantum chain redrawn again.

This construction looks more natural when we draw it with $T^\dagger$ factored out of w (Fig. 6b). Reversing rows and columns of $T^\dagger$ turns 6b into the equivalent form 6c. Let's then take this as our picture of a chain prepared in a pure state; notice that the states S and S' are identical. More generally:

Prepared quantum chain: A parallel pair of identical unitary chains whose initial vectors are the same and whose final variables are linked. (Fig. 7).

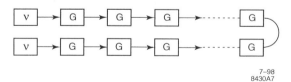

7–98
8430A7

Fig. 7. General diagram for a prepared quantum chain.

Suppose that when we measure x, the future measurement of y is indeterminate, in the sense that all outcomes are possible. This indeed is our experience of time, which we express by saying that there does not exist at time t(x) a record of the outcome of the measurement at t(y). Thus it makes no sense to speak at t(x) of the subset of records for which y has a certain value. But this implies that x and x' are linked at time t(x) by the matrix $TT^{\dagger} = 1$, which means that recording the value of x at t(x) also records the value of x' at t(x). Since we must sum over all values of y to evaluate probabilities at t(x), this avoids the paradox of Fig. 5.

Now what happens when we measure y? In our earlier example, this measurement unlinked x and z, making them in fact independent. However, in the present case, x' is already on record as having the same value as x, so it of course cannot become independent. How do we show this in our diagram? Simply by drawing a line between x and x' (Fig. 8a). Recall that a recording device is a laboratory object having (white) inputs whose output at the final time is a 1-1 function of the history of its inputs.

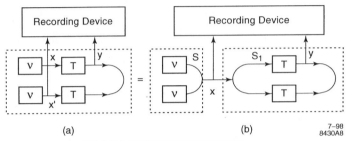

Fig. 8. a) A complete quantum measurement.
b) More detailed diagram for a complete quantum measurement.

Before y is measured, this line merely duplicates $TT^{\dagger}$, so drawing it would be redundant. But after y has acquired a definite value, this line in effect remembers the measurement of x', thus holding the fort against the disruption that would otherwise be produced by fixing y. Therefore there will be no records at all where x and x' are unequal, and the paradox of a non-null subset of the null set is avoided.

Choosing to measure x draws a line between x and x' as well as a line from x to the recorder. This is the link theory version of the projection postulate.

Probe x: Link x to x' and link x to the recorder (strictly speaking, this is a complete probe; we'll come to incomplete probes later.)

Projection postulate for a complete measurement: To make a complete measurement of the state of a prepared quantum chain at time t(x), insert a link between x and x' and measure the linked variable x, i.e. *probe x*.

The choice of whether or not to probe x can be thought of as a double-pole switch. Just what do we mean by the term 'switch'? By a *single-pole switch* we mean a three-variable component Sw[s,x,y] where s is a binary variable with values ON and OFF, Sw[ON,x,y] is the identity matrix, and Sw[OFF,x,y] is the white matrix in which every element equals 1/n, n being the dimension of x and y. Notice that for unlinked x and y, s is white, since the sum of matrix elements in the ON condition is the same as that in the OFF condition. By a *double-pole switch* we mean a pair of single-pole switches whose s's are linked.

The crucial question now arises whether this switch input satisfies axioms 1) **and** 2), i.e. whether it is white and causal. If so, then the dotted box qualifies as a laboratory object. To say that the switch is white means that the probability P of the process is unchanged by drawing the line. By the summation theorem, P is the sum of the joint distribution on x and x'. Drawing the line eliminates all off-diagonal terms in this sum. But since $TT^\dagger = 1$, these off-diagonal terms are all 0 anyway, so the line doesn't affect P. This takes care of 1); we'll come to 2) shortly.

Looking at Fig. 8b, we see that closing the switch breaks the diagram into two parts separated by x. The second part reduces to a circle, as far as x is concerned, so its presence at x is white. Thus the state S of the link from v to x is unchanged by the measurement of x; it is still $|v><v|$, and the probability distribution on x is still the diagonal of S, in accordance with Born's rule. But what is state S_1? Clearly it has the same diagonal as S. But, since $TT^\dagger=1$, its off-diagonal elements are all 0. S_1 is then the mixed state having sharp components weighted by the probability distribution on x. Thus we see that this change of state from S to S1 satisfies the standard theory of measurement, and is indeed nothing else but our old friend the collapse of the wavefront! Of course there is now no longer a wavefront to collapse, so we'll have to find another name for it. Whatever we call it, though, it's the crucial event that links the micro to the macro, and it is brought about by the *passive* act of recording the state of x. Our probe does not disturb what it sees; its effect comes from seeing alone.

That seeing alone is sufficient to collapse the wave function was a point made by HPN in the context of the double-slit experiment performed with low-velocity cannon balls, armor plate and mechanical detectors. [19] This point needs a little further discussion here in anticipation of both more abstract *and* more physical discussions of the measurement problem which are in preparation. "Seeing" the low-velocity cannon ball go through one slit or the other requires illumination, and reflection of at least one photon from the cannon ball itself, together with reflection of enough light from the two slits to distinguish which it went through. Hence there is actual energy transfer from the ball to the observer. This energy transfer has to be recorded before this particular case can be used to enter as a record in the compilation of single slit cases in contrast with double slit cases. The act is "passive" but not devoid of physical content and (pace von Neumann and Wigner) does *not* require a *conscious* observer to "collapse the wave function", as we hope the analysis made in terms of the abstract theory presented above has made clear. More generally, making the "quantum system" into a "laboratory object" requires conscious experimenters to set up the apparatus and to interpret the results after they are recorded, but this in no way distinguishes it from the protocols of classical experimentation where no question of quantum effects arises. When such processes are studied in enough detail to observe the progressive (and predicted) "randomization" of mesoscopic quantum states — as has recently been done [1] by preparing a Rydberg atom in a coherent superposition of a $n_{Bohr}=50$ and a $n_{Bohr}= 51$ state — the quantum *theoretical* treatment must include many more degrees of freedom than are usually included when discussing quantum measurement and the projection postulate. We believe that it might prove instructive to see how much of the analysis of such systems depends on how they are "carved up" between classical and quantum degrees of freedom and where considerations of experimental accuracy have to enter.

Closing the "record" switch does not change the observed probability distribution on x, and in this sense it does not disturb the measurement. However, it does affect the state of y, which would be $TST^\dagger$ with the switch open but is $T S_1 T^\dagger$ with the switch closed. From the point of view of the laboratory, closing the switch causes a change in the state of y. Inside the dotted box, though, there is nothing resembling cause and effect; indeed, nothing even happens inside the box, since closing the switch merely duplicates an existing link.

Measuring y is just like measuring x, the only difference being that S_1 is mixed (Fig. 8b). Thus we see that throwing the switch at y doesn't affect the measurement of x, or the whiteness of the x switch, so the switch inputs satisfy 2), the causality axiom. This shows that the dotted box is a classical object, and the sequence of measured variables is in fact a Markov process.

So far we've only been considering complete measurements. An incomplete measurement of x is represented by a functional box f(x) going from x to the recorder, where f is not 1-1. Recall that, in a complete measurement, connecting x to x' means that by recording x at t(x) we have also recorded x' at t(x). However, in the case of a partial measurement, it's not x but only f(x) that gets recorded at t(x), so this connection should go from f(x) to f'(x') rather than from x to x' (Fig. 9).

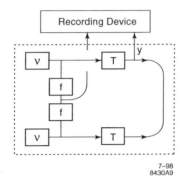

Fig. 9. Partial quantum measurement.

This arrangement satisfies the standard general rule S' $=\sum_k P_k SP_k$ for the change of state due to a partial measurement, where the P_k are the projectors of the measured observable.

What about preparation? As remarked, a preparation is always equivalent to a selection based on a measurement. But we can also represent a preparation as an input to an unprepared system, i.e. a system in the white state (fig. 10a); clearly such an input is always white.

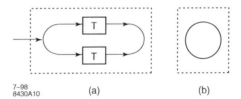

Fig. 10. a) An unprepared quantum system
b) An unprepared and unmeasured quantum object.

The white state, whose matrix is 1, has a unique role in quantum theory, since it is unchanged by measurement (i.e., if S=1, then $\sum_k P_k SP_k = \sum_k P_k = S$). The diagram of an unprepared and unmeasured quantum object is simply a circle (Fig. 10b); we can think of it as representing pure entropy. Perhaps this is the only quantum diagram that makes sense in isolation from the classical world.

Our conclusion that the "collapse of the wavefront" requires an open future seems to reintroduce an absolute arrow of time. But this is an illusion, since an unprepared system is symmetrical with respect to past and future, whether it is measured or not. Time's arrow comes not from quantum openness but from classical preparation, which we have conventionally placed at the beginning of the process, but which could just as well occur at the end, or at both the beginning and end, since it is equivalent to a selection of records.

The topic of measurement, especially partial measurement, deserves a much fuller treatment than we have given it here. However, even the present abbreviated account calls for a few more remarks concerning macroscopic matter:

In our analysis above, we have treated quantum and laboratory objects as essentially different kinds of things. And yet sooner or later we must face the fact that laboratory objects are made out of atoms, and try to render an account of just what this means. In standard quantum mechanics, when you assemble quantum objects you always get other quantum objects, and this indeed does create insoluble problems, since there's no way to "bend" quantum-ness into anything else. The present approach doesn't have that problem, of course, since it treats quantum-ness as a mode of analysis rather than an intrinsic property of things, and there is no reason why the appropriate mode of analysis should be the same at different scales. But we can hardly let things go at that, since there is a huge body of empirical knowledge about the relationship between micro and macro that needs to be translated into our present conceptual scheme.

In the last century, the laws of mechanics were thought to form a closed system that completely governs physical change, all other physical facts being relegated to the category of "accidents" of the world state. How, in such a scheme, is one to deal with the undoubtedly lawful character of thermodynamic events like heat flow? The nineteenth-century answer, which never quite satisfied anyone, was to impose a secondary layer of "laws among the accidents", so-to-speak, these being somehow based on the laws of probability. With the advent of quantum mechanics this idea becomes pretty far-fetched, since the equation governing quantum mechanical change has, except for the factor i, exactly the same form as the diffusion equation, which is one of the alleged laws among the accidents.

Let us hope that this classical two-layer concept of law is now well behind us. How, then, do we describe the relationship of micro to macro without it? Let's start at the level of process. Remember, in a process as such there are no states until you decide how to carve it up, only correlated variables, so it literally makes no sense to ask whether a process is quantum or causal. It does, however, make sense to ask whether or not its joint distribution contains negative probabilities.

Let's pretend for the moment that it doesn't. What we are then given is a huge number of micro-variables associated with individual atoms plus a much smaller number of macro variables associated with regions large enough to contain many atoms, the latter often closely corresponding to sums or averages of the micro variables. The macro-variables can be directly and repeatedly measured without disturbing them, and they seem to satisfy pretty well our axioms for laboratory objects, at least for simple inorganic objects. We implicitly assume that spatial boundaries inside the object correspond to separation boundaries in our present sense—one sometimes speaks of this assumption as locality. Applying our fundamental theorem makes it possible to define states at these boundaries, thus turning the macro-behavior of the object into a composition. This is essentially how an scientist carves something up when he wants to make a computer model of it, and there is no doubt that it works very well.

The reason it works is that macro-behavior has a good deal of autonomy from what we assume about the behavior of the unmeasured micro-variables, and for most purposes we can ignore quantum effects completely, at least under average terrestrial conditions. It might be supposed that this is simply a consequence of locality plus the complexity of the processes the macro-variables are averaging over, and indeed it would be, if the micro-probabilities were always positive.

However, with negative micro-probabilities the situation is quite different. Consider a chain whose generator is periodic, i.e., whose n'th power is the identity for some n. With only positive probabilities allowed, such a generator would have to be deterministic. But this is not true with negative probabilities, and physicists are well acquainted with periodic unitary transformations (indeed, any self-adjoint unitary transformation has period two.) Now a probabilistic chain of period n is of course separable at every variable, and would thus satisfy locality if it were strung through a material object. However, its first and last variables are perfectly correlated, just as if they were directly linked! From the point of view of common sense, this is non-locality with a vengeance.

Clearly we need a new assumption that keeps such non-local "virtual links" from playing a significant role in macro-behavior. Physicists already have a name for what is probably the right assumption: decoherence. Since this concept is defined in terms of wave functions, it needs to be reformulated in the language of probability. And it undoubtedly needs to be generalized to deal with matter at low temperatures, and perhaps with organic matter as well. Though all this could become technically complicated, it looks straightforward enough.

There is a much more serious problem which is not solved by decoherence, however, and that is to explain why negative probabilities remain confined to the micro-realm. What prevents them from "breaking out" into the everyday world and playing havoc with logic? We avoided this problem in the case of quantum measurements by assuming that making a measurement replaces certain "virtual links" due to quantum coherence by actual links. Something of the same sort must apply to the relationship between a macro-variable and the quantum average to which it corresponds. That is, a macro-variable, by being linked to the world at large, actualizes the virtual link on the quantum level between an average variable and its time-reversed dual. To put it more simply, macro-regions are self-measuring. Notice that measuring an average of many quantum micro-variables extracts only a minute fraction of the information in a pure joint state of these micro-variables, so these "macro-measurements" are essentially non-disturbing.

To summarize: If we assume decoherence plus the actuality of virtual quantum links from averages to their duals, macro-matter becomes classical.

But before we break out the champagne to celebrate the final triumph of logic and common sense, there are a few more questions. Why doesn't the time symmetry of the microscopic level carry over to the macro-level, at least to the extent of showing up as independent macroscopic boundary constraints on past and future? To what extent does our reasoning in terms of states bear on the process itself rather than on how we choose to carve it up? Might it be that hanging on to our causal framework, even for everyday objects, exacts a large price in complexity, thus blinding us to simple and essential forms in nature? And finally, why should we assume that we can always find ways to keep the irrationality of negative probability out of our experience?

3.8 Things to Come

This has been a long paper, but we fear it has only scratched the surface of its subject. We've concentrated here on the basics rather than on results. We believe this was the best course, since these basics take some getting used to. But in the long run, its results that win the prize. The history of science is littered with the remains of seductive theories that led nowhere, von Neumann's quantum logic being a recent case in point. The first place to look for results in link theory is obviously physics, about which we've said relatively little, so here, in conclusion, is a brief prospectus on several new items of physics that we see on the horizon.

First, some general remarks about space. So far, we've mostly been drawing diagrams rather than maps; in the dialogue between composition and extension, composition has had by far the louder voice. But the world doesn't present itself to raw experience as a neat arrangement of interchangeable parts—first we have to carve it up. And before we can carve it up, we have to map it, or "mark the carcass", as we put it earlier. The places where we mark are the places that separate the information in the extended structure, the boundaries through which correlation flows, and which, when rendered "immobile" by fixing the values of their variables, break the whole into independently variable parts. The question naturally arises as to how the structure of the probabilistic separating relation $p(A\&B\&C)p(B) = p(A\&B)p(B\&C)$ is related to more familiar geometric structures like metrics and topologies and linear orders.

Consider, for instance, a Markov process. We always start out by giving its variables in a certain order, but is this order contained in the probabilities or is it arbitrary? We saw that the arrow is arbitrary, so we are really talking about the between-ness relation. Let's define $B(x,y,z)$, read "y is between x and z", to mean $p(x,y,z)p(y) = p(x,y)p(y,z)$. We know that separation is equivalent to the Markov property, which holds for any x,y and z in the given order. Thus the question is whether $B(x,y,z)$ always excludes $B(y,x,z)$ and $B(x,z,y)$; let's call this the order property. If a process has the order property, then the linear order of its variables is indeed intrinsic to that process itself.

There are some obvious instances of Markov processes in which the order property fails. For instance, a sequence of independent variables is a Markov process, but since any one of the variables separates any two others, the variables can be rearranged in any order; intrinsically, they are a mere heap. At the opposite extreme is the case where the variables are perfectly correlated, i.e., where the transition matrices are permutation matrices. If you fix any variable in such a process, you fix all the others too. But, as we saw in Section 1, a fixed variable is independent of all others, so, again, any variable separates any other two. It may seem odd to think of Laplace's perfect Newtonian clockwork universe as a mere disordered heap of states, but there it is!

The true opposite of the independent variable case is the differential process given by a system where $G = 1 + Hdt$. If H is not 0 and there are no negative transition probabilities, such a process is always ordered. Hdt doesn't have to be truly infinitesimal to guarantee this, only rather small. If G is constant in time, the probabilities not only define an order but a metric, at least up to a scale constant. The local order of a continuous process is rather curious. Since it approaches the deterministic case in arbitrarily small neighborhoods, it is, so-to-speak, locally disordered. That is, it would require arbitrarily many repetitions of the process to statistically determine the order of points that are arbitrarily close together; this is rather suggestive of the time-energy uncertainty principle.

536

The same kind of reasoning can be applied to higher-dimensional continuous processes, whose probability distributions in general contain the intrinsic geometry of the space that is used to define them, or at least its topology. Keith Bowden[20] has suggested that if we apply suitable perturbing random fields to classical fields, we can recover their spaces from the resulting joint probability distributions. It would seem natural in a "probability space" for the local structure to be quite different from the global structure, which might have some bearing on super-string theory.

4 CONCLUSION

In this paper we have shown that by extending the probability formalism to include negative case counts and extending the formalism of Markov chains to the more general Markov processes which can have boundary conditions in both past and future we can construct a covering theory which describes both classical stochastic processes and the core laws of quantum mechanics (Born's rule and the Schroedinger equation) in the same mathematical framework. We show that by using care in defining a laboratory object, such objects have the familiar stochastic properties encountered in experimental physics but can contain "internal quantum structure" that reproduces the weird behavior required by "quantum mechanical measurement theory". The articulation of this philosophical and mathematical framework to include physics proper, and with it a "physical interpretation of quantum mechanics" lies beyond the horizon of this paper. Nevertheless, we claim with confidence that by "factoring out" the philosophical and mathematical problems associated with the conventional approach to quantum mechanics, we have taken an important step toward constructing a sensible physical theory of quantum phenomena. We also believe that our "covering law" approach to the problem could allow us to describe new phenomena which lie beyond the quantum horizon if such are encountered in a controlled, scientific environment.

References

[1] S. Haroche, "Entanglement, Decoherence and the Quantum/Classical Boundary" and references therein, *Physics Today*, **51** , July 1998, pp 36-42.

[2] H.P.Stapp, *Mind, Matter and Quantum Mechanics*, Springer-Verlag, Berlin, 1993·

[3] H.P.Noyes, "STAPP'S QUANTUM DUALISM: the James/Heisenberg Model of Consciousness", SLAC-PUB-6440, February 18, 1994 (also in *Proc. ANPA WEST 10*).

[4] G.Santayana, *Skepticism and Animal Faith* Dover, New York, 1923.

[5] G.W.Mackey, *Mathematical Foundations of Quantum Mechanics*, W. A. Benjamin, New York, 1968, pp 107-109.

[6] L.H.Kauffman and H.P.Noyes, *Physics Letters*, **A 218**, 139-146 (1996).

[7] R.P. Feynman and A.R. Hibbs, *Quantum Mechanics and Path Integrals,* McGraw Hill, New york, 1965, Problem 2-6, pp34-36.

[8] L.H.Kauffman and H.P.Noyes, *Proc. Roy. Soc.(London)*, **A 452**, 81-95 (1966).

[9] B.Russell, *A History of Western Philosophy* (Simon and Schuster, New York, 1945, pp 707-708).

[10] B.Russell, *Mysticism and Logic*, Anchor Doubleday, Garden City, New York, 1957, pp 174-202; originally published in Proceedings of the Aristotelian Society, 1912-1913.

[11] G.H. von Wright, *Causality and Determinism*, Columbia University Press, New York, 1974, pp 1-2.

[12] H.Hoffding, *History of Modern Philosophy*, Dover, New York, 1955.

[13] S.Hawking, as quoted in the New York Times, October 1, 1995: "...Stephen Hawking... Having ridiculed the concept [of time travel] for years, Hawking now says that it is not just a possibility but one on which the government should spend money." Hawking also gives a quick summary of the reasons for his U-turn in the preface to a new book called *The Physics of Star Trek* by astronomer Lawrence Krauss

[14] L.Krauss, *The Physics of Star Trek*, Harper, New York, 1998.

[15] M.Kaku, *Hyperspace*, Anchor, Garden City, New York, 1995

[16] T.Etter and H.P.Noyes, "Process, System, Causality and Quantum Mechanics..." SLAC-PUB-7890, a longer version of the current paper; quant-ph 9808011.

[17] Feller, W., [1950], *Probability Theory and its Application*, Wiley, New York.

[18] W.D.Blizard, "Negative Membership", *Notre Dame Journal of Formal Logic*, **31** (3), 346-368, 1990.

[19] H.P.Noyes, "An Operational Analysis of the Double Slit Experiment", in *Studies in the Foundations of Quantum Mechanics*, P. Suppes ed., Philosophy of Science Association, East Lansing, Michigan,1980, p 103.

[20] K.Bowden, private communication to TE, 1996.

Reflections on PSCQM
Tom Etter (2000)

Section 1. The significance of PSQM today

The first draft of PSCQM [1] is now more than four years old, and a good deal has happened since its was written. The present commentary will summarize some of these new developments. First, however, I would like to reflect a bit from the perspective of hindsight on the place of this paper within what I see as a larger ongoing transformation of the scientific worldview.

Prior to the twentieth century, no one but a few speculative philosophers questioned the primacy of space, time and matter as irreducibly fundamental concepts. But thanks to the uncertainty principle and the quantum "paradoxes" of amplitude cancellation, there are a growing numbers of scientists today who believe we should be looking for definitions of space, time and matter in terms of even more fundamental concepts. There is an analogy here to the way heat, temperature and entropy came to be defined in terms of energy and probability. Entropy is particularly relevant to our present topic, since its definition extends beyond physics into probability theory in general. What PSCQM has shown is there is a similar extension into probability theory of the basic quantum formalism of state, transformation and observable. In short, the "core" of quantum mechanics belongs to mathematics as well as to physics.

Let's look briefly at how entropy overflowed the bounds of physics. First of all, note that its definition in statistical mechanics as summation $-\sum p \ln p$ refers only to probability, not to space, time or matter, and thus makes sense when applied to any probability distribution whatsoever, physical or not. Entropy becomes a physical concept only when we add to its abstract definition a physical definition of p. In classical physics, p is simply decreed to be proportional to volume in phase space. In quantum mechanics, however, p is more integral to the "machinery" of mechanics itself, since it is built into the very idea of observing a mechanical system.

The best-known general fact about physical entropy is the second law of thermodynamics. This law doesn't belong to mechanics, which is symmetrical in past and future. We do, however, see it universally manifested in the everyday physical world as an aspect of the *large-number* behavior of things that move or otherwise behave independently. The statistical explanation of the second law has been formalized within the mathematics of so-called *Markov chains*, which are abstract "processes" having nothing per se to do with space, time or matter. It is just this broader scope of the abstract second law that makes it such a plausible explanation of the physical second law. By simply extrapolating Markovian behavior to atoms that obey the laws of mechanics, it turns out that we can not only explain the physical second law but many other observed connections among entropy, energy and heat.

Historically, entropy began as a physical concept, and its statistical nature was not fully recognized for many years. The central concepts of quantum mechanics today, namely state, transformation and observable, have a position in the mainstream of physics today similar to that of entropy in the early nineteenth century.

These quantum concepts are of course physical concepts. They are, however, subject to two basic laws that don't in themselves have anything to do with matter in motion, namely $D' = T^{-1}DT$ and $<Q> = trace(QD)$ where D and D' are successive states, T is a transformation, and Q is any observable. It is these two laws about the operators D, T and Q that constitute what in PSCQM we have called the *quantum core*. A useful way to characterize the quantum core is as that part of quantum physics needed for the logical design of quantum computers. The variables of a quantum computer, like those of an ordinary computer, are simply bits, and the design of a quantum computer as such has only to do with how these bits relate to each other and nothing to do with space or matter.

Might it be that the quantum core, like entropy and the second law, really belongs to a wider domain of nature than physics? This is what PSCQM set out to show, and what it did in fact show is the mathematical fact that *the laws of the quantum core, in a generalized form, are universal laws governing Markov processes*. There is a remarkable parallel here with the second law. There is also a curious complementarity. The second law is asymmetrical in past and future, and indeed is often said to define the arrow of time. The quantum core laws, on the other hand, are time symmetrical, and this in fact turns out to be their defining property in the Markovian setting. All this will be spelled out further below; suffice to say here that the general category of Markov processes allows for harmonious hybrids of second-law and quantum processes. These are set within a covering theory that permits a much larger range of Markovian structures, some of which even have two kinds of entropy, one increasing and the other decreasing with time.

In order for the abstract second law to become a physical law it had to be interpreted as a large-number law governing physical things. The same is of course true of the quantum core. For the second law, those physical things are atoms. But atoms won't do for the quantum core, since it is the precisely the quantum behavior of things like atoms we are trying to explain by large-number laws. Recall that I began by remarking that people are starting to question the primacy of space, time and matter, and to ask whether these should be replaced by more fundamental concepts. If we hope to turn the abstract quantum core into real physics, this is no longer an idle question.

I also began by saying that the larger picture is not just about the future of physics but about the future of science itself. The science you and I learned in school took space, time and matter for granted, as it did the temporal arrow defined by the mathematics of Markov chains. The primacy of space, time and matter, classically conceived, underlies not only the specific content of the sciences but the whole of scientific method. If it is abandoned, what will replace it? What kind of science will our descendents learn in the classrooms of 2100?

It's time to get more specific about Markov processes and Markov chains. Just what are these things? Unfortunately, the answer varies quite a bit from one author to the next. I have to confess that even our Markovian terminology in PSCQM is not entirely consistent, so I'll try to clean it up somewhat here. We'll start out with the definitions in Feller's classic textbook [2], and then introduce some further distinctions that he omits.

Feller defines what he calls a *general Markov process* as a succession of variables, continuous or discrete, on which there is a joint probability distribution satisfying what I'll call the *Markov separation law*. The usual informal statement of this law is that the past only influences the future via the present. This statement is misleading in one important respect, though, in that it appears to mathematically distinguish past from future, whereas what it actually asserts is *symmetrical* in past and future. A better, though mathematically equivalent, statement of the separation law is that the correlation of the past with the future depends only on the correlation of the past with the present and of the present with the future. Feller's qualification "general" is redundant, so I'll refer to his general Markov processes simply as Markov processes unless their generality needs to be emphasized. The definition of Markov process in PSCQM is in one respect more general than Feller's in that it allows for negative as well as positive probabilities, a point to which we'll return in the next section. This doesn't bear on what follows here, however, so we for now can continue to think of probabilities in the usual positive sense.

A *Markov chain,* on the other hand, is defined as the repeated application of a *transition matrix* of conditional probabilities to an initial probability distribution. More exactly, it is the joint probability distribution that results from this repeated application. It is easily shown that, in the absence of other conditions, this joint probability distribution is always a (general) Markov process. The converse is not true, however; not every Markov process is a Markov chain, a fact that is crucial to our present topic.

How do these two definitions differ? For one thing, chains are more specialized, as we just remarked. However, there is a more fundamental difference. A Markov *process* is a joint probability distribution given simply *as such*, whereas a Markov *chain* is a joint probability distribution given as a *composition* of simpler parts. To make an analogy to integers, the difference is like that between an integer as such and an integer given as a product of its prime factors. Or to switch the analogy to vectors, it's like that between a vector as such and a vector as a superposition of basis vectors. This distinction here is subtle, but extremely important.

There is another important distinction to be made, which is that between a Markov composition of *arbitrary* parts and a Markov composition whose parts are all *alike*. If the parts of the latter, except for the first, are transition matrices, it is a Markov chain. To pursue the analogy to numbers, the difference between a general Markov composition and a Markov chain is like that between a number, any number, factored into its primes, and a number that is given as a power of a single prime. Any Markov process can be factored into *some* collection of transition matrices, but only a Markov chain can be factored into a collection of transition matrices that are all *alike*. To put it less formally, the law of change in a general Markov process can be *variable* while that in a Markov chain is *fixed*.

A useful way to characterize a Markov process is to think of it as a computer with random inputs. Clearly such a computer is a Markov process, since, if we are given its state at time *t*, then the probability of its states at time *t* + 1 is independent of its history prior to *t*. If the computer is a *closed system*, i.e. if its random inputs are from a constant random source and it has no other inputs, then its transition probabilities are constant, i.e., it's a Markov chain. If on the other hand it is an *open system*, i.e. its transition probabilities depend on variable random sources or inputs from other systems, then it is a general Markov process. The same considerations apply to processes running inside a computer, which includes all real-time simulations.

Given the importance of computer modeling in today's science, it's hardly an exaggeration to say that, for most scientists, to explain something means to describe it in a way that could in principle be turned into a real-time computer simulation. This belief, which I'll call *computerism*, usually does not rise to the level of an explicit statement; it's just one of those things that "goes without saying". It's a funny thing about things that go without saying, though, which is that when you actually say them carefully, and then take a close look at what you have said, they sometimes turn out to be wrong!

Is computerism wrong? That's not something I'll take sides on here. However, I have observed that many people hold onto computerism simply because they can't imagine any other possibility. Here is where a proper understanding of Markov processes makes a big difference. It turns out that computers are only a tiny island in the vast sea of formal possibilities encompassed by the general concept of a Markov process. The quantum is another tiny island. As mentioned, there are also hybrid forms that belong to neither island. The important point is that by no stretch of imagination can the encompassing expanse of Markovian forms be located on Computer Island alone. Quantum structures can't be located there, even quantum computers can't be located there, and most of the remaining expanse isn't even in sight.

Which brings us to the future of science. Physical science grew up in close collaboration with engineering, and for the most part shares with engineering a view of the world as something to be taken apart into functional units. To this the engineer adds the art of reassembling functional units into useful functional wholes; this is called technology. The abstract skeleton of a functional part is a *transition matrix*, also sometimes called a *transfer function*, representing the functional dependence of a set of *outputs* on a set of *inputs*. In the deterministic or "causal" case, the actual values of the outputs are a function of the values of the inputs, while in the more general case it is only the probabilities of these values that are a function of the inputs. The *generality* of engineering consists in its being to able to use a small variety of functional parts and design principles to assemble a large variety of useful complex structures.

Here is where I see the broader significance of PSCQM. I believe its chief accomplishment was to mathematically extend the basic conception of lawful change that underlies current scientific practice. This extended lawfulness retains Markovian separability, but no longer requires that we separate things into functional parts. To put it another way, it no longer requires that the internal variables be inputs connected to outputs. The links between parts, and even between past and future, can now have a two-way information flow. This is easy to say, and it turns out to be rather easy to formulate mathematically, but it also turns out to be very hard to digest. Indeed, most of the work since PSCQM has involved trying to digest it. We have studied numerous examples, which provided numerous surprises, and a lot of work has gone into grounding the mathematics at a more fundamental level – we'll come to this in the next section.

Major changes in science are foreshadowed by movements in the culture at large. A variety of cultural movements in modern times, ranging from the counterculture of Woodstock to the arcane isms of Continental philosophy, share a strong discontent with the technocratic narrowness of science as it stands. The broad message here is that nature, including human nature, has many ways of *being* besides *using things*. A world that is nothing but *functionality* is a world fit only to be *used*. The world of the engineer is an abstraction geared to a particular mode of activity, not the world we live in.

But the world of the engineer is also an enormous intellectual achievement, and there is the problem. It is romantic folly to think that throwing away this achievement would return us to some imagined idyllic state of nature. I would like to think that PSQM offers a hint of a less foolish path. It clearly describes radical alternatives to functional composition that are none-the-less accessible to the engineer's mathematical tools. It also shows how these can simply explain some of the more puzzling laws of physics. This is certainly not The Answer, but it does offer hope that there may be ways to steer the intellectual power of science into a better partnership with our real human nature.

Section 2. Further developments: Link Theory

Not long after the original PSCQM was finished I discovered a simpler way to formalize its mathematics that has proved to be much easier to understand and use. The name "link theory", which came up a few times informally in PSCQM, has been transferred to this new formalization. Link theory was first presented at PhysComp '96, and soon after became part of the research agenda of the theory group headed by Dick Shoup at Interval Research, Inc. After the demise of Interval, I continued to work on some of its more abstract aspects at Hewlett-Packard in collaboration with Jim Bowery of the E-Speak project. More recently, I have rejoined Dick Shoup in continuing our Interval work at the newly formed Boundary Institute. Apart from the original idea of link theory, the developments I will now describe took place at one or the other of these three institutions.

PSCQM uses the standard language of probability theory: random variables, sample space, marginals etc. In terminology it follows Feller, though not always in notation. This kind of language was designed for quite different purposes than link theory, however, and is a bit awkward at best. Link theory, so far, has been entirely finitistic. Issues of convergence, continuity etc. don't arise, so the measure-theoretic generality of standard probability language is wasted and just muddies the waters. For present purposes, it's better to go back to the Eighteenth Century conception of probability as defined by case counts, as in calculating the odds in a game of chance.

By a *case* will be meant the assignment of a value to each of a set of variables. For instance, $\{ x = 2, y = 1, z = 3 \}$ would be a case of x, y and z. We say that several variables are *uncorrelated* or *independent* if all cases are allowed; to put it another way, if knowing the values of some of the variables gives no information about the values of the others.

Laplace defined *probability* as the number of *favorable* cases divided by the total number of cases. For instance, the probability of getting a four in rolling a pair of dice x and y is the number of cases where x and y add up to four, i.e. 3, divided by the total number of cases when x and y are independent, which is 36; thus it is 1 in 12. This of course assumes that the dice are unbiased. We can handle biased dice by imagining that there is a more numerous set of equally probable cases of which the observed cases are unequal fractions. This makes case counting adequate at least for the mathematics of combinatorial probability theory, and its intuitive transparency compared to measure theory is a great advantage.

It turns out to be very useful to represent the correlation among a set of variables by writing their allowed cases as the rows of a table whose columns are labeled by these variables. In logic, where there are only two values 0 and 1, these are known as *truth tables*. The truth table of an AND gate, for instance, has three columns labeled x, y and z, where x and y are the inputs and z the output, and four rows 000, 010, 100, and 111. The probability of a statement S about x, y, z etc. is the number of rows for which S is true divided by the total number of rows. If S is the statement $z = 1$ about the AND gate, for instance, we go down the table and pick out the single case where $z = 1$, showing that the probability of S is ¼.

In this example, S only involved the single variable z, which means we need only to look at the z column to count the cases "favorable" to S. More generally, when we are working with probabilities that only involve a subset of our variables, we can simply erase the columns headed by the others. The resulting table will, in general, contain duplicate rows. Instead of writing out all of these duplicates, we'll adopt the shorthand of writing each row only once, followed by its *count*. Such *count* tables are the basic notational building blocks of link theory in its new form. Here are the rows of the count table of our

AND gate if we are only considering the variables x and z: 00|2, 10|1, 11|1. Remember, this is just an abbreviation; the rows of the actual table with y hidden are 00, 00, 10, 11.

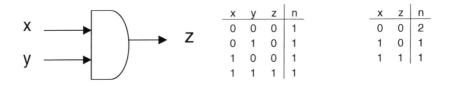

x	y	z	n
0	0	0	1
0	1	0	1
1	0	0	1
1	1	1	1

x	z	n
0	0	2
1	0	1
1	1	1

Fig 2.1. Count table of an AND gate $\{x, y, z\}$ and of its subtable $\{x, z\}$

The count column of a count table assigns to each row a case count proportional to its probability. Normalizing the count column by the total count gives the actual probability distribution on the variables of the chosen subtable. In practice it is usually better to work directly with case counts and to postpone normalization until our combinatoric calculations are finished. Thus the count table itself will take on the basic role of the sample space in standard probability theory.

We've considered independent variables; now let's turn to independent *subtables*. Consider a table T with four columns x, y, z, w. Assume there are no duplicate rows in T. Let T_1 be the subtable of T consisting of columns x and y, and T_2 be the subtable consisting of columns z and w. We say that T_1 and T_2 are *independent* in T if the rows of T are the pairs of *distinct* rows of T_1 and T_2. Notice that each distinct row of T_1 occurs in T as many times as there are distinct rows of T_2, and vice-versa. Here, for instance, is a truth table T in which T_1 and T_2 are independent NOT gates:

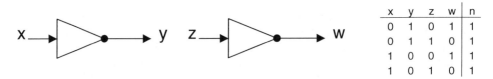

x	y	z	w	n
0	1	0	1	1
0	1	1	0	1
1	0	0	1	1
1	0	1	0	1

Fig. 2.2. Truth table of two independent NOT gates.

Recall that in Section 1 we distinguished between a joint probability distribution given simply as such, and a joint probability distribution given as a composition of simpler parts. In our present context, the counterpart of the former is a count table given simply as such, as remarked above. What, then, is the counterpart of the latter? How do we take apart a table?

For the simple table in Fig 2.2, the answer is obvious. The parts are two NOT gates, and to combine them into T, we take every combination of their rows. This operation is called *outer product*, and it applies to any two tables given separately. If the component tables are count tables, we multiply their separate row counts to get the row counts of *T*. It is easily shown that every table can be uniquely factored in this way into a product of "prime" tables.

In a Markov process, however, there are no independent parts, except in the most degenerate cases. How, then, do we take apart a Markov table?

Here is where link theory departs fundamentally from standard practice. Recall that the parts of a Markov chain, as defined by Feller, are *transition matrices* of conditional probabilities. We could follow standard practice in our table notation by introducing a new kind of table, call it a "transition table", whose rows are *conditional* cases. This would considerably complicate the logic of table analysis, however. There is a much better way, called *linking*.

The basic idea of linking is simplicity itself. Given a table *T* with variables *x* and *y*, to *link* *x* and *y* means to create a new table *T′* from *T* by requiring that *x* and *y* be equal. That is, *T′* is the table that results from discarding all of the rows of *T* in which the values of *x* and *y* disagree.

Let's create a table *T′* by linking the variables *y* and *z* of *T* in Fig. 2.2:

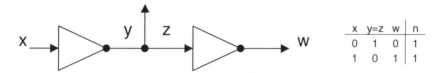

x	y=z	w	n
0	1	0	1
1	0	1	1

Fig 2.3. Two linked not gates

In effect, we are wiring the output of the first NOT gate to the input of the second. This turns the table of the outer two variables *x* and *w* into a straight "wire". Notice, incidentally, that we would get the same table *T′* by linking *x* and *w*. There is a lesson to be learned in passing here, which is that de-linking is not always unique. We'll encounter more subtle and important cases of this non-uniqueness in Section 3.

It's easy to extrapolate from this example to the general case of wiring up elementary logic gates into a logic diagram. A link is simply a wire from one gate terminal to another. Certain terminals are chosen as inputs, others as outputs. The rest are ignored; these are the so-called "hidden" variables. A computer is an initial state plus a repeated logic diagram, one copy for each clock tick, with some of the outputs from the copy at t linked to some of the inputs to the copy at $t+1$. Here is a link diagram of a very simple computer, whose logic diagram consists of a single AND gate:

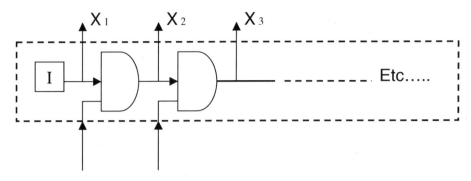

Fig 2.4. Simple open computer

Notice that we have "exposed" all of the variables in this diagram by running them outside of the dotted box. Since there are inputs from the external world, this computer must be thought of as an open system. As such it is a deterministic system. Its case table shows its inputs as independent variables and its outputs as functions of its inputs.

Had these inputs been left dangling inside the dotted box, they would be treated as hidden variables and thus erased from the case table:

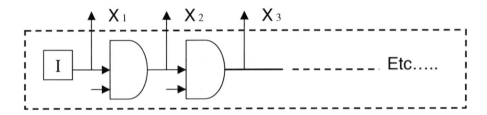

Fig. 2.5. Computer with "random inputs" as a Markov process

The outputs are no longer a function of the inputs, and are thus undetermined. The case table has well-defined case counts, however, so its outputs have a well-defined joint probability distribution. It's easy to show that this distribution is a Markov process (we must of course supply an initial state to the first upper input.)

We now come to a very important point. Remember, we carefully distinguished a Markov process from a Markov chain, which is a Markov process that has been "taken apart" into copies of a transition matrix. Fig. 2.5 shows a Markov process taken apart, not into copies of a transition matrix, but into copies of a partial case table. Be aware: this is a fundamentally different way to take things apart!

Let's go back to the analogy to numbers. What are the parts of a number? Its prime factors? That's one good answer. But an equally good answer is its decimal digits.

Neither of these is the *right* answer, but each can be right in the appropriate context. There is one important difference between the two, however, which is that the first only applies to *whole* numbers while the second also apply to fractions and even to irrational "quantities", to use the Greek term. Indeed, for the Greeks, irrational quantities were geometric rather than arithmetic entities; it was only after the invention of infinite decimal representations that it made sense to speak of irrational quantities as numbers.

As the reader may have guessed, the Markov process of Fig. 2.5 is a Markov chain. More accurately, we can reinterpret its AND gate with *y* hidden as a *transition matrix*. We can in fact do this for any case table representing a correlation between input and output variables. Indeed the very concepts of "input" and "output" rest on interpreting cases *conditionally*; we say of a NOT gate, for instance, that *if* the input is 0, *then* the output is 1, and *if* the input is 1 *then* the output is 0. Only certain kinds of case tables can be reinterpreted as transition matrices, however. Here is where linking shows its essentially greater generality, since there is no restriction on what kind of case tables can be "chained" by linking,

The result of either kind of chaining is a Markov process, and a *lawful* Markov process in the sense that the rule connecting adjacent states is constant. However, linked chains are a much larger class of lawful structures than Markov chains. Perhaps a better way to put it is that link analysis *extends* the concept of Markov chain, just as infinite decimal notation extends the concept of number. We don't yet have a word for this new kind of chain, so let me here coin one: *parachain*. This term has a precedent in the term *paracausality*, which I used in a 1977 paper for a somewhat more general concept.

Special kinds of parachains were studied as long ago as the 1960's, the earliest examples that I know of being presented in papers by the astrophysicist Helmudt Schmidt and myself [3]. It is only since link theory, however, that the concept has received a clear definition at an elementary level.

Do parachains encompass quantum processes? Not quite. The class of Markov chains, and indeed of Markov processes, must be extended a little further. To see what this involves, let's consider a real-world problem.

Imagine that killer bees have spread over a large part of the US and are continuing to spread. Where did they come from? We can trace them back to Texas, but we'd like to trace them back to the port or border town in Texas where they first made their unwelcome appearance. We have no empirical evidence about this early phase, so what we are confronted with is a problem in statistical retrodiction.

To tackle this problem, we first observe how the bees are continuing to spread and use this information to construct a *diffusion equation*, which is a certain kind of Markov chain. If this equation describes an unchanging law, we can use it to calculate the bee distribution at any time by simply running it forward or backward. Now running it forward means successively applying the Markov transition matrix to an initial state. But running it backward means successively applying the *inverse* of this transition matrix to a *final* state, assuming that it does have an inverse (most transition matrices do). The "logic" of this process is just like that of the forward process in that we can think of each entry in the inverse matrix as the proportion of bees starting in one place that end in another. There is one difference here, though, in that some of these "proportions" in the inverse matrix will be negative!

Just what is a negative swarm of bees? That's actually a deep question, to which we'll briefly return in Section 3. But then, just what is a negative pile of dollars? Suffice to note here that negative bees, like negative dollars, solve practical problems, and also enter our mathematics very easily and smoothly. The way we introduce them into link theory is by giving the their cases a negative sign; when we abbreviate duplicate rows to create a count table, those rows with negative signs subtract from the count. As we'll see, this allows both probabilities and probability amplitudes in quantum mechanics to be defined as ratios of case counts. This is spelled out in detail in PSCQM, where we also show, following Mackey, that complex amplitudes can be defined in a quantum mechanics with only real amplitudes by imposing a certain symmetry on its operators.

Section 3. Further developments: Relational logic and Markov cycles

At Interval we applied link theory to a variety of things ranging from puzzles and problems in AI to quantum computers. We also programmed a link calculator, which has since become indispensable in studying examples. Actually, we made several link calculators, the first of which was just a simple Microsoft Access macro. It turns out that the Link Theory operations of linking, hiding columns and creating count tables are in fact standard operations in relational databases, something I hadn't realized at first. It was this primitive Access calculator that first confirmed the validity of link diagrams as a way of representing quantum computers.

The simplicity of the Access calculator points to the deep roots of Link Theory in the logic of relations, a topic I have spent a good deal of time looking into over the past several years. Two long papers emerged from this work, each taking its departure from what Russell and Whitehead in *Principia Mathematica* called *Relation-Arithmetic*. The first, called *Structure Theory* [4], fixed a flaw in the *Principia* exposition that had prevented *Relation-Arithmetic* from dealing properly with relational composition. This made it possible to formulate Link Theory at a very abstract level. The second, called *Relation-Arithmetic Revived* [5], was written at Hewlett-Packard as part the theoretical work that Jim Bowery and I were doing on the design of transactional languages for the Internet. It went much further in developing another idea briefly introduced in *Structure Theory*, which is that the theory of relations, and indeed the whole of mathematics, can be formulated in a language whose only primitive predicates are *identity relations*. This new work appears to have good long-range prospects for putting link theory on a deeper logical foundation, but is outside the scope of the present paper.

What I want to present here in conclusion is a brief account of some surprising newly discovered facts about Markov processes that have emerged from our theoretical work at the Boundary Institute.

I pointed out in the last section that Markov chains are only a small fraction of the full range of orderly Markov processes, i.e., of Markov processes that have a constant dynamical law. I showed that one obtains a larger class of orderly Markov processes by construing their "dynamical parts" as linked two-variable probability distributions rather than as transition matrices. It turns out that there is another important way to generalize Markov chains, which is to allow independent boundary conditions on both their past and their *future*. It must be emphasized that, regarded as joint probability distributions, these new "parachains" are still Markov processes.

In standard Markov theory, where the matrices represent *conditional* probability distributions, it would make no sense to place a condition on the future, since that would put it on the wrong end of the *if-then* arrow. However, in a link representation there *is* no if-then arrow. There is indeed conditionality, but it operates in a very different way: the conditions on a link composition are its essentially timeless *links*.

Let me spell this out a bit. In probability theory, to *apply* a condition means to disallow certain cases. Translated into the language of link theory, this means discarding certain rows. When we construct a Markov process in link theory, we start out by forming the outer product of a collection of two-variable tables, which are the dynamical components, together with a one-variable table, which is the initial state. The result is a table with many independent parts and an exponentially larger number of rows. To create the *interdependence* that turns our table into a Markov process, we shorten this large list of rows by linking the successive parts, which means applying *conditions* of the form $Y = X$ that discard every row in which the "output" Y and "input" X are unequal.

As mentioned, a Markov *chain* in link theory is a link composition in which the (repeated) transition matrix is replaced by a (repeated) table. The relationship between this matrix and this table is very simple. Let M_{ij} be the matrix. The corresponding count table $T(i, j)$ is a table with columns i and j which has a row for every pair of values of i and j whose count is M_{ij} (this ignores normalization, but in link theory only the ratios of counts are significant.) This relationship is illustrated in Fig 3.1 for the simple two-column partial table Fig 2.1

The column indices x and z have been relabeled i and j, where i is the horizontal matrix index, j the vertical index. Note that we have added a second row with count 0 to make the table *complete*, i.e. so that it will have a row for every pair of values. Since there are only two values, this gives four rows. In the more general case, we get an n-by-n matrix, where n is the number of distinct values that occur in *either* column. To create the matrix, we simply copy the count of each row into the matrix position labeled by the i and j values of that row.

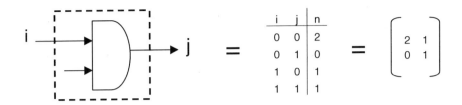

Fig. 3.1 Matrix M_{ij} of an AND gate with one input hidden

The matrix M of a Markov chain is a transition matrix, which means that it has the special property that its columns all have the same sum. A link table need not be similarly constrained, however, i.e., its corresponding *transformation* matrix need not be a transition matrix; this is why the Markov processes represented by a repeated link table are of a more general form than Markov chains.

It turns out that there is a very simple connection between linking and matrix multiplication [7]. When we link two tables and then hide their linked variables, the matrix of the resulting table is the product of the matrices of the two tables. This correspondence between linking and matrix multiplication is crucial for the application of link theory to standard science and probability theory. Here is a comparison of the two operations for a pair of NOT gates:

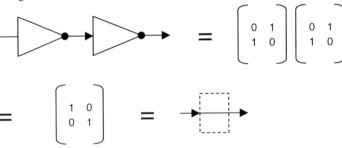

Fig 3.2 Multiplying the matrices of two NOT gates

The interchangeability of matrices and tables makes it possible to give meaning to the concept of a *final* boundary condition on a Markov chain. Here is how this works:

Start with a Markov chain given by an initial state vector I and a transition matrix T. Make I into a one-column table and T into a two-column table and replace products by links; this yields a link representation of the same Markov process. To apply a final condition given by a vector F, simply make F into a one-column table and link it to the last variable of the T series. Compositions created in this way also constitute an essentially larger class than Markov chains; they are also *parachains*, to use the term coined in

Section 2. Here is a simple example involving two AND gates. The boxes labeled 0 and 1 represent state vectors with sharp values 0 and 1. It's an interesting exercise to work out its count table and compare that to the count table of the example in fig. 2.5.

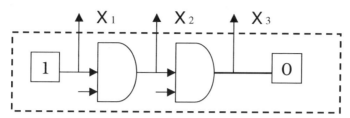

Fig.3.3 Simple finalistic parachain

Do such finalistic parachains exist in nature? Since "final causes" are generally rejected by today's science, the reader's first reaction is probably a simple no. However, leaving metaphysics aside, the empirical issues here are much more subtle than they might at first seem, as the following mathematical argument will show.

The time-evolution of a Newtonian system with n degrees of freedom is determined by an initial state that must, in general, be given by $2n$ parameters. In the most familiar cases, these n degrees of freedom are positions and orientations, the other parameters being their rates of change, or *velocities*. For *perfectly damped* systems, however, such as a capacitor discharging into a resistor, the velocity parameters are redundant. Indeed, a perfectly damped system can be *defined* as one for which knowing the rates of change of its degrees of freedom supplies no additional predictive information. Aristotelian physics, stripped of its metaphysics, is simply the physics of perfectly damped systems. As such, it persisted well into modern time as a respectable competitor to kinetic theory; Newton, for instance, proposed an Aristotelian explanation of Hooke's law [6].

In around 1730 the French priest Maupertuis brought final causes into physics with his principle of least action, which says that if we are given the values of the n degrees of freedom of a Newtonian system at the beginning and end of a certain period of time, we can calculate its time-evolution over that period by minimizing a certain integral. Notice the resemblance to our parachain example above, where an initial and final vector together with the law of the process gives the time-evolution of the probability vector. In the special case of an Aristotelian (perfectly damped) physical system, the final n parameters are of course redundant. Similarly, in the special case of an ordinary Markov chain represented as a parachain, its final boundary condition is redundant, and it can be shown that this final condition must always have the special form of a table whose counts are all equal (i.e. it must be a *white* vector; see PSCQM).

The resemblance between these two kinds of finality becomes even closer when we look at certain kinds of continuous parachains, most notably random walks for which we are given both the initial and final positions. It turns out that it is by minimizing a certain

552

integral with the dimensions of *information* that we get the expected trajectory of such a doubly conditioned random walk. This actually leads in the limit to the laws of Newtonian mechanics for the walker if we identify dispersion rate with mass [3], and it turns out that there are features of this situation suggestive of both quantum mechanics and relativity. That's another story, however. For the present, the following are the essential points:

Aristotelian physics is the special case. The general case is Newtonian physics.

The theory of Markov chains is the special case. The general case is the theory of Markov parachains.

In around 1750, final causes made a hasty and somewhat embarrassed retreat from physics when Lagrange showed that they could always be replaced by initial velocities. In essence, the reason for this is very simple. To calculate the trajectory of a general Newtonian system with n degrees of freedom we need $2n$ independent parameters, but these parameters can belong to the state of the system at any time so long as they are independent. In particular, we can have n position parameters at the beginning and n at the end. But we don't have to specify when that end comes! Even if the process ends after an infinitesimal time dt, our $2n$ parameters still remain independent and thus give us enough information to calculate the trajectory for all time. Thus, as Lagrange so famously remarked, we can start with initial positions and velocities and predict the exact course of events as far into the future as we wish. The principle of least action is completely equivalent to this formulation that puts both boundary conditions at the beginning.

What I have recently shown is that Lagrange's theorem on replacing final causes by initial velocities has a precise analogue in parachains. The mathematical details can be found on the Boundary Institute web site [7]; here I'll give a very informal sketch.

It turns out that in any Markov process there is a purely probabilistic analogue of an evolving "dynamical" state, which is simply the matrix of joint probabilities on a pair of adjacent variables X_t and X_{t+1}. We'll call this the *digram state* G_t at time t. In link theory language, G_t is the two-column table on X_t and X_{t+1} that results from hiding all other columns in the process table.

We'll come to the actual dynamical law for G_t shortly, but first I would like to reflect a bit on how this new result bears on the discussion of "computerism" in Section 2. Recall that computerism was defined as the belief that to explain something means to describe it in a way that could in principle be turned into a real-time computer simulation. In practice this means using the concept of a Markov chain as the paradigm of lawful process. Now a Markov chain is a perfectly damped system as defined above, since knowing how a state has changed tells us nothing more about the states to come. The "parameters" of a computer state are simply the computer's bits; knowing their "velocities", i.e. whether or not they have changed since the last tick, add no predictive information. Thus computerism, right or wrong, must be seen as a throwback to Aristotelian thinking.

Our history books usually tell us that Galileo and Newton refuted Aristotelian physics. That's an oversimplification. One can almost always cook up ad hoc Aristotelian gadgetry to explain the observed facts, as illustrated by Newton's explanation of Hooke's law. That's why Aristotelian physics lasted so long. The eventual triumph of Newtonian over Aristotelian physics was not so much the triumph of truth over falsehood as of conceptual simplicity and unity over overextended common sense. Today's computers have greatly increased our ability to deal with what we perceive to be complex situations without our having to find new ways of simplifying them. This is not necessarily a good thing. It's interesting to speculate what the history of physics might have been if present-day computers had existed in the sixteenth and seventeenth centuries.

The essence of the Newtonian revolution in physics was to bring velocity into the concept of state. Parachains bring velocity into the states of those systems studied by the information sciences. Therefore, when I say we should study parachains, I am not so much calling for a new revolution as enlisting in an old revolution that still has important unfinished business.

Let's now turn to the dynamics of digram states. To understand the digram dynamical rule we must introduce a new algebraic concept: the *circle product*. In link theory terms, the circle product $A \circ B$ is the *linked cycle* consisting of a pair of two-column tables A and B, as shown in Fig 3.4A and 3.4B.

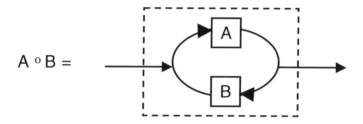

Fig 3.4A. Circle product

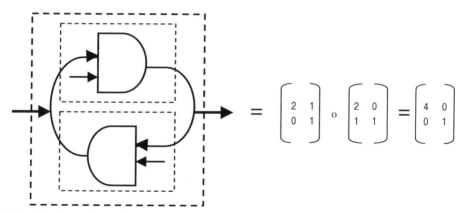

Fig 3.4B. Circle product of two AND gates

In working with the circle product it's important to keep track of the arrow directions – ingoing is column 1, outgoing is column 2 – since reversing arrows will usually create a different table. One consequence of this directionality is that $A \circ B$ is usually not equal to $B \circ A$. We'll write A^* for A with arrows reversed. Circle algebra is neither commutative nor associative; instead, we have the two rules: $B \circ A = A^* \circ B^*$ and $A \circ (B \circ C) = (A \circ B) \circ C^*$. Despite these peculiarities, it's a very useful tool for analyzing digram states and their relationship to link concepts such as the density matrix, as we'll soon see. One hint of this comes from the fact that, for unitary U, $U \circ U$ is the matrix of transition probabilities connecting a preparation to a measurement in quantum mechanics.

The example in 3.4 shows the circle product of the AND gate x-z subtable shown in Fig 2.1 times itself. Let's call this subtable A. In matrix form, the "circle square" $A \circ A$ of A is gotten by multiplying the elements of A by the corresponding elements of A^*, which is the transpose of A. This is the general rule for $A \circ B$ if A and B are square matrices of the same size.

A *Markov cycle* is defined as a joint probability distribution that can be given by a cycle of linked two-column tables. A Markov process can then be redefined as a Markov cycle in which at least one box is the *empty box E*, where E is the matrix whose elements are all 1's. Digram dynamics is most naturally formulated for Markov cycles; it can then be easily applied to Markov processes and Markov chains as special cases.

The empty box E has an important role in circle algebra, where it is the *right identity*, i.e. $A \circ E = A$. However, because of non-associativity it is not the left identity; rather we have $E \circ A = A^*$. Nevertheless, it can be used to define the *circle inverse* A^- of A as the table satisfying $A \circ A^- = A^- \circ A = E$. We'll also sometimes write the circle inverse as E / A, which makes certain formulas more understandable. The circle inverse of a matrix is gotten by first inverting all of its elements and then taking the transpose.

The basic formula of digram dynamics, which allows one to calculate successive digram states, can be derived as a theorem about three-box Markov cycles. In Fig 3.5 we can think of the cycle ABW as a process that separates from the world W at x, remains isolated at y and returns W at z. The digram state G_1 is the subtable on x and y that results from hiding z, while G_2 is the subtable on y and z that results from hiding x. G_2 is derived from G_1 by the formula $G_2 = B \circ (B^{-1}(A^- \circ G_1)A)$.

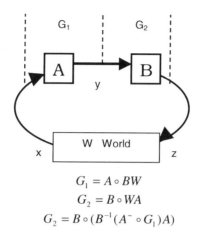

$$G_1 = A \circ BW$$
$$G_2 = B \circ WA$$
$$G_2 = B \circ (B^{-1}(A^{\sim} \circ G_1)A)$$

Fig 3.5. The digram transition formula

The most important thing to note about this formula is that it doesn't explicitly contain the world box W. It only refers to the two isolated boxes A and B and their digram states. This means it can be applied to any two consecutive boxes in an isolated series by simply dumping the remaining boxes into W:

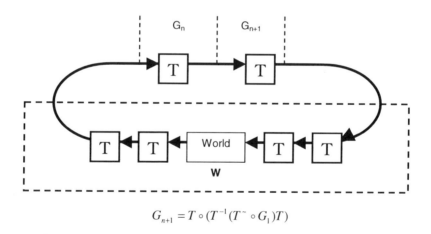

$$G_{n+1} = T \circ (T^{-1}(T^{\sim} \circ G_1)T)$$

Fig 3.6. Digram dynamics for a parachain

This formula, apart from being rather long, does not easily fit into standard mathematics because of its mixture of the two kinds of product. Worse yet, it does not gracefully go to the limit with G_n as an infinitesimal. This makes it awkward to apply to continuous processes, though using a finite G that "slides" along the time axis can in fact do the trick. Fortunately, there is better way, which is to convert to *density matrices*.

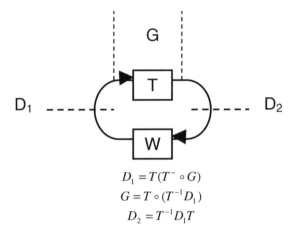

$$D_1 = T(T^{\sim} \circ G)$$
$$G = T \circ (T^{-1}D_1)$$
$$D_2 = T^{-1}D_1 T$$

Fig 3.7. Density matrix dynamics

A density matrix, as defined in PSCQM, is the matrix that results from removing a link in a link diagram and hiding all but the two unlinked variables. It is this definition that leads to the combinatorial interpretation of quantum mechanics. Rather surprisingly, it turns out that, given the transformation matrix T, the density matrix can be defined in terms of the digram matrix and vice versa; the conversion formulae are $D = T(T^{\sim} \circ G)$ and $G = T \circ (T^{-1}D)$. The density matrix dynamical law $D_2 = T^{-1}D_1 T$ is much more manageable and has a simple infinitesimal form, where T is written as $1 + A dt$. The great advantage of the digram law is that it refers to potentially observable quantities – G is an *actual* joint probability distribution that might well be observed as a statistical distribution – whereas D is a *counterfactual* distribution that can only be observed by breaking a connection. However, by using the above conversion formulas we can have the best of both worlds, using finite G's in the context of measurement and working with infinitesimal T's in the context of smooth change.

This has been a very condensed summary of digram dynamics; as mentioned, a more expanded version can be found in [7]. In conclusion, let me make one final point about Newton vs. Aristotle.

We first defined a Markov process in Section 1 as a process for which the past plus the present contains no more predictive information than the present alone. We have just been looking at examples for which this doesn't at face value seem to be true, since in a parachain, which is a Markov process, the digram state on X_{n-1} and X_n definitely has more predictive power than the probability distribution on X_n alone. The resolution of this seeming paradox lies in the fact that the transformation matrix is to some extent arbitrary.

557

Any Markov chain, even a parachain, can be analyzed as an initial state to which we apply a succession of transition matrices T_1, T_2, T_3 etc. Given all of these transition matrices, then the states prior to any time t indeed do not supply predictive information for the states after t. However, and here is the key point, in a parachain all of these transition matrices are different, whereas the digram transformation matrix T, which is not a transition matrix, remains fixed. The digram dynamical law is given by T alone, but we would need a second dynamical law governing the changing T_i if we insisted on using transition matrices to calculate the changing single-variable states. Neither way of describing the chain dynamics is "truer" than the other, but the way that keeps T constant is essentially simpler.

Galileo and Newton described a falling object as being subject to a constant force that produces a constant acceleration. To do so requires that velocity belong to its state. The Aristotelian account of a falling body, which also makes force into the agency of change, identifies the state with position alone. Since the velocity of the falling object changes, for this to work the force pushing it down must change too. There is no logical contradiction in such an account, and indeed Kepler at one point tried to calculate the "forces" that push the planets around in their orbits. But the Newtonian account, with its constant force that operates to change velocity, has the virtue of much greater simplicity and unity in this and a wide diversity of other situations. If we take the "force" of change in a Markov process to be the transformation matrix T, having that T remain constant is much simpler and cleaner than having to deal with a changing T_i, as we must if we ignore the "velocity" component of G. My biggest hope for the new revolution, or rather for the continuation of the Newtonian revolution, is that its way of analyzing statistical change can make the kind of difference in our understanding of the life world that the Newtonian way made in our understanding of matter in motion.

References.

1. *Process, System Causality and Quantum Mechanics* by Tom Etter and H. Pierre Noyes, Physics Essays Dec. 1999, and Chapter 15 in this book.
2. *An Introduction to Probability Theory and its Applications* by William Feller, John Wiley & Sons, 1950
3. *On the Occurrence of some Familiar Processes Reversed in Time* by Tom Etter, 1960, www.boundaryinstitute.org
4. *Structure Theory* by Tom Etter, 1999, www.boundaryinstitute.org
5. *Relation Arithmetic Revived* by Tom Etter, 1999, www.boundaryinstitute.org
6. *Statistical Physics and the Atomic Theory of Matter* by Stephen G. Brush, Princeton University Press, 1958, p.20
7. *Digram States in Markov Parachains* by Tom Etter, www.boundaryinstitute.org

BIT-STRING PHYSICS PREDICTION OF η, THE DARK MATTER/BARYON RATIO AND Ω_M *

H. Pierre Noyes

Stanford Linear Accelerator Center

Stanford University, Stanford, CA 94309

Abstract

Using a simple combinatorial algorithm for generating finite and discrete events as our numerical cosmology, we predict that the baryon/photon ratio at the time of nucleogenesis is $\eta = 1/256^4$, $\Omega_{DM}/\Omega_B = 12.7$ and (for a cosmological constant of $\Omega_\Lambda = 0.6 \pm 0.1$ predicted on general grounds by E.D.Jones) that $0.325 > \Omega_M > 0.183$. The limits are set not by our theory but by the empirical bounds on the renormalized Hubble constant of $0.6 < h_0 < 0.8$. If we impose the additional empirical bound of $t_0 < 14\ Gyr$, the predicted upper bound on Ω_M falls to 0.26. The predictions of Ω_M and Ω_Λ were in excellent agreement with Glanz' analysis in 1998, and are still in excellent agreement with Lineweaver's recent analysis despite the reduction of observational uncertainty by close to an order of magnitude.

Contributed paper presented at DM2000

Marina del Rey, California, February 23-25, 2000.

First Afternoon Session, Thursday, February 24

*Work supported by Department of Energy contract DE–AC03–76SF00515.

The theory on which I base my predictions is unconventional. Hence it is easier for me to show you first the consequences of the predictions in comparison with observation, in order to establish a presumption that the theory *might* be interesting, and then show you how these predictions came about.

The predictions are that (a) the ratio of baryons to photons was $\eta = 1/256^4 = 2.328... \times 10^{-10} = 10^{-10}\eta_{10}$ at the time of nucleogenesis, (b) $\Omega_{DM}/\Omega_B = 127/10 = 12.7$ and (c) $\Omega_\Lambda = 0.6$. Comparison of prediction (a) with observation is straightforward, as is illustrated in Figure 1.

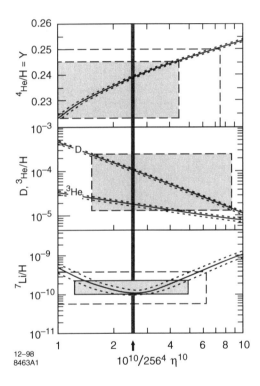

Figure 1: Comparison of the bit-string physics prediction that $\eta = 256^{-4}$ with accepted limits on the cosmic abundances as given by Olive and Schramm in [19], p. 119.

Comparison with observation of prediction (b) that the ratio of dark to baryonic matter is *not* straightforward, as was clear at DM98; I suspect that is matter will remain unresolved at this conference (DM2000). However, according to the standard cosmological model, the baryon-photon ratio remains fixed *after* nucleogenesis. In the theory I am relying on, the same is true of the of the dark matter to baryon ratio. Consequently, *if* we know the Hubble constant, *and* assume that only dark and baryonic matter contribute, the normalized matter parameter Ω_M can *also* be predicted, as we now demonstrate.

We know from the currently observed photon density (calculated from the observed 2.728 $^\circ K$ cosmic background radiation) that the normalized baryon density is given by [18]

$$\Omega_B = 3.67 \times 10^{-3} \eta_{10} h_0^{-2} \tag{1}$$

and hence, from our prediction and assumptions about dark matter, that the total mass density will be 13.7 times as large. Therefore we have that

$$\Omega_M = 0.11706 h_0^{-2} . \tag{2}$$

Hence, for $0.8 \geq h_0 \geq 0.6$ [7], Ω_M runs from 0.18291 to 0.32517. This clearly puts no restriction on Ω_Λ.

Our second constraint comes from integrating the scaled Friedman-Robertson-Walker (FRW) equations from a time after the expansion becomes matter dominated with no pressure to the present. Here we assume that this initial time is close enough to zero on the time scale of the integration so that the lower limit of integration can be approximated by zero [21]. Then the age of the universe as a function of the current values of Ω_M and Ω_Λ is given by

$$\begin{aligned}
t_0 &= 9.77813 h_0^{-1} f(\Omega_M, \Omega_\Lambda) \; Gyr \\
&= 9.77813 h_0^{-1} f(0.11706 h_0^{-2}, \Omega_\Lambda) \; Gyr
\end{aligned} \tag{3}$$

where

$$f(\Omega_M, \Omega_\Lambda) = \int_0^1 dx \sqrt{\frac{x}{\Omega_M + (1 - \Omega_M - \Omega_\Lambda)x + \Omega_\Lambda x^3}} . \tag{4}$$

For the two limiting values of h_0, we see that

$$\begin{aligned}
h_0 &= 0.8, \quad t_0 = 12.223 f(0.18291, \Omega_\Lambda) \; Gyr \\
h_0 &= 0.6, \quad t_0 = 16.297 f(0.32517, \Omega_\Lambda) \; Gyr .
\end{aligned} \tag{5}$$

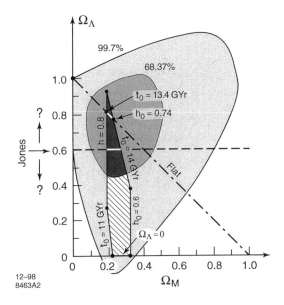

Figure 2: Limits on $(\Omega_M, \Omega_\Lambda)$ set by combining the Supernovae Type Ia data from Perlmutter, et al. with the Cosmic Ray Background Experiment (COBE) satellite data as quoted by Glanz [4] (dotted curves at the 68.37% and 99.7% confidence levels) compared with the predictions of bit-string physics that $\eta_{10} = 10^{10}/256^4$ (cf. Fig. 1) and $\Omega_{\text{Dark}}/\Omega_B = 12.7$. We accept the constraints on the scaled Hubble constant $h_0 = 0.7 \pm 0.1$ [6] and on the age of the universe $t_0 = 12.5 \pm 1.5 \ Gyr$ (solid lines). We include the predicted constraint $\Omega_\Lambda > 0$). The Jones estimate of $\Omega_\Lambda = 0.6$ is indicated, but the uncertainty was not available in 1998.

The results are plotted in Figure 2. We emphasize that these predictions were made and published over a decade ago when the observational data were vague and the theoretical climate of opinion was very different from what it is now. The figure just given was presented at ANPA20 (Sept. 3-8, 1998, Cambridge, England) and given wider circulation in[14]. The calculation (c) that $\Omega_\Lambda = 0.6$ was made by Jones before there was any observational evidence for a cosmological constant, let alone a positive one[3]. The precision of the relevant observational limits has improved

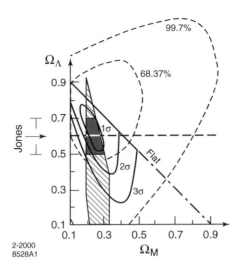

2-2000
8528A1

Figure 3: Limits on $(\Omega_M, \Omega_\Lambda)$ set by combining the Supernovae Type Ia data from Perlmutter, et al. with the Cosmic Ray Background Experiment (COBE) satellite data as quoted by Glanz [4] (dotted curves at the 68.37% and 99.7% confidence levels) and more recent much improved limits according to Lineweaver[8] (solid curves at $1\sigma, 2\sigma, 3\sigma$) compared with the predictions of bit-string physics that $\eta_{10} = 10^{10}/256^4$ (cf. Fig. 1) and $\Omega_{DM}/\Omega_B = 12.7$. We accept the constraints on the scaled Hubble constant $h_0 = 0.7 \pm 0.1$ [6] and on the age of the universe $t_0 = 12.5 \pm 1.5 \; Gyr$ (solid lines). The Jones estimate of $\Omega_\Lambda = 0.6 \pm 0.1$ is included.

considerably since DM98. A recent analysis of this new data suitable for our purposes has been made by Lineweaver[8]. His one, two and three σ contours are plotted in comparison with the previous observational limits and our (unchanged) earlier predictions in Figure 3. Note how dramatically the regions of uncertainty have shrunken in two years. It is gratifying that our *prior* predictions are still close to the center of the allowed region, indicating that it will take a lot more work to show that they are wrong!

The theory I am using has a long history[12], starting with the discovery of the *combinatorial hierarchy* in 1961 [20] and the first publication of the work on this idea by Amson, Bastin, Kilmister and Parker-Rhodes in 1966[1]. The theory is unusual in that it starts from minimal assumptions about what is needed for a physical theory and tries to let the structure of the theory grow out of them. My own preferred choice of basic assumptions are that a physicist must (a) be able to tell something from nothing, (b) be able to tell whether things are the same or different, and (c) must assume a basic arbitrariness in the universe which underlies the stochastic effects exhibited by quantum events. I further assume that we should use the simplest possible mathematical structures to model and develop these concepts. (a) is simply modeled by bit multiplication; (b) is simply modeled by bit addition (addition modulo 2, XOR, symmetric difference,...) or, as it is referred to in the ANPA program, *discrimination*.

The third requirement, together with the usual scientific assumption that we can keep *historical records* and examine them at later times, is accomplished by constructing a computer model called *program universe*[9, 16, 13] which yields a growing universe of ordered strings of the integers "0" and "1". Here we remind the reader of how we use *discrimination* ("$\oplus$") between ordered strings of zeros and ones (*bitstrings*) defined by

$$(\mathbf{a}(W) \oplus \mathbf{b}(W))_w = (a_w - b_w)^2; \quad a_w, b_w \in 0, 1; \quad w \in 1, 2, ..., W \qquad (6)$$

to generate a growing universe of bit-strings which at each step contains $P(S)$ strings of length S. The algorithm is very simple, as can be seen from the flow diagram in Fig. 4. We start with a rectangular block of rows and columns containing only the bits "0" and "1". We then pick two rows arbitrarily and if their discriminant is non-null, adjoin it to the table as a new row. If it is null, we simply adjoin an arbitrary column (Bernoulli sequence) to the table and recurse to picking two arbitrary rows. That this model contains arbitrary elements and (if interpretable in terms of known aspects of the practice of physics) an historical record (ordered by the number of TICK's, or equivalently by the row length) should be clear from the outset. The forging of rules that will indeed connect the model to the *actual* practice of physics is the primary problem that has engaged me ever since the model was created.

Program universe provides a separation into a conserved set of "labels", and a

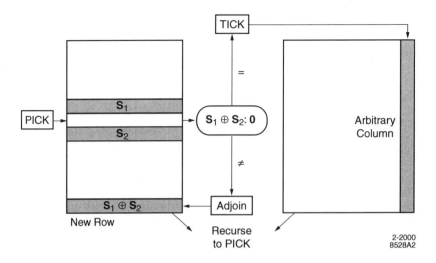

Figure 4: Flow diagram for constructing a bit-string universe growing by one row or one column at a time (see text).

growing set of "contents" which can be thought of as the space-time "addresses" to which these labels refer. To see this, think of all the left-hand, finite length S portions of the strings which exist when the program TICKs and the string-length goes from S to $S+1$. Call these *labels* of length $L = S$, and the number of them at the critical tick $N_0(L)$. Further PICKs and TICKs can only add to this set of labels those which can be produced from it by pairwise discrimination, with no impact from the (growing in length and number) set of content labels with length $S_C = S - L > 0$. If $N_I \leq N_0(S_L)$ of these labels are *discriminately independent*, then the maximum number of distinct labels they can generate, no matter how long program universe runs, will be $2^{N_I} - 1$, because this is the maximum number of ways we can choose combinations of N_I distinct things taking them $1, 2, ..., N_I$ times. We will interpret this fixed number of possibilities as a representation of the quantum numbers of systems of "elementary particles" allowed in our bit-string universe and use the growing content-strings to represent their (finite and discrete) locations in an expanding space-time description of the universe.

This label-content schema then allows us to interpret the events which lead to TICK as four-leg Feynman diagrams representing a stationary state scattering process. Note that for us to find out that the two strings found by PICK are the same, we must either pick the same string twice or at some previous step have produced (by discrimination) and adjoined the string which is now the same as the second one picked. Although it is not discussed in bit-string language, a little thought about the solution of a relativistic three body scattering problem Ed Jones and I have found [15] shows that the driving term ($^{>\atop-}{}^{<}_{-}$) is always a four-leg Feynman diagram ($> - <$) plus a spectator ($-$) whose quantum numbers are *identical* with the quantum numbers of the particle in the intermediate state connecting the two vertices. The step we do not take here is to show that the labels do indeed represent quantum number conservation and the contents a finite and discrete version of relativistic energy-momentum conservation. But we hope that enough has been said to show how we could interpret program universe as representing a sequence of contemporaneous scattering processes, and an algorithm which tells us how the space in which they occur expands.

Short-circuiting and reordering the actual route by which my current interpretation of this model was arrived at, we note that the two basic operations in the model which provide locally novel bit-strings (Adjoin and TICK) are isomorphic, respectively, to a three-leg or a four-leg Feynman diagram. This is illustrated in Fig. 5. Note that the internal (exchanged particle) state in the Feynman diagram is *necessarily* accompanied by an identical (but distinct) "spectator" somewhere else in the (coherent) memory.

We do not have space here to explain how, in the more detailed dynamical interpretation, the three-leg diagrams conserve (relativistic) 3-momentum but not necessarily energy (like vacuum fluctuations) while the four-leg diagrams conserve both 3-momentum and energy and hence are candidates for potentially observable events. We are particularly pleased that the observable events created by Program Universe necessarily provide *two* locally identical but distinct strings (states) because these are the starting point for a relativistic finite particle number quantum scattering theory which has non-trivial solutions[15]. But we *do* need to explain how this interpretation of program universe does connect up with the work on the combinatorial hierarchy.

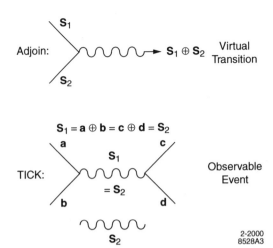

Figure 5: Interpretation of the Adjoin and TICK operations of Program Universe as Feynman diagrams(see text).

At this point we need a guiding principle to show us how we can "chunk" the growing information content provided by the discriminate closure of the label portion of the strings in such a way as to generate a hierarchical representation of the quantum numbers that these labels represent. Following a suggestion of David McGoveran's [10], we note that *we can guarantee that the representation has a coordinate basis and supports linear operators by mapping it to square matrices.*

The mapping scheme originally used by Amson, Bastin, Kilmister and Parker-Rhodes [1] satisfies this requirement. This scheme requires us to introduce the multiplication operation $(0 \cdot 0 = 0 = 0 \cdot 1 = 0 = 1 \cdot 0, \, 1 \cdot 1 = 1)$, converting our bit-string formalism into the *field* Z_2. First note, as mentioned above, that any set of n discriminately independent (*d.i.*) strings will generate exactly $2^n - 1$ discriminately closed subsets (*dcss*). Start with two d.i. strings $\mathbf{a}$, $\mathbf{b}$. These generate three d.i. subsets, namely $\{\mathbf{a}\}, \{\mathbf{b}\}, \{\mathbf{a}, \mathbf{b}, \mathbf{a} \oplus \mathbf{b}\}$. Require each dcss $(\{ \quad \})$ to contain only the eigenvector(s), of three 2×2 *mapping matrices* which (1) are non-singular (do not map onto zero) and (2) are d.i. Rearrange these as strings. They will then generate seven dcss. Map these by seven d.i. 4×4 matrices, which meet the same criteria (1) and (2) just given. Rearrange these as seven d.i. strings of length 16. These generate $127 = 2^7 - 1$ dcss. These can be mapped by 127 16×16 d.i. mapping matrices, which, rearranged as strings of length 256, generate $2^{127} - 1 \approx 1.7 \times 10^{38}$ dcss. But these cannot be mapped by 256×256 d.i. matrices because there are at most 256^2 such matrices and $256^2 \ll 2^{127} - 1$. Thus this *combinatorial hierarchy* terminates at the fourth level. The mapping matrices are not unique, but exist, as has been proved by direct construction and an abstract proof [2]. It is easy to see that the four level hierarchy constructed by these rules is *unique* because starting with d.i. strings of length 3 or 4 generates only two levels and the dcss generated by d.i. strings of length 5 or greater cannot be mapped.

Making physical sense out of these numbers is a long story [12], and making the case that they give us the quantum numbers of the standard model of quarks and leptons with exactly 3 generations has only been sketched [11]. However we do not require the completely worked out scheme to make interesting cosmological predictions. The ratio of dark to "visible" (i.e. electromagnetically interacting) matter is the easiest to see. The electromagnetic interaction first comes in when we have constructed the first three levels giving 3+7+127 =137 dcss, one of which is identified with electromagnetic interactions because it occurs with probability $1/137 \approx e^2/\hbar c$. But the construction must first complete the first two levels giving 3+7=10 dcss. Since the construction is "random" and this will happen many, many times as program universe grinds along, we will get the 10 non-electromagnetically interacting labels 127/10 times as often as we get the electromagnetically interacting labels. Our prediction of $M_{DM}/M_B = 12.7$ is that naive.

The $1/256^4$ prediction for N_B/N_γ is comparably naive. Our partially worked out scheme of relating bit-string events to particle physics [11, 12], makes it clear that photons, both as labels (which communicate with particle-antiparticle pairs) and as content strings will contain equal numbers of zeros and ones in appropriately specified portions of the strings. Consequently they can be readily identified as the most probable entities in any assemblage of strings generated by whatever pseudo-random

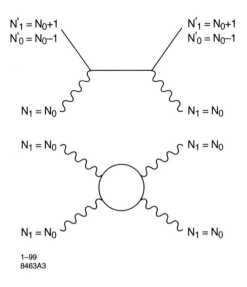

Figure 6: Comparison of bit-string labeled processes after the label length is fixed at 256 interpreted as baryon ($N_1' = N_0 + 1$) photon ($N_1 = N_0$) and photon-photon scattering. Here N_1 and N_0 symbolize, respectively, the number of ones and zeros in the label part of the string (which is of length 256). Program universe guarantees that, in the absence of further considerations, the content part of the strings will have an equal number of zeros and ones with very high probability as the string length (universe) grows.

number generator is used to construct the arbitrary actions and bit-strings needed in actually running program universe. This scheme also makes the simplest representation of fermions and anti-fermions contain one more "1" and one less "0" than the photons (or *visa versa*). (Which we call "fermions" and which "anti-fermions" is, to begin with, an arbitrary choice of nomenclature.) Since our dynamics insures conventional quantum number conservation by construction, the problem — as in conventional theories—is to show how program universe introduces a bias between "0" 's and "1" 's once the full interaction scheme is developed.

Since program universe has to start out with two strings, and both of these cannot be null if the evolution is lead anywhere, the first significant PICK and discrimination will necessarily lead to a universe with three strings, two of which are "1" and one of which is "0". Subsequent PICKs and TICKs are sufficiently "random" to insure that (at least statistically) there will be an equal number of zeros and ones, apart from the initial bias giving an extra one. Once the label length of 256 is reached, and sufficient space-time structure ("content strings") generated and interacted to achieve thermal equilibrium, this label bias for a 1 compared to equal numbers of zeros and ones will persist for 1 in 256 labels. But to count the equilibrium processes relevant to computing the ratio of baryons to photons, we must compare the labels leading to baryon-photon scattering compared to those leading to photon-photon scattering. This requires the baryon bias of 1 to appear in one and only one of the four initial (or final, since the diagrams are time symmetric) state labels of length 256 involved in that comparison; the two relevant diagrams are illustrated in Fig.6, which assumes that the above mentioned interpretation of the strings causing observable TICK's as four leg Feynman diagrams has been satisfactorily demonstrated. As a trivial example take the baryon-antibaryon-photon vertex to be $\mathbf{B} \oplus \bar{\mathbf{B}} \oplus \gamma = \mathbf{0}$ with $\mathbf{B} = (1110)$, $\bar{\mathbf{B}} = (0010)$ and $\gamma = (1100)$. We conclude that, in the absence of further information, $1/256^4$ is the program universe prediction for the baryon-photon ratio at the time of big bang nucleosynthesis.

Since Jones' paper [3] is still in preparation, I am at liberty here only to quote:

> From general operational arguments, Ed Jones has shown how to start from $\sim N$ Moncktons and self-generate a universe with $\sim N'$ baryons which—for appropriate choice of N—resembles our currently observed universe. In particular it must necessarily have a positive cosmological constant characterized by $\Omega_\Lambda \sim 0.6 \pm 0.1$.

We note further that Jones' general arguments a) are completely compatible with *program universe* and b) do not in themselves fix the value of N. Further, the estimate given above, which was made before and independent of the calculations reported in the last section, fell squarely in the middle of the region allowed in 1998 (see Fig. 2), and continues to do so despite the remarkable progress that has been made since

DM98 (see Fig.3). Clearly, pursuing the combination of these two lines of reasoning could prove to be very exciting.

References

[1] T. Bastin, "On the Scale Constants of Physics", *Studia Philosophica Gandensia*, **4**, 77 (1966).

[2] T. Bastin, H. P. Noyes, J. Amson and C. W. Kilmister, *Int'l. J. Theor. Phys.*, **18**, 445-488 (1979).

[3] E. D. Jones, private communications to HPN starting in 1997; this fact, and a seminar on the type Ia supernovae data by Goldhaber at SLAC were primary motivations for HPN to attend DM98 and make the presentation already cited[14] that summer at ANPA20 (Sept., 1998).

[4] J. Glanz, *Science*, **280**, 1008, 15 May 1998.

[5] J. Glanz, *Science*, **282**, 1247, 13 September 1998.

[6] D. E. Groom, et al., "Astrophysical Constants", in [19], p.70.

[7] C. J. Hogan, "The Hubble Constant", in [19], pp 122-124.

[8] C. H. Lineweaver, "A Younger Age for the Universe", *Science*, **284**, 1503-7 (1999).

[9] M. J. Manthey, "Program Universe" in SLAC-PUB-4008, June 1986 (Part of Proc. ANPA 7), pp 101-110.

[10] D. O. McGoveran, private communication to HPN November 19, 1998. McGoveran has tried to get this point across to this author, and to others committed to the ANPA research program, for several years.

[11] H. P. Noyes, "Bit-String Physics, a Novel 'Theory of Everything' ", in *Proc. Workshop on Physics and Computation (PhysComp '94)*, D. Matzke, ed.,94, Los

Amitos, CA: IEEE Computer Society Press, 1994, pp. 88-94 and SLAC-PUB-6509. Aug 1994.

[12] H.P.Noyes, "A Short Introduction to BIT-STRING PHYSICS", in *Merologies, Proc. ANPA 18*, T.L.Etter, ed; [available from ANPA c/o K.Bowden, Theoretical Physics Research Unit, Birkbeck College, Malet St., London WC1E 7HX]; SLAC-PUB-7205, (June 1997) and hep-th 970702. This reference contains reasonably complete citations of the earlier literature.

[13] See [12], Sect. 2, pp 28-32, for a brief discussion of *program universe* and a guide to the literature.

[14] H. P. Noyes, "Program Universe and Recent Cosmological Results" SLAC-PUB-8030 (Jan 99), [gr-qc/9901022]; it also appeared in the Proceedings of the 20th annual meeting of the Alternative Natural Philosophy Association, *Aspects II*, K.G. Bowden, Ed. pp 192-214 [available from ANPA c/o K.Bowden, Theoretical Physics Research Unit, Birkbeck College, Malet St., London WC1E 7HX].

[15] H.P.Noyes and E.D.Jones, "Solution of a Relativistic Three Body Problem", *Few Body Systems* (in press) and SLAC-PUB-7609(rev), June 1998, and hep-th 971077.

[16] H.P.Noyes and D.O.McGoveran, *Physics Essays*, **2**, 76 1989), and SLAC-PUB-4528, Oct 1998.

[17] K. A. Olive and D. N. Schramm(dec.), "Big-bang Nucleosynthesis", in [19], pp 119-121.

[18] K. A. Olive,"Big-bang Cosmology", in [19], pp 117-118.

[19] Particle Data Group, "Review of Particle Properties", *Phys.Euro. J.* **C 3**, 1-794 (1998).

[20] A. F.Parker-Rhodes, "Hierarchies of Descriptive Levels in Physical Theory", Cambridge Language Research Unit, internal document I.S.U.7, Paper I, 15 Jan-

uary 1962; reprinted, together with comments by John Amson in K. Bowden, ed. *Int.J. General Systems*, **27** Nos. 1-3(1998), pp 57-80.

[21] J. R. Primack, "Dark Matter and Structure Formation", in *Formation of Structure in the Universe*, Proc. of the Jerusalem Winter School 1996, A. Dekel and J. P. Ostriker, eds. Cambridge University Press (in Press) and astro-ph/9797285v2 25 Jul 1997,

APPENDIX

In order to underpin our claim that we can model a finite particle number version of relativistic quantum mechanics with particle creation, etc. using bit-strings we give on the next page the predictions of coupling constants and mass ratios calculated using our theory. As in any mass, length, time theory we are allowed three empirical, dimensional constants which are measured by standard techniques to connect our abstract theory to measurement. These we take to be the mass of the proton m_p, Planck's constant $\hbar$ and the velocity of light c. Everything else is calculated. Agreement with observation, given on the next page, is not perfect; we believe it is impressive. For more detail see[12].

A tentative bit-string representation of the quantum numbers of the (three generation) standard model of quarks and leptons is given on the following page (Fig. 7).

"BIT-STRING PREDICTED" VERSUS EXPERIMENTAL VALUES
OF QUANTUM NUMBERS

predicted values   *experimental values*

$$G_N^{-1}\frac{\hbar c}{m_p^2} = [2^{127}+136]\times[1-\frac{1}{3\cdot 7\cdot 10}] = 1.693\ 31\ldots\times 10^{38}$$
$$1.693\ \mathbf{58}(21)\times 10^{38}$$

$$\alpha^{-1}(m_e) = 137\times[1-\frac{1}{30\times 127}]^{-1} = 137.0359\ \mathbf{674}\ldots$$
$$137.0359\ 895(61)$$

$$G_F m_p^2/\hbar c = [256^2\sqrt{2}]^{-1}\times[1-\frac{1}{3\cdot 7}] = 1.02\ \mathbf{758}\ldots\times 10^{-5}$$
$$1.02\ 682(2)\times 10^{-5}$$

$$sin^2\theta_{Weak} = 0.25[1-\frac{1}{3\cdot 7}]^2 = 0.2267\ldots$$
$$0.2259(46)$$

$$\frac{m_p}{m_e} = \frac{137\pi}{<x(1-x)><\frac{1}{y}>} = \frac{137\pi}{(\frac{3}{14})[1+\frac{2}{7}+\frac{4}{49}](\frac{4}{5})} = 1836.15\ \mathbf{1497}\ldots \qquad (7)$$
$$1836.15\ 2701(37)$$

$$m_\pi^\pm/m_e = 275[1-\frac{2}{2\cdot 3\cdot 7\cdot 7}] = 273.12\ \mathbf{92}\ldots$$
$$273.12\ 67(4)$$

$$m_{\pi^0}/m_e = 274[1-\frac{3}{2\cdot 3\cdot 7\cdot 2}] = 264.2\ \mathbf{143}\ldots$$
$$264.1\ 373(6)$$

$$m_\mu/m_e = 3\cdot 7\cdot 10[1-\frac{3}{3\cdot 7\cdot 10}] = 207$$
$$206.768\ 26(13)$$

$$G_{\pi N\bar{N}}^2 = [(\frac{2M_N}{m_\pi})^2-1]^{\frac{1}{2}} = [195]^{\frac{1}{2}} = 13.96\ldots \quad exp. = 13.3(3),\ or\ greater\ than\ 13.9$$

574

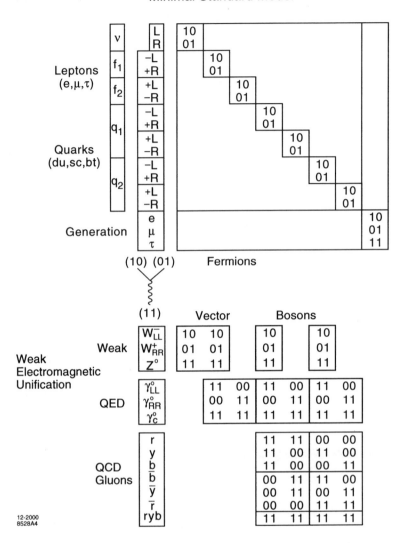

Figure 7: Skeleton of a label scheme for labels of length 16 which conveys the same quantum number information as the standard model of quarks and leptons.

Comment on "*Bit-String Physics Prediction of η, the Dark Matter/Baryon Ratio and* Ω_M"
H. Pierre Noyes (2000)

As we go to press, the '"BOOMERANG" results, and other cosmological observations now generally accepted, reduce the "area of confusion" in the observed $(\Omega_M, \Omega_\Lambda)$ plane down to a size comparable to the uncertainty in the theoretical prediction given by Bit-String Physics combined with that given by Ed Jones' calculation. Fortunately for us our predicted region is still almost exactly centered on the observational result, as can be seen by comparing Fig.3 above as compared with Fig. 15.3 in Kolb and Turner's "Pocket Cosmology"[1].

The one unknown constant in Ed Jones' microcosmology, which he uses to estimate the cosmological constant density Ω_Λ, is the number of plancktons which the universe contains at the Planck time $[\hbar/Gc]^{\frac{1}{2}}$. We add to what is given above the interpretative postulate that the Planck time in Program Universe corresponds to the first step in constructing the content portion of the strings which lies beyond the labels of length 256. This starts when all $2^{127}+136$ labels which can be produced by the combinatorial hierarchy construction have been realized. Then any PICK will provide the proto-quantum-number description of a process which conserves baryon number, charge, and lepton number, as argued in Ch.13. The number of possible distinct processes will be the square of the number of labels, i.e. $N = (2^{127} + 136)^2 \approx 2^{254}$. This turns out to be approximately the number of particles present at the Tev scale needed for Jones' calculation of the cosmological constant to be self-consistent. Whether this is a coincidence, an interpretative requirement we wish to impose on Program Universe, or a number that our interpretation of program universe predicts using more general considerations depends on the outcome of a more detailed investigation that Ed Jones and I are now pursuing. I stress the fact that Ed Jones' calculation arrives at this number from arguments of self-consistency and in itself requires only conventional physical reasoning.

Now that the cosmological observational data have brought the uncertainty in the $(\Omega_M, \Omega_\Lambda)$ plane down to a size comparable to our *a priori* prediction, it is time for us to see if we can also calculate other observational parameters — such as the "acoustic pulse" in the cosmic background radiation spherical harmonic expansion, etc. This, and refinements of our estimates of $(\Omega_M, \Omega_\Lambda)$ will require a more detailed bit-string dynamics than we currently possess. We hope that younger people interested in our approach will have the time and energy to pursue what we believe we have shown to bee a promising and fruitful new approach into the interfaces between cosmology, particle physics and natural philosophy.

References

[1] E.W.Kolb and M.S.Turner, "The Pocket Cosmology", in REVIEW OF PARTI-CLE PROPERTIES, Particle Data Group, *Euro.Phys.J.*, **15**, No. 1-4 (2000), pp 125-132.

Acknowledgments

My hearty thanks go to my friends and collaborators John Amson, Ted Bastin, Lou Kauffman, Clive Kilmister and David McGoveran for not only giving me permission to reprint some of our joint work here but also for generously writing comments about how they view this work today. This is the more true because, as you have seen above, no two of us agree about how to evaluate this work! I also wish to thank David Mcgoveran for permission to reprint excerpts from his own work which help to clarify and extend our collaborative efforts. More generally, I wish to express my gratitude to the members of the Alternative Natural Philosophy Association and of ANPA West for keeping me on my toes intellectually and broadening and deepening my understanding of where a finite and discrete approach to natural philosophy might give new insight into old problems. This includes in particular my editor, Hans van den Berg, whose advice and help have been critical in bringing this enterprise to a successful conclusion.

My particular thanks are owed to Carl Dennis whose generous financial support for aspects of the work which SLAC could not be expected to fund made the effort enormously easier, and to Enrique Zeiger who himself supported and enlisted others to support me in a similar way. My gratitude also goes out to the contributers to Enrique's fund.

With these exceptions, all my work which went into this volume was supported by the Stanford Linear Accelerator Center. I am grateful to my colleagues here for their forbearance as my research, over the years, increasingly became focused in this idiosyncratic direction. The funds which SLAC used came from The U.S. Research and Development Administration for Ch.1, from the Department of Energy under contract number EY-76-C-03-051 for Ch.2 and under contract number DE-AC03-76SF-00515 for the rest of the volume. We also acknowledge permission to reprint articles from several journals for the main ingredients of several chapters as follows: *Proc. of ANPA* for excerpts in Ch.6, and for the main ingredients of Ch.'s 8, 14; *Physics Essays* **A** for Ch.'s 5,6,15; *World Scientific* for Ch. 9; *Science Philosophy Interface* for Ch.10; *Proc.R.Soc.Lond.* **A** for Ch.11; *Physics Letters* **A** for Ch.12.

SERIES ON KNOTS AND EVERYTHING

Editor-in-charge: Louis H. Kauffman *(Univ. of Illinois, Chicago)*

The Series on Knots and Everything: is a book series polarized around the theory of knots. Volume 1 in the series is Louis H Kauffman's Knots and Physics.

One purpose of this series is to continue the exploration of many of the themes indicated in Volume 1. These themes reach out beyond knot theory into physics, mathematics, logic, linguistics, philosophy, biology and practical experience. All of these outreaches have relations with knot theory when knot theory is regarded as a pivot or meeting place for apparently separate ideas. Knots act as such a pivotal place. We do not fully understand why this is so. The series represents stages in the exploration of this nexus.

Details of the titles in this series to date give a picture of the enterprise.